Synthetic Aperture Radar

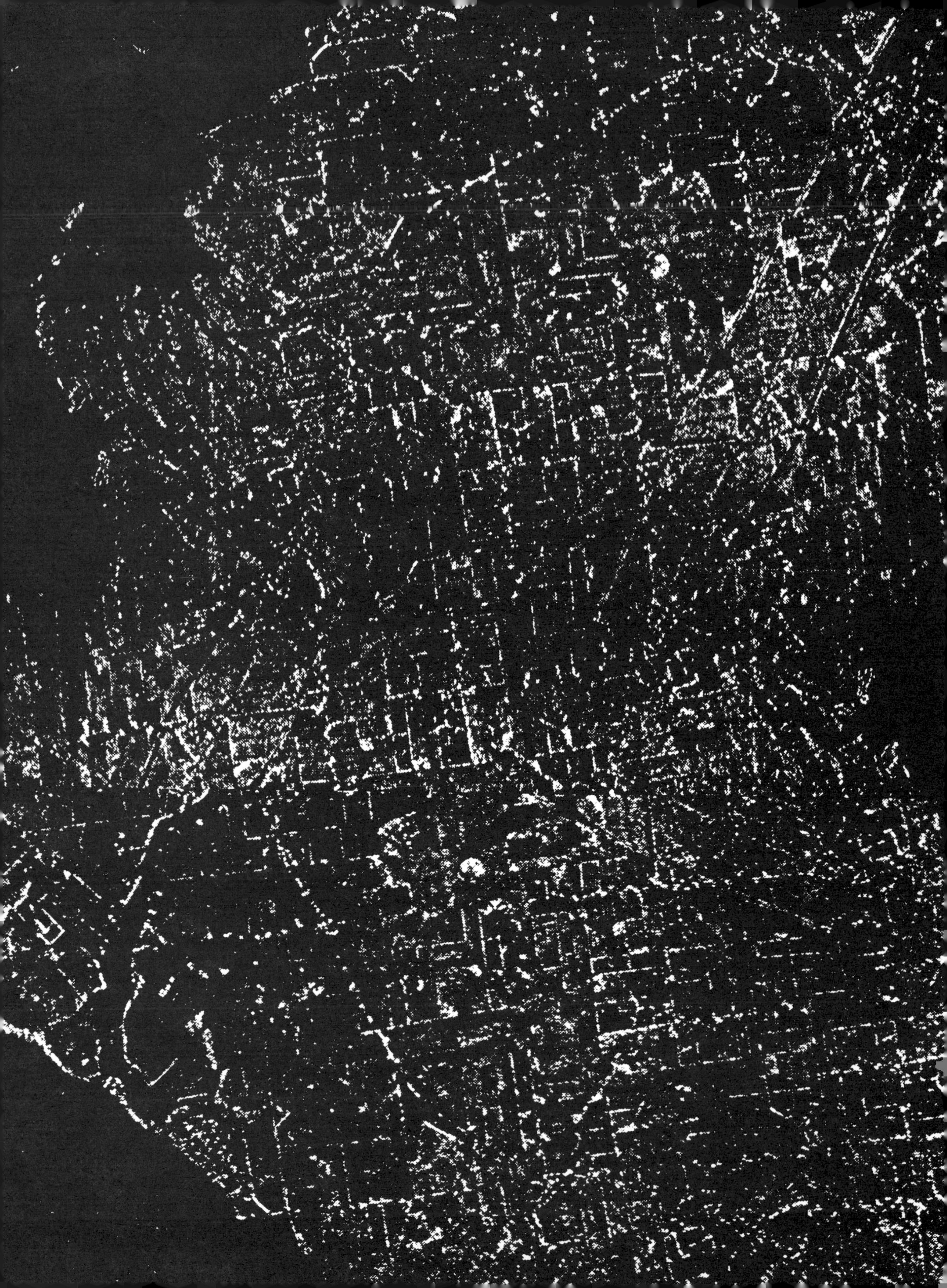

Synthetic Aperture Radar

John J. Kovaly

Artech

Consulting Engineer
Raytheon Company
Missile Systems Division
Bedford, Massachusetts

Synthetic Aperture Radar

TABLE OF CONTENTS

Cover and design facing title page are details from a synthetic aperture radar image of Munroe, Michigan area taken by the Radar and Optics Laboratory, University of Michigan at Ann Arbor.

Library of Congress Cataloging in Publication Data
Main entry under title:

Synthetic aperture radar.

(The Artech radar library)
Includes bibliographical references.
1. Synthetic aperture radar -- Addresses, essays, lectures. I. Kovaly, John J., 1928-
TK6592.S95S95 621.3848'9 76-42314
ISBN 0-89006-056-8

ARTECH HOUSE, INC.
610 Washington Street
Dedham, Massachusetts 02026

Printed and bound in the United States of America

Library of Congress Card Catalog Number: 76-42314
Standard Book Number: 0-89006-056-8

PREFACE

The first experimental demonstration of a synthetic aperture radar was made by the Control Systems Laboratory of the University of Illinois early in 1953. Eight years later, the first journal publication appeared in the open literature to describe this most elegant of electronic sensor systems. Thus far, published discussions of the different aspects of this subject have been limited primarily to various conference proceedings, transactions and journals. There is a critical need for books and reference materials in the technical community. Only two books have been published on synthetic aperture radar. One by R.O. Harger was completed in 1970; more recently in 1974, the Russian book by A.P. Reutov and B.A. Mikhaylov was translated into English.

My colleagues C.W. Sherwin, J.P. Ruina, and R.D. Rawcliffe (whose team did much of the original research and development on synthetic arrays, built the first flyable system and made the first synthetic aperture radar map) summarized the early development when they said, "As is known by everyone who has been associated with military technological developments, the discovery and early application of new concepts and ideas, even though they may be of general interest outside of their classified military application, are often not documented in journal publications. The most that can be done is to search the classified literature as carefully as possible and, at such time that the principles involved may be properly declassified, describe their development, giving reference to the original authors." The authors, whose works are a part of this book, have made this search and have documented their theoretical and practical experiences on synthetic aperture radar.

This book was first drafted in 1968. Since then, it has been updated twice, in 1974 and now in 1976; throughout it has maintained its basic structure. However, Chapter III on performance has been expanded to include discussions on moving targets in synthetic imagery and the effects of hard-limiting in synthetic aperture processing. Chapter V has been added to include papers on motion compensation in synthetic aperture radars, without which superfine radar imagery could not be attained. Finally, Chapter VII has been modified to include the results and real imagery obtained from the expanded applications of synthetic aperture radars – general ground mapping, mapping for geological and environmental purposes, and mapping of the Moon in order to detect subsurface geologic structures.

This book has been prepared to satisfy the assumed interest and needs of radar systems engineers, electronic development engineers, program managers, and military users. For those individuals unfamiliar with the subject, this collection of reprints saves them the burdensome and time-consuming effort involved in a literature search. For those engineers already involved in this technology, this edition can be used as an immediate reference. Lastly, it is hoped that this book will aid in a better understanding of synthetic aperture radar and will stimulate wider interest in and further applications of this fundamental concept.

John J. Kovaly
Dover, Massachusetts

INTRODUCTION

Radar had its biggest impact during the second World War. Since that time, the major innovations in the field of radar have been limited – the synthetic aperture, pulse compression, and digital signal processing. This book is on Synthetic Aperture Radar (SAR), a class of high resolution radar which obtains fine angular resolution by coherent processing of backscattered Doppler histories. This introduction serves as a synopsis of the fundamentals which are covered in more detail by the individual chapters.

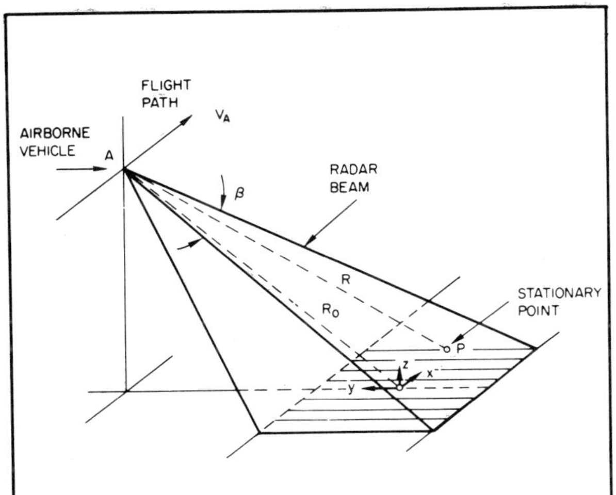

Figure 1 **Side-Looking Radar System Geometry**

Figure 1 defines the geometry of a side-looking radar system; the assumption is that an airborne vehicle carries on-board a coherent radar. The physical aperture of the radar generates an RF beam whose beamwidth is β. It illuminates the shaded patch on the ground with RF energy as the aircraft moves in the direction of the flight path. A point, P, on the ground, is shown just entering the beam; its range at that time is R. As the aircraft moves, the point P traverses the RF beam. It arrives at the point of closest approach when it is broadside at a range R_o. It continues on through the beam and has its original range, R, as it exits from the beam.

A block diagram of the coherent radar in the aircraft is shown in Figure 2. The coherent radar remembers the phase difference from transmission of a pulse to reception of a pulse; its stable local oscillator generates the carrier frequency, f_C. The transmitted pulse has a frequency modulation about the RF carrier which varies linearly from some high frequency (at the beginning of the pulse) to some lower frequency (at the end). The physical aperture radiates this energy, in the form of a beam, and receives the backscattered signal. The backscattered signal contains the carrier frequency, with a Doppler shift in frequency which is, in reality, its phase as a function of time. The receiver section conditions the signal. In the coherent detector the Doppler received signal is compared with the sum of a replica of the carrier frequency and a recording offset frequency, f_o. The recording frequency, f_o, is used to shift the Doppler spectrum away from zero frequency. The output of the coherent detector is the Doppler spectrum centered about f_o. After coherent detection the information enters a correlator and finally is applied to a display for visual observation.

The top view of Figure 3 shows, from the Pythagorean theorem, that the slant range to point P is time varying. The Doppler shift in frequency is equal to the time varying range multiplied by the ratio of $2/\lambda$. After going into the detector and making the proper substitutions, the detected frequency is the offset frequency plus a time varying frequency. The coefficient of the time variable is to be recognized from the physics of circular motion as the centripetal acceleration of the point along the radar beam.

The output of the coherent detector is called the Doppler frequency history; its characteristics are described in Figure 4. This frequency history for each scatterer is generated from the time when the backscatterer enters the beam to the time when it leaves the physical aperture beam — an interval of time, T. Each scatterer has the same Doppler history which goes from some high frequency to some low frequency. The total bandwidth which is generated over that time is the centripetal acceleration of the ground point multiplied by the ratio of T/λ.

Thus far, the discussion has described the received radar signal characteristics when the radar's beam is at a right angle to the aircraft's velocity vector. Similar signal characteristics are obtained when the radar beam is directed at a squint angle, θ_S, relative to the aircraft's velocity vector. Figure 5 describes the Doppler frequency history of a stationary ground object located at a squint angle.

Similar concepts are used to describe the signal characteristics of the transmitted pulse; they are summarized in Figure 6. Conventional pulse radars generate a pulse τ microseconds long with amplitude A. To obtain fine range resolution, the RF carrier incorporates a linear change in frequency from some frequency, f_1, to another frequency, f_2, over the pulse length, τ. Mathematically, the linear FM transmitter pulse has the following waveform:

$$f(t) = \cos\left(2\pi f_c t - \frac{2\pi\alpha t^2}{2}\right)$$

$$-\frac{\tau}{2} \leqslant t \leqslant +\frac{\tau}{2}$$

Elsewhere, $f(t) = 0$

Where:

f_C = Carrier Frequency

α = Rate of Frequency Sweep $(\Delta f)_R/\tau$

$(\Delta f)_R$ = Swept Spectrum Bandwidth

τ = Uncompressed Pulsewidth

The Detected Frequency is:

$$f_{DET}(t) = f_o - \alpha t$$

The detected frequency on a pulse-to-pulse basis, neglecting Doppler shifts for the time being, has the same form as the previously described Doppler frequency history. When this signal is passed through a correlator or a pulse compression device (see Figure 7) the output of the one dimensional correlator is the envelope function shown (with its characteristic sin x/x function). Because of the frequency coding of the pulse and the proper phase matching in the correlator, an accentuated response is formed where the height of the mainlobe is the original amplitude multiplied by the square root of the time bandwidth product. The width of this main pulse at its 4 dB point and the distance between the first two nulls are related to the time bandwidth product. For uniform weighting of the amplitude of the transmitted pulse, the first sidelobe of sin x/x function is 13.3 dB down. Sophisticated amplitude weighting techniques in the transmitter or receiver are used to reduce these sidelobe levels at the expense of broadening of the mainlobe width.

High resolution through synthetic aperture generation (Doppler frequency histories) and pulse compression (FM transmitted pulse) are linearly independent operations which can be combined, as in Figure 8. The three orthogonal coordinates are range (pulse), azimuth (Doppler), and amplitude. The figure shows that the output of the coherent detector appears as a series of pulses; each pulse has a duration which may be on the order of micro-

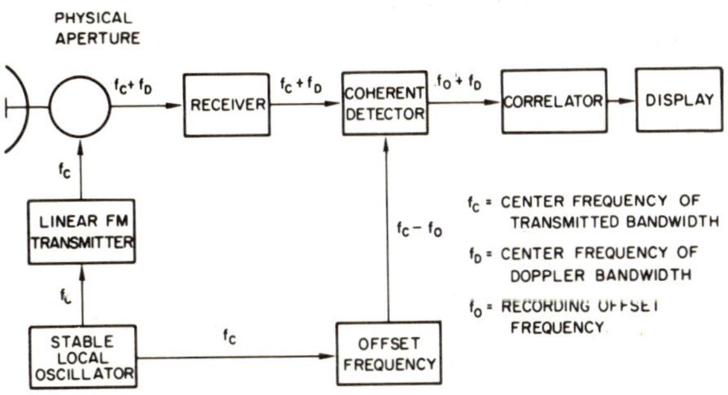

Figure 2 Coherent Radar System

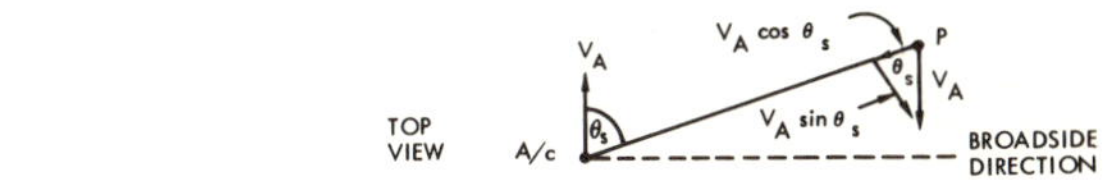

$$f_D(t) = \frac{-2V_A \cos\theta_s}{\lambda} - \frac{2V_A^2 \sin^2\theta_s}{\lambda R_o}(t - t_o)$$

$$f_{DET}(t) = f_{RCVR} - f_{REF}$$

$$f_{DET}(t) = f_o \frac{-2V_A^2 \sin^2\theta_s}{\lambda R_o}(t - t_o)$$

Figure 5 Doppler-Frequency History of Stationary Ground Object (Squint Angle)

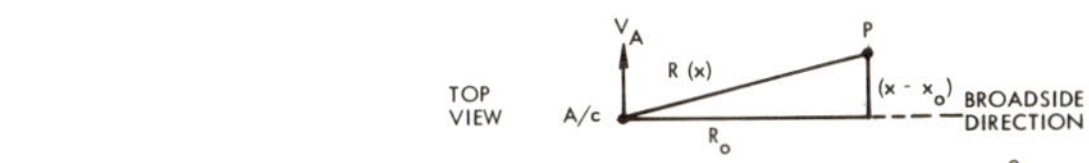

$$R(x) = \sqrt{R_o^2 + (x - x_o)^2} \approx R_o + \frac{(x - x_o)^2}{2R_o}$$

BUT $(x - x_o) = V_A(t - t_o)$

$$R(t) = R_o + \frac{V_A^2 (t - t_o)^2}{2R_o}$$

$$f_D(t) = \frac{-2\dot{R}(t)}{\lambda}$$

$$f_D(t) = \frac{-2V_A^2 (t - t_o)}{\lambda R_o}$$

$$f_{DET}(t) = f_{RCVR} - f_{REF}$$

$$f_{DET}(t) = f_o \frac{-2V_A^2 (t - t_o)}{\lambda R_o}$$

Figure 3 Doppler-Frequency of Stationary Ground Object (Broadside)

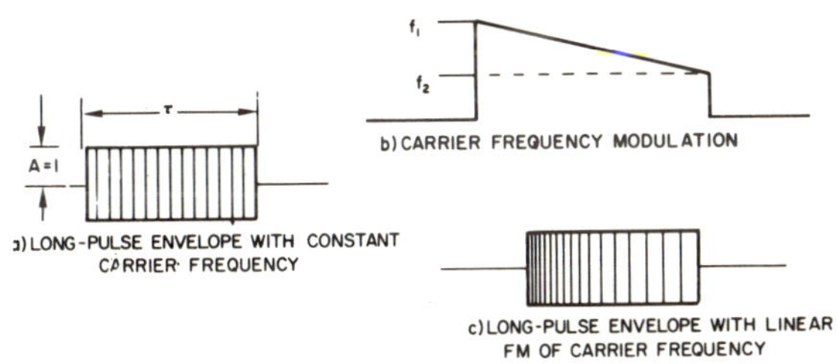

Figure 6 FM Transmitted Pulse

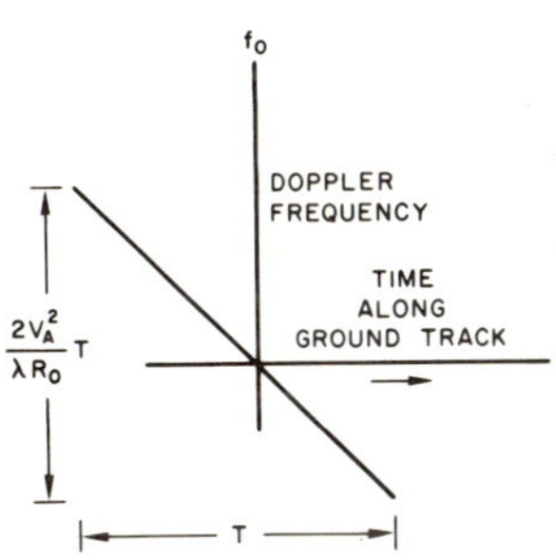

WHERE:

V_A = AIRBORNE VEHICLE VELOCITY

λ = CARRIER WAVELENGTH

R_o = SHORTEST SLANT RANGE TO GROUND SCATTERER

T = TIME SCATTERER STAYS IN PHYSICAL BEAMWIDTH

f_o = RECORDING OFFSET FREQUENCY

t_o = TIME WHEN SCATTERER PASSES THROUGH CENTER OF BEAM

Figure 4 Doppler Frequency History

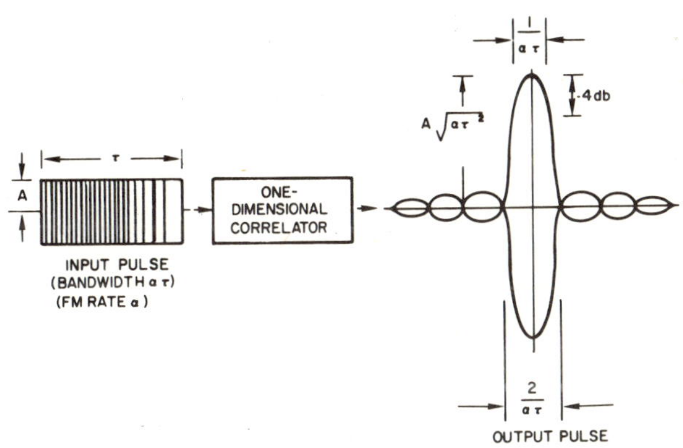

Figure 7 One-Dimensional Correlator

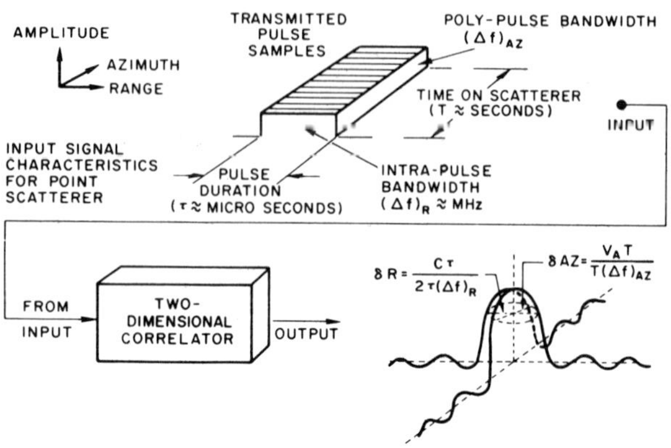

Figure 8 **Two-Dimensional Correlator**

seconds and a bandwidth which is on the order of megahertz. These pulses are the backscattered signals received by the radar and are stored in the correlator over a period of time, T, which is on the order of seconds. Over T a Doppler bandwidth is generated. With this bulk of information as the input to a two-dimensional correlator, the output is a bi-variant sin x/x distribution. In the range dimension, the resolution is δR; in the orthogonal direction, the azimuth resolution is δA_z. Of interest are the similar forms of resolution in the two directions. Each has a numerator which is related to a velocity – in range, the

	BEAM–SHARPENING	PULSE COMPRESSION
CODING	PULSE–TO–PULSE	WITHIN A PULSE
DIMENSION	AZIMUTH	RANGE
TIME SCALE	SEVERAL SECONDS	MICROSECONDS
BANDWIDTH	FEW HUNDRED HZ/S (DOPPLER BANDWIDTH)	MHZ/S
CODE GENERATION	LINEAR FREQUENCY MODULATION (AIRCRAFT FLIES STRAIGHT PATH)	LINEAR FREQUENCY MODULATION (ELECTRONICALLY)

Table 1 **Comparison of Beam-Sharpening with Pulse Compression**

RANGE DIRECTION	AZIMUTH DIRECTION
1. TIME WIDTH τ_R OF IMAGE AT -4 dB LEVEL $\tau_R = 1/(\Delta f)_R$ WHERE $(\Delta f)_R$ IS TRANSMITTING FM BANDWIDTH	1. TIME WIDTH τ_{Az} OF IMAGE AT -4 dB LEVEL $\tau_{Az} = 1/(\Delta f)_{Az}$ WHERE $(\Delta f)_{Az}$ IS DOPPLER BANDWIDTH
2. MULTIPLY BOTH SIDES OF ABOVE EQUATION BY C/2 WHERE C IS VELOCITY OF LIGHT (FACTOR OF 2 REQUIRED FOR TWO-WAY RANGE): $\frac{C\tau_R}{2} = C/2\,(\Delta f)_R$	2. MUTLIPLY BOTH SIDES OF ABOVE EQUATION BY AIRBORNE VEHICLE VELOCITY V_A: $V_A\ \tau_{Az} = V_A/(\Delta f)_{Az}$
3. HERE $\frac{C\tau_R}{2}$ IS DEFINED AS THE RANGE RESOLUTION δR: $\delta R = C/2\,(\Delta f)_R$	3. HERE $V_A\ \tau_{Az}$ IS DEFINED AS THE AZIMUTH RESOLUTION δAz: $\delta Az = V_A/(\Delta f)_{Az}$

Table 2 **Comparison of Derivations for Range Resolution (δR) and Azimuth Resolution (δAz)**

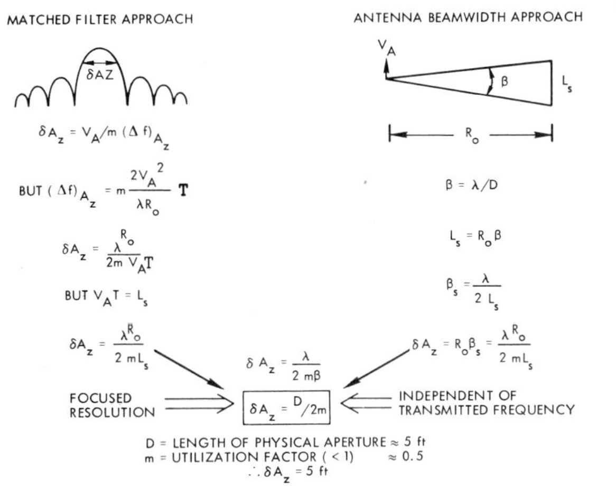

Figure 9 **Synthetic Aperture Radar Azimuth Resolution-Focused Case**

velocity of light; in azimuth, the velocity of the aircraft. Each resolution function is inversely proportional to a bandwidth – in the range dimension, the RF bandwidth; in the azimuth dimension, the Doppler bandwidth. Fundamentally, then, resolution is obtained in both dimensions by generating a bandwidth. At this point in the system – when this resolution function has been manifested – a synthetic aperture has been generated. Comparisons of the two degrees of resolution are shown in Tables 1 and 2. Workers in the field have used the expression "Beam-Sharpening" for the increased resolution provided by the synthetic aperture.

Depending upon the system requirements, the full degree of resolution can be obtained by processing all of the Doppler bandwidth and by compensating for the phase difference on a pulse-to-pulse basis. If the phase variations are not compensated for, an unfocused or lower level of resolution is obtained. Two approaches for describing the focused SAR azimuth resolution are shown in Figure 9. The more erudite matched filter approach starts with the relationships of vehicle velocity and Doppler bandwidth. An alternate approach begins with the resolution of the physical antenna. A constant term, m, (called the utilization factor) is introduced in both of these approaches. It can have a value between 0 and 1, depending upon the amount of Doppler bandwidth to be processed. The focused azimuth resolution is independent of transmitted frequency and is dependent only on the size, D, of the physical aperture and the utilization factor, m. If D were 5 feet and m were 0.5 (or 50% utilization), then the azimuth resolution would be 5 feet.

On the other hand, Figure 10 derives the SAR azimuth resolution for the unfocused case. Here the data are stored without compensating for phase differences. The phase is permitted to run out until the phase difference at the edge of the integration time (relative to the middle of that time) is 90°. Simple geometry shows that, in this case, the azimuth resolution is one half of the square root of the product of λ and R_0.

A mathematical comparison of the three degrees of fine azimuth resolution is as follows:

Conventional Case

$$\Delta X_{CONV} = \frac{\lambda R}{D}$$

Unfocused Synthetic Antenna

$$\Delta X_{UNF} = \frac{1}{2}\sqrt{\lambda R}$$

Focused Synthetic Antenna

$$\Delta X_{FOC} = \frac{D}{2m}$$

A numerical comparison for a 5 foot aperture is plotted in Figure 11. Figure 12 shows the resolution functions for both the focused and unfocused case. In the focused case, the sin x/x variation has nulls and sidelobes which are very well defined; in the unfocused case, the nulls are filled in, and the first sidelobe level is about 9 dB down from the main lobe (compared to 13.3 dB down in the focused case).

Over the years, in line with the need for obtaining improved resolutions in other applications, different forms of mapping have been devised. One relatively old type, the strip map in Figure 13, operates in a squint mode which is characterized by the squint angle (the angle between the velocity vector and the direction in which the physical aperture is pointing). The advent of phased arrays and digital processing and a better understanding of what composes a synthetic aperture have enabled mappers to use the squint mode of operation. To the right of the mapping type in the figure is the fundamental azimuth resolution equation with the constant parameters identified. Pulse compression in the range direction is assumed in all mapping modes. The second type of mapping mode, characterized as Doppler beam sharpening, is akin to the Planned Position Indicator (PPI) type of mapping. The

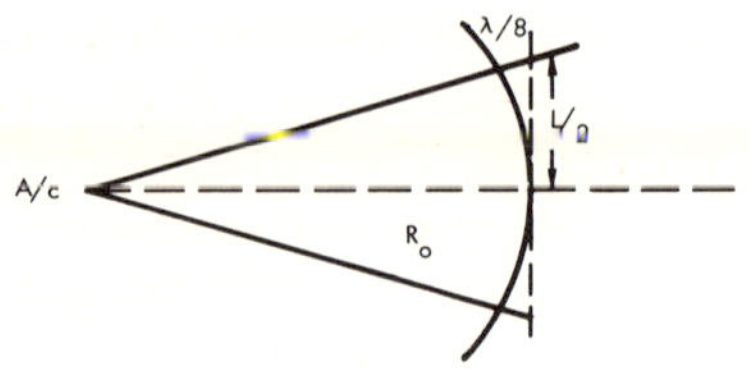

$$(R_o + \lambda/8)^2 = R_o^2 + (L/2)^2$$

$$R_o^2 + \frac{\lambda R_o}{4} + \frac{\lambda^2}{64} = R_o^2 + L^2/4$$

$$\therefore L = \sqrt{\lambda R_o}$$

$$\delta A_z = \frac{\lambda R_o}{2L} = \frac{\lambda R_o}{2\sqrt{\lambda R_o}}$$

DEFOCUSED RESOLUTION ⇨ $\delta A_z = 1/2 \sqrt{\lambda R_o}$ ⇦ RANGE & TRANSMITTED FREQUENCY DEPENDENT

Figure 10 **Synthetic Aperture Radar Azimuth Resolution-Unfocused Case**

Various forms of high resolution maps can be configured; they are depicted in Figures 13 and 14. In the early days of synthetic aperture mapping, the investigators pointed their physical aperture at an angle 90° from the velocity vector where the largest Doppler bandwidth is obtained; this method is the most convenient for generating surveillance maps.

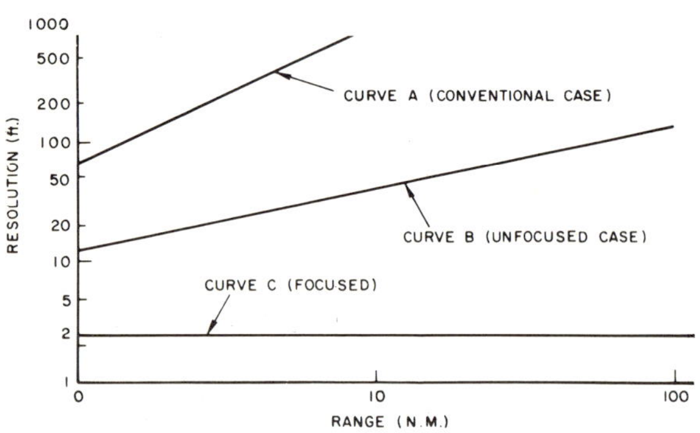

Figure 11 **Azimuth Resolution for Three Cases (A) Conventional, (B) Unfocused, and (C) Focused**

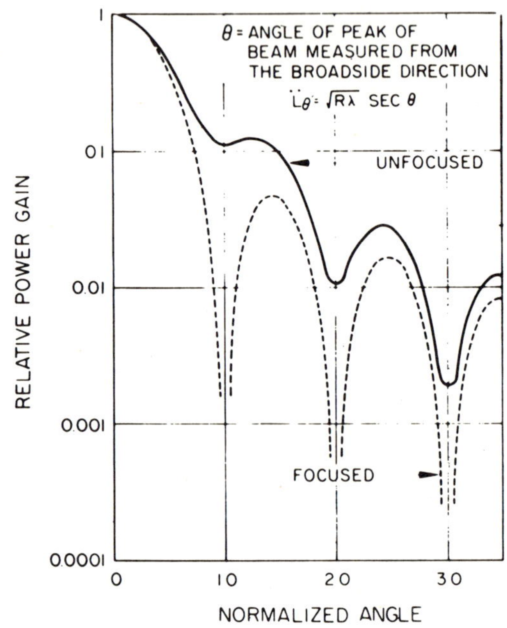

Figure 12 **Comparative Power Gain Patterns for Focused and Unfocused Synthetic Arrays**

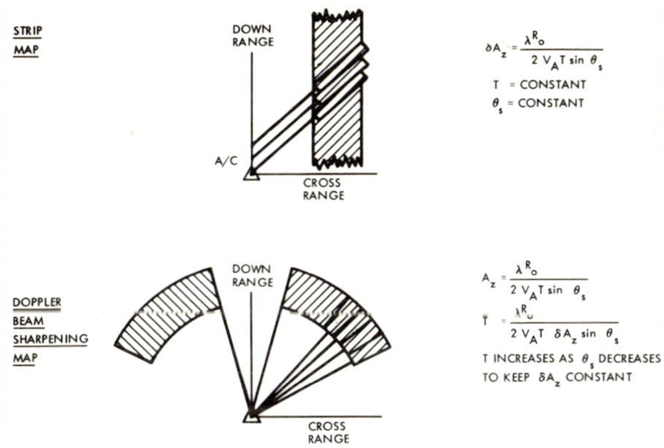

Figure 13 **Mapping Modes**

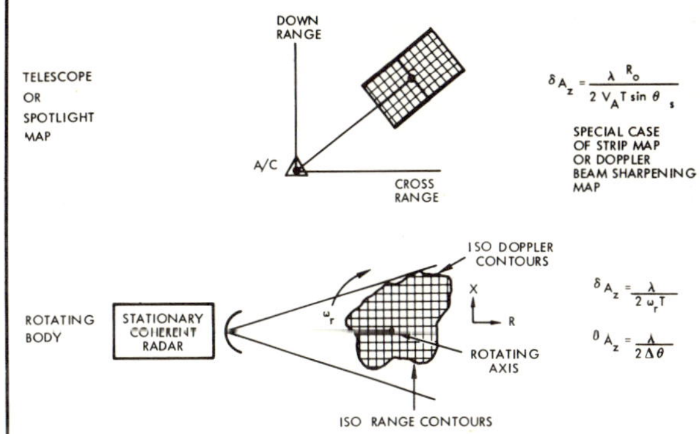

Figure 14 **Mapping Modes**

Doppler beam sharpening map can be made on either side of the velocity vector but not directly over it. In this case, the squint angle is varied; however, as the beam gets closer to the velocity vector and the squint angle decreases, the integration time increases. Some systems use two physical apertures, one on each side of the vehicle. If a phased array is used and employs its high beam switching rate capability, then strip maps and Doppler beam sharpening maps can be made.

Figure 14 shows another type of mapping mode, called the telescope or spotlight map. In this case, a very high resolution map of a particular patch on the ground over an extended period of time is produced. It is similar to the strip map; the physical aperture diameter, time, etc., are adjusted to give the right size map.

Early investigators knew that the aircraft's motion in sweeping the physical beam across the ground was paramount to generating a synthetic aperture. However, there is a more fundamental point to be made, as shown in the lower part of Figure 14 – the coherent radar remains stationary while the object to be mapped rotates within the physical beam. As the object rotates, it changes its aspect angle relative to the radar. This type of mapping has been used for mapping Venus and the Moon; it is particularly useful when dealing with a Venusian planet, where the surface is hidden by an atmosphere too difficult to penetrate optically. In creating a radar map in such a case, the range coordinate is orthogonal to the beam and the isodops are shown parallel to the axis of rotation. Azimuth resolution is inversely proportional to the angular velocity of the rotating object; conversion of the angular velocity provides a more fundamental form of the azimuth resolution which is dependent upon the change in aspect angle of the object with respect to the radar. Thus, it is not necessary to have aircraft move in order to generate a Doppler bandwidth which contains the information for improved azimuth resolution.

Signal processors are the "heart" of mapping systems which capitalize on the use of the synthetic aperture. The more common processor types are listed below:

- Range-Gated Filter Bank
- Recirculating Delay Line
- Integration on Photographic Film
- Electronic Storage Tube Integrator

} *Unfocused*

- Optical System
- Digital Signal Processing

} *Focused*

In early developmental efforts, great emphasis was placed on trying to determine the best manner in which to store the returned data required to process the Doppler bandwidth. The first four types of processors can be focused but, historically, range-gated Doppler filter banks, circulating delay lines, integration on film, and electronic storage tubes were commonly unfocused systems. Since 1960, primarily focused systems of the optical and digital signal processing types have been developed.

Figure 15 both summarizes the essentials of optical type processing and introduces new terminology. A single sweep is generated across the cathode ray tube; the output of the radar's coherent detector is used to modulate the intensity of this sweep. The film moves across the tube; the speed of the film is directly proportional to the aircraft's velocity. The lines are packed on the film so that there is some overlap. The phase history of the backscattered energy is stored on this film; the phase history within the pulse is stored along the range dimension. When stored on the film, these phase histories have been characteristically called radar holograms, owing to their similarity to optical holograms. A signal film from the image film is produced in the optical processor, which typically is constructed with the elements shown in the lower diagram of Figure 15. A coherent source of light (collimated to form a plane wave) shines on the signal film which contains the radar hologram; the conical lens has a varying focal length to compensate for the phase variations in range. Because there is a different focal length for each range element, a cylindrical lens is used to bring all of the range information into focus in one particular plane. Attempts have been made to accomplish the processing in real time; characteristically, though, the film is processed on the ground.

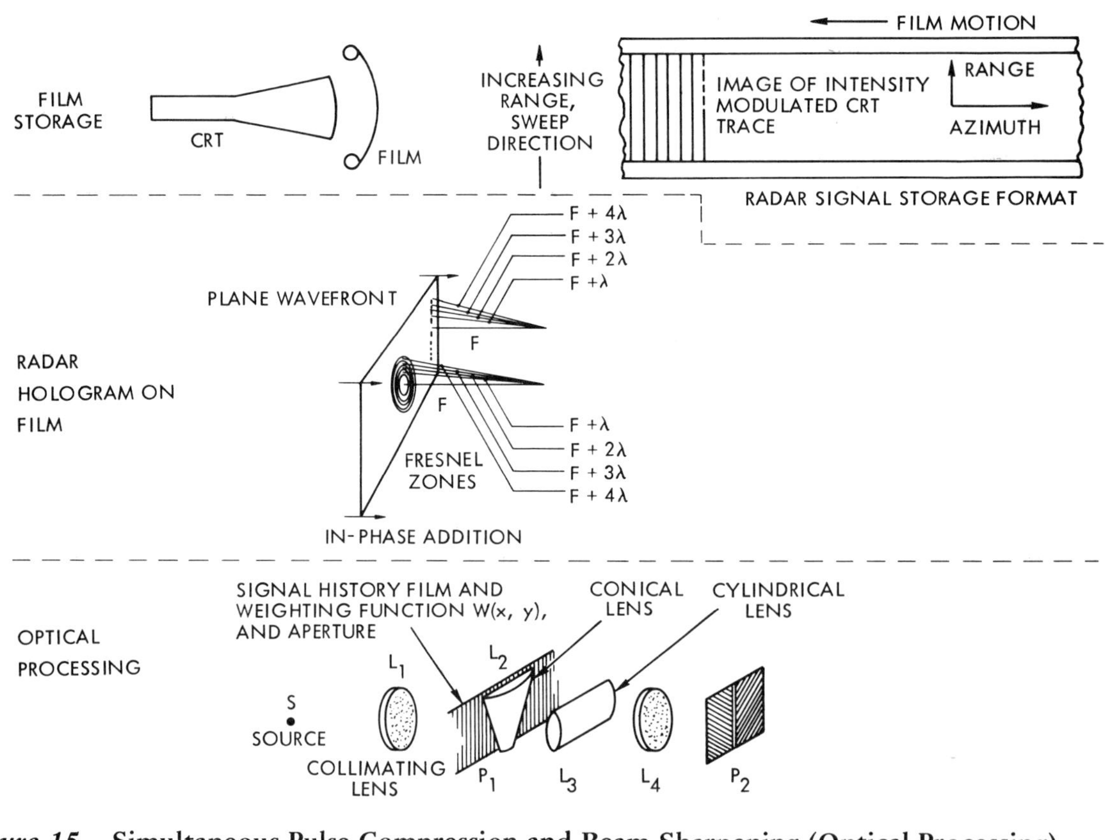

Figure 15 **Simultaneous Pulse Compression and Beam Sharpening (Optical Processing)**

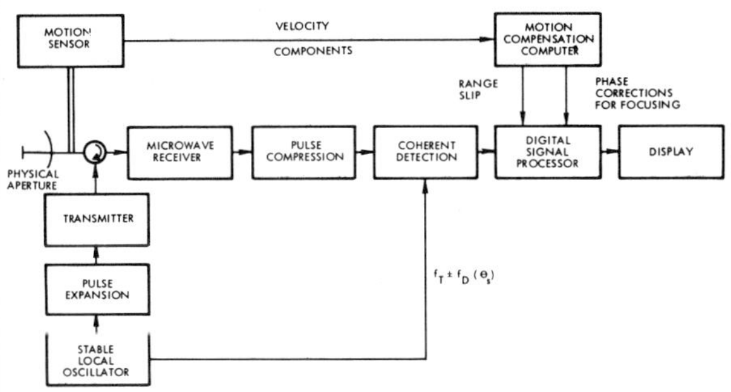

Figure 16 **High Resolution Radar**

Figure 16 is a block diagram of a high resolution radar which employs a real time digital signal processor and which incorporates two compensation loops that provide motion compensation for corrections for azimuth and range slip. These corrections are required in ultra high resolution radars in order to keep the individual points resolvable over long integration times and at squint angles other than 90°. Figure 17 is an expanded block diagram of the real time digital signal processor. In-phase and quadrature video is converted into digital format with A/D converters; correct strobing of the A/D converters compensates for range slip and phase corrections, computed by the motion compensation computer, focus the azimuth phase histories. A roughing or pre-summing filter is used to reduce the amount of data which enters the main memory. The Doppler bandwidth may be larger than what is needed; to save storage in the main memory, the roughing filter passes only the Doppler band of interest. After storage, a Fast Fourier transform is used to accomplish the frequency analysis. Real time processing is available today.

The main sources of error which limit the performance of high resolution radars are shown in Table 3. An expanded description of the errors (and their causes) which limit azimuth resolution, motion limits, and distortion specifications are listed in Table 4.

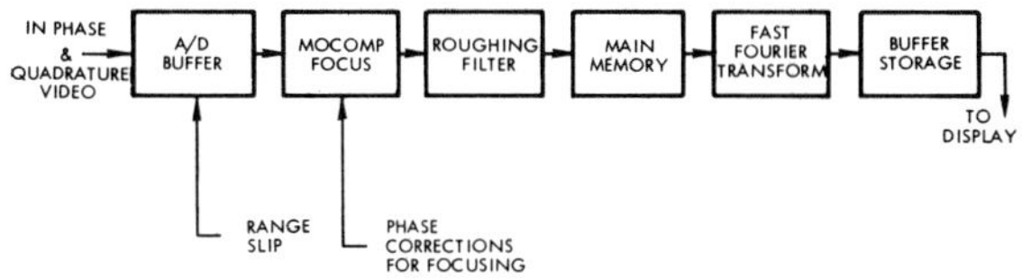

Figure 17 **Digital Signal Processor (Real Time)**

ERROR SOURCE	PERFORMANCE	LIMITATIONS
TROPOSHERIC FLUCTUATIONS	LENGTH OF SYNTHETIC APERTURE	< 1000 ft
MOTION COMPENSATION	FOCUSING	> FEW MILLI g's
OSCILLATOR INSTABILITY	RANGE	> 50 nmi

Table 3 **Sources of Phase Errors**

DISTORTIONS	CAUSES	MOTION LIMITS	DISTORTION SPECIFICATIONS
NON-UNIFORM BRIGHTNESS ACROSS MAP	OFFSET BETWEEN MAP CENTER AND REAL BEAM CENTER	LOW-FREQUENCY VELOCITY ERROR	BRIGHTNESS VARIATIONS ≤ 6 dB
DEFOCUS OF MAP CENTER	"PAIRED ECHOES" NEAR FULL NULL OF AZIMUTH RESOLUTION CELL	MID-FREQUENCY VELOCITY ERROR	AMPLITUDE OF SPURIOUS SIGNAL ≥ 26dB BELOW PEAK RESPONSE
INCREASE IN PEAK AND RANDOM FAR-OUT SIDELOBE LEVELS	"PAIRED ECHOES" BEYOND NULL OF AZIMUTH RESOLUTION CELL	HIGH-FREQUENCY VELOCITY ERROR	DISCRETE RESONANCES 26 dB BELOW RANDOM ERRORS 30dB BELOW

Table 4 **Motion Compensation**

Realism is introduced when real high resolution maps are finally imaged, as manifested by the many examples of high resolution radar imagery contained in the various papers in this book. Table 5 lists the dates, the organizations, and the conditions of early synthetic aperture mapping. Although the University of Michigan is recognized as the leading authority in this area, the first synthetic aperture map was produced by the University of Illinois in 1953. The University of Michigan has made a significant impact in bringing synthetic aperture radars to their fruition because of its optical processing facilities. From July 1953 (when the first map was made) into the sixties, pulse compression work in radar and synthetic aperture research proceeded in parallel fashion; only later did efforts combine both technologies to produce high resolution two dimensional maps.

The more common uses of synthetic aperture radars have been combat surveillance, precision navigation, aerospace surveillance, and missile guidance. The advantages of SAR are capability to be used at either day or night; atmospheric or cloud penetration; surface penetration; photographic quality; large area coverage; physical aperture and spacecraft (size) compatibility; and its usefulness for collecting geological, geophysical, and geomorphological information.

Acknowledgments

I am most grateful to Dr. C.W. Sherwin for the technical leadership of the coherent radar group at the Control Systems Laboratory of the University of Illinois, where his direction led us to design, build, and test the first experimental synthetic aperture radar and to produce the first synthetic aperture radar map.

My deep appreciation is to be acknowledged to Lorraine Bick, the Head Librarian of the Bedford Laboratories at the Raytheon Company Missile Systems Division, and to my secretary, Beatrice Traversy, for their patient and understanding assistance on the many tasks that required their talents to complete this book. But especially, we wish to express our thanks to the publishers and authors for their permission to reprint the papers in this book.

CHAPTER I
Development of Synthetic Aperture Radar

1

The advent of synthetic aperture radar was heralded by three significant events. The earliest statement of the principle that high angular resolution can be obtained by means of frequency analysis of the signals from a coherent moving radar system was reported by Carl Wiley of Goodyear Aircraft Corporation in June, 1951. The first experimental demonstration of a synthetic aperture radar and the production of the first synthetic aperture strip map were made by a group at the University of Illinois and reported on 8 July 1953. And a high resolution airborne mapping radar combat-surveillance system which employed the synthetic aperture technique achieved success on 23 August 1957, developed by a team from the University of Michigan. At this time, the capability of the synthetic-antenna technique was recognized.

The first paper in this section gives an account of the development of the first combat surveillance system, the AN/UPD-1 (XPM-1); it was the first reporting of the use of the synthetic aperture technique in the open literature. The system was composed of an airborne side-looking coherent radar and a ground van housing radar-data processing equipment. Cutrona, Vivian, Leith, and Hall describe the team efforts of a university and industry group which finalized the fabrication and flight testing of their airborne mapping system which employed the fundamental SAR concept. One of the key elements of their system was the application of a Doppler navigator and a gyroscope mounted on the physical antenna base, the combination of which yielded drift angle information to maintain the radar antenna at an angle perpendicular to the ground track. The paper includes high resolution radar imagery obtained by this system.

The second paper by Sherwin, Ruina, and Rawcliffe describes the very early developments in synthetic aperture radar systems. The first experimental coherent airborne radar is described and the very basic laboratory frequency analysis of raw radar data to yield the first synthetic aperture radar image is reported. This experimental system relied upon the pilot's ability to fly the aircraft "straight" on the ground track, with the physical antenna pointed at a *squint angle* relative to that track. In spite of their rudimentary experimental methods, the group at the University of Illinois was able to make the first experimental demonstration. The paper includes further experimental tests of coherent radar and some early developments in signal-processing. The results are described of having used three different processors — a re-entrant delay line, photographic film, and a storage tube.

An editorial survey paper by Holahan (1963) lists some of the other developers of synthetic aperture radar. The reader is referred to this paper for those organizations that followed in the footsteps of the teams at the University of Illinois and the University of Michigan.

A High Resolution Radar Combat-Surveillance System*

L.J. Cutrona, W.E. Vivian, E.N. Leith, and G.O. Hall
(Cutrona, Leith, and Hall — Institute of Science and Technology, University of Michigan, Ann Arbor; Vivian — Conductron Corporation, Ann Arbor)

IRE Transactions on Military Electronics, Vol. MIL-5, No. 2 (April 1961)
Pages 127-131

*Received by the PGMIL 21 February 1961. This work was conducted by Project MICHIGAN, under Department of the Army Contract DA-36-039 SC-78801, administered by the US Army Signal Corps.

***Summary*—An account is given of the development of the AN/UPD-1 (XPM-1) system. This airborne mapping radar, by synthesizing an extremely long antenna which expands in length in direct proportion to radar range, provides a linear resolution in the azimuth direction that is constant for all radar ranges.**

I. Introduction

On April 19, 1960 at the Washington, D. C., National Airport, the U. S. Army unveiled a new type of combat-surveillance radar system which is designated as the AN/UPD-1 (XPM-1). The principal elements of the system are an airborne side-looking coherent radar and a ground van housing radar-data processing equipment. The unique feature of the system is the data processing that produces the equivalent effect of an extremely long antenna which expands in length in direct proportion to radar range. This, in principle, can provide a linear resolution in the azimuth direction that is constant for all radar ranges.

This combat-surveillance system represents a dividend from a continuing research program in radar at the Institute of Science and Technology of The University of Michigan. This radar research is a part of Project MICHIGAN, which is sponsored by the U. S. Army Combat Surveillance Agency of the U. S. Army Signal Corps. Four models of the AN/UPD-1 (XPM-1) experimental systems were produced through separate funding for Project MICHIGAN. The airborne portion of the system was subcontracted to Texas Instruments, Inc., of Dallas, who in turn subcontracted to General Precision Laboratories and Kearfott, Inc., for Doppler inertial subsystems. The ground vans containing the radar-data handling and processing equipment were fabricated by the Institute of Science and Technology of The University of Michigan.

The AN/UPD-1 system is a good example of how a team effort of a university and industry can provide experimental models of new devices for early tactical evaluation. In less than ten months from the first successful demonstration with a breadboard system, a subcontract was in force and work was under way on the construction of the experimental systems.

It is the purpose of this article to trace the development of this system from its inception, to discuss the basic concepts of the system and its components, and to present some of the high-resolution radar photographs obtained with it.

II. The Fundamental Idea

In the summer of 1953, scientists from all parts of the country met at the University with representatives of the U. S. Army and personnel from The University of Michigan. This summer study, known as Project WOLVERINE, was for the purpose of recommending research and development programs leading to an improved capability for securing information about the battle area.

The basic idea behind the high-resolution radar was advanced in a subgroup considering the application of radar to combat surveillance. The group recognized that the major shortcoming of radars was in angular resolution. In combat surveillance the objects of military interest are against a background of the surrounding terrain; therefore, unless a radar has fine resolution in both range and azimuth, the radar return from a target of military interest is likely to go undetected because of the large clutter return from the background. With conventional techniques, fine resolution in range was readily attainable with short-duration radar pulses. Since with conventional techniques angular resolution is directly related to the radar antenna beamwidth, the only avenues for improvements in azimuth resolution were through an increase in antenna size, an increase in operating frequency, or a combination of both. Although the largest feasible size for an airborne antenna and the highest practical radar frequency (dictated by available peak power sources and acceptable attenuation for various weather conditions) were to be used, it was evident that the angular resolution was inadequate for the intended surveillance functions at radar ranges beyond a few miles.

A new idea for producing the equivalent effect of an extremely long antenna became the subject of discussion by Dr. C. Sherwin of the University of Illinois, W. Hausz of the General Electric Company, J. Koehler of the Philco Corporation and Dr. Louis Cutrona of the University of Michigan. It was thought that an extremely long antenna could be synthesized by using the forward motion of the aircraft to carry a side-looking radar antenna to positions which could then be treated as though they were the individual antenna elements of a linear antenna array.

The fundamental idea can best be grasped by consideration of Fig. 1. For the physical antenna array shown in Fig. 1(a), the individual transmitting and receiving elements are dipoles. While each dipole has a

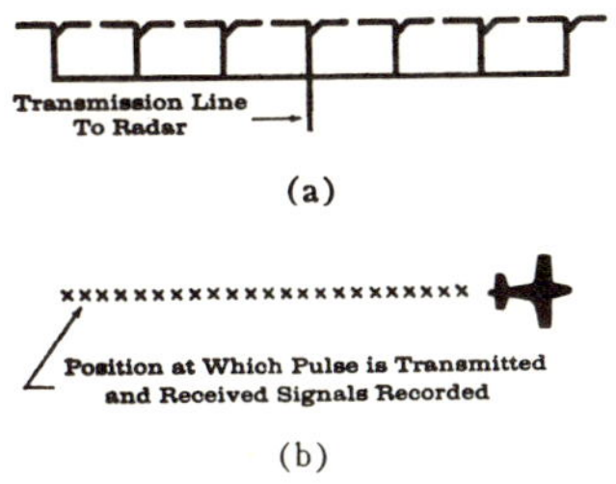

Fig. 1—(a) Physical array of dipoles. (b) Synthetic-array generation.

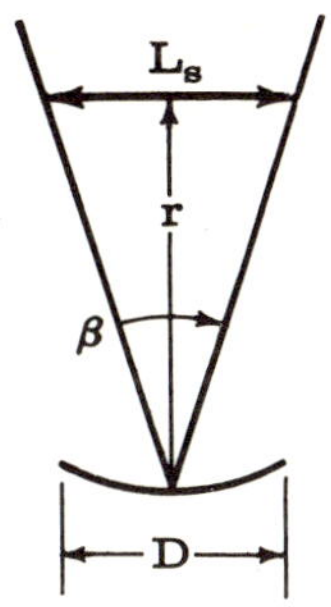

Fig. 2—Synthetic-antenna geometry.

broad radiation pattern, the assemblage of dipoles is made to produce a narrow antenna beam by making the electrical lengths of the individual transmission lines such that signals arriving in phase at the dipoles are added in phase at the main transmission line to the radar. Since transmission and reception at each of the dipoles is simultaneous, this fixed adjustment of phases serves to maintain the desired beam pattern.

Fig. 1(b) depicts the generation of the synthetic array. In this case the individual elements of the array, as indicated by the x's, do not exist simultaneously. Starting with the position at the extreme left, a radar pulse is transmitted and the return signals recorded. A short time later the aircraft has carried the side-looking antenna to the second position where another pulse is transmitted and the received signals recorded. In this way a signal is recorded for each of the positions of the synthetic array. To achieve the effect of a linear array such as Fig. 1(a), the return signals must contain phase information which is preserved in the recording. Then by appropriate data processing it should be possible to retrieve the stored data and combine it in proper phase to synthesize the desired effect of a narrow-beamed antenna.

It was evident that the radar would have to have excellent transmitter frequency stability and a stable frequency reference for use in comparing the phase of the return signals with the phase of the transmitted pulses. The method of signal storage (recording) had to preserve the range information so that the data processing could be accomplished separately for each element of radar range.

The well-known formula for antenna beamwidth showed that, at least in theory, the synthetic-antenna concept had a great potential for fine angular resolution. The half-power beamwidth β of the physical side-looking antenna (Fig. 2) is

$$\beta = k \frac{\lambda}{D} \text{ radians} \qquad (1)$$

in which

λ is the wavelength of the radar,
D is the length of the physical aperture, and
k is the illumination factor (greater than unity).

The distance across the antenna beam for a radar range r is βr, which represents the amount of data that can be collected between half-power points for this range. If this amount of data is processed at each range, this would represent a synthetic antenna length

$$L_s = \beta r \text{ feet} \qquad (2)$$

if r is expressed in feet. The phase information obtained for each of the element positions of the synthesized antenna is based on the round-trip distance between the side-looking antenna and echoing objects; for the physical array the phase depends on the one-way path length. As a consequence of this difference, the phase difference between elements of equally spaced elements of a synthetic array is twice that of a physical array with the same spacing; a synthetic antenna has half the beamwidth of a physical antenna of a given length. That is,

$$\beta_s = k_1 \frac{\lambda}{2L_s} \text{ radians} \qquad (3)$$

in which k_1 is the illumination function for the synthesized antenna. Using (1) this can be expressed in terms of the size of the physical antenna and radar range as

$$\beta_s = \frac{k_1}{k} \frac{D}{2r} \text{ radians.} \qquad (4)$$

This indicates that the synthetic beamwidth β_s is not only independent of frequency, but also decreases with radar range.

To obtain a measure of resolution, the usual assumption was made that targets could be resolved in angle if they were separated by one antenna beamwidth. For the beamwidth β_s of the synthetic antenna, the distance across the beam at any radar range r is

$$\Delta X = \beta_s r = \frac{k_1}{k} \frac{D}{2} \text{ feet} \qquad (5)$$

if the length of the physical antenna aperture D is expressed in feet.

In theory, then, the resolution in the azimuth direction for a synthetic antenna radar is independent of range, independent of radar frequency, and smaller than the physical length of the actual side-looking antenna carried by the aircraft. This assumes the generation of a synthetic antenna of a length equal to the distance across the radar beam at each radar range.

III. The Research and Development Program

When Project MICHIGAN was inaugurated in the fall of 1953, a small group undertook an evaluation of this new concept for radars. Analytic expressions were obtained for the received data and the theoretical azimuth resolution was verified with filter theory. Expressions were also obtained for the operations to be performed by the data processor. Because of the achievable lengths of synthetic antennas, it was not feasible to neglect the curvature of the phase front of received signals. It would be impossible to approach the theoretical angular resolution unless the antenna were focused. Since the curvature of the phase front varies with radar range, the data processor needed to be one which applied a focusing function that varied with range.

As this analysis progressed, the group became convinced that the synthetic-antenna technique held promise and that a breadboard system should be built for experimental evaluation. Photographic film was selected as the storage medium for the received radar signals. With film as a storage medium, three dimensions were available for recording: distance across the film, distance along the film, and the exposure. Film had the further advantage of providing a permanent record which could be processed many times using different processing parameters. Digital processing techniques for generating the synthetic antenna did not appear promising because of the number of bits of information to be handled and the need for focusing functions which differed for each segment of radar range. After a period of basic laboratory experimentation an analog data-processing technique was evolved which was not only simple, but which also provided simultaneous focusing of the radar data at all radar ranges.

By the end of 1954, a great many of the components for the radar system had been proven by laboratory experimentation and microwave components for the radar were on order. It was decided to build a preliminary model of the coherent radar in the laboratory and to check its performance with a roof-mounted antenna before proceeding with the fabrication of an airborne breadboard. In this way enough experience could be gained to finalize a design before an aircraft became available. A stabilizing cavity and a crystal-controlled reference at the IF frequency were selected as the means for obtaining the necessary radar frequency stability and coherence. With this scheme, the prime frequency reference is a continuously operating klystron stabilized by a high Q cavity. The second reference at the IF frequency, also continuously operating, is used to single-sideband modulate the RF frequency. A chain of pulsed klystron amplifiers is then used to produce the desired RF pulse shape and to boost the power level for transmission. On reception, the prime frequency reference serves as the local oscillator for heterodyning the received signals.

The laboratory breadboard was operating with a roof-mounted antenna by the fall of 1955. Standard laboratory oscilloscopes were used as an output display and a modified scope camera was used for recording. In the spring of 1956 work was started on the fabrication of the components for an airborne breadboard system.

A C-46F aircraft was obtained on bailment from the U. S. Air Force in the fall of 1956. Aircraft overhaul, modification, and radar equipment installation were completed in July of 1957. Figs. 3 and 4 show external and internal views of the aircraft with its installed equipment. The radome houses the side-looking radar antenna and the antenna for Doppler navigation radar.

Fig. 3—The C-46F.

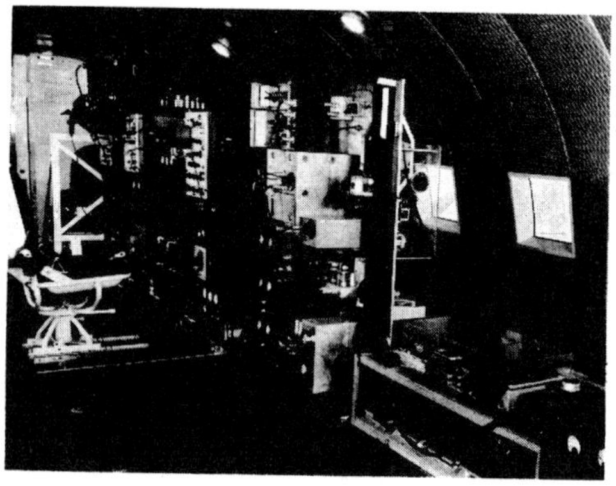

Fig. 4—Breadboard system in the C-46F.

In the foreground of the interior view is a 14-channel Ampex tape recorder which is used to obtain records of aircraft accelerations, antenna orientation, and various voltages within the radar system. Tape records were extremely valuable for post-flight evaluation, for trouble-shooting equipment, for formulating modification requirements, and for improving flight-operation procedures. The first rack of the radar proper houses the coherent transmitter, which is followed in turn by the main control and operators' panel, power supplies, the receiver, and two radar signal displays with camera recorders. The components for the Doppler navigator are mounted in a separate frame at the extreme forward position in the cabin.

For this breadboard system, standard laboratory Tektronix 535 oscilloscopes with P-11 phosphor cathode-ray tubes were used. The recording cameras, which were greatly modified Fairchild scope cameras, continuously advanced 35-mm film in a direction perpendicular to an intensity-modulated range trace on the display. The rate of advance of the film was controlled by ground-speed information obtained from the Doppler navigator. The repetition rate of the radar was also ground-speed controlled to maintain constant film exposure. The radar antenna was maintained perpendicular to the ground track through a combination of drift-angle information from the Doppler navigator and a

Fig. 5—The AN/UPD-1 system.

Fig. 6—Interior of the L-23D.

Fig. 7—Data processor.

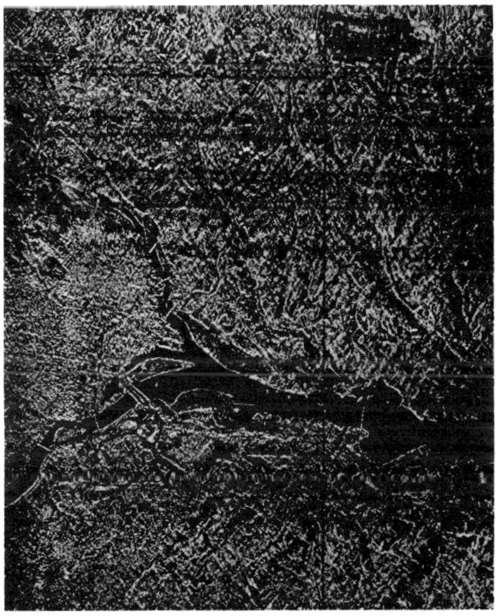

Fig. 8—AN/UPD-1 map of Washington

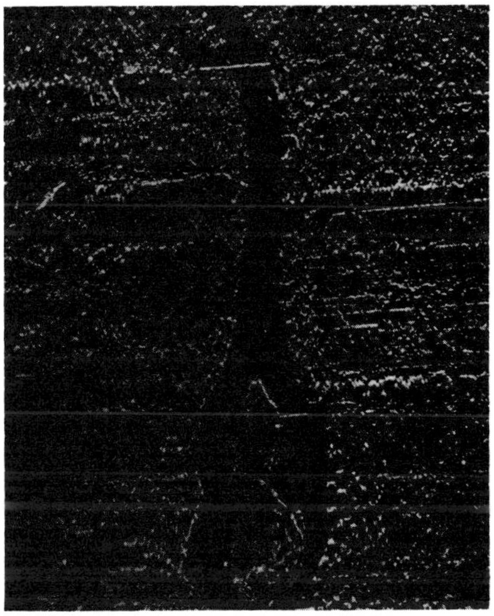

Fig. 9—AN/UPD-1 map of Detroit.

gyro mounted on the antenna base. The navigator established the long-time average while the gyro corrected for short-time yawing.

Success was achieved on August 23, 1957. This was the fifth flight on which data were recorded. After this one successful flight, which demonstrated high angular resolution with the synthetic-antenna technique, it was two months before another successful flight was achieved. As with most R&D programs there was a period in which everything seemed to go wrong. However, there were some results which were encouraging.

Equipment reliability was improved and a number of fruitful flights were made in the Willow Run area. In the spring of 1958 high-resolution strip maps were obtained of the Fort Huachuca area and of Tucson, Arizona.

By this time, the capability of the synthetic-antenna technique was recognized and steps were taken to initiate the procurement of experimental models mounted in U. S. Army aircraft.

The research and development program with the C-46 "flying radar laboratory" is being carried on and the results of the experimentation will be made available for the next generation of high-resolution radars.

IV. The AN/UPD-1 (XPM-1) Systems

In 1958, at the request of the U. S. Army Combat Surveillance Agency, specifications were prepared and a subcontract let to Texas Instruments, Inc., for four experimental models of the airborne portion of the high-resolution radar system installed in L-23D aircraft. The ground portion of the system was fabricated at Willow Run Laboratories and installed in U. S. Army expansible vans. Work started on July 1, 1958.

The first of the four models of the airborne equipment arrived at Willow Run Laboratories in the fall of 1959. Successful high-resolution operation was achieved in the Dallas area, in Michigan, and in the area around Washington, D. C. On April 19 a press conference was held in Washington and some high-resolution radar photographs were cleared for release to the press. Two complete systems were on display at the Washington National Airport.

Fig. 5 shows the aircraft and the ground processing van. As in the C-46F aircraft, the radome houses both the side-looking radar antenna and a Doppler navigator antenna. The van houses, in addition to film-development equipment and a data processor, an enlarging viewer to display high-resolution strip-map photographs at standard map scale. Fig. 6 shows the equipment in the L-23D. The complete weight of the radar including Doppler-inertial auxiliaries is 700 lbs. Fig. 7 is a photograph of the data processor in the processing van.

Figs. 8 and 9 are high-resolution radar photographs made with the AN/UPD-1 systems.

Some Early Developments In Synthetic Aperture Radar Systems*

C.W. Sherwin, J.P. Ruina, and R.D. Rawcliffe
(Sherwin and Rawcliffe — Aerospace Corporation, El Segundo, California [formerly at Coordinated Sciences Laboratory (Control Systems Laboratory), University of Illinois, Urbana] ; Ruina — Advanced Research Projects Agency, Washington, DC [on leave from the University of Illinois])

IRE Transactions on Military Electronics, Vol. MIL-6, No. 2 (April 1962)
Pages 111-115

*Received by the PGMIL 25 January 1962. This program was made possible by support extended to the University of Illinois Control Systems Laboratory jointly by the Department of the Army (Signal Corps and Ordinance Corps), Department of the Navy (Office of Naval Research), and the Department of the Air Force (Office of Scientific Research, Air Research and Development Command) under Signal Corps Contract DA-36-039 SC-56695.

I. Introduction

RECENTLY Cutrona, *et al.*,[1] described a high-resolution radar system in which very high angular resolution was obtained by the processing of the signals from a moving coherent radar system. The purpose of this paper is to describe some of the early developments which laid the foundations of this and other systems which are based upon the principle: frequency analysis of coherent radar signals yields very high angular resolution of targets illuminated by the radar beam. We also wish to describe briefly some of the experiments in the data-processing of such radar signals. These experiments were carried out at the University of Illinois, starting in 1952.

II. Early Development of the Concepts

As is known by everyone who has been associated with military technological developments, the discovery and early application of new concepts and ideas, even though they may be of general interest outside of their classified military application, are often not documented in journal publications. The most that can be done is to search the classified literature as carefully as possible and, at such time that the principles involved may be properly de-classified, describe their development, giving reference to the original authors.

As far as we can determine, the earliest statement of the principle that high angular resolution can be obtained by means of frequency analysis on the signals from a coherent moving radar system was reported by Carl Wiley of Goodyear Aircraft Corporation in June, 1951. Wiley noted that if v is the velocity of a coherent radar system of frequency F_0 with a broad physical beamwidth, and θ_1 is the angle between v and the line from the radar to a fixed reflecting point P_1, then the Doppler shift f_1 in the frequency of the reflected signal is

$$f_1 = (2v/c)F_0 \cos \theta_1. \tag{1}$$

If a second reflecting point P_2 is located at a slightly different angle θ_2, then its signal will have the Doppler shift f_2. The two Doppler frequencies f_1 and f_2 will differ by

$$\Delta f = (2v/c)F_0 \sin \theta\Delta\theta. \tag{2}$$

If both targets are known to be fixed relative to the frame in which the radar is moving, the measurement of f_1 and f_2 yields the two angles θ_1 and θ_2. If the observation of the two signals extends over a time interval ΔT, then a difference $\Delta f \cong 1/\Delta T$ can be distinguished and the effective angular resolution between the two targets can be obtained from (2). In the event that a target point is moving, an additional Doppler shift is added as a result of the radial component of the target's motion; therefore unless this additional velocity is known, the angular position of the target cannot be determined accurately.

Wiley also noted that, with proper allowance for the harmonic properties of the frequency spectrum, a pulsed radar system could be employed. If many different range intervals were to be frequency-analyzed at one time, new developments were needed in frequency-analysis systems and he proposed a possible "comb" filter which might perform this task. Wiley also pointed out the advantages of looking in a direction exactly normal to the flight path. Here sin θ is unity; the Doppler frequency approaches zero, but the rate of change of the Doppler frequency with θ is a maximum.

Independent of Wiley's theoretical work, a program of experimental observations on the nature of radar reflections of ground targets, as observed by an ordinary noncoherent airborne X-band radar system, was under way at the Control Systems Laboratory at the University of Illinois. These experiments were aimed at studying the noncoherent Doppler signals resulting from the presence of both moving and nonmoving targets. In these experiments an X-band radar of beam width $\Delta\theta \cong 4°$ was mounted in a low-flying aircraft O moving at the velocity v and pointed at a fixed angle θ with respect to the ground track of the aircraft (see Fig. 1). The signals from a particular range interval P, of 0.5-μsec duration, located at a fixed range R, were "stretched" or "box-carred" in order to augment their low-frequency components, and then recorded on tape. Sample tapes of terrain signals were selected. These were made into loops for convenience in repetitive reproduction and the signals were frequency-analyzed. In a report dated

[1] L. J. Cutrona, W. E. Vivian, E. N. Leith, and G. O. Hall, "A high-resolution radar combat-surveillance system," IRE Trans. on Military Electronics, vol. MIL-5, pp. 127–131; April, 1961.

March 21, 1952, Kovaly, Lyman, Prothe and Sherwin of the University of Illinois observed that certain terrain samples gave distinctive frequency spectra containing sharp lines. They noted that these sharp frequencies must be due to a small number of strongly reflecting fixed targets located inside the illuminated ground patch P. Thus, in Fig. 1, two point-targets A and B have slightly different radial velocities with respect to the radar at O. They would produce a single modulation frequency in the magnitude of the reflection from the illuminated patch P. In a section entitled "Last minute reflections," Kovaly, *et al.*, noted that this new, experimentally observed effect could provide the basis for radar systems with greatly improved angular resolution.

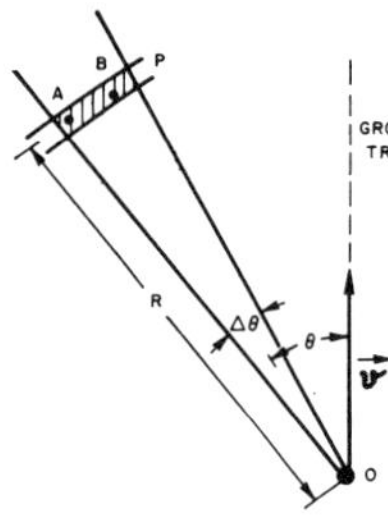

Fig. 1—Method of observing ground targets from a moving aircraft O. The radar beam position θ and the range R of the illuminated patch P are fixed with respect to the aircraft.

In a subsequent report also in March 1952, following shortly upon the initial discovery of the basic effect, Sherwin applied the principle to a pulsed, coherent radar system in which a relatively broad physical beam was held at a fixed angle with respect to the velocity vector of the aircraft, as in Fig. 1, and each range interval, as defined by the pulse length, was to be frequency-analyzed. He showed that with beams directed at 30° with respect to the ground track of a 600-mph aircraft, a high angular resolution could readily be obtained. In April, 1952 Sherwin reported that the resolution could be explained in geometrical terms; *i.e.*, the angular resolution was approximately equivalent to that of a physical antenna whose length was equal to the distance travelled by the aircraft during the time the data-processing system was able to store the phase information. Thus if the signals from one target excited a filter of width Δf so that the signal memory time was $\Delta T = 1/\Delta f$, then the synthetic antenna had an equivalent aperture of approximately $v\Delta T$. In a physical array, n equally spaced dipoles transmit signals in phase (thus the beam is normal to the array), all n dipoles receive signals from a target, and all these signals are added in the antenna itself. In the synthetic antenna, n signals are transmitted in sequence (one for each radar pulse), received in sequence, and stored in such a manner that their phases and amplitudes are preserved. Finally, during the signal-processing operation, the stored signals are added together, thus reconstructing the effective aperture.

III. The Principle of Focusing at all Ranges Simultaneously

In April, 1952 Sherwin also reported that a synthetic antenna could be made to focus at any selected range simply by producing the proper systematic phase shifts in the stored signals from that range as they were read-out for processing. Upon read-out, the phases of the stored signals were to be shifted at just the rate that would have occurred if the aircraft had been flying in a circle with its center at the target. Since the coherent signals from each range interval were to be separately stored and processed, the synthetic antenna could be made to focus at all ranges simultaneously, a feat which is impossible with geometrical antennas. He also noted that random deviations from the ideal straight-line physical flight path could be sensed with an accelerometer and compensated for by an appropriate phase shift in the received signal before it is stored. Thus, in principle, radar systems with very high effective resolution could be designed. One could conceive of having an antenna whose effective aperture D was hundreds or even thousands of feet long, but still have the limit on resolution be determined only by the ratio λ/D where λ is the wavelength.

IV. The First Experimental Demonstration

During the ensuing year a group at the University of Illinois constructed an X-band radar which used an echo box to obtain a coherent reference signal. This equipment, which was installed in a C-46 aircraft provided by and operated by the U. S. Air Force, was used to study reflected signals from both the ground and sea surface as received from a 0.5-μsec interval at a fixed range of about 10,000 ft. The signals from the selected range interval were stretched out with a "box-car" circuit and were tape-recorded for later analysis in the laboratory. In a report dated July 8, 1953, Kovaly, Newell, Prothe and Sherwin produced a strip map (see Fig. 2) of a section of Key West, Fla., which was generated entirely by the frequency analysis of the signal from a single range interval in a nonscanning radar beam whose geometrical width was 4.13°. The effective angular resolution of the system was about 0.4°, due partly to the width of the filter in the frequency analyzer and partly to the inherent frequency modulation of the target as it moved through the range gate. [Because of the geometrical arrangement employed (see Fig. 2, right-hand side) a target enters the range gate at one azimuthal angle and leaves it at a slightly different angle.] As can be seen from Fig. 2. land-water boundaries with the correct geometrical forms are clearly visible, and individual buildings are resolved. Thus the feasibility of the technique was demonstrated experimentally for the first time.

The same basic equipment was used to study sea "clutter," and many distinctive properties of radar echoes

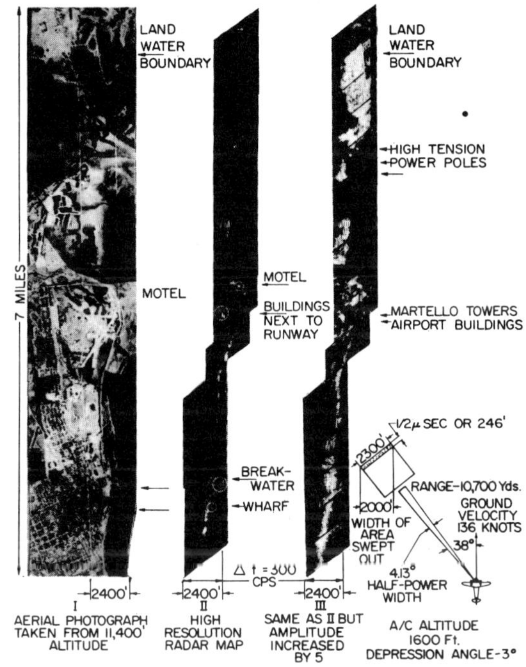

Fig. 2—Strip maps (right) produced early in 1953 of a section of Key West, Fla. These radar maps were made entirely by means of frequency analysis, entirely inside the natural geometrical beamwidth of the nonscanning radar beam. On the left is shown an optical photograph of the same region, demonstrating the close correspondence between the two methods of observation.

from the sea surface were observed.[2,3] Thus, for example, the motion of waves and spray was clearly observable. Radar echoes from snorkels,[2] arising from the uniform motion of the submarine, appeared as sharp, narrow frequency bands which in general were displaced from the broader spectrum sea-clutter signals.

V. Further Extension of the Theory

During the summer of 1953 the above theoretical analyses and the experimental observations were communicated to the members of the summer study, Project WOLVERINE, at the University of Michigan. These results formed the basis of the program which culminated in the Surveillance System described by Cutrona, *et al.*[1] In early 1954 Cutrona and Vivian reported that, for a side-looking system with optimum data-processing, the ultimate focused resolution was 1) of a linear dimension approximately equal to the physical size of the illuminating aperture, 2) independent of range, and 3) independent of wavelength. These results are published in the 1961 report.[1]

VI. Comparison of Real and Synthetic Aperture Radar Systems

There are a number of significant differences between physical and synthetic apertures.

One feature of the synthetic antenna of effective length D is that its angular resolution is actually twice that of a physical dipole array of the same aperture D. In the physical array the beam pattern is determined by the *one*-way phase shifts and in the synthetic array by the *two*-way phase shifts. The extra information obtainable from the synthetic array (permitting the higher resolution) arises from the fact that the information from each dipole is separately transmitted, stored and processed.

Consider the most simple case (see Fig. 3). On the left [in Fig. 3(a)] is a physical array of ordinary dipoles, uniformly illuminated. All the dipoles radiate and receive simultaneously, each one receiving from the target the energy which was originally radiated from all of the others. The radiated and received antenna patterns are identical, and their half angle to the first null is given by $\theta_n = \lambda/D$.

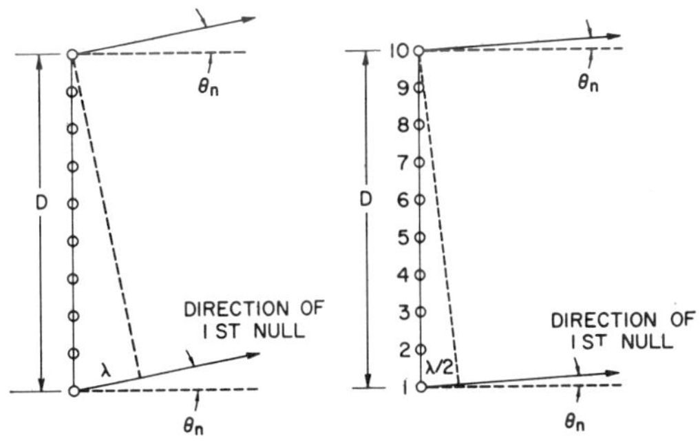

(a) (b)

Fig. 3—The comparison of the angular resolution of the physical and the synthetic dipole arrays. Uniform illumination. (a) Physical dipole array $\theta_n = \lambda/D$. (b) Synthetic antenna $\theta_n = (\lambda/2)/D$.

For the synthetic antenna [in Fig. 3(b)] we consider the case of a single dipole moving with a uniform velocity and transmitting single pulses at regular intervals. Let the radar have such a pulse recurrence frequency and velocity that the points where the transmissions occur are located at exactly the same positions as for the physical array. Each dipole transmission of the synthetic array has a very broad physical radiation pattern. Here each dipole receives from the target *only* the radiation which was transmitted by itself. The received signal, including both amplitude and phase information, is assumed to be stored at each dipole.

[2] J. J. Kovaly, G. S. Newell, W. C. Prothe and C. W. Sherwin. "The Observations of Snorkels and Sea Clutter Using Coherent Airborne Radar," Control Systems Lab., University of Illinois, Urbana, Ill., Rept. R-27; November 20, 1952.

[3] B. L. Hicks, N. Knable, G. S. Newell, J. P. Ruina and C. W. Sherwin, "The spectrum of x-band radiation backscattered from the sea surface," *J. Geophys. Res.*, vol. 65, pp. 825–837; March, 1960.

After all ten signals are stored they are read back simultaneously, which corresponds to directing the synthetic array exactly normal to the flight path. (This follows since, if a distant point target were located exactly normal to the flight path, the stored signals would all have the same phase and amplitude and upon simultaneous addition would yield the maximum sum.) If however, a distant target is assumed to be located in the direction θ_n in Fig. 3(b), the signal received and stored at dipole 1 is out of phase with that stored at dipole 10. If reproduced simultaneously the sum of these signals will be (substantially) zero, producing the first null in the synthetic array at $\theta_n = \lambda/2D$. Thus for a given effective aperture D, the synthetic array has only half the angle between nulls as for the physical array of the same aperture D. In between the nulls of the synthetic antenna, however, the sum signal varies as $\sin \theta/\theta$ rather than as $\sin^2 \theta/\theta^2$ as in the physical array.

The ability to focus the synthetic antenna at all ranges at once also gives the synthetic array greater information-collecting efficiency than that of a single physical array. Again this property arises from the fact that the information from each dipole and from each range element is separately and optimally processed.

The requirement of a physical array that it be mechanically rigid has its parallel in the requirement for the synthetic array that the RF phase in successive pulse transmissions be preserved. However, deviations of the elements of a synthetic array from a straight line by virtue of nonlinear aircraft motion can in principle be sensed and corrected by properly varying the phase of the RF signal before processing.

The spacing between elements in a synthetic array is determined by the pulse repetition frequency of the radar and the distance the aircraft travels between pulses. When both the radar range and aircraft speed are high, the spacing between elements may far exceed λ, thereby causing azimuth ambiguities. The flexibility available in determining the spacing in a real antenna array may not be available here.

Another difference to note is that in a physical array the linearity of the summing process is not an issue. The signals from the array elements are summed at RF in a waveguide or cable and the process is one of extremely wide dynamic range and linearity. However, this is not the case in the synthetic array where the data processor may be limited in both dynamic range and linearity, thereby degrading the effective antenna pattern.

VII. Further Experimental Tests of Coherent Radar and Some Early Developments in Signal-Processing

The realization of a practical coherent high-resolution radar primarily involved the development of a suitable data-processing device. The heart of a data processor is a storage element which may be a delay line, a photographic film, a storage tube, resonant filters, magnetic drums, or tape. In about 1954, several other projects were started, all of them aimed at developing practical data-processing equipment. The earliest work was done at the Goodyear Aircraft Corporation where the emphasis was placed on a filter-bank processor. The Philco Corporation developed a re-entrant delay-line processor which was later used in flight tests at the University of Illinois. The University of Michigan started the development of a processor, referred to in Cutrona, *et al.*,[1] which employed photographic storage and was designed ultimately to permit the much desired all-range focusing.

In this section we describe the experimental results obtained by using three different processors in the program carried out at the Control Systems Laboratory at the University of Illinois: 1) a re-entrant delay line, 2) photographic film, and 3) a storage tube. This program was reported in detail in April, 1959 by Ruina and Rawcliffe, and only a brief outline is given here.

In these experiments, a "coho-stalo" type of coherent radar[4] was employed. In this system the memory of the transmitted phase is preserved in a coherent phase-locked IF oscillator which uses a very stable local oscillator as a reference. With this method coherent operation of the radar could be obtained to ranges much beyond the 2 or 3 miles permitted by the simple echo box used in the first feasibility experiments.

Since frequency analysis at a multiplicity of range intervals was now possible, the new experiments were all performed with the synthetic beam directed exactly normal to the flight path. It is essential to understand in the following discussion that for this condition of operation the property of the signals from a target have a distinctive behavior with time. The physical beam is assumed to be directed normal to the flight path. Consider a point target as it first appears in the physical beam and there produces an observable signal. At this stage the target is slightly forward of the exact normal to the flight path and, in fact, has a small radial velocity approaching the aircraft. Thus the signal frequency is slightly higher than the transmitted frequency of the constant-amplitude coherent reference oscillator. The sum of the IF signal and the reference oscillator modulates (with the Doppler frequency) about a constant-amplitude IF reference level set by the coherent reference signal itself. This sum signal when rectified is called the "bipolar video." As the target approaches the exact normal to the flight path, its Doppler signal becomes lower and for a short time, while the target crosses this line, the sum of its signal plus the reference has a nearly constant amplitude (the "zero-beat" condition). When the target gets aft of the normal it recedes from the radar and again develops a Doppler modulation in the radar output.

If now an integrating device is constantly summing the output signal, whenever the target is *off* the normal line to the flight path the integrator sum will be very nearly constant and is due only to the presence of the reference

[4] R. McConnell, "Radar Systems Engineering," M.I.T. Radiation Lab. Ser., McGraw-Hill Book Co., Inc., New York, N. Y., vol. 1, pp. 626–638; 1947.

signal. However, during the normal-crossing or "zero-beat" interval, the sum will tend to deviate from the standard sum. For example, if the target signal happens to have such a range that it is in-phase with the reference, then the sum signal from the integrator will be greater than the reference level. Similarly if the target signal happens to be out of phase with the reference, the sum signal will be below the reference level. A target in exact quadrature phase is not observed. By rectifying the sum signal and presenting it as intensity modulation on a display, one then obtains the high-resolution map.

The first experiments were with a re-entry delay line which was developed at the Philco Corporation.[5] Its application to the high-resolution problem was suggested by D. E. Sunstein and B. Steinberg. This delay line has a 60-db insertion loss which was compensated for by a wide-band FM amplifier. Even so, only about 100 signals could be added before the distortion of the initial $\frac{1}{2}$-μsec pulse became so extensive that range resolution suffered noticeably. An effective antenna length at X band of about 35 ft was obtained by using a line with a delay of approximately 1400 μsec and an aircraft flying at 250 ft/sec. Fig. 4 shows a typical map taken with this radar. On the left the delay-line integrator is operating, giving the high angular resolution. On the right the map is obtained from the ordinary noncoherent signal which is limited in angle only by the physical width of the beam from the 5-ft aperture antenna (3.5°). Here the delay line gives an improvement of a factor of 7 in angular resolution. This is, of course, far below the theoretical focused maximum at this range (5 ft at about 20 miles, or about 1/300°), but it does demonstrate the feasibility of the delay-line processor.

The long steady signal at the "zero-beat" point can also be sensed by direct integration of the "bipolar video" on a photographic film, a possibility first pointed out by W. Hausz of the Electronics Laboratory of the General Electric Company in October, 1952. As long as the target is producing an oscillating amplitude bipolar video signal, and if the film is linear in its response, then a uniform gray level is produced. However, when the target passes through the zero-beat condition the film usually receives either an unusually high or an unusually low integrated exposure (the finite spot size of the cathode-ray tube display provides the necessary overlapping for the proper integration) identifying the exact spot at which the target reached a point normal to the flight path.

In spite of the bipolar or unrectified nature of the high-resolution signals, when this technique was used in the Illinois flight tests, Ruina and Rawcliffe were able to demonstrate recognizable high-resolution maps recorded directly on film. To demonstrate this effect it was necessary to develop cathode-ray tubes with a highly linear response and to use carefully controlled film-processing. The quality was seriously limited by the lack of uniformity of the phosphors in the cathode-ray tube.

[5] D. E. Sunstein, U. S. Patent No. 2,736,021; February 21, 1956.

Fig. 4— Demonstration of the improved angular resolution using a delay-line integrator on a coherent X-band radar. On the left the integrator is processing about 100 pulses of $\frac{1}{2}$ microsecond duration, producing an effective aperture of 35 ft. On the right the noncoherent signals are limited in angle only by the 5-ft physical aperture of the radar. Aircraft at 8500-ft altitude. 13–23-mile range interval. Showing region near Peoria, Ill., May 12, 1958.

The Illinois group also developed an electronic storage tube integrator. Here the bipolar video is recorded on the storage surface until it is loaded to capacity. The integration process occurs on the storage surface. The tube is then read-out, full-wave rectified (to remove the bipolar form), and presented on an ordinary display. With the storage tube used, an effective aperture of approximately 90 ft for targets at ranges of 30 miles was obtained. The actual flight tests using the Illinois storage tube system were performed with the University of Michigan aircraft and radar equipment.

The storage tube processing method was potentially the best of the three tried by the Illinois group. It has not, however, been carried beyond the feasibility test quoted above.

Because of the great difficulty in storing and reproducing exact analog data, an analysis was performed by Bitzer[6] which showed that phase shifts stored in quantized form produced tolerable degradation in the equivalent aperture. 12° steps were calculated to produce generally unimportant changes in either the main lobe or the sidelobes, but 90° steps produced large sidelobes. These results, however, have not been tested on synthetic aperture systems.

ACKNOWLEDGMENT

The research program described here extended over a period of a number of years and it was the result of the contributions of many people. The authors particularly wish to acknowledge the invaluable contributions of E. M. Lyman, N. Knable, G. S. Newell, J. J. Kovaly, W. C. Prothe, D. L. Bitzer, D. Skaperdas, J. Gooch, and H. V. Krone.

[6] D. L. Bitzer, "Steering and Focusing Antenna Beams by the Use of Quantized Phase Shifts," Control Systems Lab., University of Illinois, Urbana, Ill., Rept. No. I-84; 1959.

CHAPTER II
Comparative Synthetic Aperture Radar Theory

The advancement of a new principle or evolutionary concept such as the synthetic aperture and its application in the synthetic aperture radar requires a degree of understanding which can be obtained only by comparison to similar and known principles and concepts. This chapter includes those papers which have made the most significant comparisons of synthetic aperture radar; as such, they have made its principles of operation more understandable.

In the first paper, L.J. Cutrona and G.O. Hall compare the azimuth resolution of the more common physical aperture with the improved resolution which can be obtained with the synthetic aperture. The azimuth resolutions of a focused and an unfocused synthetic aperture are included. Three expressions for achieving azimuth resolution are shown to be dependent on the basic parameters of λ, the wavelength of the radar signal transmitted; D, the horizontal aperture of the antenna; and R, the radar range. Most significantly, the case of the focused synthetic aperture is shown to provide the fine resolution which is independent of the transmitted frequency and is dependent only on the size of the physical aperture. For other comparisons of the synthetic aperture with antenna theory the reader is referred to Bickmore (1966) and Collin and Zucker (1969); sideways looking airborne radar antennas are discussed in Cary and Thraves (1971).

H.L. McCord, in the second paper, shows that vector addition, cross-correlation, and filtering are three equivalent approaches for deriving synthetic array patterns and for analyzing processing techniques. The three formulations are shown as derived from one basic equation. The systems designer of a synthetic aperture radar may use the vector addition approach to derive stability requirements, cross-correlation to investigate processing techniques, and the Doppler filtering formulation to arrive at the range-angle ambiguity limits.

The optical analog of the synthetic aperture radar is described by W.M. Brown and L.J. Porcello in the third paper. They demonstrate that the radar data film which is obtained by a synthetic aperture radar is an analog of the all-optical holographic technique. This paper provides a good base for the understanding of the optical processors which are instrumental in providing the most detailed radar images. For more reading on the microwave holographic aspects of synthetic aperture radars, the reader is referred to Kock (1970 and 1971).

In the fourth paper, J.J. Kovaly shows the similarity between improved range and azimuth resolutions. The technique of frequency coding the transmitted pulse to obtain range resolution which is smaller than the pulse is shown to be similar to the generation of a Doppler frequency in synthetic aperture to obtain azimuth resolution better than the beamwidth of the physical aperture. The two operations are linearly independent and are used in modern synthetic aperture radars.

The reader is referred to Rihaczek (1969) and Cutrona (1970) to learn about the properties of the ambiguity function which can indicate the azimuth and range resolution of a synthetic aperture radar.

A Comparison Of Techniques For Achieving Fine Azimuth Resolution*

L.J. Cutrona and G.O. Hall
(University of Michigan, Ann Arbor)

IRE Transactions on Military Electronics, Vol. MIL-6, No. 2 (April 1962)
Pages 119-121

*Received by the PGMIL 15 January 1962. This work was conducted by Project MICHIGAN, under Department of the Army Contract DA-36-039 SC-78801, administered by the US Signal Corps.

***Summary*—In the discussion of techniques for achieving azimuth resolution, it is instructive to compare the achievable resolution for three cases:**

1) the conventional case for which

$$\text{Res} = k\frac{\lambda R}{D}$$

2) the unfocussed synthetic antenna case for which

$$\text{Res} = k\sqrt{\lambda R}$$

3) the focussed synthetic antenna case for which

$$\text{Res} = kD$$

where λ is wavelength, D is aperture of physical antenna used, and R is range, and the k's are constants of proportionality.

These cases are compared. It is shown for unfocussed synthetic antennas that there is a maximum length which depends on range and wavelength. This restriction on length can be removed by focussing.

A number of degrees of freedom exist for the design of a synthetic antenna which are not available for a real linear array. Among these is the ability to focus simultaneously at all ranges.

I. Introductory Remarks

THERE HAVE BEEN a number of important developments in airborne radar which affect the azimuth resolution capability of such equipments. This paper will concern itself chiefly with a discussion of some aspects of the azimuth resolution technique together with a comparison of these techniques with more conventional means for achieving azimuth resolution.

The application of the notion of the generation of a synthetic antenna for achieving fine angular resolution was discussed by Cutrona, *et al.*[1] In that paper two radar maps using the synthetic antenna technique for azimuth resolution improvement were shown and a derivation was given which indicates that a linear resolution transverse to range equal to half the horizontal antenna aperture is theoretically achievable. This expression for the resolution was derived by making an analogy between the synthetic antenna and a linear array. No comments were made regarding considerations which need to be taken into account in order to synthesize an antenna aperture of sufficient length to give a resolution equal to half the horizontal antenna aperture.

It is a purpose of this paper to discuss the factors which are pertinent for achieving a synthetic antenna length sufficient for achieving maximum resolution. The processing which is implied in this case will be defined as focussing the antenna.

Consideration will also be given to the case in which focussing is not used. In this case there is a limitation to the length of synthetic aperture which can be synthesized and, therefore, also a limit to the fineness of resolution which can be achieved in the unfocussed case.

A major objective of this paper will be to compare the resolution performance capability of three means for achieving azimuth resolution, namely:

1) *The conventional technique:* In this technique azimuth resolution depends upon the width of the radiated beam.
2) *The unfocussed synthetic antenna technique:* In this case the synthetic antenna length is made as long as the unfocussed technique permits.
3) *The focussed synthetic antenna technique:* In this case the synthetic antenna length is made equal to the linear width of the radiated beam at each range.

As is shown in the sections which follow, the linear transverse resolution for the *conventional case* is given by

$$\text{Resolution}_{conv} = \frac{\lambda R}{D}. \tag{1}$$

For the *unfocussed case*, the resolution is given by

$$\text{Resolution}_{unf} = \tfrac{1}{2}\sqrt{\lambda R} \tag{2}$$

whereas for the *focussed case*, the resolution is given by

$$\text{Resolution}_{foc} = \frac{D}{2}. \tag{3}$$

In the above expressions λ is the wavelength of the radar signal transmitted, D is the horizontal aperture of the antenna, and R is the radar range.

Fig. 1 is a plot of the resolution for each of these cases as a function of radar range. This plot is for an antenna aperture of 5 ft and a wavelength of 0.1 ft.

[1] L. J. Cutrona, W. E. Vivian, E. N. Leith and G. O. Hall, "A high-resolution radar combat-surveillance system," IRE Trans. on Military Electronics, vol. MIL-5, pp. 127–131; April, 1961.

II. Conventional Azimuth Resolution

The conventional technique for achieving azimuth resolution has been that of radiating a narrow beam. In this case the resolution of a target depends upon whether the target is included within the radiated beam. While some techniques exist for resolving targets somewhat less than a beamwidth apart, it will be assumed for the purposes of this paper that the resolving capability of an antenna is equal to the width of the radiated beam at each range.

The computation of the linear azimuth resolution for the conventional case is well known. The appropriate expression is obtained by noting that the width of the radiated beam in radians is given by the ratio λ/D whereas the linear width of the beam at range R is the product of this beamwidth by range. These considerations lead to the result already written as (1).

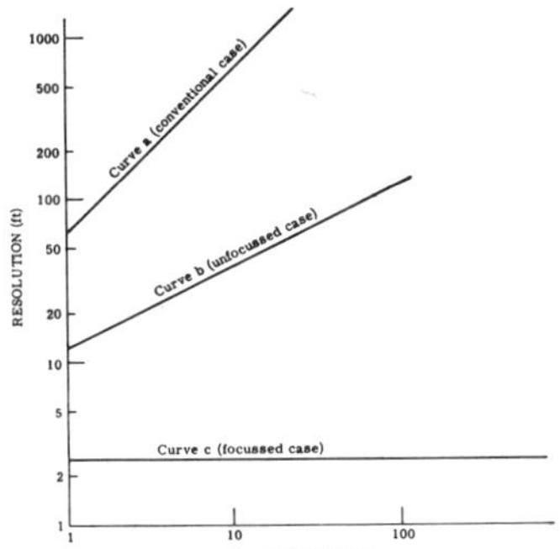

Fig. 1—Azimuth resolution for three cases: (a) conventional, (b) unfocussed, (c) focussed.

It is noted that for a given wavelength and antenna aperture the fineness of resolution deteriorates in direct proportion to range. For a specified antenna aperture and range the linear resolution can be improved by using a shorter wavelength. Conversely, with the wavelength fixed, an increase in antenna aperture is required to improve resolution.

A consideration not evident from the above discussion but pertinent from antenna theory is that (1) applies only to the far-field pattern of an antenna. The beginning of the far field occurs at a distance R_{min} for which

$$R_{min} \simeq \frac{D^2}{\lambda} \cdot \tag{4}$$

It will be noted by substitution of (4) that the finest resolution achievable using the conventional technique is given by

$$\text{Minimum Conventional Resolution} = D. \tag{5}$$

III. The Unfocused Synthetic Antenna

The simpler of the synthetic antenna techniques is that which generates an unfocussed synthetic aperture. In this case the coherent signals received at the synthetic array points are integrated with no attempt made to shift the phases of the signals before integration. This lack of phase adjustment imposes a maximum length upon the synthetic antenna length which can be generated. This maximum synthetic antenna length occurs at a given range when the round-trip distance from a radar target to the center of the synthetic array differs by $\lambda/4$ from the round-trip distance between the radar target and the extremities of the synthetic antenna array.

The pertinent geometry is shown in Fig. 2. In this figure R_0 represents the range from a radar target to the center of the array, L_{eff} represents the maximum synthetic antenna length for which the distance from the target to the extremities of the synthetic antenna does not exceed $R_0+\lambda/8$.

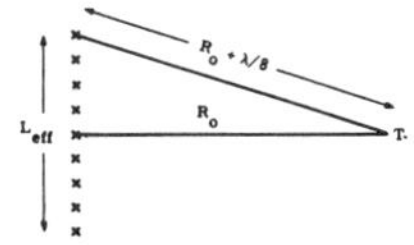

Fig. 2—Geometry for unfocussed synthetic antenna.

It is evident from this geometry that

$$\left(R_0 + \frac{\lambda}{8}\right)^2 = \frac{(L_{eff})^2}{4} + R_0^2. \tag{6}$$

If this expression is solved for L_{eff}, subject to the assumption that $\lambda/16$ is small compared to R_0, the result is

$$L_{eff} = \sqrt{R_0\lambda}. \tag{7}$$

Because of coherence, the relative phase of signals received from a target depends on the round-trip distance to it. Consequently, the beamwidth of the synthetic antenna is half that of a conventional array of the same length, *i.e.*,

$$\beta = \frac{\lambda}{2L_{eff}} \text{ radians}. \tag{8}$$

Multiplying this beamwidth by range results in the resolution given by (2).

As shown by (2) the transverse linear resolution is independent of the antenna aperture size. For the unfocussed case fineness of resolution is increased by the use of shorter wavelengths, but in comparison with the conventional case, the improvement in fineness of resolution varies as the square root of λ rather than directly as λ.

It will also be observed that the resolution as a function of range deteriorates as the square root of range rather than directly as range for the conventional case. A plot of (2) is given in Fig. 1.

IV. The Focussed Case

An expression for the resolution achievable in the focussed case was given in Cutrona, *et al.*[1] This expression has been given as (3). It is significant to note that the azimuth resolution achievable for this case depends only upon the synthetic antenna aperture and that in contradistinction to the conventional case finer resolution is achievable by using small rather than large antennas.

Also significant for this case is the fact that the achievable resolution for a given antenna size is independent both of the range and of the wavelength used. A graph of (3) is also shown on Fig. 1. Inasmuch as the resolution achievable in this case is a constant, this curve is a straight line parallel to the range axis.

In order to achieve the resolution indicated by (3), a synthetic antenna length given by (9) is required, namely

$$L_{eff} = \frac{\lambda R}{D}. \tag{9}$$

The considerations used in arriving at (7) indicated that unless additional processing were applied to the signals, antenna lengths such as those implied by expression (9) could not be achieved. The processing required is that of an adjustment of the phases of the signals received at each point of the synthetic antenna which makes these signals co-phase for a given target. If this is done, the restrictions which limited the maximum antenna length to that given by (7) are no longer pertinent, and the new limitation on the length of synthetic antenna achievable becomes simply the linear width of the radiated beam at the range of the target. It will be noted that (9) gives an expression for the maximum length in this case.

In some cases a resolution coarser than $D/2$ is sufficient. Then a fraction α of the maximum focussed synthetic antenna length can be used. Eq. (9) becomes

$$L_{eff} = \frac{\alpha\lambda R}{D} \tag{9a}$$

and the achievable resolution is

$$\text{Resolution}_{foc} = \frac{D}{2\alpha}. \tag{3a}$$

For those situations for which the synthetic antenna length given by (9a) is less than or equal to the synthetic antenna length for the unfocussed case as given by (7) only a limited improvement in resolution is achievable for the focussed case. However, if a resolution finer than that given by (2) is desired, then focussing must be used, and in this sense the use of focussing, in effect, removes the restriction on synthetic antenna length which would otherwise apply.

V. Summarizing Remarks

The discussions in Sections II, III and IV have indicated that the resolution capabilities of conventional unfocussed and focussed antenna systems differ markedly in their characteristics. It has been shown that the fineness of resolution, other parameters being fixed, varies directly as λ for the conventional case, as the $\sqrt{\lambda}$ for the unfocussed case, and is independent of λ for the focussed case.

The dependence of fineness of resolution on antenna size has been shown to improve for larger apertures in the conventional case, to be independent of aperture in the unfocussed case, and to improve with smaller apertures for the focussed case.

The dependence of the fineness of resolution on range has been shown to deteriorate directly as range in the conventional case, as the $\sqrt{R}$ in the unfocussed case, and to be independent of range for the focussed case. These results are summarized in graphical form in Fig. 1.

It is evident from these figures that as the requirement for finer azimuth resolution increases, one may be led to consider, in turn, the conventional, the unfocussed and the focussed cases in that order.

The Equivalence Among Three Approaches To Deriving Synthetic Array Patterns And Analyzing Processing Techniques*

H.L. McCord
(Aerospace Group, Hughes Aircraft Company, Culver City, California)

IRE Transactions on Military Electronics, Vol. MIL-6, No. 1 (January 1962)
Pages 116-119

*Received 2 December 1961

***Summary*—The equivalence of the vector addition, cross-correlation, and filtering approaches to deriving synthetic array patterns and analyzing processing techniques is demonstrated here. A mathematical model is defined which establishes a geometry, a transmitted and received signal, and a general synthetic array weighting function. A preliminary analysis of this model derives the signals received at the flight-path points where radar pulse transmission and reception occurred, and the effects at these flight-path points of the synthetic array weighting function. The weighted, received signal expression is then summed over the array, and the resultant shown to be a vector addition, a cross-correlation integral, or a filtering operation depending on the changes of variable and formal manipulations employed.**

I. Introduction

A SYNTHETIC array radar consists basically of a coherent radar whose antenna becomes a new element of the synthetic array with each succeeding transmission, and of a data-processing system whose function is to store and to sum coherently the data collected (with successive pulses) by the antenna. In order to understand the principles behind the operation of a synthetic array radar system, it is helpful to consider the means by which a conventional linear antenna forms a beam. Signals from reflecting objects are received essentially simultaneously by the elements of the array, a vector summation of the signals is performed by the feed system of the array, and the resultant is sent to the receiver. The signals received by each of the elements in the array differ in phase. In the direction of the main lobe, these signals will be essentially in phase, while, for other directions, there is a phase progression along the array. Thus, the magnitude of the vector sum changes as the antenna is scanned and this variation gives the familiar pattern of major and minor lobes.

There is no fundamental reason why signals incident upon different elements of the array must be received simultaneously. The vector integration can take place just as well in a data processor used in conjunction with one element of the linear array which successively assumes the positions of the several array elements. If successive signals are collected by a small antenna and stored with both amplitude and phase preserved, they can be vectorially summed at the end of the sampling time. The result will be a narrow beam which is substantially equivalent to that achievable with a large, linear array. Thus, by using synthetic array techniques, a single, small antenna can have an effective beamwidth as narrow as that which could be obtained by a long linear array.

Typically, the worker in the synthetic array field chooses an analytical approach convenient to the problem at hand. For example, stability requirements may be derived from the vector addition point of view; a processing technique may be investigated from the cross-correlation approach; and the range-angle ambiguity limits may be determined from the Doppler filtering formulation, all in one synthetic array system design. The equivalence of the vector addition, cross-correlation, and filtering approaches to deriving synthetic array patterns and analyzing processing techniques, which is usually taken as a lemma, is demonstrated here.

II. Model and Preliminary Analysis

Consider the basic geometry of Fig. 1. The length of the line x_1T is

$$x_1T = \sqrt{R^2 + (x_1 - X)^2} \tag{1}$$

which is, for $(x_1 - X)/R \ll 1$, well approximated by

$$x_1T \approx R\left[1 + \frac{(x_1 - X)^2}{2R^2}\right] \tag{2}$$

so that

$$p = x_1T - R \approx \frac{(x_1 - X)^2}{2R}. \tag{3}$$

The length p is approximately the difference in line lengths between x_1T and XT.

Fig. 2 shows the geometry associated with the aircraft flight path and target. The aircraft is assumed to transmit and receive radar pulses at the flight-path points indicated; a formulation of the resultant signals from the target in both the X and t dimensions follows.

Let the transmitted signal be

$$f(t) = \operatorname{Re} A(t)e^{i\omega_0 t} \tag{4}$$

where

$A(t)$ = envelope, typically pulse modulation
ω_0 = carrier frequency, radians per second

and t is zero when x of Fig. 2 is zero. (It is to be noted that the x-coordinate system is moving.) The received signal at any position along the array from a particular target may then be written

$$s(x, t) = \operatorname{Re} kA(t - \alpha)e^{i\omega_0(t-\alpha)}F(x) \tag{5}$$

where

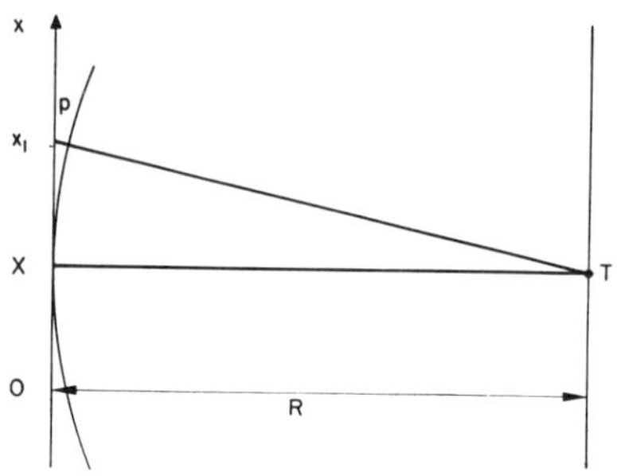

Fig. 1—Basic geometry.

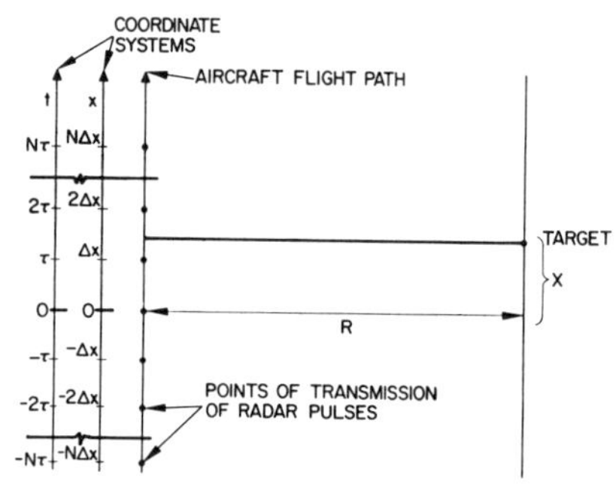

Fig. 2—Flight path and target geometry.

k = constant—a function of range, reflectivity, etc.,[1]

α = round-trip time delay to target from the array position of interest,

$F(x)$ = complex synthetic array weighting function (actually applied by the data-processing following reception).

Using (2),

$$\alpha = \frac{2}{c}\left[R + \frac{(x - X)^2}{2R}\right] \tag{6}$$

where X is the position of the target with respect to $x=0$ (the array center), and c is the velocity of propagation. By substituting (6) into (5), an expression for the signal received at any position along the array from a particular target results:

$$s(x, t) = \operatorname{Re} kA\left\{t - \frac{2}{c}\left[R + \frac{(x - X)^2}{2R}\right]\right\} e^{i\omega_0 t} \cdot e^{-i(2\omega_0 R/c)} F(x) e^{-i(\omega_0/Rc)(x-X)^2}. \tag{7}$$

Of particular interest are the signals received at the various flight-path points where pulse transmission and reception occurred. Substitution of $x = n\Delta x$ into (7) gives

$$s(n\Delta x, t) = \operatorname{Re} kA\left\{t - \frac{2}{c}\left[R + \frac{(n\Delta x - X)^2}{2R}\right]\right\} e^{i\omega_0 t} \cdot e^{-i(2\omega_0 R/c)} F(n\Delta x) e^{-i(\omega_0/Rc)(n\Delta x-X)^2}. \tag{8}$$

The successive returns from each range of interest are summed separately, so that the array output for this target is

$$E(X, t) = \sum_{n=-N}^{N} s(n\Delta x, t) \tag{9}$$

or[2]

$$E(X, t) = \operatorname{Re}\left\{kA_R e^{i\omega_0(t-2R/c)} \sum_{n=-N}^{N} F(n\Delta x) \cdot e^{-i(\omega_0/Rc)(n\Delta x-X)^2}\right\}. \tag{10}$$

Letting

$$S(X) = kA_R \sum_{n=-N}^{N} F(n\Delta x) e^{-i(\omega_0/Rc)(n\Delta x-X)^2} \tag{11}$$

then

$$E(X, t) = \operatorname{Re}\left\{ |S(X)|\, e^{i\omega_0(t-2R/c+\theta_s/\omega_0)}\right\}$$

or

$$E(X, t) = |S(X)| \cos \omega_0\left(t - \frac{2R}{c} + \frac{\theta_s}{\omega_0}\right) \tag{12}$$

where $|S(X)|$ and θ_s are defined by the relation

$$S(X) = |S(X)| e^{i\theta_s}.$$

Thus, the amplitude of the synthetic array output in this case (k a constant) is the magnitude of $S(X)$. In the following sections, (11) will be viewed as an addition of vectors, a cross-correlation, and a filtering operation.

III. Vector Formulation

Consider the following vector, $r(n\Delta x, X)$:

$$r(n\Delta x, X) = kA_R F(n\Delta x) e^{-i(\omega_0/Rc)(n\Delta x-X)^2}. \tag{13}$$

If it is understood that $r(n\Delta x, X)$ is rotating at ω_0 radians/sec, and that $-2\omega_0 R/c$ is the reference, or zero, phase for this vector, and if it is assumed that subsequent operations on this vector are carried out independently for each range, then (13) is the vector representation of (8).

The sum of the $2N+1$ vectors can be written

$$R(X) = kA_R \sum_{n=-N}^{N} F(n\Delta x) e^{-i(\omega_0/Rc)(n\Delta x-X)^2}$$

which is identical to (11).

Eq. (11) shows how the vector sum $S(X)$ of the signals received at the array elements varies with the displacement, X, of the target from the array center. Moving the target displacement X through a number of values will

[1] The effect of the real antenna on the amplitude and phase of the received signal has been assumed negligible. Such effect can be accounted for by a suitable modification of $F(x)$.

[2] A_R denotes $A(t-2R/c)$; the quantity $(n\Delta x - X)^2/2R$ is negligible.

result in a variation in $S(X)$, the magnitude of which gives the antenna pattern.

The complex synthetic array weighting function could be written

$$F(n\Delta x) = W(n\Delta x)e^{i\phi(n\Delta x)}$$

where $W(n\Delta x)$ is the amplitude weighting function and $\phi(n\Delta x)$ is the phase weighting function, however, no loss of generality obtains from not making this substitution. Typically, the phase weighting function, $\phi(n\Delta x)$ could be:

$$\phi(n\Delta x) = \frac{\omega_0}{Rc}(n\Delta x)^2$$

in order to focus the array for targets at range R and to get maximum response at $X=0$.

If

$$W(n\Delta x) = 1$$

and

$$\phi(n\Delta x) = \frac{\omega_0}{Rc}(n\Delta x)^2,$$

then (11) becomes

$$S(X) = kA_R e^{-i(\omega_0/Rc)X^2} \sum_{n=-N}^{N} e^{i(2\omega_0/Rc)n\Delta x X}$$

or

$$S(X) = kA_R e^{-i(\omega_0/Rc)X^2} \frac{\sin (2N+1)\frac{\omega_0}{Rc}\Delta x X}{\sin \frac{\omega_0}{Rc}\Delta x X}.$$

The magnitude of this last equation is recognized as the familiar pattern associated with equally weighted arrays. It has a maximum of $(2N+1)kA_R$, a first null at $X=(\pi/2N+1)(Rc/\omega_0\Delta x)$, and it is periodic at intervals of $X=\pi Rc/\omega_0\Delta x$. This periodicity results in spurious main lobes, which are usually suppressed by the real antenna pattern.

IV. Cross-Correlation Formulation

Eq. (11) is in the form of a cross-correlation, with X the variable. The signal return, $kA_R e^{-i(\omega_0/Rc)}(n\Delta x - X)^2$ is being cross-correlated with the antenna weighting function, $F(n\Delta x)$, to give the function $S(X)$. It is instructive to change from x coordinates to time coordinates. Referring to Fig. 2 and making the changes of variable, $\Delta x = v\tau$ and $X = v\beta$, (8) may be rewritten:

$$s(nv\tau, t) = \operatorname{Re} kA\left\{t - \frac{2}{c}\left[R + \frac{v^2}{2R}(n\tau - \beta)^2\right]\right\}e^{i\omega_0 t} \cdot e^{-i(2\omega_0 R/c)}F(nv\tau)e^{-i(\omega_0 v^2/Rc)(n\tau-\beta)^2}. \tag{14}$$

If, again, the processing is carried out independently for each range, then the envelope, or slowly varying portion, of (14) is

$$\langle s(nv\tau, t)\rangle = kA_R F_1(nv\tau)e^{-i(\omega_0 v^2/Rc)(n\tau-\beta)^2}. \tag{15}$$

Defining the following functions:

$$\langle s_1(n\tau, t)\rangle = \langle s(nv\tau, t)\rangle$$
$$F_1(n\tau) = F(nv\tau)$$
$$S_1(\beta) = S(v\beta),$$

then (15) can be written

$$\langle s_1(n\tau, t)\rangle = kA_R F_1(n\tau)e^{-i(\omega_0 v^2/Rc)(n\tau-\beta)^2}. \tag{16}$$

The sum of the $2N+1$ envelope values is

$$S_1(\beta) = kA_R \sum_{n=-N}^{N} F_1(n\tau)e^{-i(\omega_0 v^2/Rc)(n\tau-\beta)^2} \tag{17}$$

which is (11) with a change of variable.

A formal manipulation of (17) gives

$$S_1(\beta) = \int_{-\infty}^{\infty} kA_R F_1(\lambda)e^{-i(\omega_0 v^2/Rc)(\lambda-\beta)} \cdot \sum_{n=-N}^{N} \delta(\lambda - n\tau)d\lambda. \tag{18}$$

If the finite time cross-correlation function

$$R_{xy}(\beta) = \int_{-\infty}^{\infty} x(t)y(t+\beta)dt$$

is compared with (18), it is seen that $S_1(\beta)$ is analogous to $R_{xy}(-\beta)$, and the functions

$$A_R F_1(\lambda) \sum_{n=-N}^{N} \delta(\lambda - n\tau) \quad \text{and} \quad ke^{-i(\omega_0 v^2/Rc)(\lambda-\beta)^2}$$

are being cross-correlated to give the result, $S_1(\beta)$. The function

$$ke^{-i(\omega_0 v^2/Rc)(\lambda-\beta)^2}$$

is associated with the signal, while the function

$$A_R F_1(\lambda) \sum_{n=-N}^{N} \delta(\lambda - n\tau)$$

is associated with the synthetic array weighting function $-F_1(\lambda)$ being applied in the data processor and

$$A_R \sum_{n=-N}^{N} \delta(\lambda - n\tau)$$

being applied by the modulation of the radar and its linear motion.

V. Doppler Filtering Formulation

To show the equivalence of cross-correlation and filtering, it is necessary to investigate the Fourier transform of of (18).

The Fourier transform and its inverse are defined

$$\hat{f}(\omega) = \int_{-\infty}^{\infty} f(t)e^{-i\omega t}dt \tag{19}$$

and

$$f(t) = \frac{1}{2\pi}\int_{-\infty}^{\infty} \hat{f}(\omega)e^{i\omega t}dw. \tag{20}$$

The Fourier transform of (18) is

$$\hat{S}_1(\omega) = \int_{-\infty}^{\infty}\int_{-\infty}^{\infty} kA_RF_1(\lambda)e^{-i(\omega_0 v^2/Rc)(\lambda-\beta)} \cdot \sum_{n=-N}^{N} \delta(\lambda - n\tau)e^{-i\omega\beta}d\lambda d\beta \tag{21}$$

making the change of variable, $\lambda - \beta = r$

$$\hat{S}_1(\omega) = \int_{-\infty}^{\infty}\int_{-\infty}^{\infty} kA_RF_1(\lambda)e^{-i(\omega_0 v^2/Rc)r^2} \cdot \sum_{n=-N}^{N} \delta(\lambda - n\tau)e^{-i\omega(\lambda-r)}d\lambda dr$$

which can be written

$$\hat{S}_1(\omega) = \left\{\int_{-\infty}^{\infty} A_RF_1(\lambda) \sum_{n=-N}^{N} \delta(\lambda - n\tau)e^{-i\omega\lambda}d\lambda\right\} \cdot \left\{k\int_{-\infty}^{\infty} e^{-i(\omega_0 v^2/Rc)r^2}e^{i\omega r}dr\right\}. \tag{22}$$

The two bracketed factors on the right side of (22) are the Fourier transforms of the complex time array weighting function and the received signal function, respectively. The output spectrum $\hat{S}_1(\omega)$ can be thought of, then, as the result of filtering the signal spectrum with the transform of the time array weighting function.

ACKNOWLEDGMENT

Some earlier unpublished work by Dr. H. V. Hance first suggested to the writer that the vector addition, cross-correlation, and filtering formulations could be rigorously derived from one basic equation.

An Introduction To Synthetic Aperture Radar

W.M. Brown and L.J. Porcello
(University of Michigan, Ann Arbor)

IEEE Spectrum (September 1969)
Pages 52-62

One way of achieving fine-resolution terrain imagery using airborne, side-looking radar is to boost frequency; another is to decrease along-track tracking. Neither is attractive. But they would have had to do were it not for coherent wave radar—upon which synthetic-aperture radar is premised. Using coherent radar, resolution can depend, not on the width of the beam, but on Doppler frequency shift. The azimuthal resolution of side-looking radar can therefore be of the same order of magnitude as that for range resolution. The key to converting the theoretical groundwork into a "full-bodied" system is an appropriate data-processing scheme. And the simplest scheme for working, processing, and deciphering the data is optical.

Although the amount of information carried by an electromagnetic signal is constrained, depending on the technicians' resourcefulness, there are limits and *limits*. Were it not for some electronic "trickery," therefore, the range resolution in the direction of flight of present-day, side-looking, airborne radar would be relatively poor—and the future, despite increasingly sophisticated and costly components, would not hold much promise for improvement.

One useful technique for overcoming the physical constraints that limit the more "straightforward" type of radar systems is called synthetic-aperture radar. Using it, along-track resolution does not depend exclusively on how narrow the beam is. Rather, resolution is provided by Doppler processing the received echoes. In this way, greatly improved performance—an order of magnitude or more—is possible.

The synthetic-aperture concept

The genesis of the synthetic-aperture concept appears to have been the work of Carl Wiley of the Goodyear Aircraft Corp. in the early 1950s; the development of his ideas is retraced in a paper by Sherwin *et al.*[1] Wiley observed that a one-to-one correspondence exists between the along-track coordinate of a reflecting object (being linearly traversed by a radar beam) and the *instantaneous Doppler shift* of the signal reflected to the radar by that object. He concluded that a frequency analysis of the reflected signals could enable finer along-track resolution than that permitted by the along-track width of the physical beam itself.

This "Doppler beam-sharpening" concept was not only exploited by Goodyear, but by a group at the University of Illinois. An experimental demonstration of the beam-sharpening concept was carried out by the Illinois group in 1953 through use of airborne coherent X-band pulsed radar, "boxcar" circuitry, a tape recorder, and a frequency analyzer.

One major problem, recognized quite early, was implementation of a practical data processor that could accept wide-band signals from a storage device and carry out the necessary Doppler-frequency analysis at each

FIGURE 1. Antenna of side-looking radar illuminates terrain strip indicated by area in color.

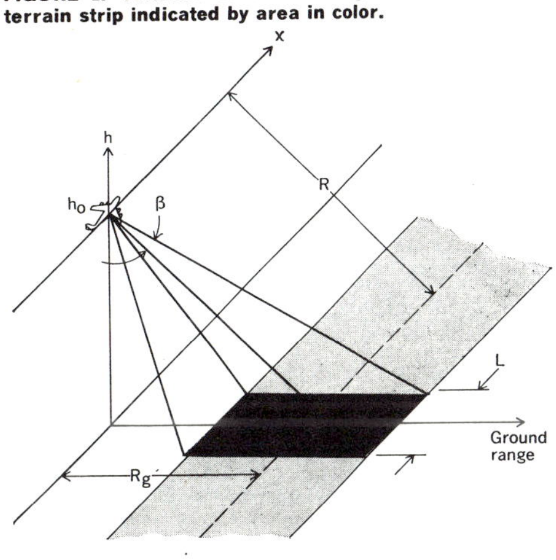

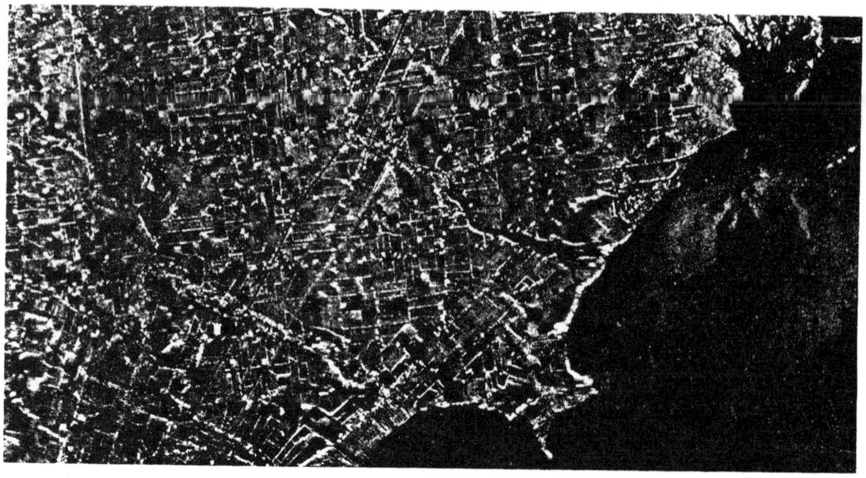

FIGURE 2. Here is an example of side-looking radar's imaging capabilities.

resolvable slant range. During 1953 a summer study was held at the University of Michigan. Known as Project Wolverine, it served as a point of departure for expanded efforts in, and diverse approaches to, fine-resolution, terrain-imaging radar.

The basic geometry of the side-looking, terrain-imaging radar is shown in Fig. 1. The airborne radar radiates through an antenna whose vertical pattern is chosen to illuminate a terrain strip properly—the colored region in Fig. 1—parallel to the flight path. The antenna beam, usually narrow in the along-track (azimuthal) dimension, is not rotated, but scans the terrain as a result of the forward motion of the aircraft. The radar system generates a two-dimensional image of the reflectivity distribution of the terrain strip, with one dimension proportional to slant range and the second to along-track coordinate position. Some examples of side-looking radar imagery are shown in Fig. 2. The slant-range dimension is vertical, whereas along-track is horizontal.

A side-looking radar achieves slant-range resolution through the use of pulsing and time-delay sorting, as illustrated in Fig. 3. If the radar transmits a very short pulse, reflected by a target at slant range R, then the round-trip propagation time between the radar and this target is given by

$$\Delta T = \frac{2R}{c} \tag{1}$$

where c is the propagation speed of the radar wave. The reflections from targets at different ranges will, naturally, arrive at the receiver with different time delays. If the transmitter pulse is very short, say of duration τ, then the returns from targets at sufficiently different ranges will be nonoverlapping in time. Specifically, the required separation is

$$\Delta R_s \geq \frac{c\tau}{2}$$

Accepting this as a measure of slant-range resolution ρ_R, then

$$\rho_R \approx \frac{c\tau}{2} \tag{2}$$

and the corresponding ground-range resolution is

$$\frac{c\tau}{2} \sec \psi$$

where ψ is the depression angle of the line of sight to the target with respect to local horizontal, as shown in Fig. 3.

In a simple, pulsed radar, generation of a pulse of duration τ requires a transmitter bandwidth of the order of

$$W \approx \frac{1}{\tau}$$

and preservation of the range resolution $c\tau/2$ requires that the receiver and display also have bandwidth $W \approx 1/\tau$. Obviously, the key to achieving fine range resolution is wide-band radar transmitters and receivers—as amply demonstrated in pulse-compression technology.* The bandwidth sets the fundamental constraint on range resolution:

$$\rho_R \approx \frac{c}{2W} \tag{3}$$

By way of example, a radar with a bandwidth of 1 GHz provides a theoretical range resolution of

$$\rho_R \approx 15 \text{ cm}$$

Now the along-track or azimuthal cross section of the

* In a pulse compression radar, we radiate "dispersed" pulses with time–bandwidth product $\tau W \gg 1$ (rather than $\tau W \approx 1$ as in the case above) and use a data processor to "compress" the received pulses to a time duration $1/W$.

antenna pattern has a half-power angular width of β radians. The corresponding along-track beam width at range R is

$$L \approx \beta R \tag{4}$$

If the distance L is accepted as measure of the along-track resolution ρ_x of this radar, then the only recourse for achieving fine resolution at long range is to make β very narrow. But an antenna aperture with along-track dimension D, operating at its diffraction limit at wavelength λ, yields a half-power angular beam width of

$$\beta \approx \frac{\lambda}{D} \text{ radians} \tag{5}$$

therefore,

$$\rho_x \approx \frac{\lambda R}{D} \tag{6}$$

To keep ρ_x small as R increases, D must be increased and/or λ decreased. Each of these options becomes unattractive beyond certain limits. Large-D antennas are incompatible with airborne operation; operation at very short wavelength leads to weather limitations; and, finally, construction of a diffraction-limited airborne antenna with $D \gtrsim 10^3\lambda$ can be costly as well as difficult. By way of example, a radar with $D \approx 10^3\lambda$ realizes, at slant range $R = 10$ km, an along-track resolution of

$$\rho_x \approx 10 \text{ meters}$$

The simple radar system of Fig. 1, therefore, has the potential to achieve fine range resolution, but appears to be constrained to relatively poor azimuthal resolution at long operating ranges. The synthetic-aperture technique can improve azimuthal resolution to the point where ρ_x is commensurate with ρ_R.

If the radar emits a sinusoid of frequency ν_0 Hz, and if the angle γ (see Fig. 4) changes sufficiently slowly, then the reflected signal reaching the receiver is a sinusoid with slowly varying frequency $\nu = \nu_D + \nu_0$ where the instantaneous Doppler shift ν_D is given by

$$\nu_D = \frac{2v}{c} \nu_0 \cos \gamma \tag{7}$$

When $\theta = \frac{\pi}{2} - \gamma$ is small compared with one radian:

$$\nu_D \approx \frac{2v}{\lambda R} (x_0 - x) \tag{8}$$

where $\lambda = c/\nu_0$. Therefore, at any range R, the Doppler shift ν_0 is a linear function of $(x - x_0)$, and a frequency analysis of the return displays the radar reflectivity of the terrain as a function of $(x_0 - x)$.

The synthetic-aperture technique, although first conceived from the Doppler viewpoint, can also be approached from an entirely different, but mathematically equivalent, viewpoint. Following from Fig. 1, observe that a reflector at range R is illuminated by the radar while the latter moves through a distance of

$$L = \beta R$$

If the physical antenna is regarded as one element of a linear array (extending in the direction of flight), occupying in time sequence all the elemental positions germane to the array, then, intuitively, it should be possible to "synthesize" an aperture of length L by suitably storing the received signal before processing the data. At first glance, one would expect this synthetic aperture to have a "synthetic angular beam width" of approximately

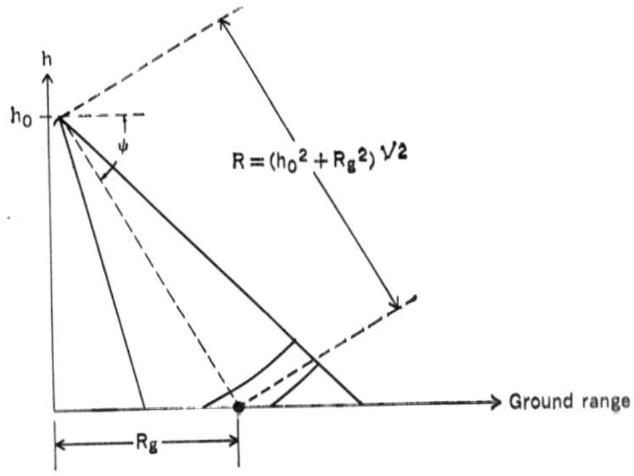

FIGURE 3. Behind side-looking radar resolution are pulsing and time-delay sorting.

FIGURE 4. Coherent radiation enables the radar beam, through processing, to be effectively narrowed—hence the appellative, synthetic-aperture radar.

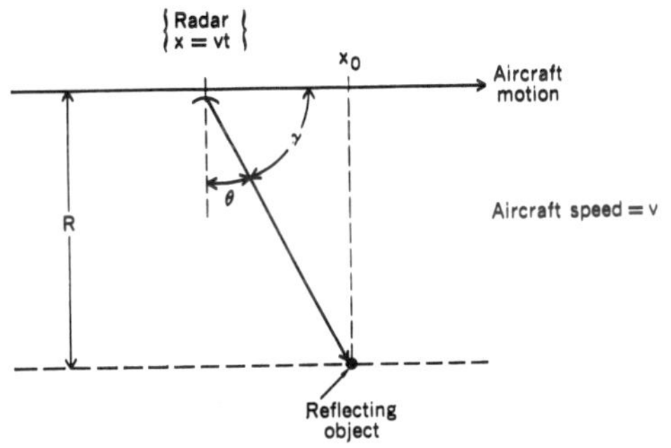

$$\beta' \approx \frac{\lambda}{L} \text{ radians} \tag{9}$$

and a corresponding along-track width $R\beta'$. Substituting for β', $R\beta'$ equals D.

Radar observation of a rotating target

The motion of a terrain-imaging, synthetic-aperture radar past a target gives essentially arbitrary resolution in the along-track dimension (perpendicular to the range dimension). It should be expected that other forms of relative motion between radar and rigid target should lead to similarly fine resolution capability. In fact, if a stationary, coherent radar illuminates a target field consisting of a rotating rigid body, as in Fig. 5, it is possible to form a two-dimensional image of the object's radar reflectivity distribution by carrying out a delay-Doppler (or range-Doppler) analysis of the reflected signals.

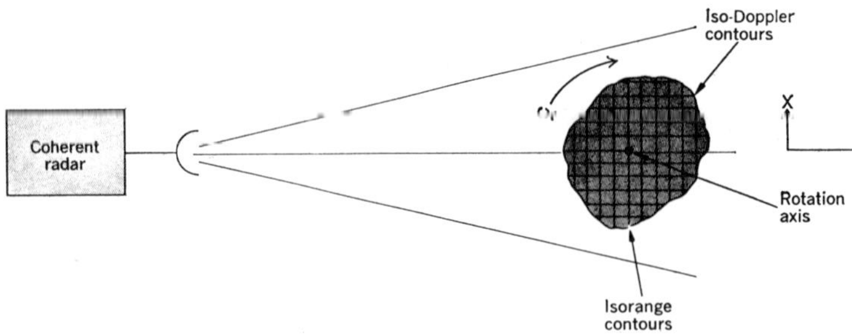

FIGURE 5. Using a range-Doppler technique, it is possible to get a two-dimensional image of the radar reflectivity from a rotating body.

Assume that the angular velocity of the body is ω_r rad/s about an axis perpendicular to the line of sight, with the radar far from the body. Then, the radial velocity of a point at "cross-range" position x on the body is $\omega_r x$, and the Doppler shift of radar returns from this point is

$$\nu_D = \frac{2\omega_r x}{\lambda} \tag{10}$$

If the radar returns are sorted according to Doppler frequency, the scatterers are resolved in the cross-range dimension. If the radar also provides resolution in range, a two-dimensional (radar) image of the body can be obtained. And, if appropriate coherent processing is applied to the signals received over a time T, the obtainable Doppler resolution is about T^{-1} Hz. Since the Doppler shift corresponding to the radial velocity $\omega_r x$ is $\nu_D = 2\omega_r x\lambda^{-1}$, a Doppler resolution of $\Delta\nu_D = T^{-1}$ implies a cross-range resolution of

$$\rho_x \approx \frac{\lambda}{2\omega_r T} = \frac{\lambda}{2\Delta\theta} \tag{11}$$

where $\Delta\theta$ is now the change in total aspect angle during the coherent processing time. The rotating-target-field version of the synthetic-aperture concept has been utilized by radar astronomers to obtain radar images of planets.[2,3]

For planetary radar, the contours of constant range are annuli about the point on the planetary surface closest to the earth (subradar point); the contours of constant Doppler are the intersections of the planetary surface with a family of planes—parallel to the rotation axis and to the line joining this axis and the radar. (An ambiguity normally exists between the upper and lower hemispheres of the planet, but this can often be resolved by auxillary methods.) Figure 6 shows, first, a radar image of a lunar crater generated at a wavelength of 3.8 cm by the Haystack radar and an electronic data processor and, second, an optical image of the same region.

Reconsider now the side-looking, terrain-imaging, synthetic-aperture radar situation. Consider a point target at an angle θ from broadside. The Doppler frequency associated with such a target is $\nu_D = \lambda^{-1} 2v \sin \theta$. A θ-interval β radians wide is well illuminated, and the corresponding Doppler bandwidth is $B_d \approx \lambda^{-1} 4v \sin (\beta/2)$; data with such a bandwidth can be time-resolved to about $\rho_t \approx B_d^{-1}$. In the along-track dimension, $x = vt$, so the corresponding along-track resolution is $\rho_x \approx vB_d^{-1}$ or

$$\rho_x \approx \frac{\lambda}{4 \sin \left(\frac{\Delta\theta}{2}\right)} \tag{12}$$

where $\Delta\theta = \beta$ is the change in aspect angle over which a target is observed by the side-looking radar. If in (12) we let $\Delta\theta = \lambda/D$ and $\sin \Delta\theta/2 \approx \Delta\theta/2$, resolution for the side-looking case is $D/2$. Actually, (12) is the most appropriate cross-range resolution formula for both the side-looking and rotating-target-field cases. In addition, consider a physical aperture and recall that the diffraction-limited angular beam width of approximately λ/D yields an azimuthal linear resolution at range R of about $R\lambda/D = \lambda/\Delta\theta$. Here* $\Delta\theta = D/R$ is again the aspect-angle interval over which the physical aperture views the target field. Observe the general property that a system collecting data over an aspect-angle interval $\Delta\theta$ can yield cross-range resolution as in (12) that agrees with (11) for small $\Delta\theta$.

Converting theory to practice

The remarkable property—that one can enjoy along-track resolution

$$\rho_x \approx D \tag{13}$$

independent of range and of wavelength, that can be improved by *reducing* the dimension D—was first reported by Cutrona *et al.*[4] The key is the use of an appropriate data-processing scheme. In the long-antenna context, requirements are: to *store* the reflected data, to *amplitude-weight and phase-shift* the returns from sequential locations of the physical antenna, and to *coherently sum* these weighted returns as for a physical array. In fact, as L becomes large, one finds it necessary to *focus* the synthetic aperture—i.e., to insert a phase shift, quadratic in distance along the aperture, to compensate for the curvature of the reflected wavefront—in order to realize

* The $\lambda/\Delta\theta$ differs from (11) by a factor of two because two-way propagation is involved in both synthetic-aperture situations discussed. Two-way propagation doubles phase shifts and halves resolutions. The same factor of two was ignored earlier when D was derived as the along-track resolution for the side-looking radar.

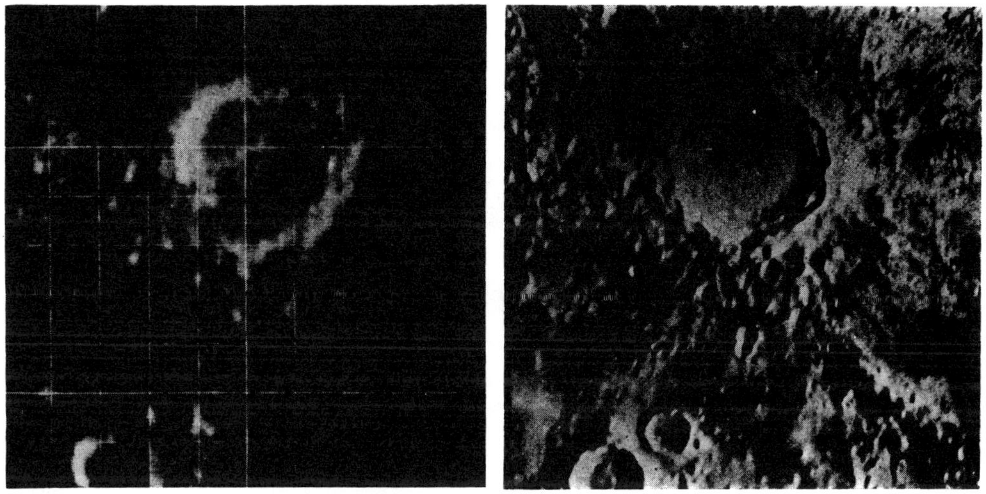

FIGURE 6. Radar image, left, compares favorably with optical image of a moon crater. (Rectified Lunar Atlas, University of Arizona Press, 1963; courtesy G. H. Pettengill.)

the ultimate along-track resolution. This focusing operation, which is range dependent, complicates the data processing to some degree, and makes it somewhat more difficult than a simple, range-dependent, Doppler-spectrum analysis. Systems failing to provide this focusing operation are known as *unfocused synthetic-aperture radars*, and have an along-track resolution capability of approximately

$$\rho_x \approx \sqrt{\lambda R_s} \quad (14)$$

Systems that provide focused processing are termed *focused synthetic-aperture radars* and have the resolution potential

$$\rho_x \approx D \quad (15)$$

Various forms of signal processors were attempted by early workers in the field. Goodyear placed initial emphasis on the development of a range-gated filter bank. But it turned out that it was impractical, at that time, to incorporate the focusing operation into this device. The Illinois group pursued three distinct processing approaches, again all limited to the unfocused case: (1) a recirculating delay line, developed at the Philco Corp.; (2) integration on photographic film; and (3) an electronic storage tube integrator. The Michigan group suggested an optical system signal processor for synthetic-aperture generation. (As we shall show, an optical system, illuminated with coherent light, can serve as a versatile analog computer for performing signal-processing operations.) The potential of a coherent optical system as a general-purpose signal processor was soon recognized and a coherent optical data processor that permitted generation of a synthetic aperture properly focused at each slant range was subsequently devised by Cutrona *et al.*[5,6] Finally, in 1957, a synthetic-aperture radar using a coherent optical data processor was successfully demonstrated.

Optical processors were subsequently applied to pulse compression and to range-Doppler radars such as those used in planetary mapping, and to the problem of automatic shape recognition. Some of the optical techniques devised in the context of radar data processing were appropriately modified by E. N. Leith for application to wavefront reconstruction—initially suggested by Gabor[7]—and formed the basis of modern holography.

System configuration

To employ optical processing, it is customary to generate a photographic transparency on which the unprocessed data are stored. In the case of a synthetic-aperture radar, this transparency has the properties of an array of one-dimensional off-axis holograms. As a result, the optical data processor—the heart of a synthetic-aperture radar—can be described as a viewer in which a one-dimensional wavefront reconstruction process takes place, without losing the range resolution of pulsed radar. The simplified block diagram of Fig. 7 shows the interrelation of a radar, a processor, and an associated storage device, which collectively constitute a synthetic-aperture radar. A stable oscillator serves as a common reference source for both the transmitter and receiver; i.e., the radar is coherent. Therefore, the phase delay associated with a round-trip propagation signal can be determined by synchronously demodulating the received signal with respect to the transmitter output. This permits the physical antenna to sequence observations rather than to make a full set of spatially diverse observations of the reflected wave at one instant, as with a physical array system. Pulsing the radar need not upset the fundamental coherence provided by the stable source.

Data processing

The synthetic-aperture, terrain-imaging, data-processing concept might be straightforward in principle. But

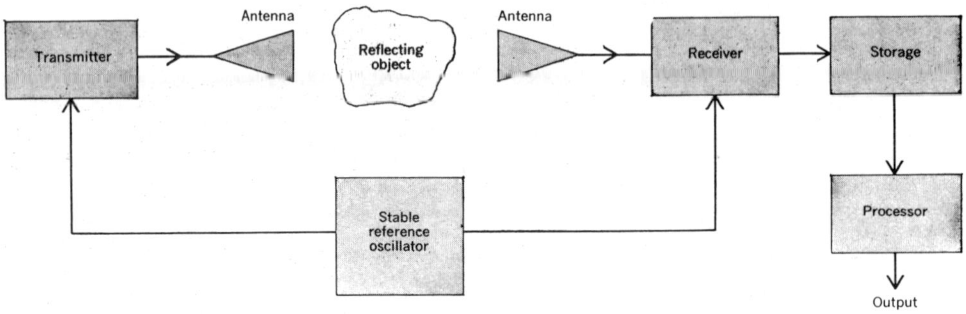

FIGURE 7. These pieces collectively constitute a synthetic-aperture radar.

one critical, early question was whether the necessary processor could be implemented in a form neither prohibitively cumbersome nor expensive for large-scale tactical use.

The output of the radar receiver typically has a bandwidth of many megahertz, and the storage device associated with the processor must accept data at a commensurate rate. It was recognized quite early that photographic film is (but for the nuisance of chemical development) well suited to the task of storing the data. The stored data could be read out optically as a viewable image. It made sense, therefore, to consider processing using optical elements alone. (See box.)

In a simple, optical, holographic system, an object is illuminated with coherent light as shown in Fig. 11 (A); each scattering center on the object generates a spherical wave that impinges on a plane P. A collimated, reference wave also passes through P_1 and a film, in plane P, records the intensity distribution of the sum of the reference and scattered waves. This recorded interference pattern, or hologram, has the appearance of a section of a Fresnel zone plate* of focal length f'—the distance from the scattering center to P. The portion of this recorded zone plate pattern is a function of the size and location of the aperture at P and the offset angle α between the reference beam and the scattered wave's propagation vector.

A recorded portion of the zone plate, when it is removed from the origin is, inferentially, called an "off-axis" zone plate. Inserting a "spatial offset frequency" into the pattern is equivalent to translating the origin of the pattern; this offset frequency is controlled by the angle α. If the photographic transparency is developed and then reilluminated with a collimated, coherent light beam, the recorded zone plate gives rise to a pair of spherical waves—one diverging and one converging—each of focal length f', as illustrated in Fig. 11(B). The two waves correspond, respectively, to a virtual and a real image of the scattering center. The propagation directions of the waves are offset by the angle α with respect to the hologram. A third wave associated with the average gray level, or optical bias of the hologram, continues along the optic axis.

A complex scattering object, consisting of a spatial superposition of scattering centers, gives rise to a corresponding superposition of reconstructed waves. The virtual image from the hologram produced by the complex scattering centers appears to be an accurate three-dimensional facsimile of the object itself. One might, therefore, anticipate that optical elements can be used to display the two-dimensional spectrum of a light distribution, appropriately phase-shift and/or amplitude-weight portions of the spectrum by changing the gray scale (through suitable filtering), and then examine the (modified) light distribution. Alternatively, it should be possible to display a one-dimensional spectrum in each of many, independent channels; to modify the spectrum in each channel; and to examine the modified light distribution.

In order to apply these optical filtering properties to radar data processing, the radar data must be stored in an appropriate format.

In its simplest form, the radar emits short pulses of microwave energy, of duration τ second, periodically at at rate of ν_s pulses per second. The reflected signal has the form of Fig. 12; R_1 is the shortest slant range illuminated by the radar and R_2 is the maximum slant range illuminated. The signal emerging from the synchronous demodulator in the receiver is bipolar with respect to some arbitrary bias level; the sum of signal plus bias, a monopolar waveform, is used to intensity-modulate a cathode-ray tube whose beam is swept repetitively in a straight line. To map the full slant-range interval illuminated, the nth sweep is started at time $T_0 + (n - 1) \frac{2R_1}{c}$ and is continued for a duration $\frac{2}{c}(R_2 - R_1)$. The intensity-modulated sweep is then photographed onto the radar data film that moves in a direction perpendicular to the sweep direction, as shown in Fig. 13. The variable y_1 is then a measure of slant range and x_1 corresponds (after scaling) to the along-track position of the air-

* A conventional Fresnel zone plate is a film transparency with a radially symmetric intensity transmission $I(x,y)$ given by

$$I(x,y) = |A_0 + A_1 \cos[\mu(x^2 + y^2)]|^2$$

or a nonlinear function of I. If $I(x,y)$ is precisely as given in the equation, the zone plate has a focal length

$$f' = \pm \frac{\pi}{\mu\lambda}$$

where λ is the wavelength used to illuminate the zone plate. The ($\pm$) signs denote conjugate foci—i.e., an incident collimated beam gives rise to a real and a virtual point image. If the plate is a nonlinear function of I, then higher-order conjugate foci, corresponding to harmonics generated by the nonlinearity, also appear.

craft—hence the position of the physical antenna element. Ideally, the intensity modulation of the CRT and the development of the film are both controlled so that, after development, the light-amplitude transmissivity of the film is a linear function of the bipolar video signal amplitude.

If the data film is illuminated with a normally incident, coherent light beam, the light emerging from the film is uniform in phase (assuming the film itself does not corrupt the phase) and is amplitude modulated as a function of x_1 and y_1. For any y_1 = constant, the x_1-dimension displays the elemental signal received at each point along the synthetic aperture for the range corresponding to y_1.

The optical analog to electronic methodology

In a physical radar array, elemental returns would be amplitude-weighted (to control sidelobe levels), phase-shifted (to provide beam steering and/or to focus the array), and then coherently summed. In an optical system, amplitude-weighting can be accomplished by inserting a shaded transparency with uniform phase thickness adjacent to the data film. Phase-shifting is attained by inserting appropriately shaped pieces of glass adjacent to the data film. If for any given y_1 corresponding to a slant range R the array is to be focused at R, a phase shift quadratic in x_1 (with coefficients a function of R) must be inserted across the aperture to compensate for the quadratic phase delay that occurs when a spherical wave, emanating from the reflecting object, is observed along the synthetic aperture. Such a quadratic phase correction is provided by a simple plano-convex lens placed in proximity to the data film at the appropriate y_1. At each y_1, the lens must have an x_1-dimension focal length proportional to y_1. This is shown in Fig. 14(A). A lens that continuously satisfies this requirement for

The Fourier property

The most significant property of a coherently illuminated optical system is the Fourier transform property. If coherent illumination of wavelength λ_L is incident on a spherical lens, then complex light amplitude distributions $\hat{A}_1(x_1,y_1)$ and $\hat{A}_2(x_2,y_2)$ in the front and back focal planes of the lens are quite nicely related by a Fourier transformation. This is illustrated in Fig. 8. For simplicity, assume the illumination to be linearly polarized; by "complex light amplitude," we mean the magnitude and phase (relative to some arbitrary reference phase) of the corresponding linearly polarized electric field intensity. Employing the coordinate system of Fig. 8, this can be expressed as

$$\hat{A}_2(x_2,y_2) \sim \int\int \hat{A}_1(x_1,y_1) \exp[-j\alpha(x_1x_2 + y_1y_2)]dx_1\,dy_1 \quad (16)$$

where

$$\alpha = -\frac{2\pi}{\lambda_L f_1} \quad (17)$$

and f_1 is the focal length of the lens.

In other words

$$\hat{A}_2(x_2,y_2) \sim \mathfrak{F}_{x,y}\{\hat{A}_1(x_1,y_1)\} \quad (18)$$

where $\mathfrak{F}_{x,y}\{\cdot\}$ denotes a Fourier transform with respect to both x and y. This interesting relation follows directly from the Kirchhoff diffraction integral and simply says that the amplitude spectrum of a two-dimensional function displayed at plane P_1 can be observed at plane P_2. If two such stages are cascaded with the lenses separated by $2f_1$, as illustrated in Fig. 9, then

$$\hat{A}_3(x_3,y_3) \sim \mathfrak{F}_{x,y}\{\mathfrak{F}_{x,y}[(\hat{A}_1x_1,y_1)]\} \quad (19)$$

which renders $\hat{A}_3$ proportional to $\hat{A}_1$ in reversed coordinates:

$$\hat{A}_3(x_1,y_1) \sim \hat{A}_1(-x_1,-y_1) \quad (20)$$

It should be apparent, therefore, that a function $f(x,y)$ can be inserted at P_1; the spectrum at P_2 can then be modified by inserting slits, stops, shaded transparencies, and/or optical phase shifters, and finally observed at P_3.

A companion relationship exists for cylindrical lenses having curvature in one dimension only. In this case, a Fourier transform is effected in only one dimension; the illumination simply diverges in the other dimension. Using a cylindrical lens and a spherical lens of equal focal lengths in the combination of Fig. 10: in the x-dimension, a Fourier transform relation exists between $\hat{A}_1$ and $\hat{A}_2$, whereas in the y-dimension, $\hat{A}_2$ is an image of $\hat{A}_1$ with the usual inversion of coordinates.

Therefore,

$$\hat{A}_2(x_2,y_2) \sim \int \hat{A}_1(x_1,y_1)\Big|_{y_1=-y_2} \exp[-j\alpha x_1x_2]\,dx_1 \quad (21)$$

This represents a multichannel one-dimensional spectrum analyzer, in which the channels are stacked in the y-dimension.

all values of y_1 is conical, as illustrated in Fig. 14(B). Since the synthetic-aperture length L is also proportional to R, and hence to y_1, tapering the lens to a point at $R = 0$ is consistent with the required processing.

The coherent light emerging from the film–conical lens combination now has the property that the wave emanating from each y_1 (or range channel) is collimated in the x_1-dimension—being either uniform in phase or having a phase linear in x_1. But it has no focal properties in the y_1-dimension. Integration with respect to x_1 can be handled via a Fourier transform (see box below), evaluated at $\omega_x = 0$

$$\left(\text{i.e., } \mathfrak{F}\{f(x)\}\Big|_{\omega=0} = \int_{-\infty}^{\infty} f(x)e^{-j\omega x}dx\Big|_{\omega=0} = \int_{-\infty}^{\infty} f(x)\,dx\right)$$

To image in the y-dimension for the range resolution inherent in the data film calls for a sphere–cylinder lens combination, as shown in Fig. 15. A view at P_2 through a vertical slit shows the output of the synthetic aperture for all displayed ranges for one value of the along-track dimension. Note that the slit is displaced from the origin because a bias is needed in order to record a bipolar function and therefore record the radar data in a spatial, bandpass form purposefully to keep its spectrum away from dc. (There can be no negative intensities available from the CRT trace.) A small x-dimension carrier frequency is employed and the x-dimension at P_2 is then evaluated at the corresponding off-axis position.

Input–output correspondence

It should be apparent, then, that a radar data film made with the optical conversion recording scheme of Fig. 13 is an analog of the all-optical holographic technique.

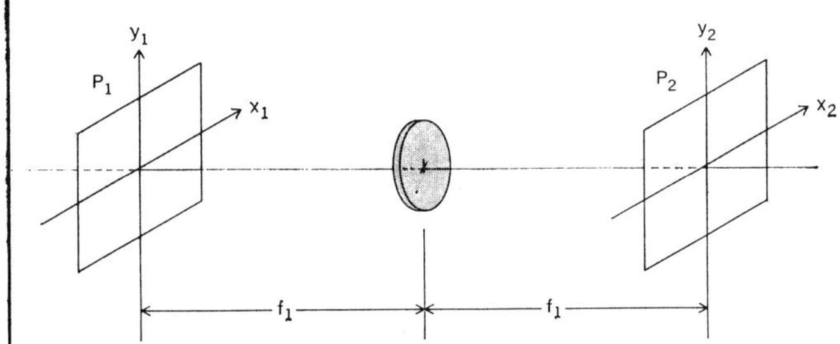

FIGURE 8. Light distributions at the front and back planes of the lens are related by a Fourier transform.

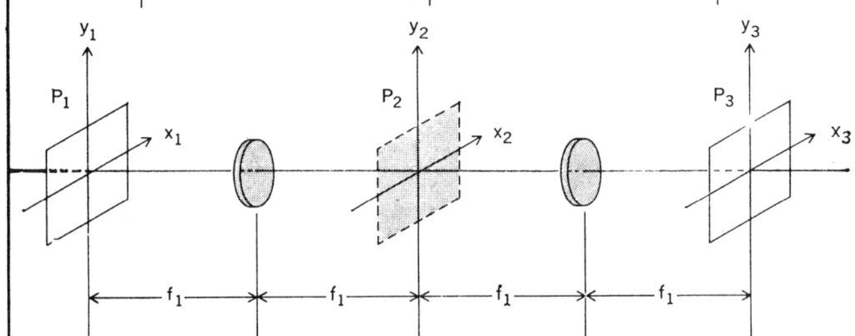

FIGURE 9. A succession of two transformations, achieved with this double lens arrangement, reverses the coordinates of the original amplitude distribution.

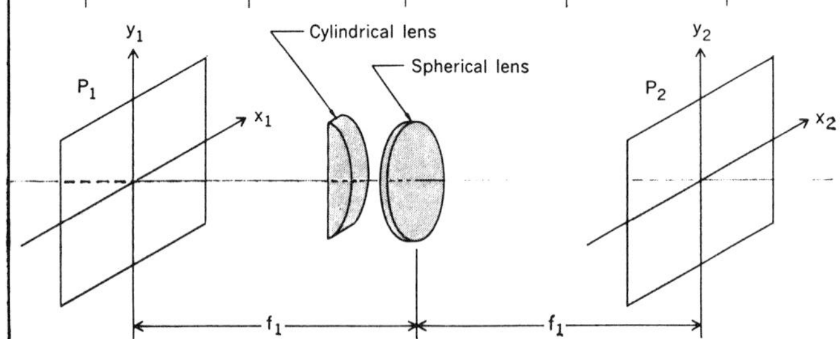

FIGURE 10. Information from each of the coordinates can be processed differently. Here, A_2 is a Fourier transform of A_1 in the x-direction and simply inverted in the y-direction.

FIGURE 11. Top diagram depicts conventional optical setup for producing a hologram. Hologram (bottom) can, in turn, reproduce both a virtual and real three-dimensional image.

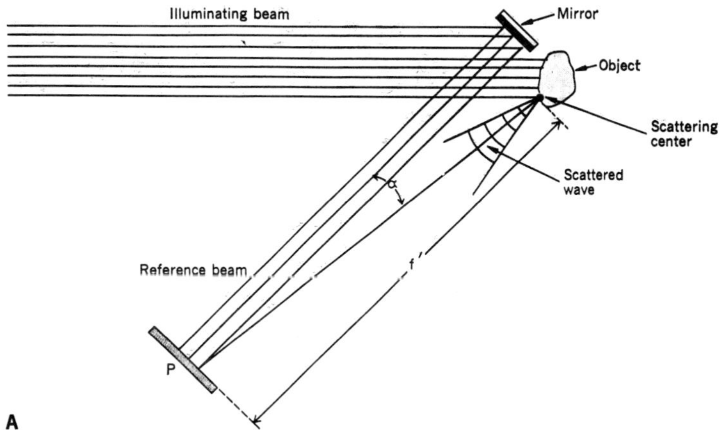

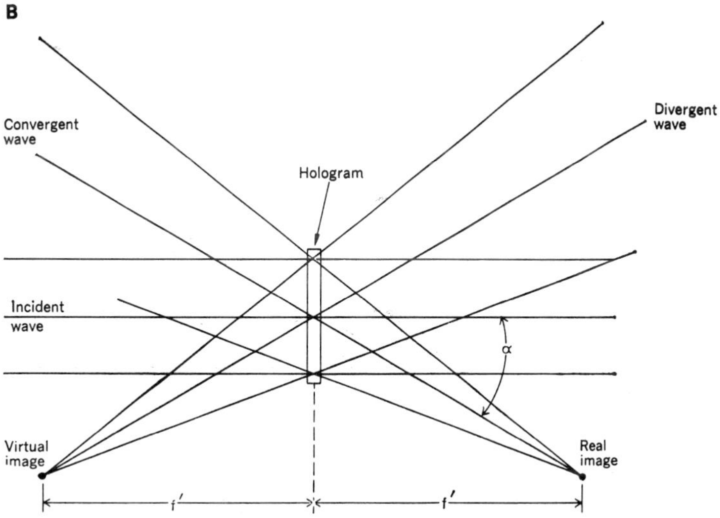

Each scattering object on the ground produces a one-dimensional Fresnel zone plate at the value of y_1 corresponding to the range of the object, and with a focal length proportional to this range. Reflectors, at the same range but with different along-track coordinates, appear as zone plates with centers displaced in the x-direction; their radar cross sections are manifested in the modulation intensity of the zone plate.

When this collection of one-dimensional holograms is illuminated by a coherent, collimated beam of light, a one-dimensional wavefront reconstruction takes place: The spatial pattern associated with each reflector generates a real and virtual image, each having an x-dimension focal length proportional to range. This is illustrated in Fig. 16. An image of the target field focused in the x-dimension only would be observed by appropriately examining one of these two planes. However, the target responses observed in either of these planes are not focused in the range dimension; furthermore, the tilt angle of the plane may make such observation inconvenient. A special viewer could serve to erect the tilted plane, and to bring the plane of x-focus and that of y-focus into coincidence. The optical system, Fig. 15, does exactly this: the conical lens erects the tilted plane of x-focus, and moves it to infinity; the sphere then moves the plane of x-focus to P_2; and the sphere–cylinder combination images the y-focused structure of plane P_1 at P_2. However, the one-dimensional holograms are of an off-axis reference beam variety (a consequence of using an offset frequency in the recording process), and the energy associated with the virtual image does not spatially overlap that associated with the bias term or with the real image. The desired image may, therefore, be viewed through a slit in P_2.

The data processing required for a rotating-target radar (or, for that matter, for any range-Doppler radar) is some-

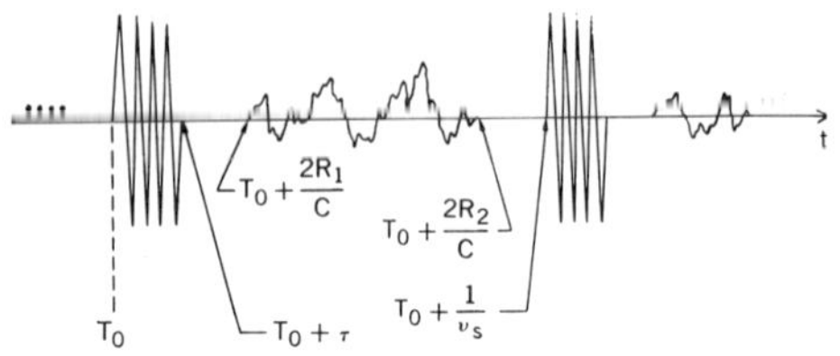

FIGURE 12. Delay relationships exist between outgoing and incoming side-looking radar pulses. R_1 and R_2 represent the nearest and farthest slant range probed.

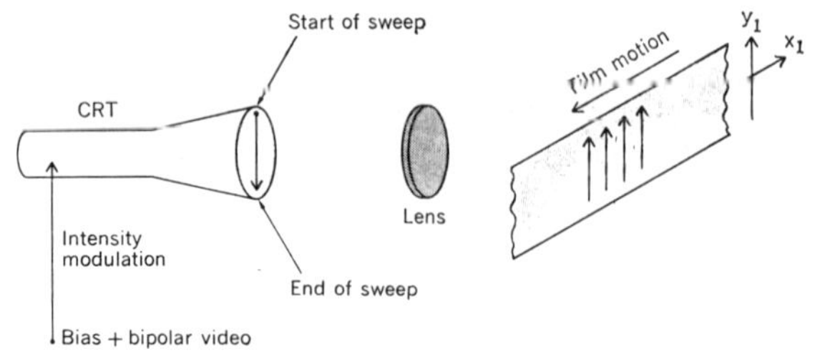

FIGURE 13. Radar signals are sent through a biased CRT, and then recorded on film for subsequent optical processing.

what simpler than just described. In such cases, a conventional image is obtained for range and a Fourier transform is formed for the orthogonal dimension. Optical compression of the range pulse may easily be incorporated in either the range-Doppler or the airborne terrain-imaging case. This is done by optically synthesizing the "matched filter" associated with the range-pulse waveform, and inserting this into the optical channel. The processor becomes more complicated (for both the range-Doppler case and the airborne terrain-imaging case) when, over the integration time used by Doppler analysis, the range to points in the target field changes by a distance comparable to or greater than the range resolution; i.e., when the "range-walk" problem is significant. These problems can be solved, but the solutions are beyond the scope of this article.

Ambiguity constraint

Finally, a remark on the ambiguity problem associated with synthetic-aperture side-looking radar—and one that does put a constraint on its usefulness.

If the vertical cross section of the physical antenna pattern illuminates a slant-range interval of ΔR and if a periodic pulse train is transmitted, range ambiguities can be avoided, provided that the interpulse period satisfies the relation

$$T \geq \frac{2\Delta R}{c} \tag{22}$$

However, the Doppler bandwidth of the received signals imposes an upper bound on the interpulse period. The highest well-illuminated Doppler frequency is $(\beta/2)2v\lambda^{-1}$, and sampling theorem dictates a minimum sampling rate of twice this highest frequency:

$$T^{-1} \geq \frac{2\beta v}{\lambda}$$

Equivalently, since $\beta \approx \lambda/D$, the minimum rate required for proper sampling of the Doppler frequencies is

$$T^{-1} \geq \frac{2v}{D} \tag{23}$$

The constraint of Eq. (23) indicates that at least one pulse must be transmitted whenever the aircraft moves a distance of $D/2$, this being approximately the azimuthal resolution limit of the system; i.e., as one might expect, the spacing of the samples in the along-trace dimension must be as fine as the achievable resolution.

FIGURE 14. A quadratic phase delay, introduced by a plano-convex lens, (top) is used to focus abscissa information for a particular slant range R. To bring entire area between two slant ranges into focus, therefore, requires a conical lens (bottom).

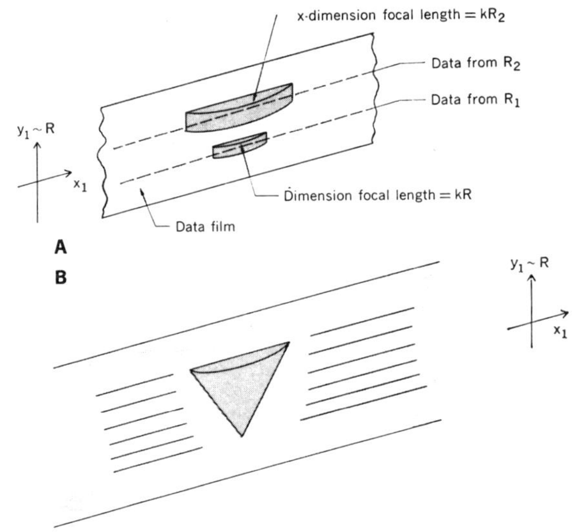

Various additional elementary conclusions can be drawn. With operation at the ambiguity limit [i.e., equality in both (22) and (23)], an image is acquired at a rate of W (two-dimensional) resolution cells per second, where W is the RF bandwidth of the radar in hertz. Also, let $\rho_x = D/2$ be the azimuthal resolution of the system; then (22) and (23) give

$$\Delta R = \frac{c}{2v} \rho_x \tag{24}$$

as the unambiguous obtainable swath width. Since c is so great, (24) usually permits generous coverage for side-looking radar. However, if these same ideas are considered for sonar, c becomes the speed of sound in water. In turn, the resulting slant-range interval of (24) is disturbingly small; several orders of magnitude in available slant-range interval are typically lost when one goes over to the sonar case.

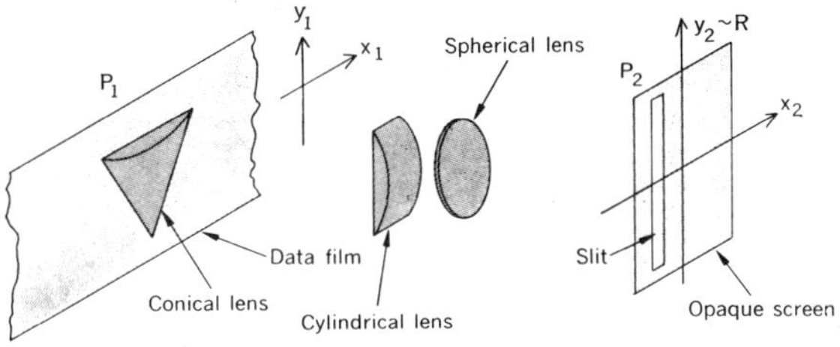

FIGURE 15. With this optical arrangement, a view through the vertical slit at P_2 displays output of the synthetic aperture for all ranges with the same along-track dimension.

FIGURE 16. A one-dimensional wavefront reconstruction generates a real and virtual image of the slant-range reflections so that they fall in a plane.

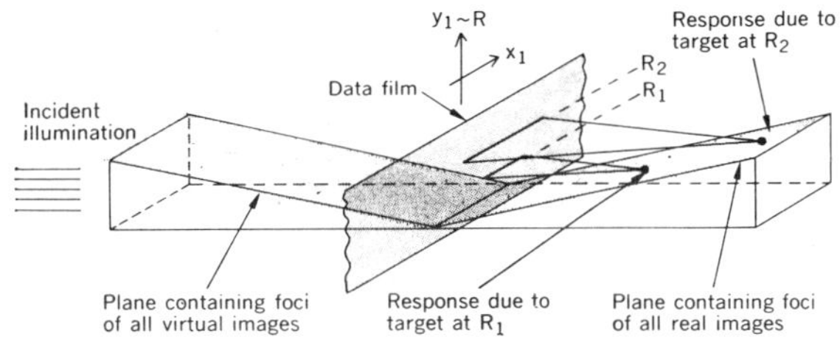

REFERENCES

1. Sherwin, C. W., Ruina, J. P., and Rawcliffe, R. D., "Some early developments in synthetic aperture radar systems," *IRE Trans. Military Electronics*, vol. MIL-6, pp. 111–115, Apr. 1962.
2. Shapiro, I. I., "Planetary radar astronomy," *IEEE Spectrum*, vol. 5, pp. 70–79, Mar. 1968.
3. Evans, J. V., and Hagfors, T., *Radar Astronomy*. New York: McGraw-Hill, 1968.
4. Cutrona, L. J., Vivian, W. E., Leith, E. N., and Hall, G. O., "A high-resolution radar combat-surveillance system," *IRE Trans. Military Electronics*, vol. MIL-5, pp. 127–131, Jan. 1961.
5. Cutrona, L. J., Leith, E. N., Palermo, C. J., and Porcello, L. J., "Optical data processing and filtering systems," *IRE Trans. Information Theory*, vol. IT-6, pp. 386–400, June 1960.
6. Cutrona, L. J., Leith, E. N., Porcello, L. J., and Vivian, W. E., "On the application of coherent optical processing techniques to synthetic-aperture radar," *Proc. IEEE*, vol. 54, pp. 1026–1032, July 1966.
7. Gabor, D., "A new microscopic principle," *Nature*, vol. 161, p. 777, May 1948.

BIBLIOGRAPHY

Brown, W. M., "Synthetic aperture radar," *IEEE Trans. Aerospace and Electronic Systems*, vol. AES-3, p. 217, 1967.

Brown, W. M., and Fredricks, R., "Range-Doppler imaging with motion through resolution cells," *IEEE Trans. Aerospace and Electronic Systems*, vol. AES-5, p. 236, Jan. 1969.

Brown, W. M., and Palermo, C. J., "Effects of phase errors on resoluton," *IEEE Trans. Military Electronics*, vol. MIL-9, pp. 4–9, Jan. 1965.

Brown, W. M., and Palermo, C. J., *Random Processes, Communications, and Radar*. New York: McGraw-Hill, 1969.

Cook, C. E., and Bernfeld, M., *Radar Signals: An Introduction to Theory and Application*. New York: Academic Press, 1967.

Cutrona, L. J., and Hall, G. O., "A comparison of techniques for achieving fine azimuth resolution," *IRE Trans. Military Electronics*, vol. MIL-6, pp. 119–121, Apr. 1962.

Develet, J. A., "Performance of a synthetic-aperture mapping radar system," *IEEE Trans. Aerospace and Navigational Electronics*, vol. ANE-11, pp. 173–179, Sept. 1964.

Greene, C. A., and Moller, R. T., "The effect of normally distributed random phase errors on synthetic array gain patterns," *IRE Trans. Military Electronics*, vol. MIL-6, pp. 130–139, Apr. 1962.

Harger, R. O., "An optimum design of ambiguity function, antenna pattern, and signal for side-looking radars," *IEEE Trans. Military Electronics*, vol. MIL-9, pp. 264–278, July/Oct. 1965.

Harger, R. O., and Crimmins, T. R., "The effect of phase errors on weighted spectra," *IEEE Trans. Military Electronics*, vol. MIL-9, pp. 298–299, July/Oct. 1965.

Heimiller, R. C., "Theory and evaluation of gain patterns of synthetic arrays," *IRE Trans. Military Electronics*, vol. MIL-6, pp. 122–129, Apr. 1962.

Klauder, J. R., Price, A. C., Darlington, S., and Albersheim, W. J., "The theory and design of chirp radars," *Bell. Sys. Tech. J.*, vol. 39, p. 745, 1960.

Kozma, A., and Kelly, D. L., "Spatial filtering for detection of signals submerged in noise," *Appl. Optics*, vol. 4, p. 387, Apr. 1965.

Leith, E. N., "Photographic film as an element of a coherent optical system," *SPSE*, vol. 6, p. 75, Mar.–Apr., 1962.

Leith, E. N., Upatnieks, J., and Haines, K. A., "Microscopy by wavefront reconstruction," *J. Opt. Soc. Am.*, vol. 55, p. 981, 1965.

Leith, E. N., and Upatnieks, J., "Recent advances in holography" *Progress in Optics*, vol. 6, E. Wolf, ed. Amsterdam: North Holland Pub. Co., 1967.

Leith, E. N., "Optical processing techniques for simultaneous pulse compression and beam sharpening," *IEEE Trans. Aerospace and Electronic Systems*, vol. AES-4, p. 879, 1968.

Leith, E. N., and Ingalls, A. L., "Synthetic antenna data processing by wavefront reconstruction," *Appl. Optics*, vol. 7, p. 539, 1968.

McCord, H. L., "The equivalence among three approaches to deriving synthetic array patterns and analyzing processing techniques," *IEEE Trans. Military Electronics*, vol. MIL-6, pp. 116–119, Apr. 1962.

Turin, G. L., "An introduction to matched filters," *IRE Trans. Information Theory*, vol. IT-6, pp. 311–329, June 1960.

Tyler, G. L., "The bistatic, continuous-wave radar method for the study of planetary surfaces," *J. Geophys. Res.*, vol. 71, p. 1559, 1966.

Vander Lugt, A., "Operational notation for the analysis and synthesis of optical data-processing systems," *Proc. IEEE*, vol. 54, pp. 1055–1063, Aug. 1966.

Radar Techniques For Planetary Mapping With Orbiting Vehicle

J.J. Kovaly
(Raytheon Company, Bedford, Massachusetts)

Annals of the New York Academy of Sciences, Vol. 187 (25 January 1972)
Pages 154-176

Introduction

From a survey [1] of planet characteristics and an estimate of the systems required to perform aerial mapping of planetary bodies, J. L. Sheehan shows that a radar system, for imagery and reflectivity measurements, will be one of the primary sensors. For the Jovian type of planets (gas giants), Jupiter, Saturn, Uranus, Neptune, and probably Pluto, which are mostly encompassed by hydrogen and helium, the resolution requirements of a radar system would be estimated from the expected size of waves in an ocean of these gases. However, a resolution of better than one meter should be used to map the terrestrial type (Earth-like), Mercury, Venus, Mars, Earth, and the moon, in order to map the topography and determine landing sites. In one case, that of Venus, a radar sensor may be the only way of seeing its surface because of its thick atmosphere.*

Extensive field measurements performed with a side-looking airborne radar AN/APQ-27 [2] have already demonstrated the many advantages derived from the use of radar for reconnaissance mapping of the Earth's surface. J. H. Simons has demonstrated that this radar imagery provides good evidence of geological structure, geomorphologic features, and cultural patterns of land uses. The results of this work have led to a recent application of the side-looking radar as an exploration tool for the petroleum industry with which to map the perpetually cloud-bound eastern slope of the Andes Mountains in Ecuador, South America.

Another advantage of the radar mapping sensor is its self-illuminating capability. An orbiting vehicle with an onboard radar mapper can map completely around the moon where optical photography is limited to areas under solar illumination.

A preliminary design of a radar system for mapping of Venus, from an orbiting spacecraft, was shown to be feasible by R. F. Schmidt.[3] This conceptual radar was designed to provide measurements of surface contour and surface area/reflectivity. Within the constraints of beamwidth limits (1 degree and 2 degrees) and pulsewidth limits (1 μs and 10 μs), this radar could provide surface resolutions of 15–20 km or better, depending on orbit altitude.

The primary purpose here is to set forth the fundamental principles associated with coherent side-looking imaging radars that have produced radar maps with photograph-like quality. High-resolution maps in two dimensions can be produced by an airborne radar through the generation of a synthetic aperture [4, 5]

* The Corps of Engineers is using a high-resolution side-looking airborne radar (SLAR), AN/APQ–97 to obtain topographical maps of large land areas in Panama and bordering Colombia that are perpetually concealed by continuous cloud cover; see Technology Week, May 22, 1967, p. 11, and Aviation Week and Space Technology, May 8, 1967, pp. 57–64.

in the along-track or azimuth direction, and pulse compression[6] in the cross-track or range direction. Similarly, a stationary coherent radar can produce a high resolution image of a rotating or moving object. Astronomers have mapped the moon and Venus with earth-stationed radars by synthetic aperture concept processing of the backscattered signals.[7, 8]

Excellent tutorial papers have been written on the separate topics of the general theory of side-looking synthetic aperture radar systems[9] and linear FM pulse-compression radar systems.[10] This paper shows that the high-resolution properties of these separate concepts are derived from identical fundamental principles. The radar system, which employs these two techniques simultaneously, can produce a symmetrical, two-dimensional, high-resolution map by equating the time-bandwith products in the two directions of azimuth and range.

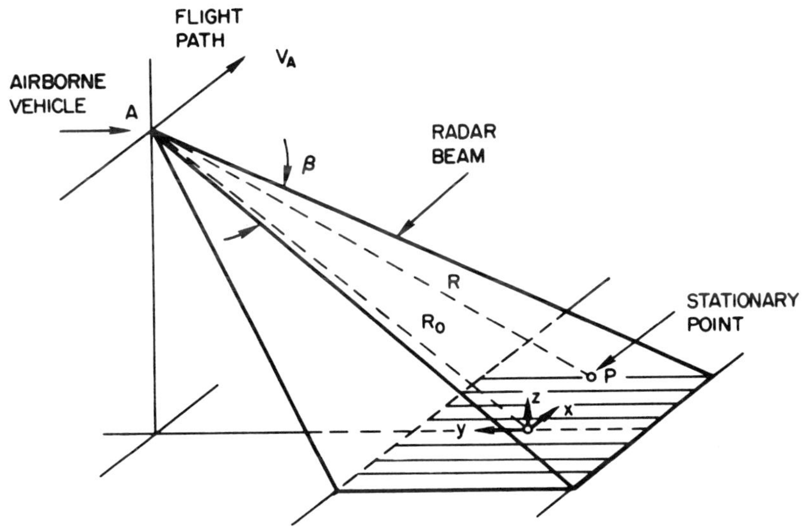

FIGURE 1. Side-looking radar system geometry.

CONVENTIONAL SIDE-LOOKING AIRBORNE RADAR (SLAR)

In general, the system geometry associated with a side-looking radar is shown in FIGURE 1. An airborne vehicle, at Point A, is traveling in the positive x direction with velocity V_A. Onboard the vehicle is a doppler radar system whose physical aperture, with beamwidth β, illuminates a ground-swath parallel to the flight path. The radar beam points in a direction orthogonal to the vehicle's flight path.

A noncoherent transmitter emits a CW pulse of length τ, which illuminates the ground. The range equivalent of this pulse length is $\frac{C\tau}{2}$, and thus, the conventional range resolution δR_{conv} is

$$\delta R_{conv} = \frac{C\tau}{2}.$$

In the cross range direction at range R_o, the physical aperture illuminates a lineal distance of $R_o\beta$. Here, the diffraction-limited aperture beamwidth, β, is equal to λ/D where λ is the wavelength of transmission and D is the aperture size. Then, the cross range or azimuth resolution δ AZ is

$$\delta Az_{conv} = \frac{\lambda R_o}{D}.$$

These two fundamental relationships for range and azimuth resolution indicate that a conventional side-looking airborne radar (SLAR) can achieve fine resolution by transmitting a narrower pulse at a higher frequency with a larger physical aperture. This approach to getting improved radar imagery is characterized as the "brute force" method.

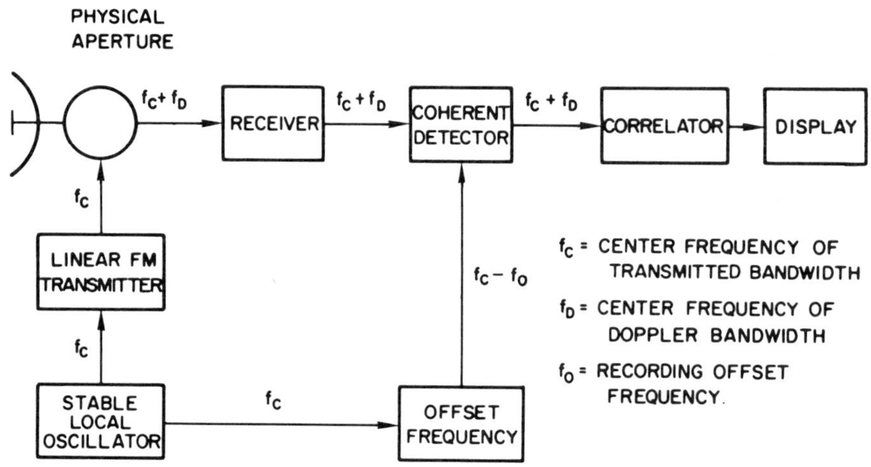

FIGURE 2. Basic block diagram of coherent radar system with linear FM transmitter.

SYNTHETIC APERTURE RADAR (SAR) WITH FM PULSE COMPRESSION

Here, the airborne vehicle carries a coherent, pulse doppler radar. FIGURE 2 is a basic block diagram of a coherent radar system [11] with a linear FM transmitter carried to the airborne vehicle. The synthetic aperture radar obtains range information by pulsing the transmitter. Fine range resolution and fine azimuth resolution is derived by storing the returned pulses over the time that a ground scatterer stays within the physical aperture beamwidth, and the processing the stored data through the correlator. Each pulse is linearly FM'ed within each pulse by the transmitter; and the sequence of pulses, gathered over a beam dwell time, are doppler FM'ed because of the airborne vehicle's motion across the ground scatterer.

The finely resolved images are displayed on a two-dimensional display to form a radar map of the ground. On this display, the range coordinate is in the cross-track direction and azimuth is the along-track direction.

In the analyses that follow, attention is focussed on the frequency histories of the returned signals at the output of the coherent detector, and how these histories are matched to the center frequency, time duration, and bandwidth of the correlator.

Doppler-Frequency History of Standard Ground Objects

Consider a stationary point reflector P on the ground at time t when it first enters the radar beam of the physical aperture (see FIGURE 1). The slant range to this point is given by the approximation:

$$R(x) = \sqrt{R_o^2 + (x-x_o)^2} \approx R_o + \frac{(x-x_o)^2}{2R_o} \qquad (x-x_o) << R_o$$

and since $(x-x_o) = V_A(t-t_o)$,
then

$$R(t) \approx R_o + \frac{V_A^2(t-t_o)^2}{2R_o}. \tag{1}$$

Now, the doppler shift in frequency for this point is:

$$f_D(t) = -\frac{2\dot{R}(t)}{\lambda} \tag{2}$$

where λ is the wavelength of the transmitted RF signal. After taking the time derivative of Equation 1, substitution in Equation 2 gives:

$$f_D(t) = -\frac{2V_A^2}{\lambda R_o}(t-t)_o \tag{3}$$

The received frequencies at the input to the coherent detector (see FIGURE 2) become:

$$f_{RCVR}(t) = f_C + f_D(t) = f_C - \frac{2V_A^2}{\lambda R_o}(t-t_o) \tag{4}$$

where f_C is the primary transmitted frequency. The reference frequency to the coherent detector is:

$$f_{REF} = f_C - f_o \tag{5}$$

where f_o is the recording offset frequency which is used to keep the doppler information band away from dc, and thus avoid frequency foldover. Basically, the coherent detector performs the function of subtracting the reference frequency (Equation 5) from the input-received frequency (Equation 4), so that the coherently-detected frequency is:

$$f_{DET}(t) = f_{RCVR} - f_{REF} = (f_C + f_D) - (f_C - f_o) = f_o - \frac{2V_A^2}{\lambda R_o}(t-t_o) \tag{6}$$

This time function of the detected frequency is commonly referred to as the "doppler-frequency history" of a point scatterer. FIGURE 3 is a plot of this history with the center of the diagram taken to be at time, t_o, when the ground scatterer is in the center of the physical aperture, and when the frequency is the offset frequency, f_o. With the scatterer at broadside, its doppler frequency falls to zero. The slope of the frequency history, or the FM rate, is $-2V_A{}^2/\lambda R_o$. Here, the quantity $-2V_A{}^2/R_o$ is recognized as the two-way radial acceleration of the point scatterer along the boresight of the physical aperture toward the aircraft. Now, the -3 db lineal azimuth width of the physical aperture is $R_o\beta$, so that the time extent over which the scatterer stays within the beam is essentially;

$$T=\frac{R_o\beta}{V_A} \tag{7}$$

The bandwith (BW) of the doppler history is the product of its slope and its time duration, so that:

$$BW=\frac{2V_A{}^2}{\lambda R_o}T \tag{8}$$

and, thus, the time-bandwidth product is:

$$(TBW)=\frac{2V_A{}^2}{\lambda R_o}T^2. \tag{9}$$

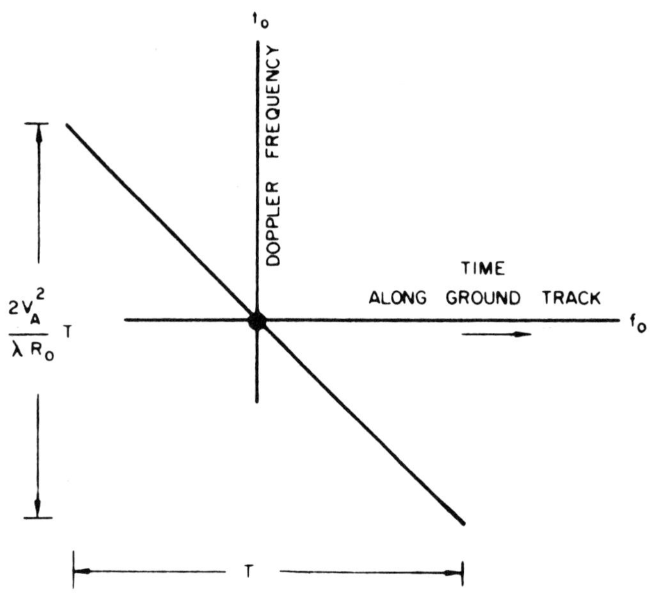

FIGURE 3. The doppler frequency history of a stationary ground scatterer. Key: V_A=airborne vehicle velocity; λ=carrier wavelength; R_o=shortest slant range to ground scatterer; T=time scatterer stays in physical beamwidth; f_o=recording offset frequency; t_o=time when scatterer passes through center of beam.

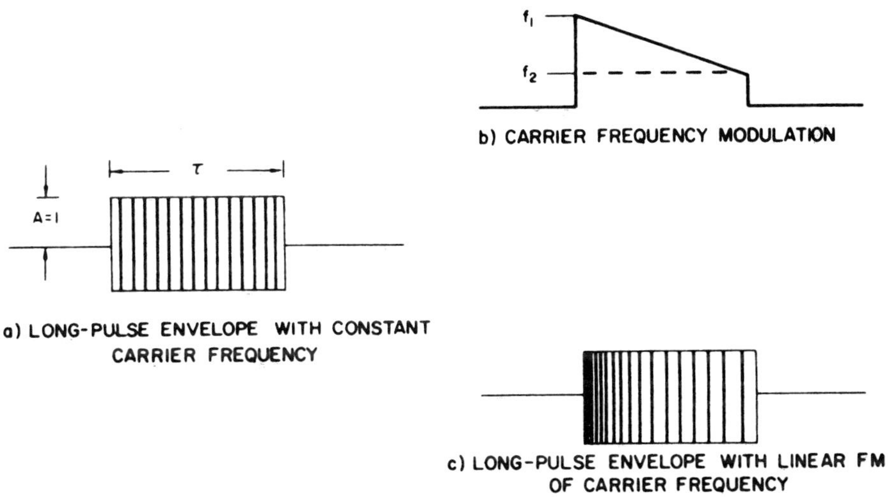

FIGURE 4. Linear FM transmitted pulse characteristics.

Linear FM Transmitted Pulse

Recall that the pulse doppler radar obtains its range information by transmitting pulses to the ground scatterer. Furthermore, in order to obtain more efficient use of the average transmitted power, and to increase range resolution, the pulse is transmitted with a linear FM sweep of its carrier frequency under the envelope of the pulse.[10] FIGURE 4 shows the long-pulse envelope and the carrier frequency modulation.

The transmitted waveform for the linear FM pulse can be written:

for

$$f(t) = \cos\left(2\pi f_C t - \frac{2\pi \alpha t^2}{2}\right) \tag{10}$$

$$-\frac{\tau}{2} \leq t \leq +\frac{\tau}{2}$$

$$f(t) = o \text{ elsewhere}$$

where

f_C = carrier frequency
α = rate of frequency sweep $(\Delta f_R/\tau)$
$(\Delta f)_R$ = swept spectrum bandwidth
τ = uncompressed pulsewidth

The frequency content of each transmitted pulse is obtained from the argument of cosine function so that:

$$f_{TRANSMITTED} = f_C - \alpha t\,. \tag{11}$$

In the absence of any doppler shifts in frequency of the returned pulses from a ground scatterer, the frequency content of each coherently-detected pulse, after mixing with the reference signal, as in Equation 6, gives:

$$f_{\substack{DET\\PULSE}} = f_0 - \alpha t \cdot \quad \textbf{(12)}$$

Comparing Equations 6 and 12 provides an insight into the similarity of the FMing functions in the two directions of azimuth and range. It is to be noted that the intentional frequency slope α for each transmitted pulse in the range direction is similar (but not equal) to the frequency slope $2V_A^2/\lambda R_0$ for the doppler frequencies of a point scatterer. The range FMing α is introduced to each pulse by a function generator in the radar transmitter, whereas the azimuth FMing is produced by the change in aspect angle of the ground scatterer as the airborne vehicle moves along the ground track. Both of these quantities are frequency rates, and only their magnitudes may be different. The frequency slope α (in the range direction) sweeps over the pulse interval, τ, whereas the frequency slope, $2V_A^2/\lambda R_0$, sweeps over the time interval T, for which the ground scatter remains within the azimuth beamwidth.

The radar pulses are transmitted at a pulse repetition frequency (PRF) consistent with: 1) at least two samples per cycle at the highest doppler-received frequency; 2) avoidance of second-time range ambiguities; and 3) sufficient energy to get a good S/N. As such, the doppler frequency history along the azimuth direction is a sampled function rather than the continuous function shown in FIGURE 3. For the purposes of this paper, this intermediate sampling process may be neglected, since reconstructed doppler histories are readily derived from the samples. However, the PRF must be high enough to sample the doppler bandwidth adequately in the azimuth angle illuminated by the radar.

Two-Dimensional Resolution

As shown previously, it takes time to sample the doppler bandwidth associated with a point scatterer during the interval it remains within the beamwidth of the physical aperture. The information within this bandwidth is required in order to derive the full resolution capability of the synthetic aperture. Therefore, the sampled data is stored on some medium e.g. film, tape, or storage tube, over this interval before processing in the two-dimensional correlator.

Now, the focusing or collapsing of this information, which lasts for T seconds in the azimuth dimension and τ seconds in the range direction, and whose frequency is changing with a rate of $2V_A^2/\lambda R_0$ cycles/second/second in the azimuth dimension, and α cycles/second/second in the range dimension, into a finely-resolved or narrow image is accomplished in the correlator.[12]

There are any number of equivalent methods [13] for describing the focusing function of the correlator, e.g. vector summation, matched filter, or cross-correlation. All of these methods can be summarized, pictorially as in FIGURE 5. The input pulse of constant amplitude † A and width τ with linear FM rate, α, is passed through the correlator which "matches" these characteristics. The output compressed pulse [14] has a sin x/x envelope with a pulsewidth $1/\alpha\tau$ when measured 4 db down from the peak amplitude. The spacing between the first zeros

† The amplitude A is an involved radar quantity that depends on radiated power level, phase shift, fourth-power range law, and average-target reflectivity.

of this envelope is $2/\alpha\,\tau$, and the peak amplitude is $A\sqrt{\alpha\,\tau^2}$. This compressed pulse is referred to, in our case, as the image of the ground scatterer.

The above general correlator output characteristics for a linear FM pulse input are directly applicable to the two orthogonal directions of range and azimuth. FIGURE 6 shows the format of the input data to the two-dimensional correlator. Note the similarities of the two-dimensional packet of input data. In the range direction, there is a time interval, τ, microsecond long with intrapulse bandwidth $(\Delta f)_R = \alpha\tau$ MHz, whereas in the azimuth direction there is T seconds of samples with polypulse bandwidth $(\Delta f)_{Az} = 2V_A^2/\lambda\, R_0\, T$ cycles. In both directions, bandwidths are generated by linear FM over time intervals given. A two-dimensional correlator e.g. an optical correlator [15] can focus the input packet in each dimension without any coupling between them. Then, as shown in FIGURE 6, the focused image in each direction is a sin x/x function. Thus, the image of the ground scatterer at the correlator output is a bivariate

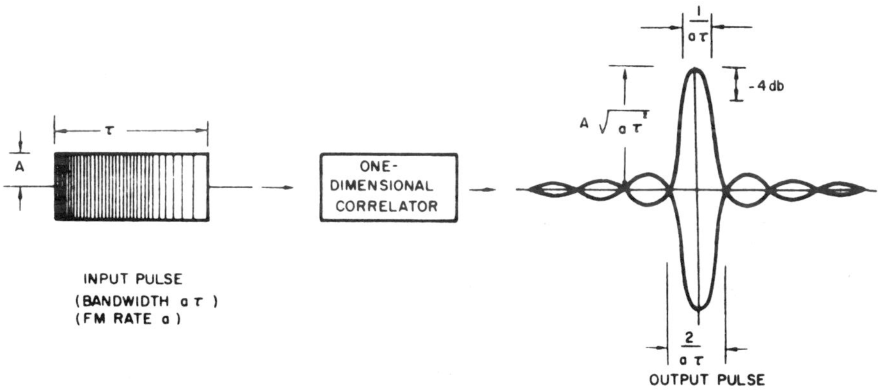

FIGURE 5. Input/output characteristics of a one-dimensional correlator for a general input pulse length τ seconds long with a linear FM rate.

sin x/x function. The resolution of this output function is the width of the image at the -4 db level down from the maximum. *It is at this point in the system and at the time when this image is formed that the combination of airborne vehicle, coherent radar with linear FM transmitter, and correlator have generated the "synthetic aperture."*

The derivation of the range resolution (δ R), and azimuth resolution (δ Az) are given in TABLE 1, for easy comparison. (Note the direct similarity of the resolution Equations 13 and 14, TABLE 1.) The resolutions in either direction are fundamentally related to the total information content [16] $2\,\tau(\Delta f)_R$ and $2\,T(\Delta f)_{Az}$, the velocities C and V_A, and then the time intervals τ and T. These resolution forms show that, as the total information content is increased, the resolution becomes smaller or greatly improved. Furthermore, the condition for making the image symmetrical in both directions is to let $\delta R = \delta Az$ and $\tau(\Delta f)_R = T(\Delta f)_{Az}$ so that $C\tau = 4V_AT$.

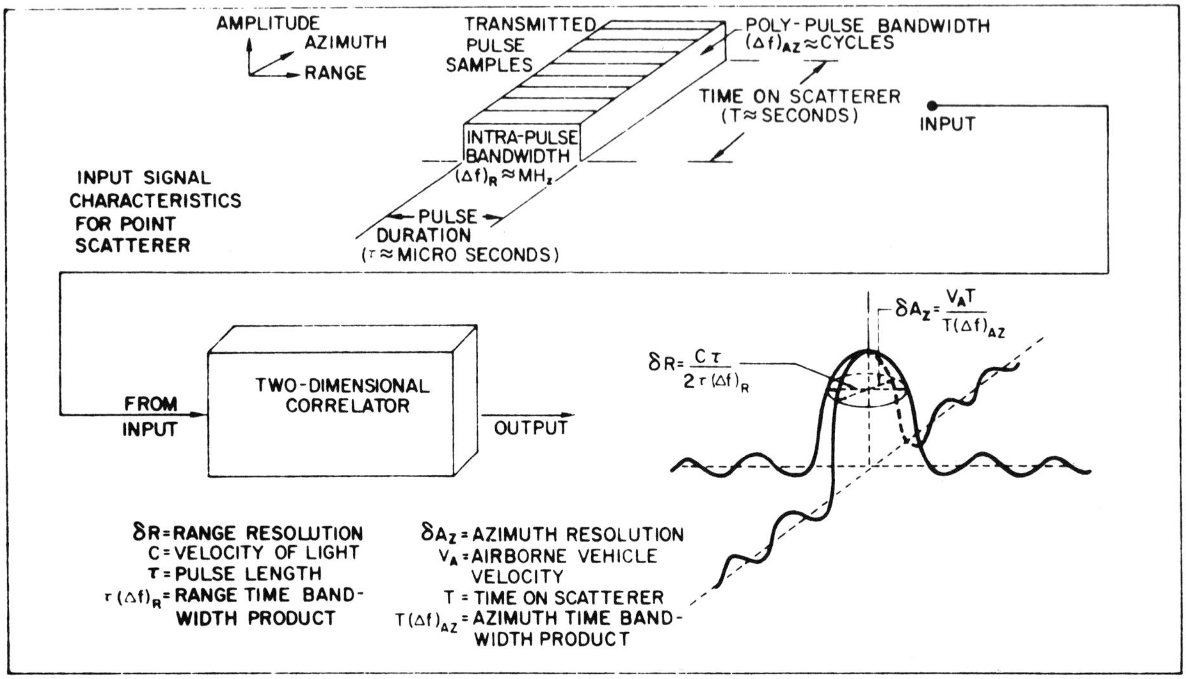

FIGURE 6. Input/output characteristics of a two-dimensional correlator.

TABLE 1

COMPARISON OF DERIVATIONS FOR RANGE RESOLUTION (δR) AND AZIMUTH RESOLUTION (δAz)

Useful Relationships

Range Direction	Azimuth Direction
1. Time width τ_R of image at -4 db level (Refer to Figure 5): $\tau_R = 1/(\Delta f)_R$ where $(\Delta f)_R$ is transmitting FM band-width.	1. Time with τ_{Az} of image at -4 db level (Refer to Figure 5): $\tau_{Az} = 1/(\Delta f)_{Az}$ where $(\Delta f)_{Az}$ is doppler bandwidth.
2. Multiply both sides of above equation by C/2 where C is velocity of light (factor of 2 required for two-way range): $\frac{C\tau_R}{2} = C/2\,(\Delta f)_R$	2. Multiply both sides of above equation by airborne vehicle velocity V_A: $V_A \tau_{Az} = V_A/(\Delta f)_{Az}$
3. Here $\frac{C\tau_R}{2}$ is defined as the range resolution δR: $\delta R = C/2\,(\Delta f)_R$	3. Here $V_A \tau_{Az}$ is defined as the azimuth resolution δAz: $\delta Az = V_A/(\Delta f)_{Az}$
4. Multiplying the numerator and denominator by pulse width τ gives range resolution in terms of range time-bandwidth product $\tau(\Delta f)_R$: $\delta R = C\tau/2\tau(\Delta f)_R$ (13)	4. Multiplying the numerator and denominator by dwell time T, during which ground scatterer stays in physical aperture, gives azimuth resolution in terms of azimuth time-bandwidth product $T\,(\Delta f)_{Az}$: $\delta Az = V_A T/T\,(\Delta f)_{Az}$ (14)

Some other familiar forms for the resolution in the azimuth direction are easily obtained. Since:

$$(\Delta f)_{Az}=\frac{2V_A^2}{\lambda R_o}T$$

substitution into Equation 14 gives the azimuth resolution in terms of the time on target T as:

$$\delta\, Az=\frac{\lambda R_o}{2\, V_A T}. \qquad \textbf{(15)}$$

Here, the product V_AT is usually considered as the length L of the generated synthetic aperture so that:

$$\delta\, Az=\frac{\lambda R_o}{2L}. \qquad \textbf{(16)}$$

Furthermore, since $L=R_o\beta$ where β is the beamwidth of the physical aperture on board the airborne vehicle, and since $\beta=\lambda/D$ where D is the actual length of the physical aperture, then:

$$\delta\, Az=\frac{D}{2}. \qquad \textbf{(17)}$$

Here, Equation 17 shows that, if all the data from a point scatterer is processed optimally, the theoretical azimuth resolution of a synthetic aperture is independent of transmitted frequency or wavelength and is dependent only on the length D of the physical aperture. In practical systems, the theoretical resolution may not be obtainable, due to the many sources of phase errors [17, 18] that may be introduced throughout the system. TABLE 2 summarizes the salient characteristics for imaging a stationary point scatterer by a focussed synthetic radar system ‡ in the azimuth or along-track direction.

RADAR MAPPING OF A ROTATING OBJECT

The previous paragraphs have shown how fine azimuth or cross-range resolution is obtained by relative motion of a moving coherent radar past a stationary object. These same benefits can be derived by another form of relative motion. Consider the case as shown in FIGURE 7. Here, the coherent radar is stationary, say at an Earth site, and the rotating body is the moon or a planet. As the body rotates with angular velocity ω_r, all the points on the body at a distance x will have a radial velocity $V\sin\theta$ toward the radar where θ is the subtended angle of a point at the radar. These points have a doppler shift in frequency equal to $2V\sin\theta/\lambda$. If the total change in aspect angle of one rotation of the body is $\Delta\theta$, then the corresponding doppler bandwidth is $\left(4V\sin\frac{\Delta\theta}{2}\right)/\lambda$. The azimuth resolution for these points is, from Item 3 in TABLE 1:

$$\delta\, Az_{rot}=\lambda/4\sin(\Delta\theta/2).$$

‡ A comparison of the azimuth resolution obtainable from a physical aperture, an unfocused SAR, and a focused SAR has been treated by L. J. Cutrona and G. O. Hall.

TABLE 2

SUMMARY OF RELATIONSHIPS FOR A FOCUSED SYNTHETIC APERTURE RADAR SYSTEM *

Useful Relationships	Theoretical Forms
Doppler History	$f_o + \frac{2V_A^2}{\lambda R_o}(t - t_o)$
Time on Target	$\frac{R_o\beta}{V_A}$
Doppler Bandwidth (BW)	$\frac{2V_A^2}{\lambda R_o} T$
Time Bandwidth Product (TBW)	$\frac{2V_A^2}{\lambda R_o} T^2$
Equivalent Forms for Azimuth Resolution	**Theoretical Forms**
Time on Target (T) Formulation	$\frac{\lambda R_o}{2V_A T}$
Synthetic Aperture Length (L) Formulation	$\frac{\lambda R_o}{2L}$
Physical Aperture Length (D) Formulation	$\frac{D}{2}$
Time-Bandwidth Product (TBW) Formulation	$\sqrt{\frac{\lambda R_o}{2\ T(\Delta f)_{Az}}}$

* Key: f_o=recording offset frequency; V_A=airborne vehicle velocity; λ=carrier wavelength; t=time; t_o=time when scatter is in beam center; β=beamwidth of physical aperture.

Side-Looking Airborne Radar (SLAR) Imagery

Some examples of the type of imagery obtained by a SLAR system are shown in Figure 8. This airborne system had two physical apertures that were mapping to both sides of the aircraft. The dark horizontal lines were unmapped areas directly beneath the aircraft. Clearly visible are the third-order basin and range provinces. Fourth-order individual mountains and valleys are discernible. The fifth- and sixth-order microreliefs and soil types are also indicated, as well as vegetation.

Synthetic Aperture Radar Imagery

The first experimental demonstration of the synthetic aperture radar concept was performed by a group at the University of Illinois. In a report dated

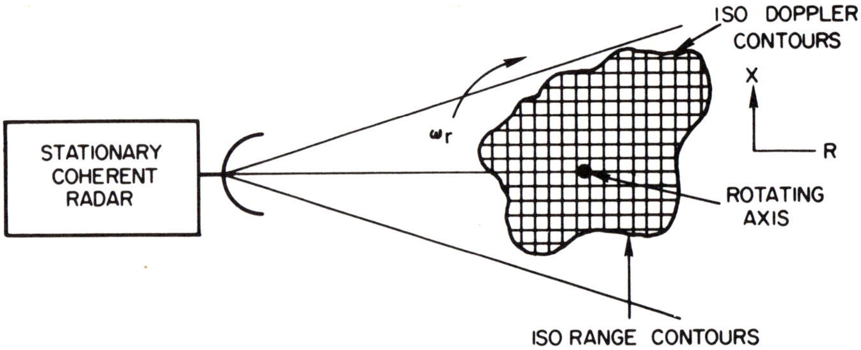

Figure 7. High-resolution radar mapping of a rotating body with the use of range-doppler technique.

July 8, 1953, Kovaly and co-workers [5] produced a strip map (see Figure 9A) of a section of Key West, Fla. With this early system, the 4.13-degree physical aperture beamwidth was used to make an effective synthetic beamwidth of 0.4 degree. Even in this first synthetic aperture radar (SAR) imagery, the land-water boundaries, with correct geometrical forms, were clearly visible, and individual buildings were resolved.

The radar system that produced this early imagery did not employ pulse compression for improved range resolution. Furthermore, the azimuth processing was unfocussed; that is, there was no phase correction to compensate for the changing doppler frequency over the integration time.[19]

Since the first SAR map was made back in 1953, many and significant advances have been made in the design of SAR systems. The group at the Radar and Optics Laboratory, the University of Michigan, Ann Arbor, Mich., have been the forerunners in advancing the state of the art in the design of this elegant electronic system. An example of the early high-resolution SAR imagery, produced by this group, is shown in Figure 9B. There the improve-

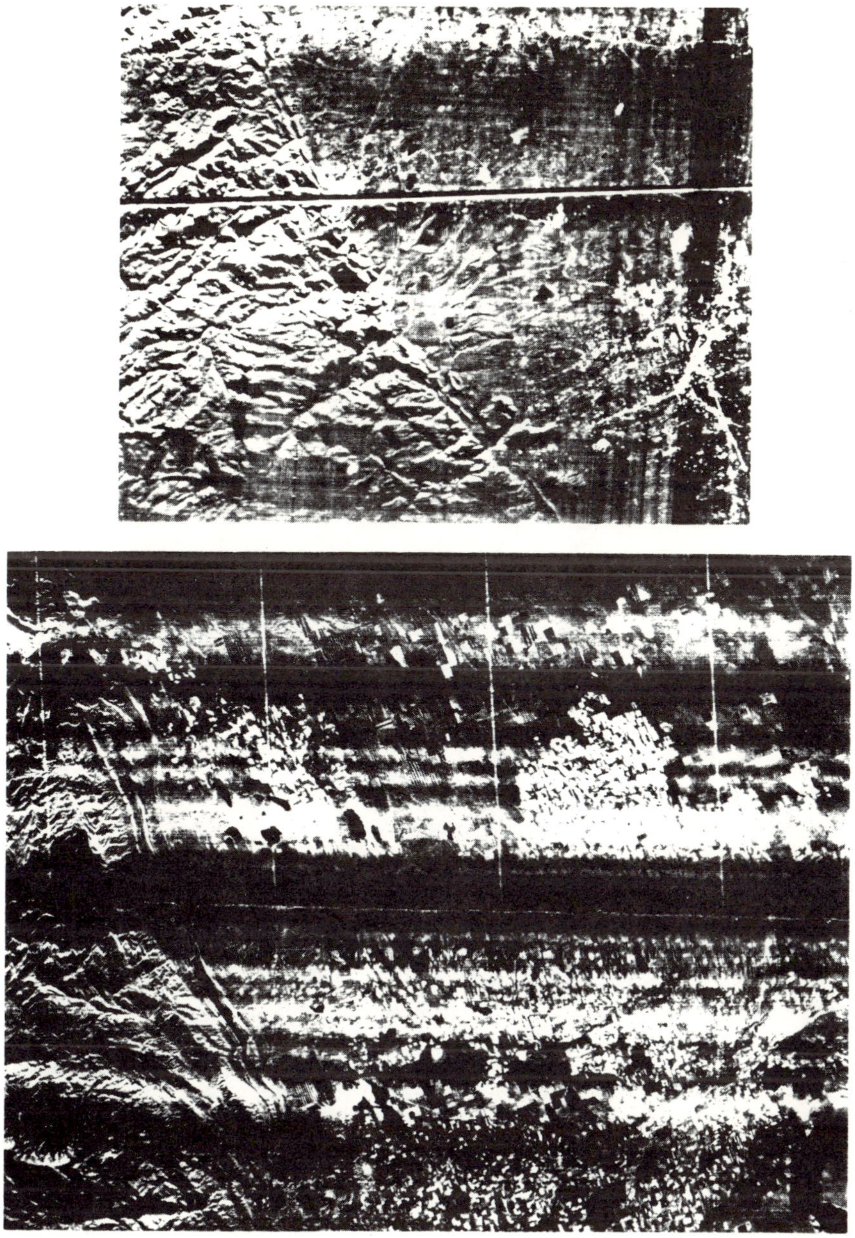

FIGURE 8. Radar imagery made by a side-looking airborne radar (SLAR). North Plateau, Wyo. (*top*); Denver, Colo. (*bottom*). (By permission of Autometric Operation, Equipment Division, Raytheon Company.)

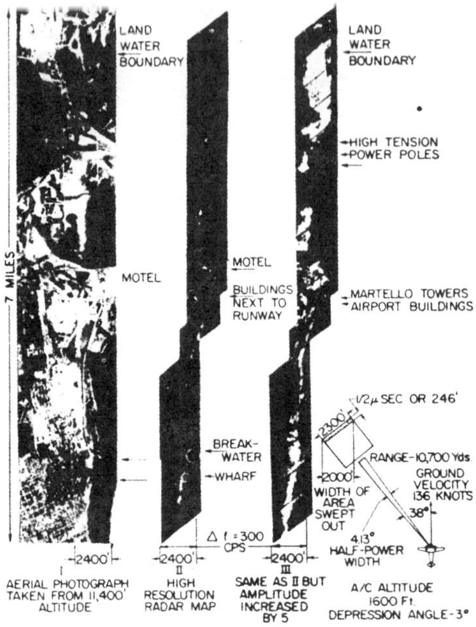

FIGURE 9A. First SAR map, an example of early synthetic aperture radar imagery. (By permission of IEEE Transactions.[5])

ment in radar imagery is clearly discernible when comparing this map with the early attempt made in FIGURE 9A.

Later, improved radar imagery was made by the University of Michigan group, as shown in FIGURES 10A and 10B.

FIGURES 10A & 10B are radar images of the Detroit, Michigan area. The water areas of Lake St. Clair and the Detroit River, with the Ambassador and MacArthur bridges, are easily identified. Of significant geological interest are the broken ice floes on Lake St. Clair. Man-made structures like buildings and roads, as well as natural terrain, are easily recognized.

The Goodyear Aerospace Corporation has contributed significantly to the production of high-resolution imagery. Some examples of their radar maps are shown in FIGURE 11. The Basin and Range province of Southwestern United States at the top. Compare this SAR image [2] with a similar geological example made with SLAR map of FIGURE 8. The well-defined shadowed areas of the mountains show the highly eroded nature of this ridge. Some other geologically interesting areas are the parallel drainage region and the sand dunes. The cultured patterns of barren and irrigated fields are vividly discernible. Figure 11 (*bottom*) shows much geological structure in the Meteor Crater, Winslow, Arizona area. Numerous lineaments are not included on the generalized Arizona geologic map. The structured signatures from this radar image were analyzed in comparison with the known structures, and a revised map was prepared. The new maps contain more detail than was appearing on the old geological maps.

FIGURE 12 is a comparison of a photographic image (left) of the moon's surface taken by the NASA Lunar Orbiter I Spacecraft and the radar image (right) of the Meteor Crater Area, Arizona. This comparison emphasizes the photo-like quality of the radar image that can be obtained by a future orbiting vehicle of some of the unmapped planets that could not be seen by an optical imaging system.

FIGURE 9B. An example of early synthetic aperture radar imagery. Compare with earlier attempt shown in FIGURE 9A. (By permission of Radar and Optics Laboratory, University of Michigan, Ann Arbor, Mich.)

Delay Doppler Radar Images from Rotating Body

Radar imagery of Venus [20] has been produced with Earth-based radars by employing the delay-doppler (synthetic aperture) imaging of a rigid rotating body. FIGURE 13 is a Venus radar map prepared by the Cornell University

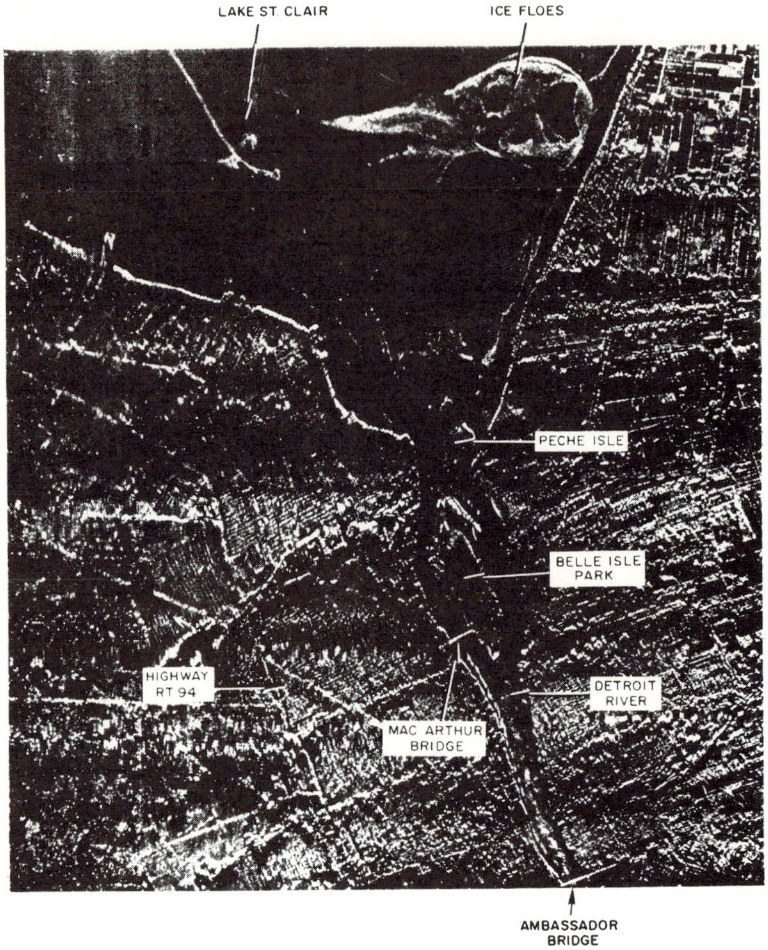

FIGURE 10A. Synthetic aperture radar imagery of Detroit area. (By permission of Radar and Optics Laboratory, University of Michigan, Ann Arbor, Mich.)

astronomers with the radio telescope, located at Arecibo, Puerto Rico. Locations of rough and smooth surfaces can be pinpointed. First-order bulge, second-order platform and third-order ranges with fourth-order mountain masses are present. Here is a clear indication of how a radar system can penetrate through the thick cloud cover around Venus, a task that the Earth optical

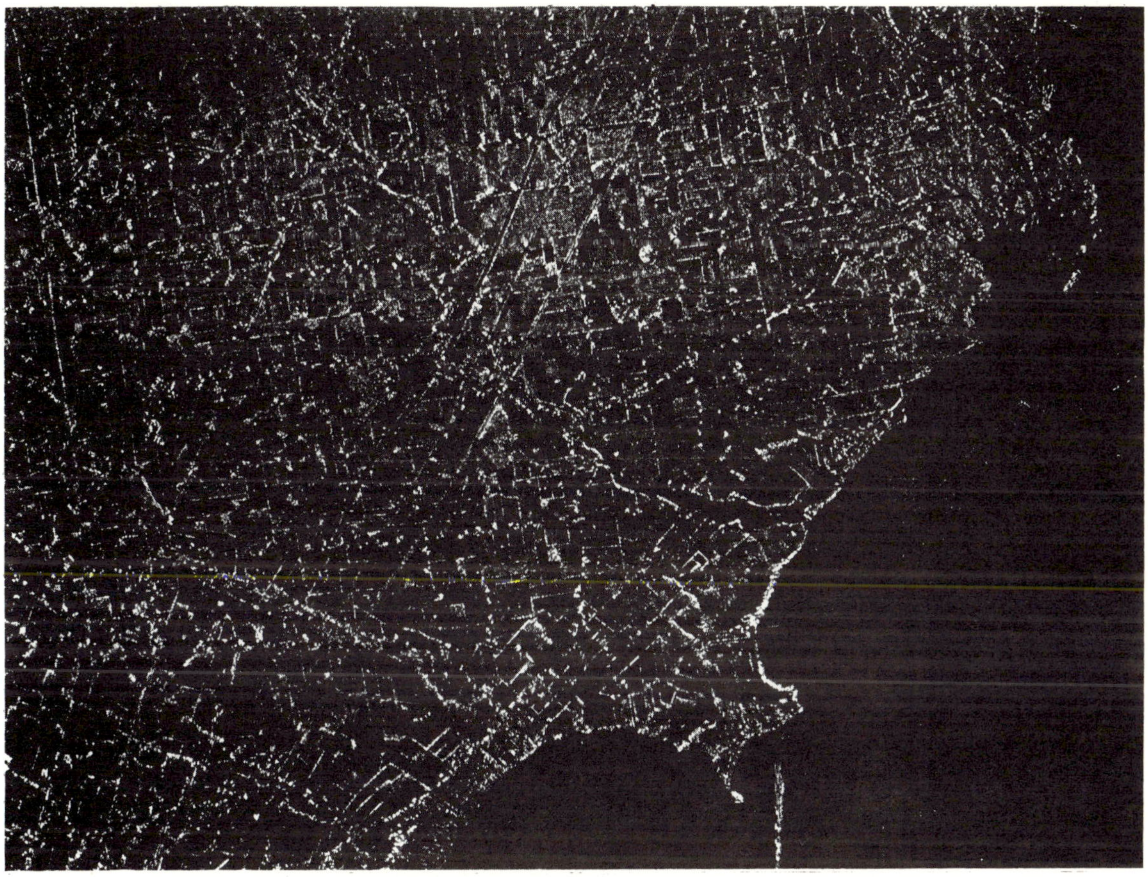

FIGURE 10B. Recent synthetic aperture radar imagery of Munroe, Mich. area. (By permission of Radar and Optics Laboratory, University of Michigan, Ann Arbor, Mich.)

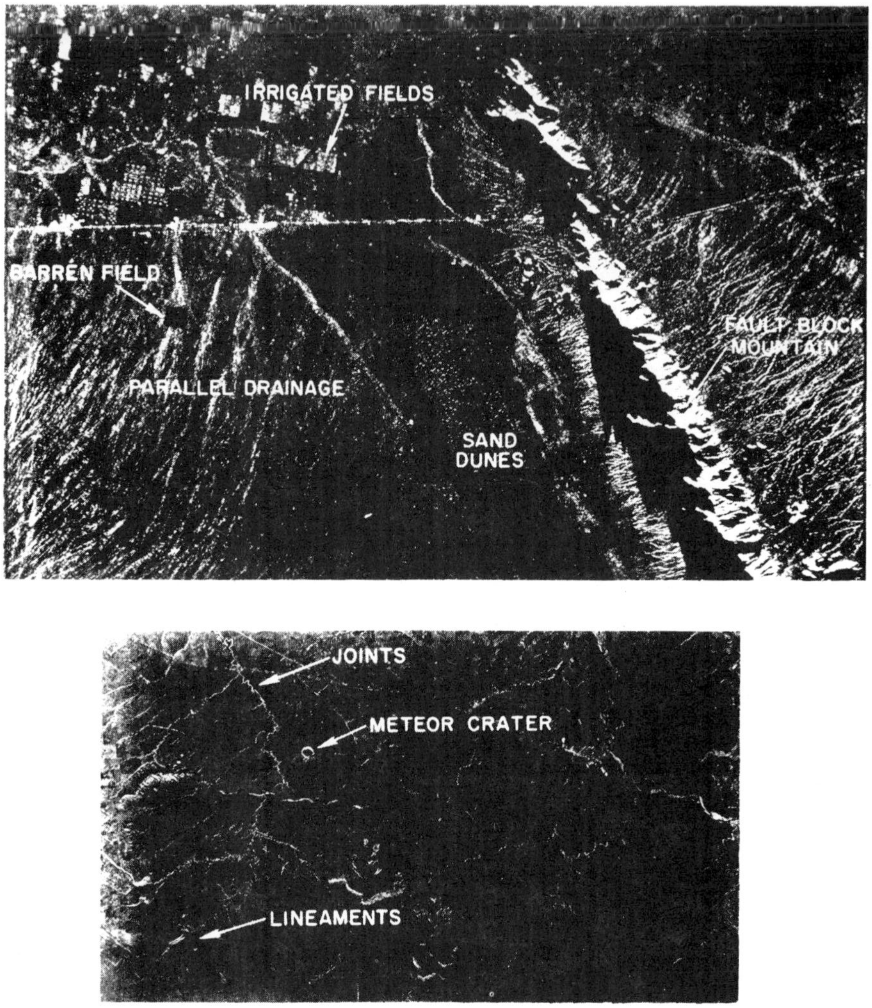

FIGURE 11. Recent synthetic aperture radar imagery of rural areas: Southwestern Arizona (*top*) and Meteor Crater Area, Ariz. (*bottom*). (By permission of Goodyear Aerospace Corp., Arizona Division.)

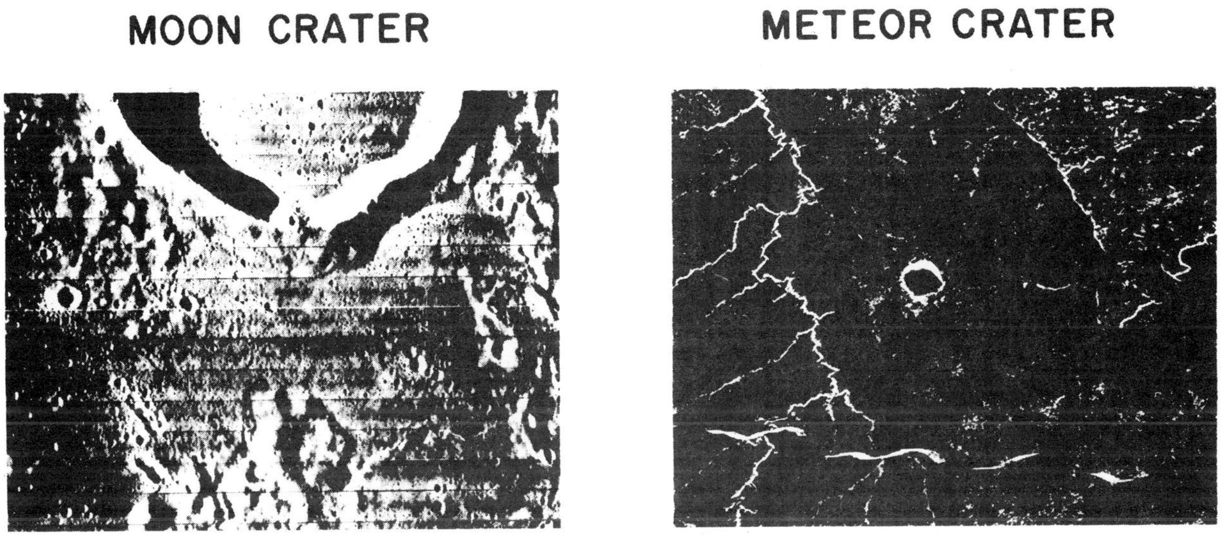

FIGURE 12. Comparison of photographic (*left*) and radar imagery (*right*) of similar terrain types. (*Left*, by permission of NASA, Langley Research Center, Hampton, Va.; *right*, by permission of Goodyear Aerospace Corp., Arizona Division.)

telescope cannot match. Similar radar maps of Venus were imaged by the group at Lincoln Laboratory of Massachusetts Institute of Technology.[21] From these radar maps, the radar astronomers say that the surface of Venus is twice as rough as the average surface of the moon, and that the average reflectivity of Venus is higher than that of the moon. In all probability, then, Venus's surface is composed of a denser material than that of the moon.

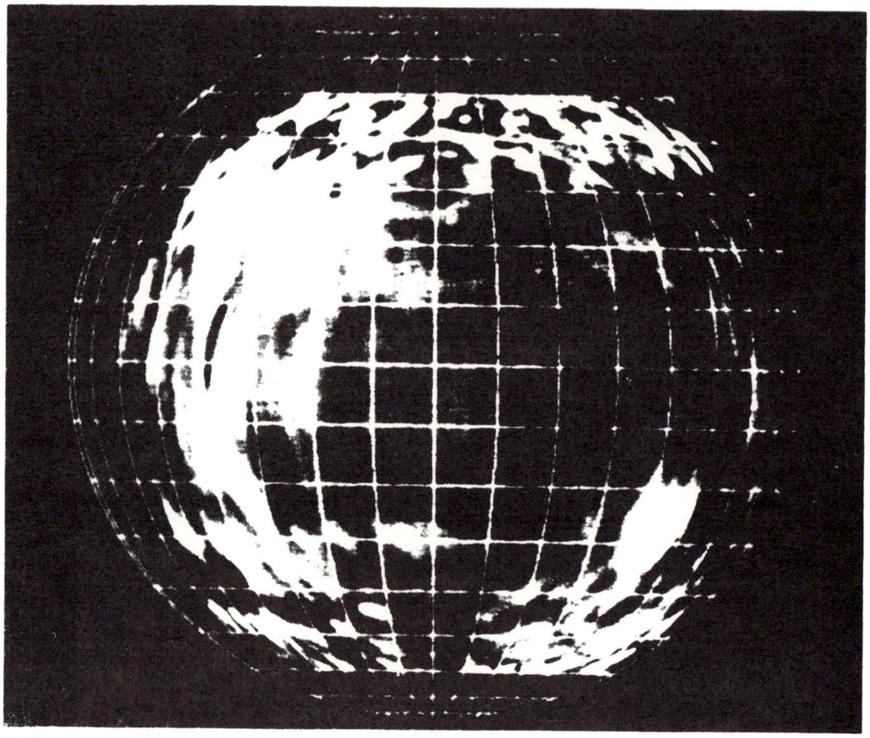

FIGURE 13. Delay-doppler radar map of Venus. (Courtesy of Cornell University Center for Radiophysics and Space Research, Ithaca, N.Y.)

CONCLUSION

With the capability of performing during day or night, good or bad weather, and in gaseous media, radar should play a significant role as one of the primary sensors for planetary aerial mapping. In addition, and far more importantly, radar systems can provide the high-resolution radar imagery that results in photo-like quality to the maps. This paper has shown that fine azimuth or along-track resolution can be achieved through the generation of a synthetic aperture. A physical aperture, whose dimensions are commensurate with the size of a spacecraft, in combination with a coherent radar and data processor, can synthesize a very long antenna, which is required to achieve a narrow beam in the azimuth dimension. In addition, fine-range resolution can be obtained by linear FM modulation of the transmitted radar pulse. Here, appropriate transmitter modulation makes for efficient use of peak-power limited tubes to achieve the necessary average power. The combination of a synthetic aperture radar, with linear FM transmitter modulation, can result in a symmetrical, two-dimensional, high-resolution response function by designing equal time-bandwidth products in both the azimuth and range direction. Through this paper, the two sophisticated electronic concepts, synthetic aperture radar and linear FM transmission, can be treated together with the same fundamental concepts.

References

1. Sheehan, J. L. 1967. Aerial mapping planetary bodies. IEEE Trans. Aerospace and Electronics Systems. AES–**3**(2): 202–206.
2. Simons, J. H. 1964. Some applications of side-looking airborne radar. Proc. Third Symp. Remote Sensing of Environment: 563–571. Infrared Physics Lab. Univ. Mich. Ann Arbor, Mich.
3. Schmidt, R. F. 1964. Radar mapping of Venus from an orbiting spacecraft. Proc. Third Symp. Remote Sensing of Environment: 51–61. Infrared Physics Lab. Univ. Mich. Ann Arbor, Mich.
4. Cutrona, L. J., W. E. Vivian, E. N. Leith & G. O. Hall. 1961. A high-resolution radar combat-surveillance system. IEEE Trans. Military Electronics MIL–**5:** 127–131.
5. Sherwin, C. W., J. P. Ruina & R. D. Rawcliffe. 1962. Some early developments in synthetic aperture radar systems. IEEE Trans. Military Electronics. MIL–**6**(2): 111–115.
6. Cook, C. E. 1960. Pulse compression—key to more efficient radar transmission. Proc. IEEE. **48**(3): 310–316.
7. Shapiro, I. I. 1968. Planetary radar astronomy. IEEE Spectrum. **5:** 70–79.
8. Evans, J. V. & T. Hagfors. 1968. Radar Astronomy. McGraw-Hill Book Co. New York, N.Y.
9. Brown, W. M. 1967. Synthetic aperture radar. IEEE Trans. Aerospace and Electronics Systems. AES–**3**(2): 217–229.
10. Bernfeld, M., C. E. Cook, J. Paoillo, & C. A. Palmieri. 1964, 1965. Matched filtering, pulse compression, and waveform design. Microwave J. Part I, Oct. 1964: 57–64; Part II, Nov. 1964: 81–90; Part III, Dec. 1964: 70–76; Part IV, Jan. 1965: 73–81.
11. Skolnik, M. I. 1962. Introduction to Radar Systems: 143–163. McGraw-Hill Book Co. New York, N.Y.
12. Cutrona, L. J., E. N. Leith, L. J. Procello & W. E. Vivian. 1966. On the application of coherent optical processing techniques to synthetic-aperture radar. Proc. IEEE. **54:** 1026–1032.
13. McCord, H. L. 1962. The equivalence among three approaches to deriving synthetic array patterns and analyzing processing techniques IEEE Trans. Military Electronics MIL–**6**(2): 116–119.
14. Klauder, J. R., A. C. Price, S. Darlington & W. J. Albersheim. 1960. The theory and design of chirp radars. Bell System Tech. J. **39:** 745–808.
15. Cutrona, L. J., E. N. Leith, C. J. Palermo & L. J. Porcello. 1960. Optical Data Processing and Filtering Systems. IEEE Trans. Information Theory. IT–**6:** 386–400.

16. WOODWOOD, P. M. 1955. Probability and Information Theory with Application to Radar. McGraw-Hill Book Co. New York, N.Y.
17. GREENE, C. A. & R. T. MOLLER. 1962. The effect of normally distributed random phase errors on synthetic array gain patterns. IEEE Trans. Military Electronics MIL–**6**(2): 120–139.
18. BROWN, W. M. & J. C. PALERMO. 1965. Effects of phase errors on resolution. IEEE Trans. Military Electronics. MIL–**9**(1): 4–9.
19. CUTRONA, L. J. & G. O. HALL. 1962. A comparison of techniques for achieving fine azimuth resolution. IEEE Trans. Military Electronics. MIL–**6**(2): 119–121.
20. CAMPBELL, D. B., R. F. JURGENS, R. B. DYCE, F. S. HARRIS & G. H. PETTENBILL. 1970. Radar Interferometric Observations of Venus at 70 cm. Cornell-Sydney University Astronomy Center, Arecibo Observatory, Arecibo, Puerto Rico. (Preprint Oct.)
21. RODGERS, A. E. & R. P. INGALLS. 1970. Radar mapping of Venus with interferometric resolution of the range-Doppler ambiguity. Radio Science **5**(2): 425–433.

CHAPTER III
Performance of Synthetic Aperture Radar

Design considerations for a synthetic aperture radar encompass the descriptions of its performance in terms of the complex interplay of many system parameters. Some of these design elements are synthetic aperture length, physical aperture size, average transmitted power, pulse repetition frequency, time, bandwidth, vehicle velocity, and vehicle altitude; characteristic performance criteria are resolution, signal-to-noise ratio, least squares target field optimization, gain patterns and, sidelobe levels. Since the synthetic aperture radar is a remote sensing system the major role of which is the generation of a radar image, its performance is measured by the fidelity of the created image of a rather complex target field. However, limitations of ideal performance can be introduced by receiver noise, system nonlinearities, phase errors, or interactions of the target field in the radar's processor.

W.M. Brown describes the performance of a SAR system when receiver noise limits performance. He derives the length of a synthetic aperture in terms of the noise level and the average transmitted power. In addition, the effects of phase errors on azimuth resolution are considered. Although Brown's paper is an excellent introduction to SAR performance in the presence of receiver noise and system phase errors, the reader can enhance his depth of understanding by delving into Brown and Palermo (1962) for more theoretical aspects of coherent systems and Brown and Palermo (1965) for phase errors on resolution. In addition, Cheng (1964) provides expressions for computing the upper bounds of the effects of phase errors on azimuth resolution of synthetic apertures and the range resolution of pulse-compression radars.

The performance of the synthetic aperture as an antenna is described by R.C. Heimiller in terms of gain patterns and sidelobe response for a focused and unfocused aperture. The gain pattern for a focused synthetic aperture is characterized, in general, by a narrow main beam with low sidelobe levels and well-defined nulls. For an unfocused aperture, the mainbeam is broad, with relatively high sidelobes and filled-in nulls. Considerations are given to the use of rectangular and exponential amplitude weighting to reduce the sidelobes of an unfocused aperture. The third paper by R.O. Harger includes the optimum design of antenna patterns and transmitted signal by minimizing all spurious responses in the ambiguity function. The solution of a nonlinear design problem for a SAR can be found in Harger and Crimmins (1968).

The next two papers by Develet and Harger treat the mapping performance of the SAR. Develet postulates an ideal SAR when mapping a point target, homogeneous terrain, and terrain of arbitrary reflectivity. His performance description includes resolution, swath width, and map signal-to-noise ratio in terms of vehicle altitude, velocity, terrain scattering coefficient, and signal wavelength. Then, Harger provides an optimum design of a SAR which is intended to classify randomly reflecting areas such as agriculture fields. He includes a set of summary design curves. The imaging of extended targets by coherent systems exhibits what is termed as *coherent break-up.*

The reader is referred to Mitchell (1974) for simulation results on studies of extended targets and their coherent radar images.

So far, this chapter has attempted to provide the fundamental methods which can be used to describe the performance of a synthetic aperture radar. The next four papers include SAR performance under certain system design requirements and particular target field characteristics. When the dynamic range of a data link is of concern, as in the case of planetary mapping by satellite, Hildebrand considers the transmission of the unfocused data with subsequent compression to achieve fine resolution. Bayma and McInnes provide a design procedure to select the aperture size and pulse repetition frequency for a terrain mapping SAR to keep the Doppler ambiguities at an acceptable level. Raney considers the effects of SAR imagery when the target field has moving targets. Characteristically, the moving target image is blurred with a reduced intensity. Finally, to reduce the dynamic range limits of electronic processing in a SAR, Steinberg studies the use of hardlimiting to produce a fixed amplitude processor input. Effects such as azimuth false targets can be formed when a large target predominates. On the other hand, statistically homogeneous clutter can serve to linearize the super-limiters performance. For the effects of relative source strength and signal-to-noise ratio on the angular resolution, the reader is referred to Steinberg (1974).

Synthetic Aperture Radar

W.M. Brown
(Institute of Science and Technology, University of Michigan, Ann Arbor)

IEEE Transactions on Aerospace and Electronics Systems, Vol. AES-3, No. 2 (March 1967)
Pages 217-229

Abstract

The general theory of side-looking synthetic aperture radar systems is developed. A simple circuit-theory model is developed; the geometry of the system determines the nature of the prefilter and the receiver (or processor) is the postfilter. The complex distributed reflectivity density appears as the input, and receiver noise is first considered as the interference which limits performance. Analysis and optimization are carried out for three performance criteria (resolution, signal-to-noise ratio, and least squares estimation of the target field). The optimum synthetic aperture length is derived in terms of the noise level and average transmitted power. Range-Doppler ambiguity limitations and optical processing are discussed briefly.

The synthetic aperture concept for rotating target fields is described. It is observed that, for a physical aperture, a side-looking radar, and a rotating target field, the azimuth resolution is λ/α where α is the change in aspect angle over which the target field is viewed.

The effects of phase errors on azimuth resolution are derived in terms of the power density spectrum of the derivative of the phase errors and the performance in the absence of phase errors.

Key Words—Aperture, radar, resolution, signal-to-noise, synthetic, system.

I. Introduction

Coherent side-looking imaging radar [1], [2] is one of the most elegant electronic systems to be developed during the last decade. This particular system development is a part of the rather extensive activities in range-Doppler radars, which in turn are a part of a broad scope of activities in the processing of analog data (these activities include various applications of pulse compression to radar and communications, coherent optical data processing with recent emphasis on holography, laser radar and communications, work in automatic shape recognition, and the like).

Manuscript received June 23, 1966; revised October 31, 1966.

This work was conducted for the U. S. Army Electronics Command under Project MICHIGAN, Contract DA-28-043-AMC-00013(e), and for the U. S. Air Force under Contract AF 33(615)-1452.

The major concern of this (primarily tutorial) paper is coherent side-looking radar; however, in the interest of obtaining a unified picture, comments are made on physical aperture systems, and on the range-Doppler imaging of rigid rotating target fields. In all of these systems the obtainable cross-range resolution on the target field is determined by the change in aspect angle over which the target field is viewed; specifically, if α is the change in aspect angle in radians over which the target is viewed, then the obtainable resolution is approximately

$$\frac{\lambda}{2 \sin \frac{\alpha}{2}} .$$

Imperfect phasing in a physical aperture and/or in Doppler-processing systems (such as in a side-looking radar) may well preclude obtaining the phase-error-free theoretical resolution; hence, the resolution limits due to phase errors will be described for the various systems under consideration. Also, optical processing and some elementary ambiguity limitations will be described briefly.

Resolution as a Performance Criterion [3]

Suppose the complex illumination on a linear aperture of length d is $H_0(x)$, then in the far field the amplitude and phase of the electric field at an angle θ from broadside (for small θ) is essentially

$$h(\theta) = \int_{-d/2}^{d/2} H_0(x) \exp\left(j\frac{2\pi x}{\lambda}\theta\right) dx.$$

Or if $2\pi x/\lambda = \omega$ and $\lambda H_0(\lambda\omega/2\pi) = H(\omega)$, then

$$h(\theta) = \frac{1}{2\pi}\int_{-\infty}^{\infty} e^{j\omega\theta} H(\omega) d\omega \qquad (1)$$

where the illumination, $H(\omega)$, is subject to the constraint

$$H(\omega) = 0 \qquad \text{for } |\omega| > B = \frac{\pi d}{\lambda} \cdot$$

Typically, we desire that the beamwidth associated with h be narrow; i.e., the spread of $h(\theta)$ in θ is the resolution corresponding to the illumination $H(\omega)$, and the first-order problem is to select H so as to optimize the resolution. The illumination function H is constrained by the physical extent of the antenna (or optical aperture, depending on the value of λ). Of course, we have arrived at a rather classic mathematics problem; namely, given that $H(\omega)$ is constrained in bandwidth, how narrow can the Fourier transform of H be. This is something of a subject in itself [3, Ch. 5]—due partly to the fact that the time (or θ) duration of h and the ω spread of H can be quantitatively defined in several interesting ways. However, for our discussion, it is enough to recall one lower bound on the time-bandwidth product of a signal.

Let the time spread (or resolution) of $h(t)$ be defined as the width of an equivalent rectangle; specifically, if the maximum value of $|h(t)|^2$ occurs at $t=0$, then the resolution is defined quantitatively as

$$T = \frac{\int_{-\infty}^{\infty} |h(t)|^2 dt}{|h(0)|^2}.$$

Now if

$$h(t) = \frac{1}{2\pi}\int_{-\infty}^{\infty} e^{j\omega t} H(\omega) d\omega$$

and $H(\omega)=0$ for $|\omega| > B$, then[1]

$$TB \geq \pi \tag{2}$$

and $T=\pi/B$ provided $H(\omega)$ is constant in $|\omega| < B$.

In terms of antenna illumination, the minimum obtainable beamwidth is $T=\Delta\theta=\pi/B=\pi/(\pi d/\lambda)=\lambda/d$ radians. which of course is the usual result; the simple observation here is that (within the definition of resolution used) the bandwidth limit (aperture limit) imposes an absolute lower bound on the angular resolution where this lower bound is trivially tied to the relation between time and frequency spreads of Fourier transform pairs.

Now at a range R, the position resolution provided by the angular resolution λ/d is $\rho=R\ \lambda/d$; however, d/R is (approximately) the angle subtended by the physical aperture d at range R. That is, if $\alpha=d/R$, then the positional resolution of the beam at any range (with these simplifying assumptions) is $\rho=\lambda/\alpha$.

The above simple remarks will be recast in a more complete imaging context in Section II; however, some additional simple observations on resolution as a performance criterion are in order. Specifically, consider a more or less arbitrary linear time-invariant system

$$y(t) = \int_{-\infty}^{\infty} h(\tau)x(t-\tau)d\tau$$

where h is the impulse response of the system. If we think of the output $|y|^2$ as the image of x, then $|h|^2$ is the image produced by a point source. In optical systems, antenna (illumination) design, and the synthesis of pulse compression networks, it is common practice to use resolution (the spread of the point "target" response $|h|^2$) as the (primary) performance criterion.

The idea is simply that if $|h|^2$ is concentrated on a short time interval, then y should "resemble" x. (Said in another way, if we strive to make h approach the δ-function, this should make y approach x.) It is somewhat interesting to compare resolution as a performance criterion with other criteria such as mean-square error, point target detection probability, etc. Suffice it to say here that if the system is constrained in bandwidth, the choice of system impulse response for small mean-square error is consistent with the choice of h for fine resolution. Specifically, if $H(\omega)=0$ for $|\omega| > B$, then

$$\begin{aligned} E(|y-x|^2) &= \int_{-\infty}^{\infty} |1-H(\omega)|^2 S_x(\omega)d\omega \\ &= \int_{|\omega|<B} |1-H|^2 S_x(\omega)d\omega \\ &\quad + \int_{|\omega|>B} S_x(\omega)d\omega \end{aligned}$$

where S_x is the power density spectrum of the input signal. Of course the H which minimizes the mean-square error is $H(\omega)=1$ for $|\omega| < B$, and this transfer function also minimizes the equivalent rectangle resolution under the same bandwidth constraint.

Actually work in pulse compression, radar waveform selection, and antenna illumination has resulted in something of a folklore on the desirable properties of the point target response h (required sidelobes etc.). For example, an $h(t)=1$ for $|t| < T$ and $h(t)=0$ for $|t| > T$ has most of the charms praised in the lore; however, if viewed in terms of the transfer function, $H(\omega)$, such a system contains rather fantastic phase distortion[2] (as well as amplitude distortion; however, compensation for the amplitude of a transfer function typically involves noise enhancement while phase compensation typically involves no noise penalty). Such phase distortion degrades equivalent rectangle resolution, mean-square error, etc., and would be intuitively viewed as offensive in an audio system. Comparable remarks can be made concerning the tapering of the amplitude frequency response of pulse

[1] As for the proof, we use $\int|h|^2=1/2\pi\int|H|^2$, and $h(0)=1/2\pi\int H$; then

$$T = 2\pi \frac{\int_{-B}^{B} |H|^2}{\left|\int_{-B}^{B} H\right|^2}.$$

The integrand in the denominator is viewed as $1\cdot H$, then the Schwarz inequality gives

$$\left|\int_{-B}^{-B} 1\cdot H\right|^2 \leq \int_{-B}^{B} d\omega \cdot \int_{-B}^{B} |H|^2 d\omega \cdot$$

Hence $T \geq (\pi/B)$ with equality when $H(\omega)$ is constant on $|\omega| < B$.

[2] The Fourier transform of a rectangular h is alternately positive and negative and negative H is the ultimate in phase distortion.

compression systems[3]—trading resolution for sidelobes. Again what is attractive in the lore can almost be viewed with horror from a transfer function distortion viewpoint.

It might be noted that resolution as a performance criterion (like concern for amplitude and phase distortion in an amplifier) can be (is) employed without specifying properties of the input (such as the power spectra of the signal and noise); this feature of resolution as a criterion has practical appeal. A somewhat more compelling appeal for resolution as a performance criterion occurs when system performance is limited by phase errors (or stochastic delays). Specifically, under such a disturbance resolution remains indicative of image quality while other popular criteria (mean-square error, etc.) tend to be inappropriate. For example, phase errors may not degrade resolution significantly, yet introduce enough distortion (misplacement of the output) to cause extravagant mean-square error. In such a case the quality of the output image may be excellent, but this fact would not be seen from, say, a mean-square error analysis.

II. Side-Looking Radar

With the major exception of ambiguity limitations, the range channel and azimuth channel of a side-looking radar can be analyzed independently. Hence, it is helpful to start with only the azimuth channel. To this end consider targets at just one range and a CW radar (it is simple enough later to pulse the radar to obtain range resolution in which case the reflecting material in a specific range resolution cell corresponds to the line of reflecting material to be considered now).

Mathematical Model

Let $A(\theta)$ be the two-way electric field pattern of the side-looking antenna in Fig. 1, and for simplicity let this pattern be real. The function $f(u)$ denotes the complex reflectivity density at position u. This reflectivity density is defined so that if $\cos \omega_0 t$ is transmitted from position x on the line of flight, then the signal returned at time t from the increment $(u, u+du)$ is

$$A(\theta)\,|f(u)|\,\cos\left[\omega_0\left(t-\frac{2s}{c}\right)-\gamma(u)\right]du$$

where $f=|f|e^{j\gamma}$. The total signal received at time t is given by summing over all incremental returns, thus

$$\left(\text{where } \theta \cong \frac{x-u}{R}\right):$$

$$g_b(t) = \int A\left(\frac{x-u}{R}\right)|f(u)| \cdot\cos\left[\omega_0\left(t-\frac{2s}{c}\right)-\gamma(u)\right]du.$$

[3] A procedure borrowed from the design of antenna illumination functions.

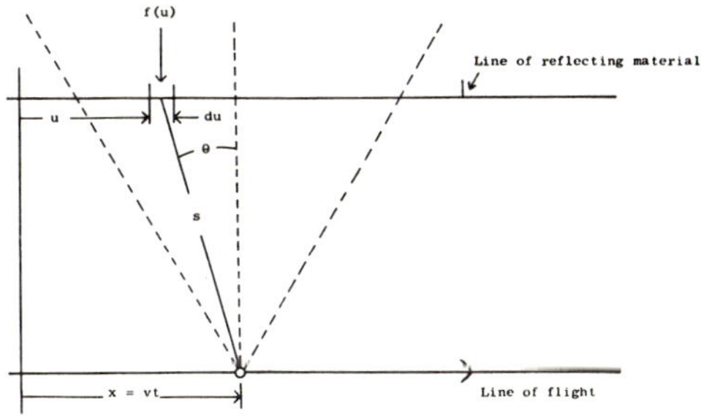

Fig. 1. Geometry of side-looking radar.

If g_b is mixed[4] with $\cos \omega_0 t$ and only the difference frequency is retained, we obtain

$$g_r(x) = \int A\left(\frac{x-u}{R}\right)|f(u)|\,\cos\left[\frac{2\omega_0}{c}s+\gamma(u)\right]du.$$

This mixer output is a function of time in that if the vehicle speed is v, then $x=vt$; however, since the basic objective of the system is to estimate (or image) a function of position (f), the data will be considered as a function of x.

If g_b is mixed in a quadrature mixer ($\sin \omega_0 t$), the signal at the difference frequency is

$$g_i(x) = \int A\left(\frac{x-u}{R}\right)|f(u)|\,\sin\left[\frac{4\pi}{\lambda}s+\gamma(u)\right]du.$$

Now the ordered pair of mixer outputs is a complex signal $g=g_r+jg_i$. It is convenient to approximate

$$s = \sqrt{R^2+(x-u)^2} \cong R + \frac{(x-u)^2}{2R}$$

and let f absorb the constant phase factor $\exp j(4\pi/\lambda)R$. Then if

$$W(x) = A\left(\frac{x}{R}\right)\exp j\frac{2\pi}{\lambda R}x^2$$

$$g(x) = W(x) * f(x). \tag{3}$$

Intrinsic Resolution

Of course the resolution of the basic data (without processing) is the spread of W; e.g., if the beamwidth of the illuminating antenna is α, then the unprocessed data provides a ground resolution of αR. On the other hand, the bandwidth of W offers the potential for much finer resolution. Specifically since W is a chirp pulse having an "instantaneous frequency" of $(4\pi/\lambda R)x$ for a point target located at $u=0$, the bandwidth of the data collected over

[4] For this analysis the carrier frequency, ω_0, will be removed, and the system analyzed in terms of complex lowpass signals; however, we could just as well treat the system in terms of real bandpass signals [3, chps. 6 and 9].

the beamwidth of the illuminating antenna is

$$2B = \frac{4\pi}{\lambda R}(\alpha R) \text{ radians/ft.}$$

In turn (2) suggests that the obtainable resolution is

$$\rho = \pi/B = \frac{\lambda}{2\alpha}\cdot \qquad (4)$$

If d is the physical aperture, it is common to take $\alpha \cong \lambda/d$; then the "theoretical resolution" is $\rho = d/2$.

Before proceeding, consider the point target response without approximating s:

$$W(x) = A\left(\frac{x}{R}\right)\exp\left(j\frac{4\pi}{\lambda}\sqrt{R^2 + x^2}\right).$$

The instantaneous frequency is

$$\frac{4\pi}{\lambda}\frac{x}{\sqrt{R^2 + x^2}}\cdot$$

Thus, data considered over a path[5] of $\pm X$, gives rise to a bandwidth of $B = 4\pi/\lambda \sin \alpha/2$ where α is the change in aspect angle corresponding to this path length. Thus the anticipated resolution is

$$\rho = \frac{\lambda}{4 \sin \alpha/2}\cdot \qquad (5)$$

If one-way propagation is considered (as occurs, say, in an ordinary coherent optical system), then $(2\pi/\lambda)s$ appears in place of $(4\pi/\lambda)s$. In this case a coherent aperture which subtends an angle of α leads to a (diffraction-limited) resolution twice that given in (5).

Basic Optimization

The above remarks on resolution were based on the bandwidth of the data received from a point target; however, it was implicit that the Doppler histories were "compressed." That is, suppose we process the received data, g, with a filter having an impulse response $K(x)$. If noise is present at the output of the quadrature mixers $(n = n_r + jn_i)$, then the final system output is

$$h = K * W * f + K * n.$$

The mathematical model is depicted in Fig. 2. For a point target f is a δ-function and the overall point target response is $h_0 = K * W$.

The first issue is that of selecting an appropriate K; here we suffer from too many available attractive theories. Specifically, K might be selected on the basis of resolution; alternatively, having introduced additive noise, a matched filter suggests itself as does the selection of K so as to minimize the mean-square error, $E(|f-h|^2)$. However, since W tends to be a pulse having a large time-

[5] Here we consider a broad illuminating beam so as to accommodate a Doppler history of any desired length.

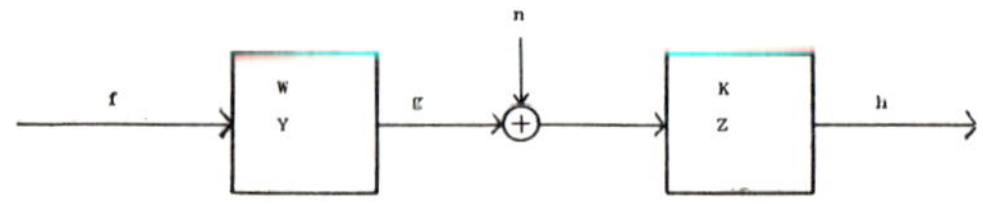

Y and Z denote Fourier transforms of W and K

Fig. 2. Model for azimuth channel of side-looking radar.

bandwidth product ("time" duration of about αR; bandwidth of about $2\alpha/\lambda$; and a product, or available compression ratio, of about $2\alpha^2 R/\lambda \cong 2\lambda R/d^2$), the major obligation of the postfilter is that it phase compensate for the prefilter. That is, if h is to resemble f, it is intuitively clear that the phase of the frequency response Z should be the negative of the phase of Y (save possibly for a linear phase factor which would merely introduce time delay).

Actually, the use of the matched filter

$$Z = \frac{\bar{Y}}{S_n},$$

where S_n is the power density spectrum of the noise, generally involves some slight sacrifice of resolution in the interest of noise rejection. Conceptually the least squares receiver is splendid, it compromises between resolution and noise rejection. That is, the least squares receiver is

$$Z = \frac{\bar{Y}S_f}{|Y|^2 S_f + S_n}\cdot \qquad (6)$$

As $S_n \to 0$, $Z \to 1/Y$ which provides a (theoretically) perfect image of f; and as $S_n \to \infty$, $Z \to (\bar{Y}/S_n)S_f$ which is about the same as the point target matched filter.

The question of optimizing the system in the presence of additive noise is not of overwhelming interest since imperfect coherence (phase errors) imposes, perhaps, more serious performance limitations. The effects of phase errors will be covered in the last Section. A rather basic question suggests itself at this stage; namely, granting that the least squares receiver (Z) is employed, how do we optimize the prefilter transfer function Y.

Let $e = h - f$, then from Fig. 2 we obtain the following, where $S = |Y|^2 S_f + S_n$:

$$S_e = \left|S^{1/2}Z - \frac{S_f \bar{Y}}{S^{1/2}}\right| + \frac{S_f S_n}{S}$$

which yields the optimum Z of (6). The corresponding mean-square error is

$$\overline{e^2} = \frac{1}{2\pi}\int \frac{S_f S_n}{S_f |Y|^2 + S_n} d\omega. \qquad (7)$$

The problem is to minimize $\overline{e^2}$ over $|Y|^2$. Of course as $|Y|^2 \to \infty$, $\overline{e^2} \to 0$; thus for the minimization some constraint is required on the prefilter transfer function, Y. The most reasonable basic constraint is on the illumination power of the physical aperture. We now proceed to interpret this constraint in terms of Y.

Recall that

$$Y(\omega) = \int e^{-j\omega x} A\left(\frac{x}{R}\right) \exp\left(j\frac{2\pi}{\lambda R}x^2\right) dx.$$

To a good approximation:

$$|Y(\omega)|^2 = \frac{\lambda R}{2} A^2\left(\frac{\lambda}{4\pi}\omega\right). \qquad (8)$$

As an approximation (8) simply states that in the direction θ from broadside we receive a Doppler of $4\pi\theta/\lambda$ radians per unit of distance traveled by the vehicle (or a Doppler frequency of $2v\theta/\lambda$ Hz), and that these Dopplers are weighted in amplitude by $A(\theta)$.

The point of (8) is that our selection of the prefilter $|Y|^2$ is equivalent to selecting the square of the two-way pattern A. Within the context of a real two-way pattern, $A(\theta)$ is the square of the one-way pattern $h(\theta)$ which is the Fourier transform of the physical illumination function $H(\xi)$ where ξ denotes position along the physical antenna. A transmitter power constraint amounts to a constraint on $\int|H|^2\leq P_1$, or by Parseval's theorem $\int|h(\theta)|^2 d\theta\leq P_2$. Thus we wish to minimize (7) with $\int|Y|d\omega\leq P$ where P is directly proportional to transmitted power.

A radar scene rich in detail is well modeled by a random process f which has a broad spectral density; thus let $S_f(\omega)=F$ for $|\omega|<B$ and $S_f(\omega)=0$ otherwise and let the noise be white, $S_n=N$. To do the problem justice, a finite aperture constraint should be placed on the physical antenna; however, rather than treat this problem (which requires numerical analysis) the central issue can be dealt with in a simple way. The central issue is the following: for any F and N (signal-to-noise ratio), does the optimum Y pass the entire bandwidth associated with S_f. For this question it is enough to consider a rectangular illumination pattern. That is, we approximate h (and hence A and Y) as a constant over some angle so that (by the power constraint) $|Y|=P/b$ on a subinterval of length b in $|\omega|<B$, and proceed to find the optimum value of b. Equation (7) gives

$$2\pi E(|e|^2) = F(2B-b) + \frac{bF}{\frac{F}{N}\frac{P^2}{b^2}+1}.$$

The optimum b maximizes

$$b - \frac{b^3}{b^2+\frac{FP^2}{N}} \quad \text{or} \quad \frac{\frac{FP^2}{N}b}{b^2+\frac{FP^2}{N}}.$$

Thus with FP^2/N fixed, the appropriate Doppler bandwidth to be covered by the radar is

$$b = P\sqrt{F/N}. \qquad (9)$$

A comparable problem occurs in the range channel of the radar; however, in this case [3, ch. 9] W is the complex modulation on the transmitted pulse and an energy constraint on the pulse gives $\int|W^2|^2\leq D/2\pi$, so that $\int|Y|^2\leq D$ and $|Y|^2=D/b$. The quantity then to be maximized is

$$\frac{b\frac{FD}{N}}{b+\frac{FD}{N}}.$$

However, this is monotone in b (for $b>0$) so that for any assumed bandwidth B in the structure of the radar field, the (least squares) optimum is $b=B$; i.e., in the range dimension of a radar, limited energy in the transmitted pulses does not call for a compromise of the attempt to obtain arbitrarily fine resolution.

Of course illumination over a broad angle thins the power density over the Doppler bandwidth, and leads to a small (physical) reception aperture. Such compound payment for fine azimuth resolution accounts for the limited resolution ambitions called for in (9).

Two-Dimensional Imaging and Ambiguity Limitations

If the transmitter is pulsed, then for each range rather than having $g(t)$ of Fig. 2 available, a sampled version of $g(t)$ is available where the PRF of the radar determines the sampling rate. The detailed derivation of the equations and mathematical model for the range-azimuth system is not necessary here; however, a few remarks on the appropriate mathematical model are worthwhile. Let y denote range, $W_2(t)$ the complex modulation on each transmitted pulse, and $W_1(y)=W_2(2y/c)$. A two-dimensional reflectivity is introduced $f_0(x, y)$, and a receiver impulse response $K_2(t)$ is applied to each "range sweep" received by the radar with $K_1(y)=K_2(2y/c)$. There is a propagation phase shift which renders

$$f(x, y) = \exp\left(j\frac{4\pi}{\lambda}y\right) f_0(x, y)$$

the effective input to the system, and the observable data (to a good approximation) become

$$g(x, y) = f(x, y) * [W_y(x)W_1(y)]$$

where we now display the range dependence of the azimuth impulse response, $W_y(x)$. Finally, $g(x, y)$ is available only at $x=mX$ where X is the distance traveled between transmitted pulses and m is an integer variable (or index on the transmitted pulses); also receiver noise accompanies each range sweep. Thus, the model takes the form shown in Fig. 3.

The fact that the azimuth filters have a slow dependence on range presents no great conceptual or analytical problems so that the sampling stands out as the major additional feature to be considered. Over a broad class of

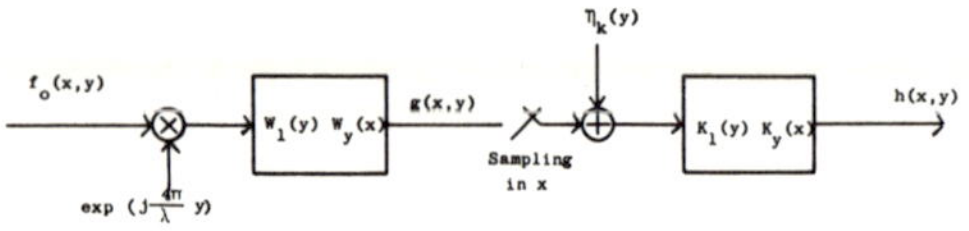

Fig. 3. Two-dimensional model.

system parameters, the sampling also introduces no appreciable effect; and in such cases it is seen that the elementary considerations discussed for the azimuth channel apply in the two-dimensional system and for that matter generalize in a rather obvious way to provide a theoretical basis for the entire system.

Transmissions other than a periodic pulse train can be considered, and fundamentally we are simply dealing with a range-Doppler radar which suggests that most of the theory developed for range-Doppler radars (usually in an "air defense" radar context) applies to the side-looking coherent radar. Such is the case, and with some manipulation the two-dimensional point target response of the side-looking radar can be expressed as the ordinary range-Doppler ambiguity function [4] of the transmitted signal (tapered by the pattern of the illuminating antenna). However, this viewpoint does not particularly suggest much of a departure from a periodic pulse train.

With a periodic pulse train, the ambiguity limit of side-looking radar involves the need to have a PRF high enough to adequately sample the bandwidth of the Dopplers in the azimuth angle illuminated by the radar. This minimum PRF sets a maximum on the unambiguous range interval which can be imaged. In detail two frequencies ω_1 and ω_2 are ambiguous (alias) for sampling period T if $\exp j\omega_1 mT = \exp j\omega_2 mT$ for all m, and this is the case if

$$\omega_1 - \omega_2 = p\frac{2\pi}{T} \tag{10}$$

where p is any integer. The Doppler frequencies are concentrated in a bandwidth (4)

$$\frac{4\pi}{\lambda R}(\alpha R) \text{ rad/ft}$$

where α is the azimuth beamwidth. In this notation such a bandwidth will not involve ambiguous frequencies if the distance traveled between transmitted pulses (X) satisfies

$$\frac{2\pi}{X} \geq \frac{4\pi}{\lambda}\alpha.$$

That is,

$$X \leq \frac{\lambda}{2\alpha} = \rho$$

where ρ is the theoretical "resolution." Of course this is eminently reasonable; if we desire an along-track resolution of ρ we must transmit at least one pulse per ρ-distance traveled by the vehicle. With $\alpha \doteq \lambda/d$, and a vehicle speed v this calls for the following PRF:

$$T = X/v \leq d/2v$$

$$\text{PRF} \geq \frac{2v}{d}.$$

If the illuminating pattern had zero azimuth response outside the angle $\alpha = \lambda/2X$, and if the elevation pattern provided zero illumination outside a range interval of length $R_0 = (c/2)(X/v)$, then the system would contain no ambiguities. However, since a finite aperture is used some range ambiguities ("second time around" echoes) and Doppler ambiguities occur (possibly negligible in level).

As for the appearance of azimuth ambiguous responses, consider a point target under sidelobe illumination at an azimuth angle for which the Doppler is centered around the frequency $2\pi p/X$. After sampling the data collected at such an angle will have (very nearly) the same phase history as that appearing as the main beam of the antenna passes over the target. Such data will be compressed by the azimuth processor to yield a "false" response (with an amplitude determined by sidelobe levels). The positions of these ambiguous responses (relative to the main beam response) correspond to the condition that the instantaneous Doppler $((4\pi/\lambda R)x)$ is alias to zero Doppler:

$$\frac{4\pi}{\lambda R}x = p\frac{2\pi}{X}$$

$$x = p\frac{\lambda R}{2X} \qquad (p\text{—any integer}).$$

The situation is depicted[6] in Fig. 4.

The design problem posed by these ambiguity considerations is dominated by the question of determining the optimum choice of the illumination (azimuth and elevation) on the physical aperture (of specified dimensions). This problem has been effectively dealt with [5].

Optical Processing

To this point we have considered the data collected by the radar, what is to be done with the data to produce an image, and have analyzed performance in a general way. Some comment on how the data processing might be implemented is in order. The two-dimensional nature of the received data[7] suggests the possibility of employing optical data processing [6]–[8] which at least is conceptually appealing. Here we will only comment briefly on the possibility of such an implementation.

In the range dimension let us transmit a short pulse so that the range sweeps recorded on film have the desired

[6] Of course the indicated Doppler ambiguities are equivalent to the grating lobes which appear in a physical linear array which has directive widely spaced ($\gg\lambda/2$) elements.

[7] The received data is a function of time alone, but is two-dimensional as a sequence of range sweeps with the range dimension occurring in fast time along the sweep and the azimuth data occurring in slow time across the sweeps.

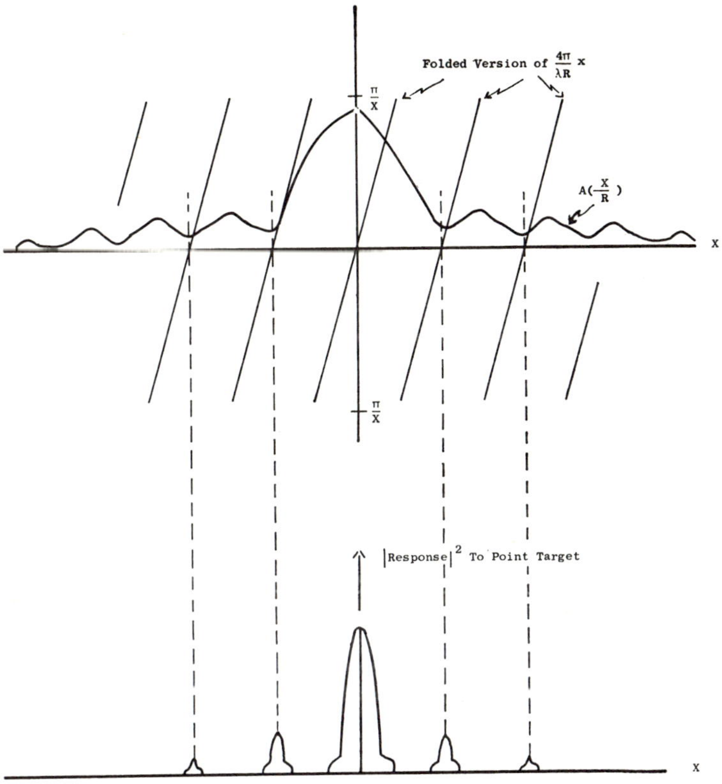

Fig. 4. Ambiguous sidelobes in azimuth response.

range resolution (i.e., the electronic receiver provides the processing depicted by K_1 in Fig. 3), then we optically synthesize only the azimuth processing. A suitable choice for the azimuth impulse response is the white noise matched filter; i.e.,

$$Z(\omega) = \overline{Y}(\omega) \quad \text{or} \quad K(x) = \overline{W}(-x).$$

Let the antenna pattern be even, then

$$K(x) = A\left(\frac{x}{y}\right) \exp - j\frac{2\pi}{\lambda y}x^2.$$

The desired output in azimuth at each range (y) is

$$\int A\left(\frac{u}{y}\right) \exp\left[-j\frac{2\pi}{\lambda y}u^2\right] g(x+u)du$$

where $g(x)$ is the data across the range sweeps. Rather than introduce the complex azimuth data in terms of real and imaginary parts, a carrier Doppler, ω_1, is inserted so that g appears as a complex modulation thus:

$$\begin{aligned} g(x) &= B(x)e^{j\beta(x)} \\ g_b(x) &= 2B(x)\cos\left[\omega_1 x + \beta(x)\right] + C \\ &= Be^{j\beta}e^{j\omega_1 x} + Be^{-j\beta}e^{-j\omega_1 x} + C \end{aligned}$$

where C is a bias which permits the recording of g_b as a positive function on ordinary film. As for the synthesis of K, a phase factor[8] $\exp(j(2\pi/\lambda y)u^2)$ can be introduced as a conical lens and the other part of the processing, $A(x/y)$, is not of much concern; however, it amounts to a wedged aperture so that we process only those Dopplers which are well illuminated. Now suppose g_b and the conical lens are placed essentially in contact in a plane wave of coherent light; let x denote the azimuth position on the film which is at the center of the conical lens, and let u denote distance from the center of the lens; then the complex amplitude of the emerging light is

[8] Since g_b contains $\bar{g}$ and g we need not use $\exp -j(2\pi/\lambda y)x^2$ which would correspond to a concave conical lens. Also, $2/\lambda y = 1/\lambda_0 f$ where λ_0 is the wavelength of the light and f is the variable focal length of the conical lens.

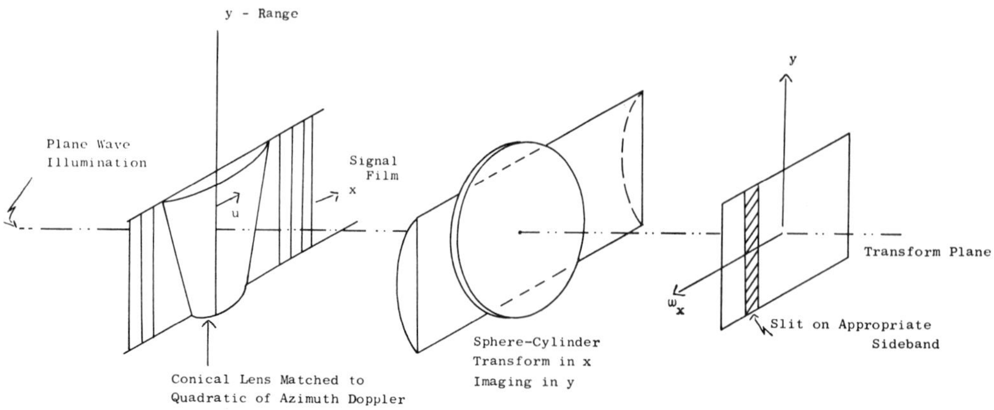

Fig. 5. Optical processing.

$$q(u) = g(x+u)e^{j\omega_1(x+u)}A\left(\frac{u}{y}\right)\exp\left(j\frac{2\pi}{\lambda y}u^2\right) + \bar{g}(x+u)e^{-j\omega_1(x+u)}A\left(\frac{u}{y}\right)\exp\left(j\frac{2\pi}{\lambda y}u^2\right) + CA\left(\frac{u}{y}\right)\exp\left(j\frac{2\pi}{\lambda y}u^2\right)$$

where u is the azimuth coordinate across the optical aperture.

If we take the Fourier transform of $q(u)$, the three terms produce outputs centered at ω_1, $-\omega_1$, and on axis (zero frequency). If ω_1 exceeds the highest frequency of gK, these three outputs are separated physically in the transform plane. Specifically, let us consider the transform

$$e^{-j\omega_1 x}\int \bar{g}(x+u)e^{-j\omega_1 u}A\left(\frac{u}{y}\right)\exp\left(j\frac{2\pi}{\lambda y}u^2\right)e^{-j\omega u}du.$$

At $\omega = -\omega_1$ (a slit in the frequency plane), the intensity of the light is

$$\left|\int \bar{g}(x+u)A\left(\frac{u}{y}\right)\exp\left(j\frac{2\pi}{\lambda y}u^2\right)du\right|^2$$

which is the desired processed radar image.

Thus with the variable focal length conical lens in place, the basic processing consists of recording the data on film with a bias and carrier offset frequency and then introducing the optical Fourier transform. The actual desired data appear at a position in the transform plane corresponding to the offset frequency. All of the above operations are in only the azimuth dimension and hence conceptually involve cylindrical lenses; however, a provision must be made to image the range dimension. This imaging involves another lens; either a cylinder acting in the range dimension, or an appropriate cylinder with a spherical two-dimensional Fourier transforming lens to provide the Fourier transform in azimuth and imaging in range. The final system takes the form shown in Fig. 5.

III. Imaging of a Rotating Target Field

The relative motion between the scene and the radar led to the ability to obtain fine azimuth resolution in a side-looking radar. Other forms of relative motion could hopefully lead to comparable results, and such is the case. A synthetic aperture effect, perhaps more basic than the side-looking radar, occurs if a radar scene simply rotates. Again, one can start by considering a line of reflecting material at some range; after the ability to obtain azimuth (or cross-range) resolution is displayed, the radar can be pulsed so as to lead to a two-dimensional imaging system. This additional synthetic aperture concept will not be developed in any detail; however, it is instructive to display another configuration in which Doppler processing provides (nearly) arbitrarily fine azimuth resolution at any range from a small aperture. The situation[9] is depicted in Fig. 6.

The amplitude and phase received from the increment at $(x, x+dx)$ is (in the CW case)

$$f(x)\exp\left(-j\frac{4\pi}{\lambda}s\right)dx.$$

The total received complex modulation as a function of time is

$$F_0(t) = \int_{-L/2}^{L/2}\exp\left(-j\frac{4\pi}{\lambda}s\right)f(x)dx.$$

[9] The range increment through the center of rotation is used; however, very little modification of the derivation is required for the other range increments.

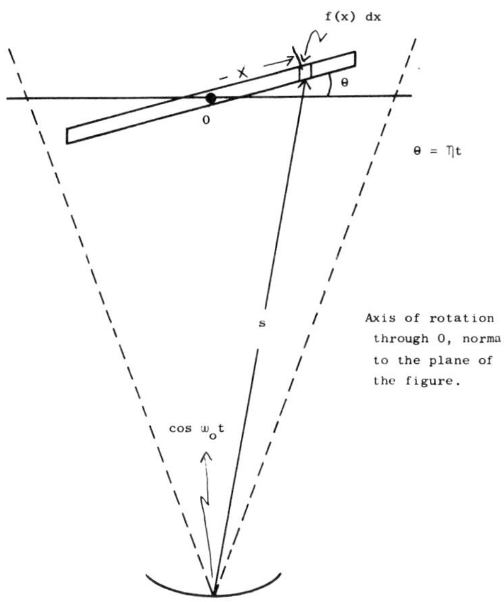

Fig. 6. Doppler processing for rotational motion.

Now $s=R+x \sin \theta=R+x \sin \eta t$ if R is large (where η is the angular velocity of the line of reflection); the constant phase associated with R is not of much concern, hence

$$F_0(t) = \int_{|x|<L/2} \exp\left(-j\frac{4\pi}{\lambda}x \sin \eta t\right) f(x)dx. \quad (11)$$

If we observe $F_0(t)$ for $|t|<T$, then we have the Fourier transform of the reflectivity over spatial frequencies

$$|\omega| < \frac{4\pi}{\lambda} \sin \eta T = B.$$

Such data on f provides a resolution capability of

$$\rho = \frac{\pi}{B} = \frac{\lambda}{4 \sin \eta T} = \frac{\lambda}{4 \sin \alpha/2}$$

where α is the change in aspect angle over which the target is viewed. Of course again we have the theoretical ability to achieve arbitrarily fine azimuth resolution at arbitrary range without the use of a large physical aperture on the radar. In this case the reflection at position x clearly gives rise to a Doppler frequency of $f_d=2\eta x/\lambda$ and a "bank of Doppler filters"[10] obviously resolves the target in the x-coordinate.

[10] The "bank of Doppler filters" amounts to taking the Fourier transform of $F_0(t)$ which is the appropriate processing on $F_0(t)$ to obtain an image of $f(x)$.

The side-looking radar can be viewed as a rotating target situation; specifically, (3) can be written thus:

$$g(vt) = \int_{|vt-u|<\alpha R/2} \exp\left[j\frac{2\pi}{\lambda R}(vt-u)^2\right] f(u)du$$
$$= \exp\left(j\frac{k}{2}t^2\right) \int \exp\left[-j\frac{4\pi}{\lambda}\left(\frac{v}{R}\right)tu\right] f_0(u)du \quad (12)$$

where $f_0(u)=\exp(j(2\pi/\lambda R)u^2)f(u)$, but f_0 and f provide the same image ($|f_0|^2=|f|^2$). Now consider a segment of the target field of length αR, and let $t=0$ correspond to when the center of this segment is closest to the radar vehicle; then if $g(vt)$ is modified by the known phase factor (which is the role of the conical lens in Fig. 5), the side-looking radar data is equivalent to that obtained by rotating the segment with an angular velocity of $\eta=v/R$. Also, the change in aspect angle in the side-looking radar case is the illumination beamwidth α.

In (11) let $\sin \eta t \cong \eta t$, then Fourier transform processing of the received data gives

$$q(\omega) = \frac{1}{2\pi}\int_{|t|<T} e^{j\omega_e t} F_0(t)dt \quad (13)$$

and $h(x)=q(4\pi\eta x/\lambda)$ is the image of f over the spatial bandwidth $|\omega|<(2\pi\alpha/\lambda)$. With white noise included in the analysis, we process $F_0(t)+n(t)$. In this case transform processing is viewed as a "bank of matched filters," which for a point target at unknown position x_0 yields the maximum likelihood estimate of x_0.

IV. Effects of Phase Errors [9], [10]

Let us first consider the simple situation of the illumination of an aperture; specifically, let

$$F(\omega) = \int e^{-j\omega t} f(t)dt \quad (14)$$

where[11] $f(t)$ denotes the illumination function and $F(\omega)$ the resulting pattern. Of course (14) and (1) depict the same situation in different variables; in (14) let ω denote pattern angle in radians, then t is related to illumination position x by $t=2\pi x/\lambda$. If the intended illumination is distorted by a phase error $\alpha(t)$, the resulting modified pattern is

$$F_m(\omega) = \int e^{-j\omega t} e^{j\alpha(t)} f(t)dt. \quad (15)$$

As suggested by (11) and (12), the effects of $\alpha(t)$ on F_m carry over to the effects of phase errors on the point target response of side-looking radar and on the imaging of a rotating target field.

[11] The symbol f was previously used for reflectivity density; in this section our few references to reflectivity density will be denoted by $\hat{f}$. Also, α now denotes the phase error process rather than the change in aspect angle.

Elementary Observations

Our major concern is with the tendency for $|F_m|^2$ to be spread in ω compared to the phase error free response $|F|^2$; however, the first-order effect of the phase errors is to shift the pattern. Thus, we are obligated to consider both the shift in position of the beam and the loss in resolution due to phase errors (and, possibly higher-order effects as well). The basic situation is displayed[12] if we approximate the phase error by the first three terms of its Taylor series

$$\alpha(t) = \alpha_0 + vt + \tfrac{1}{2}at^2$$
$$v = \alpha'(0) \quad \text{and} \quad a = \alpha''(0).$$

Then

$$F_m(\omega) = \int e^{-j(\omega - v)t} e^{j1/2at^2} f(t)dt$$

where α_0 has been omitted since it has no effect on $|F_m|^2$.

If the phase distortion were known, we would modify f to eliminate the phase errors; hence, it is reasonable to consider α a random process and then optimize f against known (or estimated) statistics of α. Now within the Taylor series approximation the pattern shift is given by $\alpha'(0)$, and $\alpha''(0)$ remains as the factor tending to spoil resolution. The basic problem is to select the illumination so as to minimize resolution (on the average). Let us defer a careful treatment of this problem, and only optimize $f(t)$ over Gaussian[13] illuminations; i.e., let

$$f(t) = \exp - \tfrac{1}{2}ct^2.$$

Then $c^{1/2}$ is indicative of the size of aperture we should strive for in the presence of phase errors. Having noted the pattern shift, the modified pattern for resolution considerations is

$$F_m(\omega) = \int e^{-j\omega t} \exp\left[-\tfrac{1}{2}(c + ja)t^2\right]dt$$
$$= \sqrt{2\pi}(c + ja)^{-1/2} \exp\left[-\tfrac{1}{2}\omega^2(c + ja)^{-1}\right].$$

Thus

$$|F_m(\omega)|^2 = \frac{2\pi}{\sqrt{c^2 + a^2}} \exp\left[-\frac{\omega^2}{2}\left(\frac{2c}{c^2 + a^2}\right)\right].$$

The corresponding equivalent rectangle resolution is

$$\hat{\rho}_0 = \sqrt{\pi}\sqrt{\frac{c^2 + a^2}{c}}. \tag{16}$$

The radius of gyration resolution is

$$\rho = \sqrt{\frac{c^2 + a^2}{2c}}. \tag{17}$$

If we now treat α as a random process, $a = \alpha''(0)$ is a random variable, and a reasonable approach is to select c so as to minimize the mean-square resolution. For either measure of resolution, this calls for the c which minimizes $(c^2 + \overline{a^2})/c$ where now $\overline{a^2}$ denotes the mean-square value of α''. The optimum c is $c = \sqrt{\overline{a^2}} = \Delta a$. With this choice of antenna illumination the rms resolution is

$$\bar{\rho} = \sqrt{\Delta a}. \tag{18}$$

As a specific interpretation of the above simple observations let $\alpha(t)$ result from geometrical distortion of a physical aperture; in this context let $r(x)$ denote the departure of the aperture from the intended straight line. Then

$$\alpha(t) = \frac{2\pi}{\lambda} r\left(\frac{\lambda}{2\pi}t\right)$$
$$\alpha' = r'$$
$$\alpha'' = \frac{\lambda}{2\pi}r''.$$

In turn, if μ is the rms value of d^2r/dx^2, then the rms resolution is

$$\bar{\rho} = \sqrt{\frac{\lambda\mu}{2\pi}} \quad \text{(radians)}$$
$$\bar{\hat{\rho}}_0 = \sqrt{\lambda\mu}. \tag{19}$$

Also, the rms displacement of the beam is equal to the rms value of dr/dx.

At this stage we could proceed with the general analysis of the effects of phase errors; however, it seems appropriate to first relate the above type of analysis to synthetic aperture radar.

Application to Synthetic Aperture Radars

First consider the case of a rotating target field. In (11) let $\sin \eta t \cong \eta t$ and let $\hat{f}(x) = \delta(x - x_0)$ so that the point target response is under consideration. Also, let $F_0(t)$ be distorted by phase errors; then the available data are

$$e^{j\alpha(t)} \exp\left(-j\frac{4\pi}{\lambda}\eta x_0 t\right).$$

It is presumed that the image of this point target is formed via

$$F_m(\omega) = \int \exp\left[-j(\omega + \omega_0)t\right] e^{j\alpha(t)} f(t)dt \tag{20}$$

where $f(t)$ essentially depicts the amount of data to be processed and $\omega_0 = 4\pi\eta x_0/\lambda$. Now within the context of a Gaussian taper on the data (which limits the effects of the phase errors), the obtainable resolution in the ω-variable is $\sqrt{\Delta a}$ where Δa is the rms value of α''. In turn the positional (radius of gyration) resolution of the image is

$$\bar{\rho} = \frac{\lambda}{4\pi\eta}\sqrt{\Delta a}. \tag{21}$$

For example, suppose the phase errors are due to a time variation in the electrical-path length (either due to physical change in the distance between the target and the radar or due to time varying propagation conditions), and let $r(t)$ denote this change in propagation path; then

[12] Simple results are also obtained if we consider a periodic phase error, e.g., $\exp jA \cos \mu t = \Sigma J_k(A)e^{jk\mu t}$ which leads to a superposition of shifted versions of the phase error free response.

[13] This is the most simplifying choice mathematically and has some intuitive appeal; however, later it will be seen to have a deeper justification.

$\alpha(t)=(4\pi/\lambda)r(t)$ and if μ is the rms value of r'' we have

$$\bar{\rho}=\frac{1}{2\eta}\sqrt{\frac{\lambda\mu}{\pi}}. \tag{22}$$

It is worthwhile to observe that within the many simplifying assumptions of this analysis, the taper which provides the minimum resolution of (21) also maximizes the output signal-to-noise ratio. Specifically, if the observed data are accompanied by white additive noise, the noise at the output is

$$n_0(\omega)=\int e^{-j\omega t}n(t)f(t)dt$$

$$E(|n_0|^2)=\frac{N_0}{2\pi}\int|f(t)|^2dt$$

where N_0 is the power density of the white additive noise. Since the mean-square noise out is independent of ω, let the position of the point target be $\omega_0=0$. Also, since $|f|^2=|e^{j\alpha}f|^2$, we have by Parseval's theorem that

$$E(|n_0|^2)=N_0\int|F_m(\omega)|^2d\omega.$$

Finally, with $\omega_0=0$ the peak signal out occurs at $\omega=0$; hence, the output noise-to-signal ratio is

$$N/S)_{\text{out}}=N_0\frac{\int|F_m(\omega)|^2d\omega}{|F_m(0)|^2}.$$

However, this is simply N_0 times the equivalent rectangle resolution. Thus within the Taylor series approximation (and Gaussian taper), processing for maximum signal-to-noise ratio and processing for minimum resolution are equivalent.

In the side-looking radar case phase errors (usually) enter as a multiplicative factor so as to modify the mathematical model of Fig. 2 thus:

$$h=[(W\cdot\hat{f})e^{j\alpha}+n]\cdot K.$$

One can easily optimize the receiver K with a least squares error criterion; however, the resulting mean-square error is not indicative of image quality. The reason for this is roughly that the phase error process, α, produces phase differences between h and $\hat{f}$ and shifts in position between h and $\hat{f}$ both of which introduce large values for $|h-\hat{f}|^2$. However, neither of these low-order effects render $|h|^2$ a poor quality image of $|\hat{f}|^2$. Thus as mentioned in the Introduction, resolution comes very much into its own as a performance criterion when we move from a concern for additive noise to a concern for phase errors (and related distortion).

Since the side-looking radar is a time[14] (space) invariant system, consider a point target at $x=0$; then if $W=A(x)\exp -j\frac{1}{2}kx^2$ and $K=B(x)\exp j\frac{1}{2}kx^2$ where $k=4\pi/\lambda R$, the point target response is

$$h(x)=\int A(u)B(x-u)\exp -j\frac{k}{2}[u^2-(x-u)^2]\cdot\exp j\alpha(u)du.$$

As for the image $|h|^2$ we can ignore the phase factor $\exp j\frac{1}{2}kx^2$; thus the response is

$$h(x)=\int\exp(-jkxu)\exp[j\alpha(u)]A(u)B(x-u)du. \tag{23}$$

A pesky detail appears because we are able to taper using A and/or B; except for additive noise considerations we can simply take $B=1$, and then (23) is in the exact form for which our resolution analysis applies. Alternatively, if the system achieves significant compression of the azimuth Doppler histories, then B is a broad function compared to h so that both sides of (23) are essentially zero for large x and $B(x-u)\cong B(u)$ for (the significant) small values of x. In any event the simple results mentioned at the beginning of Section IV apply with $\omega=kx$; specifically, (18) gives

$$\bar{\rho}=\frac{R\lambda}{4\pi}\sqrt{\Delta a}$$

where Δa is the rms value of α''. For example, if $\alpha(x)=(4\pi/\lambda)r(x)$ and μ is the rms value of r'', then

$$\bar{\rho}=\frac{R}{2}\sqrt{\frac{\lambda\mu}{\pi}}\ (ft) \tag{24}$$

which bears the expected relation to (19) and (22). Also, the previous observation about signal-to-noise ratio applies here by substantially the same derivation provided the tapering is obtained with $|K|$.

From the foregoing approximate analysis it should be clear that if the effects of phase errors on a Fourier transform are analyzed (which amounts to the physical aperture case), then by introducing appropriate scale factors the results apply to synthetic aperture systems (and other situations as well; e.g., pulse distortion due to dispersive propagation [9].

General Theory

It is analytically attractive to measure the shift in position due to phase errors by the first moment

$$\overline{\omega_\alpha}=\frac{\int\omega|F_m(\omega)|^2d\omega}{\int|F_m(\omega)|^2d\omega}.$$

If the second moment is

$$\overline{\omega_\alpha{}^2}=\frac{\int\omega^2|F_m(\omega)|^2d\omega}{\int|F_m(\omega)|^2d\omega},$$

[14] It is assumed that α and n are stationary random processes. Also, the choices of W and K involve the assumption that the phase of K is matched to the phase of W in the absence of phase errors.

then the radius of gyration of the point target response (or beamwidth in the physical aperture case) is ρ_α where

$$\rho_\alpha{}^2 = \frac{\int (\omega - \overline{\omega_\alpha})^2 |F_m|^2 d\omega}{\int |F_m|^2 d\omega} = \overline{\omega_\alpha{}^2} - (\overline{\omega_\alpha})^2.$$

We could formulate higher-order central moments of $|F_m|^2$ and be led to formulas similar to those to be derived for resolution and displacement; however, the effort is greater and the rewards modest. The problem is to calculate the mean-square value of ρ_α and $\overline{\omega_\alpha}$; these quantities are depicted in Fig. 7.

Let $f_m = e^{j\alpha}f$, then since F_m is the Fourier transform of f_m, $|F_m|^2$ is the transform of C_{f_m} where

$$C_{f_m}(t) = \int f_m(t+\tau)\bar{f}_m(\tau)d\tau.$$

It is worthwhile to note that

$$E(|F_m|^2) = E\int e^{-j\omega t}\left\{\int \exp\left[j\alpha(t+\tau) - j\alpha(\tau)\right] \cdot f(t+\tau)\bar{f}(\tau)d\tau\right\} dt$$

$$= \int e^{-j\omega t}\{R_\gamma(t)C_f(t)\}dt$$

$$= \frac{1}{2\pi}S_\gamma(\omega) \cdot |F(\omega)|^2 \qquad (25)$$

where $S_\gamma(\omega)$ is the power density spectrum of $\gamma(t) = e^{j\alpha(t)}$.

Thus as a first-order analysis of the effects of phase errors, the "expected image" is the phase error free image blurred through convolution with the power density of $e^{j\alpha}$; if phase errors are of serious concern, $|F|^2$ will behave like a δ-function and the expected response is $S_\gamma(\omega)$. The shortcoming of this observation is that the expected response provides a pessimistic estimate of performance in that the spread of $E(|F_m|^2)$ can well be due primarily to positional shift in the response (first moment). On the other hand, for large ω, $S_\gamma(\omega)$ is indicative of average sidelobe energy.

Since $|F_m|^2$ is the transform of C_{f_m},

$$C_{f_m}(t) = \frac{1}{2\pi}\int e^{j\omega t}|F_m|^2 d\omega \qquad (26)$$

$$C_{f_m}(0) = \frac{1}{2\pi}\int |F_m|^2 d\omega \qquad (27)$$

$$C_{f_m}''(0) = -\frac{1}{2\pi}\int \omega^2 |F_m|^2 d\omega. \qquad (28)$$

Hence

$$\overline{\omega_\alpha{}^2} = -\frac{C_{f_m}''(0)}{C_{f_m}(0)}.$$

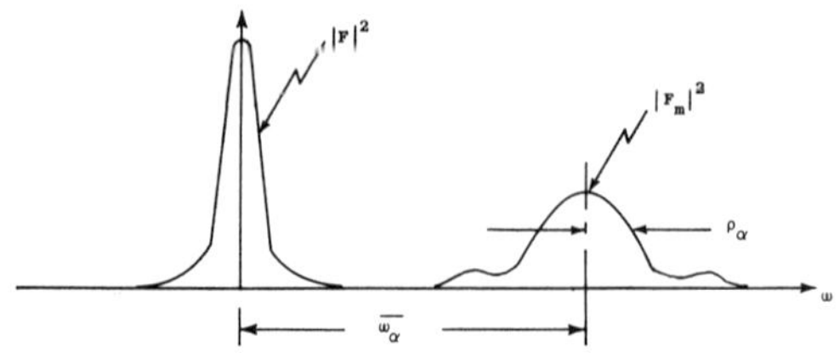

Fig. 7. Qualitative effects of phase errors.

It is easy to show that $-C_{f_m}''(t) = C_{f'_m}(t)$; also $C_{f_m}(0) = C_f(0)$. Thus we can take the expected value of $\overline{\omega_\alpha{}^2}$ (after expectation, the α subscript will be dropped):

$$C_f(0)\overline{\omega^2} = E[C_{f'_m}(0)]$$

$$f'_m = e^{j\alpha}f' + j\alpha' f e^{j\alpha}.$$

Let $E(\alpha') = 0$; then the cross terms in $E[C_{f'_m}]$ are zero and we get

$$C_f(0)\overline{\omega^2} = \int |f'|^2 + E[(\alpha')^2]\int |f|^2.$$

However, $\int |f'|^2 / C_f(0)$ is the second moment of the pattern in the absence of phase errors (which will be denoted $\overline{\omega_0{}^2}$), and $C_f(0) = \int |f|^2$; thus

$$\overline{\omega^2} = \overline{\omega_0{}^2} + E[(\alpha')^2]. \qquad (29)$$

From (26) we also have

$$jC_f(0)\overline{\omega_\alpha} = C_{f_m}'(0)$$

$$C_{f_m}'(t) = \int f_m'(t+\tau)\bar{f}_m(\tau)d\tau$$

$$C_{f_m}'(0) = \int f'\bar{f}d\tau + j\int \alpha' |f|^2 d\tau.$$

Thus

$$C_f{}^2(0)(\overline{\omega_\alpha})^2 = \left[-jC_f'(0) + \int \alpha' |f|^2 d\tau\right]^2.$$

If $E(\alpha') = 0$, the cross terms again are zero; the first term squared is seen to give rise to the square of the first moment in the absence of phase errors, $(\overline{\omega_0})^2$. The term involving α' gives rise to

$$\iint R_{\alpha'}(\tau - \theta)|f(\tau)|^2 |f(\theta)|^2 d\tau d\theta = \frac{1}{8\pi^3}\int S_{\alpha'}(\omega)|C_F(\omega)|^2 d\omega$$

since $(1/2\pi)C_F$ is the transform of $|f|^2$, where F is the pattern in the absence of phase errors. The final result is

$$(\bar{\omega})^2 = E[(\omega_\alpha)^2]$$

$$= (\overline{\omega_0})^2 + \frac{1}{2\pi}\int S_{\alpha'}(\omega)\frac{|C_F(\omega)|^2}{|C_F(0)|^2}d\omega.$$

Finally, the mean-square resolution is

$$\rho^2 = \rho_0^2 + \frac{1}{2\pi}\int S_{\alpha'}(\omega)\left[1 - \frac{|C_F(\omega)|^2}{|C_F(0)|^2}\right]d\omega \quad (30)$$

where

$$\rho_0^2 = \frac{\int \omega^2 |F|^2 d\omega}{\int |F|^2 d\omega}$$

if the phase error free pattern has zero first moment.

The problem of optimizing the system for resolution amounts to finding the phase error free pattern (F, or taper f) which minimizes ρ^2 of (30). This can be done (with numerical analysis) for any specific $S_{\alpha'}$, but seems to require considerable effort.

A worthwhile general result is available; namely, if we approximate $|C_F|^2$ for small ω by the first three terms of its Taylor series (which provides a good approximation if $S_{\alpha'}$ is concentrated at low frequencies), then the optimum $f(t)$ is found to be Gaussian

$$f(t) = \exp\left(-\tfrac{1}{2}ct^2\right)$$

and the optimum c is found to be $c^2 = E[(\alpha'')^2]$. In this case $\rho \cong \sqrt{\Delta a}$ where Δa is the rms value of α''. Of course this is in pleasant agreement with the simple analysis which resulted in (18).

What is more, for arbitrary $S_{\alpha'}$ (not necessarily concentrated at low frequencies), the said Gaussian taper renders $\rho \leq \sqrt{\Delta a}$. In addition since $\rho^2 \leq \overline{\omega^2}$, (29) easily indicates that ρ can be made less than Δv where Δv is the rms value of α'. Thus as a rather deep result we have that for arbitrary $S_{\alpha'}$, the obtainable rms resolution satisfies

$$\rho \leq \min\left(\Delta v, \sqrt{\Delta a}\right). \quad (31)$$

In summary, (30) can be used to calculate the average resolution with any phase error spectrum and taper, f. More significantly, (30) permits the selection of the optimum taper in the presence of phase errors, and (31) provides a quick estimate (often sharp) of the minimum obtainable rms resolution. Finally, the power spectrum of $e^{i\alpha}$ provides a rough estimate of remote sidelobe energy levels.

REFERENCES

[1] L. J. Cutrona, W. E. Vivian, E. N. Leith, and G. O. Hall, "A high-resolution radar combat-surveillance system," *IRE Trans. on Military Electronics*, vol. MIL-5, pp. 127–131, April 1961.
[2] C. W. Sherwin, J. P. Ruina, and R. D. Rawcliffe, "Some early developments in synthetic aperture radar systems," *IRE Trans. on Military Electronics*, vol. MIL-6, pp. 111–115, April 1962.
[3] W. M. Brown, *Analysis of Linear Time-Invariant Systems*. New York: McGraw-Hill, 1963.
[4] W. M. Siebert, "A radar detection philosophy," *IRE Trans. on Information Theory*, vol. IT-2, pp. 204–221, September 1956.
[5] R. O. Harger, "An optimum design of ambiguity function, antenna pattern, and signal for side-looking radars," *IEEE Trans. on Military Electronics*, vol. MIL-9, pp. 264–278, July–October 1965.
[6] L. J. Cutrona, E. N. Leith, C. J. Palermo, and L. J. Porcello, "Optical data processing and filtering systems," *IRE Trans. on Information Theory*, vol. IT-6, pp. 386–400, June 1960.
[7] A. B. Vander Lugt, "Signal detection by complex spatial filtering," *IEEE Trans. on Information Theory*, vol. IT-10, pp. 139–145, April, 1964.
[8] L. J. Cutrona, E. N. Leith, L. J. Porcello, and W. E. Vivian, "On the application of coherent optical processing techniques to synthetic-aperture radar," *Proc. IEEE*, vol. 54, pp. 1026–1032, August 1966.
[9] W. M. Brown and C. J. Palermo, "Effects of phase errors on resolution," *IEEE Trans. on Military Electronics*, vol. MIL-9, pp. 4–9, January 1965.
[10] J. R. Klauder, A. C. Price, S. Darlington, and W. J. Albersheim, "The theory and design of chirp radars," *Bell Sys. Tech. J.*, vol. 39, pp. 745–808, July 1960

William M. Brown (M'57–SM'63) was born in Wheeling, W. Va. on February 14, 1932. He received the B.S.E.E. degree in 1952 from the University of West Virginia, Morgantown, and the M.S.E. degree in 1955 and the D.Eng. degree in 1957 from The Johns Hopkins University, Baltimore, Md.

From 1952 to 1954, he worked in the Air Arm Division of Westinghouse. During 1954 to 1957 he was at The Johns Hopkins Radiation Lab. He then worked for a year at the Institute for Defense Analyses, Washington, D. C., and taught during the same period at George Washington University. Since 1958, he has been with the University of Michigan, Ann Arbor, where he is a Professor of Electrical Engineering and Head of the Radar and Optics Lab.

Theory And Evaluation Of Gain Patterns Of Synthetic Arrays*

R.C. Heimiller
(Aerospace Group, Hughes Aircraft Company, Culver City, California)

IRE Transactions on Military Electronics, Vol. MIL-6, No. 2 (April 1962)
Pages 122-129

*Received by the PGMIL 2 December 1961

***Summary*—The results of an analysis of synthetic array gain patterns, including sidelobe response, are summarized. Both focused and unfocused arrays are examined. The phase time history of echos from reflectors on the ground as a function of position is used to modify the equations for the broadside synthetic array for application to forward-looking antenna arrays. The dependence of the "optimal" length and resolution of an unfocused array on the look angle is derived. The gain patterns of unfocused synthetic arrays are presented for arrays of various lengths with both rectangular and exponential amplitude weighting.**

Calculations are made showing that the physical antenna gain pattern acts as amplitude weighting with resultant sidelobe suppression and illustrating that the final gain pattern *cannot* be expressed as the product of the space factor of the array times the real antenna gain pattern. However, unless the length of the synthetic array is about equal to or greater than the real antenna beamwidth times the range, the product approximation error is negligible in the region near the main lobe and amounts to a reduction in gain.

The results of this analysis are then put in the form of design curves from which such parameters as the maximum resolution of an unfocused array and the length of array necessary for a given resolution at a given frequency can be obtained.

Introduction

A TUTORIAL explanation of the principles of synthetic arrays from the different points of view of vector addition, cross-correlation and Doppler filtering is given in another article in this TRANSACTION.[1] Another interesting introduction to the subject as well as a history of the early synthetic array work was presented in an earlier issue.[2] The principles behind the operation of a synthetic array system are similar to the principles by which a conventional linear array forms a beam. In the conventional array the reflected signals are received simultaneously by the elements of the array, and the summation of the signals is performed by the feed system of the array.

In a synthetic array, the reflected signals are received at different times as the radar platform moves along. The summation of the signals takes place in a data processor as a single element successively assumes the positions of the array elements. Thus, by using synthetic array techniques, a single, small antenna can have an effective beamwidth as narrow as that which could be obtained by a long linear array. To preserve the phase relationships of the reflected signals during the integration time, a coherent radar is required.

Focused Synthetic Array

The geometry for a synthetic array beam which is canted forward of the broadside position is shown in Fig. 1.

[1] H. L. McCord, this issue, pp. 116–119.

[2] L. J. Cutrona, W. E. Vivian, E. N. Leith, and G. O. Hall, "A High-Resolution Radar Combat-Surveillance System," pp. 127–131; April, 1961 issue of these TRANSACTIONS.

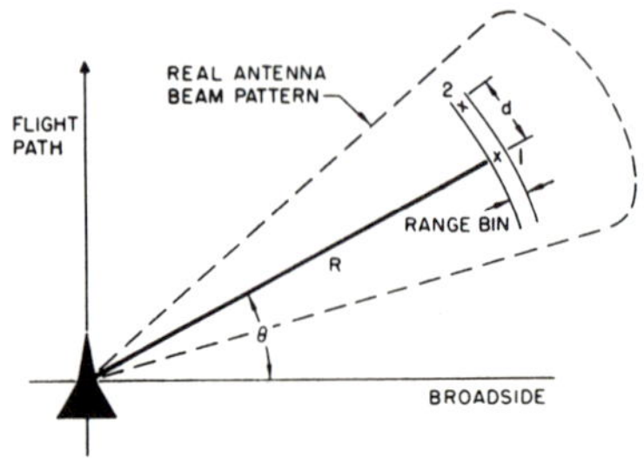

Fig. 1 Geometry for forward looking synthetic array.

CHART I

FOCUSED SYNTHETIC ARRAY RESOLUTION
(Two-Way Power Gain)

Resolution Definition	Formula
−3 db	$0\cdot44(\lambda R/L)\cos\theta$
−4 db	$0\cdot50(\lambda R/L)\cos\theta$
−6 db	$0\cdot60(\lambda R/L)\cos\theta$
Null-to-Null	$(\lambda R/L)\cos\theta$

λ = RF wavelength
R = range
L = length of synthetic array
θ = angle of beam from broadside

CHART II

SOURCES OF PHASE ERRORS
- Tropospheric Propagation
- Vehicle Motion
- Equipment Limitations
  - RF Stability
  - Processor Effects

EFFECTS OF PHASE ERRORS
- Increased Sidelobe Level
- Beam Pointing Error
- Gain Reduction
- Beam Spreading
- Beam Wander

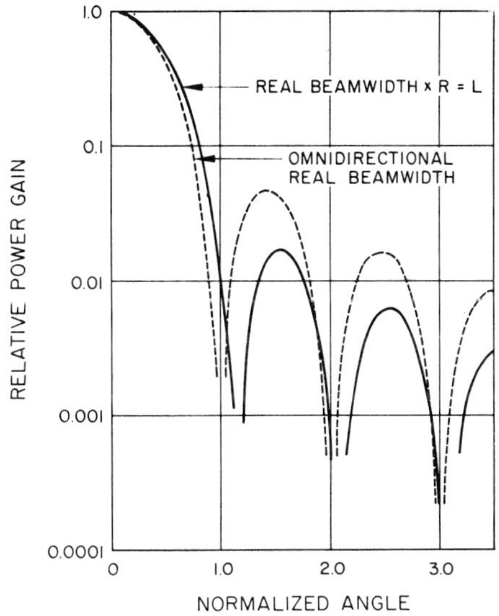

Fig. 2—Power gain patterns for a focused synthetic array illustrating effect of real beam pattern.

Using this geometry, a detailed derivation of the gain pattern of a focused synthetic array is carried out in Appendix I of this paper.

The result is that the two-way power gain of a synthetic array of length L is equivalent to the one-way power gain of a real antenna of length $2L$. The resolution formula of a focused synthetic array with rectangular weighting is given in Chart I. The theoretical resolution as calculated from these formulas may not be realized due to phase errors.

Chart II summarizes the important sources and effects of these phase errors. Another paper in this issue discusses, in detail, phase error effects.[3]

The calculation of the gain pattern in Appendix I was carried out assuming that the real antenna pattern is omnidirectional, and gives what is called the space factor of the array. To obtain the over-all pattern, this space factor is multiplied by the real antenna two-way gain pattern. However, this method is an approximation and Appendix II considers in more detail the effect of the real antenna beam pattern.

Fig. 2 shows the comparison of gain patterns for an omnidirectional real beam and for a real beam having a beamwidth equal to the synthetic array length divided by the range. For the space factor of the array, the null occurs at a normalized angle value of one, which corresponds to an actual angle of $(\lambda/2L)\sec\theta$. The effect of the physical antenna beam shape is to give an amplitude weighting which causes a slight loss in resolution but with improvement in sidelobe level.

Unfocused Synthetic Array Rectangular Weighting

If the array is not focused, that is if the quadratic phase term is not corrected, then for rectangular amplitude weighting a minimum 3-db resolution occurs at approximately $L = 1.2\sqrt{R\lambda}$. Figs. 3 and 4 show the gain patterns for the unfocused case for rectangular weighting. Actually, in practical unfocused synthetic arrays, the amplitude weighting would probably be exponential. This case is discussed later.

Fig. 3 compares an unfocused to a focused synthetic array where the length of each is $\sqrt{R\lambda}\sec\theta$. Notice that the 3-db resolution of the unfocused array is only about 8 per cent worse than that of the focused array, but that the sidelobe level is much higher. Since there are no nulls in the unfocused array gain pattern, some other definition besides null-to-null resolution is necessary to specify the resolution of the beam.

Fig. 4 shows that the resolution of an unfocused array of length $L = \sqrt{R\lambda}\sec\theta$ is much better than that of arrays of half or twice this length. If the optimum is examined more closely in the vicinity of $L = \sqrt{R\lambda}\sec\theta$, then which of the gain patterns is optimum will depend on the definition used. For example, if the criterion used was 3-db resolution, an array length $L = 1.2\sqrt{R\lambda}\sec\theta$ would be optimum. However, the sidelobe level for this length is higher than for the shorter lengths.

[3] C. A. Greene and R. T. Moller, this issue, pp. 130–140.

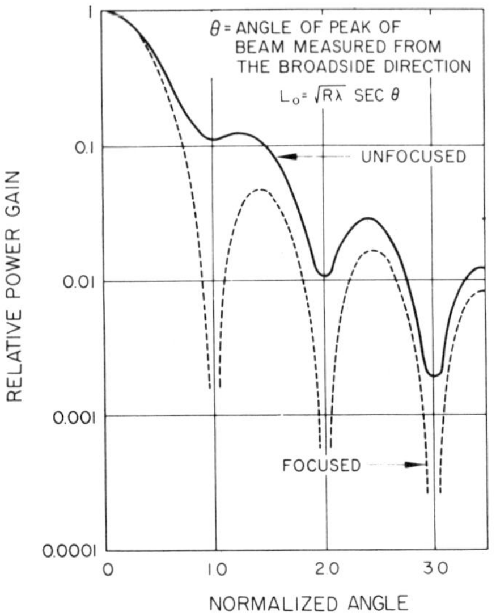

Fig. 3—Comparative power gain patterns for focused and unfocused synthetic arrays.

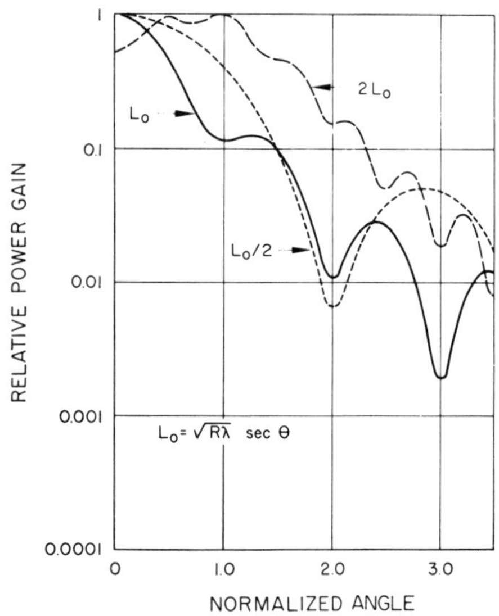

Fig. 4—Power gain for unfocused synthetic antennas with rectangular weighting.

Unfocused Synthetic Arrays—Exponential Weighting

In actual practice, unfocused synthetic arrays will have effective exponential amplitude weighting for a number of processing methods. Appendix III considers the resolution of such unfocused synthetic arrays and the results are shown in Fig. 5. The parameter γ is given by

$$\gamma = \frac{\sqrt{\lambda R_0}}{2VT \cos \theta N},$$

where VT is the interelement spacing and N is the number of pulses occurring in a time equal to the time constant of the exponential decay. The time constant then equals

$$\frac{\sqrt{\lambda R_0}}{2V \cos \theta} \frac{1}{\gamma}.$$

The resolution in feet equals

$$\sqrt{\lambda R_0}/2 \times \text{the normalized beamwidth.}$$

For

$\lambda = 0.1$ ft,

$R_0 = 10$ nautical miles $= 60{,}800$ ft

the optimum 6-db resolution $= 72$ ft and the optimum value of $1/\gamma = 2$. Note, however, that the optimum value of $1/\gamma$ is not very critical.

For the smaller values of γ, the sidelobe level stays quite high so that the 3- and 6-db resolution values do not tell the whole story. Another method of evaluating the gain patterns is comparing the main lobe and sidelobe integrated response for different resolution widths. That is, the integral of the relative two-way power gain is taken over a given interval about the center of the array pattern. This integral is called the main lobe integrated response and is a function of the width chosen. The integral outside this

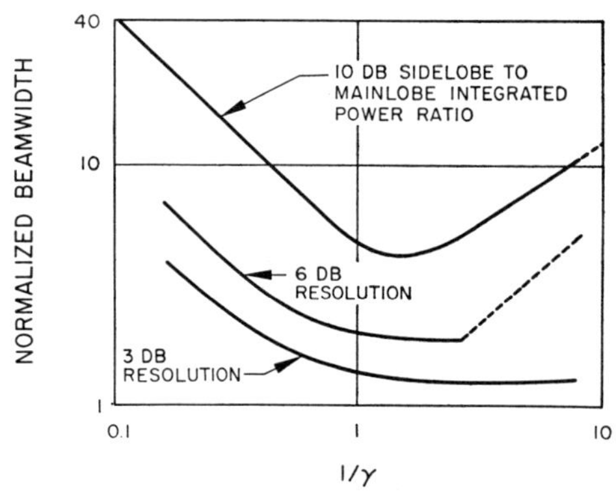

Fig. 5—Choice of time constant—unfocused array.

interval is called the sidelobe integrated response. The sidelobe to main lobe integrated response ratio for a focused synthetic array with rectangular weighting for the null-to-null width is −10 db. The "effective null-to-null width" for an unfocused array can be defined as the width which yields this ratio. This width as a function of $1/\gamma$ is shown in Fig. 5. The optimum resolution according to this definition occurs at $1/\gamma = 1.5$ and has a normalized width of 4.2. The "effective null-to-null resolution" in feet then equals

$$\frac{4.2\sqrt{\lambda R_0}}{2} = 2.1\sqrt{\lambda R_0}.$$

Resolution Curves

Figs. 6 and 7 give the 6-db azimuth resolution of a broadside synthetic array as a function of wavelength and array length for ranges of 10 and 30 nautical miles. Focused arrays with rectangular amplitude weighting and unfocused arrays with exponential weighting are considered. Figs. 8 and 9 present the corresponding curves where the resolution is defined by the "effective null-to-null width" which is discussed above. For a focused array, this corresponds very nearly to the null-to-null resolution of the array.

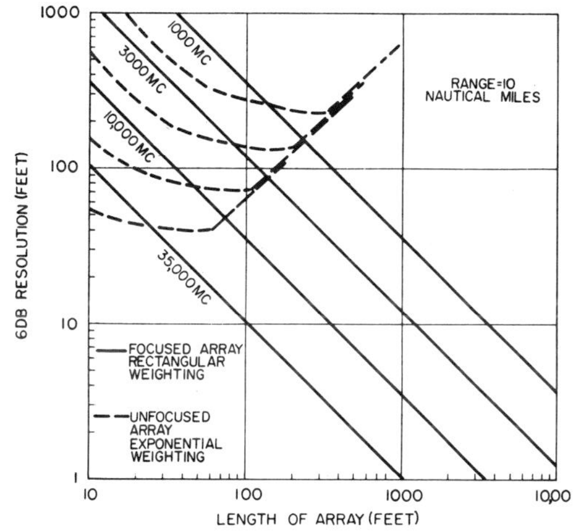

Fig. 6 6-db azimuth resolution of a broadside synthetic array.

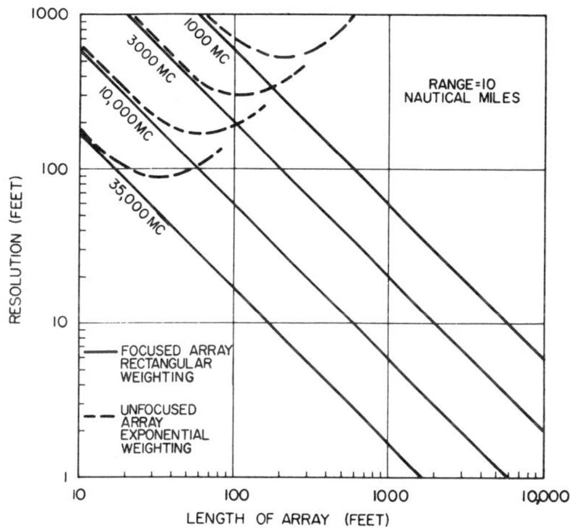

Fig. 8 Resolution for 10-db sidelobe-to-main lobe integrated power ratio broadside synthetic array.

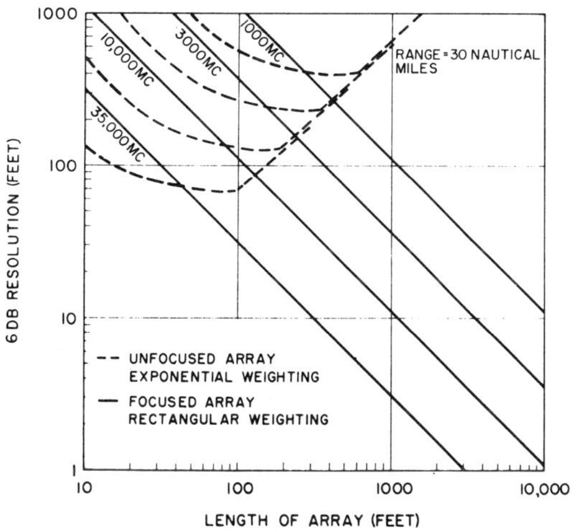

Fig. 7 6-db azimuth resolution of a broadside synthetic array.

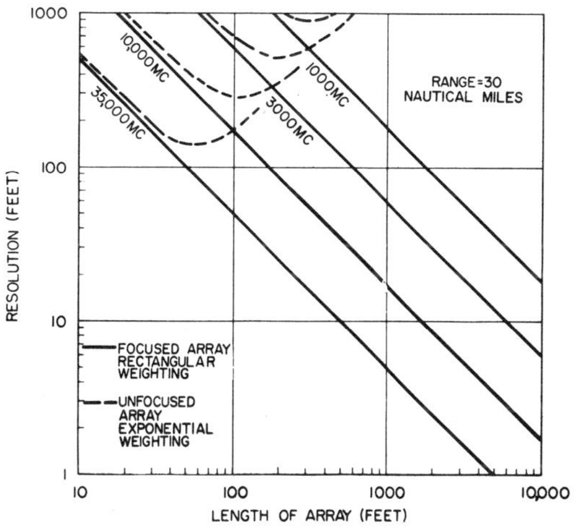

Fig. 9 Resolution for 10-db sidelobe-to-main lobe integrated power ratio broadside synthetic array.

Appendix I

The Formation of Synthetic Array Antenna Patterns

Consider the geometry of Fig. 1. Let the flight-path direction be the x axis and the broadside direction the y axis.

Define:

- t = time
- θ = angle of synthetic beam from the broadside direction
- ① = reflector on the synthetic beam axis at $t=0$
- ② = reflector off the synthetic beam axis a distance d but at the same range as reflector ①
- R = range to reflector (a function of time)
- R_0 = range at $t=0$
- V = velocity of radar platform
- ω_0 = transmitter frequency
- λ = transmitter wavelength
- x_1 = x position of reflector 1 at $t=0$
- y_1 = y position of reflector 1
- x_2 = x position of reflector 2 at $t=0$
- y_2 = y position of reflector 2
- $\Delta x = x_2 - x_1$
- α = angle between reflector 1 and reflector 2 at $t=0 \cong d/R_0$

$$R_1(t) = \text{range to reflector 1} = \sqrt{y_1^2+(x_1-Vt)^2} = \sqrt{y_1^2+x_1^2+V^2t^2-2x_1Vt} = \sqrt{R_0^2+V^2t^2-2x_1Vt}$$

$$R_2(t) = \text{range to reflector 2} = \sqrt{y_2^2+(x_2-Vt)^2} = \sqrt{y_2^2+x_2^2+V^2t^2-2Vx_2t} = \sqrt{R_0^2+V^2t^2-2Vx_2t}$$

The phase time history of reflector 1 is then

$$\phi_1(t) = \omega_0 t - \frac{4\pi R_0}{\lambda}\sqrt{1+\frac{V^2t^2-2Vx_1t}{R_0^2}}.$$

Since

$$\sqrt{1+\epsilon} = 1 + \frac{\epsilon}{2} - \frac{\epsilon^2}{8} + \cdots$$

then (including up to the 2nd power of t)

$$\phi_1(t) \cong \omega_0 t - \frac{4\pi R_0}{\lambda}\left(1 + \frac{V^2t^2-2Vx_1t}{2R_0^2} - \frac{V^2x_1^2t^2}{2R_0^4}\right)$$

$$\cong \omega_0 t - \frac{4\pi R_0}{\lambda} + \frac{4\pi Vx_1t}{\lambda R_0} - \frac{2\pi V^2t^2\left[1-\left(\frac{x_1}{R_0}\right)^2\right]}{\lambda R_0}$$

$$\frac{x_1}{R_0} = \sin\theta,\quad 1-\left(\frac{x_1}{R_0}\right)^2 = \cos^2\theta$$

$$\phi_1(t) \cong \omega_0 t - \frac{4\pi R_0}{\lambda} + \frac{4\pi Vt}{\lambda}\sin\theta - \frac{2\pi V^2t^2}{\lambda R_0}\cos^2\theta.$$

Some comment at this point may be helpful. The first term is the linear phase term from the carrier frequency, the second is a constant phase shift due to the reflector being a finite distance away, the third is a linear phase term due to the Doppler shift (which is zero for the broadside direction) and the fourth term is a quadratic phase variation (which is proportional to $\cos^2\theta$). The processor will compensate for the carrier frequency and Doppler shift terms (and for a *focused* array, the quadratic phase term). Since the quadratic phase term contains the factor $[Vt\cos\theta]^2$, the "optimal" length of an unfocused array (defined as the length for which the phase shifts at the ends of the array are 90° from the phase at the center) is proportional to $\sec\theta$.

The phase time history of reflector 2 (following the previous derivation) is approximately:

$$\phi_2(t) \cong \omega_0 t - \frac{4\pi R_0}{\lambda} + \frac{4\pi Vx_2t}{\lambda R_0} + \frac{2\pi V^2t^2\left[1-\left(\frac{x_2}{R_0}\right)^2\right]}{\lambda R_0}$$

$$\frac{x_2}{R_0} = \frac{x_1}{R_0} + \frac{\Delta x}{R_0} = \sin\theta + \frac{d}{R_0}\cos\left(\theta+\frac{\alpha}{2}\right)$$

$$\frac{x_2}{R_0} \cong \sin\theta + \frac{d}{R_0}\cos\theta$$

$$1-\left(\frac{x_2}{R_0}\right)^2 \cong 1 - \sin^2\theta - \frac{2d}{R_0}\sin\theta\cos\theta - \frac{d^2}{R_0^2}\cos^2\theta$$

$$\cong \cos^2\theta - \frac{d}{R_0}\sin 2\theta - \frac{d^2}{R_0^2}\cos^2\theta.$$

Therefore

$$\phi_2(t) \cong \omega_0 t - \frac{4\pi R_0}{\lambda} + \frac{4\pi Vt}{\lambda}\sin\theta - \frac{2\pi V^2t^2\cos^2\theta}{\lambda R_0} + \frac{4\pi Vt}{\lambda R_0}(d\cos\theta) + \frac{2\pi V^2t^2}{\lambda R_0}\left(\frac{d}{R_0}\sin 2\theta + \frac{d^2}{R_0^2}\cos^2\theta\right).$$

For small d the last term can be ignored.

$$\phi_2(t) \cong \phi_1(t) + \frac{4\pi Vt}{\lambda R_0}(d\cos\theta).$$

The last term in this equation is a linear phase change which depends on d and which provides the resolution capability of the synthetic array. For a given *distance* resolution d, the time required for a given phase shift increases as $\sec\theta$. Therefore, the length of an array required for a given resolution increases as $\sec\theta$. Since the length of an "optimum" unfocused array also increases as $\sec\theta$, this means the maximum resolution of an unfocused array does not depend on θ.

For a focused array, the processor makes the phase correction $\phi_1(t)$ to the incoming data. The residue phase

$$\Delta\phi_2(t) = \phi_2(t) - \phi_1(t) \cong \frac{4\pi Vtd\cos\theta}{\lambda R_0}.$$

The processor effectively "sums" the returns from the transmitted pulses for reflectors at a given range. The output of the processor for a focused array, assuming that the returns from a reflector are constant during the integration time, is

$$Z(d) = \sum_{n=-((N-1)/2)}^{n=+((N-1)/2)} A_n \exp\left\{j\frac{4\pi nVTd\cos\theta}{\lambda R_0}\right\},$$

where

$Z(d)$ = Voltage output of processor
N = Number of pulses integrated (assumed odd for simplicity)
T = Interpulse period
A_n = Weighting factors of the processor
$G(d)$ = Power gain pattern $= |Z(d)|^2/|Z(0)|^2$.

If rectangular weighting is used (that is, if $A_n = 1$), then the power gain pattern of the *focused* array becomes

$$G(d) = \frac{\sin^2(2\pi VTNd\cos\theta/\lambda R_0)}{\sin^2(2\pi VTd\cos\theta/\lambda R_0)}$$

$$= \frac{\sin^2(2\pi Ld\cos\theta/\lambda R_0)}{\sin^2(2\pi Ld\cos\theta/\lambda R_0 N)}$$

$$L = VTN = \text{length of synthetic array.}$$

The null-to-null resolution D is

$$D = \frac{\lambda R_0}{L}\sec\theta.$$

The 3-db resolution is $0.44\ \lambda R_0/L\sec\theta$.

The first spurious lobe will occur at that value of $d(d=d_1)$ for which

$$\frac{4\pi VT}{\lambda R_0}d_1\cos\theta = 2\pi.$$

Since $d_1/R_0 \cong \alpha_1$ for small α_1, this means

$$\frac{4\pi VT\cos\theta}{\lambda}\alpha_1 = 2\pi,$$

and

$$\alpha_1 = \frac{\lambda\sec\theta}{2VT}.$$

(Note: d_1 and α_1 denote the particular values of d and α at the spurious lobe.)

If the angular position of the first null of the real radar antenna is placed at α_1, then for a uniformly illuminated aperture $\alpha_1 \cong \sin\alpha_1 = \lambda/L_1$.

L_1 = projected length of the real radar antenna in the direction of the beam (*i.e.*, the actual physical length of the antenna if the long dimension is perpendicular to the direction of the beam or the actual physical length times $\cos\theta$ if the long dimension is parallel to the flight path).

$$\frac{\lambda}{L_1} = \frac{\lambda\sec\theta}{2VT}.$$

$$L_1 = 2(VT)\cos\theta.$$

This may be regarded as a lower limit on L. The actual restriction is

$$L_1 \geq 2(VT)\cos\theta.$$

APPENDIX II

EFFECT OF THE REAL ANTENNA BEAM PATTERN ON THE GAIN PATTERN OF A FOCUSED SYNTHETIC ARRAY

Introduction

Calculation of the gain pattern of a synthetic antenna array is usually carried out assuming that the real antenna pattern is omnidirectional and then multiplying the resultant synthetic pattern (called the space factor) by the real antenna two-way gain pattern. This section investigates this problem more closely by assuming a real antenna beam shape of a relatively simple shape and carrying out the synthetic array calculation for rectangular weighting. The results thus derived are valid in calculating the main lobe and near sidelobe gain patterns of the synthetic array but do not give results for spurious lobe response. If the antenna pattern in the direction of the spurious lobe can be expanded in sine and cosine functions, a derivation similar to the one presented here can be carried out for the spurious lobes.

Gain Pattern of Real Antenna

For the purpose of carrying out the analysis easily, the following one-way real antenna power gain pattern was assumed:

$$G(\theta) = \left(\frac{1}{2} + \frac{1}{2}\cos\frac{\pi\theta}{\beta}\right) \qquad \text{for } -\beta \leq \theta \leq \beta$$

$$= 0 \qquad \text{for } |\theta| > \beta$$

where

θ = azimuth angle measured from boresight
β = half-power azimuth beamwidth.

The two-way power gain pattern is then

$$G^2(\theta) = \left(\frac{1}{2} + \frac{1}{2}\cos\frac{\pi\theta}{\beta}\right)^2,$$

and the *two-way voltage* gain pattern is

$$A(\theta) = \frac{1}{2} + \frac{1}{2}\cos\frac{\pi\theta}{\beta} \qquad \text{for } -\beta \leq \theta \leq \beta$$

$$= 0 \qquad \text{for } |\theta| > b.$$

Formulas for Gain Pattern of Focused Synthetic Antenna

The output of the processor for a focused array was derived assuming that the returns from a reflector during the integration interval are of constant amplitude. The result is

$$Z(d) = \sum_{n=-((N-1)/2)}^{n=+((N-1)/2)} A_n \exp\left\{j\frac{4\pi nVTd\cos\theta_1}{\lambda R_0}\right\}$$

where

$Z(d)$ = voltage output of processor

N = number of pulses integrated (assumed odd for simplicity)
T = interpulse period
A_n = weighting factors of the processor
V = velocity of radar platform
λ = RF wavelength
R_0 = range to reflector at array midpoint
d = distance of reflector perpendicular to line along axis of beam (which is at the angle θ_1)
θ_1 = angle of axis of synthetic beam measured from the broadside direction.

To include the effect of the real antenna beam shape, the antenna gain as a function of n and d must be specified. Assume that for $n=0$, $d=0$ the real antenna gain is a maximum. Then for $n=0$, the voltage gain for all d is $A(d/R_0)$.

For each unit in n, the aircraft moves a distance VT and a reflector close to the axis of the synthetic array has a change in angle from the boresight of the real antenna of an amount $VT \cos\theta_1/R$. The two-way voltage gain is then:

$$A\left(\frac{d}{R_0} - \frac{nVT\cos\theta_1}{R}\right).$$

Assuming rectangular weighting in the processor (*i.e.*, $A_n = 1$ for $-(N-1)/2 \leq n \leq (N-1)/2$ and including the effect of the real antenna beam shape, the output of the processor becomes

$$Z(d) = \frac{1}{2}\frac{\sin \pi y}{\sin\left(\frac{\pi y}{N}\right)} + \frac{1}{4}\exp\left\{j\frac{\pi d}{\beta R_0}\right\}\frac{\sin\left(\pi y - \frac{\pi}{2k}\right)}{\sin\frac{1}{N}\left(\pi y - \frac{\pi}{2k}\right)} + \frac{1}{4}\exp\left\{-j\frac{\pi d}{\beta R_0}\right\}\frac{\sin\left(\pi y + \frac{\pi}{2k}\right)}{\sin\frac{1}{N}\left(\pi y = \frac{\pi}{2k}\right)}$$

where

L = length of synthetic array = NVT
D = null-to-null resolution of synthetic array (assuming that the amplitude of the return from a reflector during the integration time is constant)

$$D = \frac{\lambda R_0}{L}\sec\theta_1$$

$$y = \frac{2d}{D},$$

and

$$k = \frac{\beta R_0}{L}\sec\theta_1 = \frac{\beta R_0 \sec\theta_1}{NVT}.$$

A normalized power gain function can then be defined as

$$f(d) = \frac{|Z(d)|^2}{|Z(0)|^2}.$$

Some Calculated Results

Fig. 2 in the body of this paper shows plots of $f(d)$ with abscissa $y = 2d/D$ for two values of k ($k=\infty$, $k=1$) with the assumption that $\exp\{\pm j\pi d/\beta R_0\} = 1$ (which is valid for small y and large R_0 in a typical case). The case $k=\infty$ corresponds to the space factor derivation for an omnidirectional real beam yielding a gain pattern of the form $(\sin x/x)^2$ with null-to-null resolution D and -3 db resolution $0.44D$. The synthetic antenna length in the case $k=1$ is equal to the beamwidth of the real antenna times the range. The shape of the physical antenna beamshape gives a result similar to amplitude weighting with a slight decrease in resolution but with improved sidelobe level. The null-to-null resolution is $1.15D$ and the -3-db resolution is $0.48D$.

Some Limiting Cases

1) For $\beta \to \infty$ (which also implies that $k \to \infty$)

$$f(d) = \frac{\sin^2 \pi y}{\sin^2\left(\frac{\pi y}{N}\right)}$$

which is the result derived assuming rectangular weighting and a real antenna gain pattern which is uniform in all directions (space factor derivation).

2) Let $d/R_0 = \theta$, $k \to \infty$ which implies that the amplitude of the return from a reflector is constant during the integration period. Then

$$Z(d) = \left(\frac{1}{2} + \frac{1}{4}\exp\left\{j\frac{\pi}{\beta}\theta\right\} + \frac{1}{4}\exp\left\{-j\frac{\pi}{\beta}\theta\right\}\right)\frac{\sin\pi y}{\sin\left(\frac{\pi}{N}y\right)}$$

$$= \left(\frac{1}{2} + \frac{1}{2}\cos\frac{\pi}{\beta}\theta\right)\frac{\sin\pi y}{\sin\left(\frac{\pi}{N}y\right)}$$

$$f(d) = \left(\frac{1}{2} + \frac{1}{2}\cos\frac{\pi}{\beta}\theta\right)^2\frac{\sin^2\pi y}{\sin^2\left(\frac{\pi y}{N}\right)}$$

$$f(d) = G^2(\theta)\frac{\sin^2\pi y}{\sin^2\left(\frac{\pi}{N}y\right)}$$

or, in words, the two-way power gain pattern is equal to the space factor of the array times the two-way gain pattern of the physical antenna. *This is only true if k is large enough* (actually $k=5$ is sufficient). The calculation for $k=1$ shows that this is not true for all cases.

3) Let $\exp\{\pm j(\pi d/\beta R_0)\} \cong 1$. The effect of the real

antenna gain pattern is identical to amplitude weighting over the integration period with the weighting factors $A_n = \frac{1}{2} + \frac{1}{2}\cos m\pi/Nk$.

Appendix III

Optimum Resolution of an Unfocused Synthetic Array[4]

In actual practice unfocused synthetic arrays will effectively have exponential weighting if the processing is done either by a delay line sweep integrator or by a bank of range gates followed by single tuned filters (two practical methods). The signal is considered to be a train of equal amplitude pulses and the processing to consist of coherently integrating (or "adding") these pulses with appropriate weighting. For an unfocused array with its main beam in the direction θ, the output $Z(d)$ of the processor will be

$$Z(d) = \sum_{n=-\infty}^{n=m} a_n \cdot \exp\left\{j\left[\frac{-2\pi V^2 n^2 T^2 \cos^2\theta}{\lambda R_0} + \frac{4\pi V n T d \cos\theta}{\lambda R_0}\right]\right\}$$

where

T = interpulse period

a_n = weighting factors of the processor

and the value of m is chosen such at $Z(0)$ is maximum.

If the weighting is exponential, then $a_n = \exp(n - m/N)$ where N is the number of pulses occurring in a time equal to the time constant of the exponential decay. Then

$$Z(d) = \sum_{n=-\infty}^{n=m} \exp\left\{\frac{n-m}{N}\right\} \exp\{j \text{ etc.}\}.$$

The relative two-way power gain can be defined as

$$G(d) = \frac{|Z(d)|^2}{|Z(0)|^2}$$

$G(d)$ can be evaluated numerically by approximating the sum by an integral.

$$|Z(d)|^2 = \left|\sum_{n=-\infty}^{n=m} \exp\left(\frac{n-m}{N}\right) \exp\left\{-j\frac{\pi}{2}(an^2 - 2abn)\right\}\right|^2$$

where

$$a = \frac{4V^2T^2\cos^2\theta}{\lambda R_0}$$

$$b = \frac{d}{VT\cos\theta}$$

$$|Z(d)|^2 \cong \left|\int_{-\infty}^{m} \exp\left(\frac{x-m}{N}\right) \exp\left\{-j\frac{\pi a}{2}(x-b)^2\right\} dx\right|^2.$$

[4]This problem was originally evaluated and set up for computer evaluation by W. W. Buchman, Hughes Aircraft Company.

Let

$\mu = \sqrt{a}\,(x - b)$

$V = \sqrt{a}\,(m - b)$ (not to be confused with the V used above for velocity)

$$\gamma = \frac{1}{\sqrt{a}\,N}$$

then

$$|Z(d)|^2 = \left|\frac{1}{\sqrt{a}} \exp(-\gamma V) \int_{-\infty}^{V} \exp(\gamma\mu) \cdot \exp\left\{-j\frac{\pi\mu^2}{2}\right\} d\mu\right|^2$$

$$= \gamma^2 N^2 |H(V, \gamma)|^2$$

where

$$H(V, \gamma) = \exp(-\gamma V) \int_{-\infty}^{V} \exp(\gamma\mu) \exp\left\{+j\frac{\pi\mu^2}{2}\right\} d\mu.$$

Therefore the evaluation of the antenna gain pattern reduces to evaluating $|H(V, \gamma)|^2$. The relative two-way power gain is

$$G(d) = \frac{|H(V, \gamma)|^2}{|H(V_0, \gamma)|^2}$$

where V_0 is the value of V for the maximum of $H(V, \gamma)$.

$$V - V_0 = -\sqrt{a}\,b = \frac{-2d}{\sqrt{\lambda R}}.$$

The parameter

$$\gamma = \frac{\sqrt{\lambda R_0}}{2VT\cos\theta N}$$

where N, as stated before, is the number of pulses occurring in a time equal to the time constant of the exponential decay. NT might be called the "length" of the array. The width or resolution of these gain patterns at the 3- and 6-db points is shown vs $1/\gamma$ in Fig. 5. The 3-db resolution is maximum at $1/\gamma = 3.5$ and has a value of normalized width of 1.22. The 6-db resolution is optimum at $1/\gamma = 2$ and has a value of normalized width of 1.86. The resolution in feet equals $\sqrt{\lambda R_0}/2 \times$ normalized width. Note that the optimum value of γ is not very critical.

Acknowledgment

The author is indebted to the persons who have developed and evaluated the synthetic array principles, particularly those at the Institute of Science and Technology, University of Michigan and at the Hughes Aircraft Company. Much of this original work has not yet found its way into the open literature. Although the development given in this report is original with the author, it owes a great deal to this previous work. Some specific references are included but no attempt has been made to be comprehensive and apologies are probably due to persons whose work is not referenced.

An Optimum Design Of Ambiguity Function, Antenna Pattern, and Signal For Side-Looking Radars

R.O. Harger
(Radar Laboratory, Institute of Science and Technology, University of Michigan, Ann Arbor)

IEEE Transactions on Military Electronics, Vol. MIL-9, Nos. 3 and 4 (July/October 1965)
Pages 264-278

***Abstract*—The design of antenna pattern and transmitted signal for a coherent-side-looking (range-azimuth) synthetic antenna radar is studied. The general design criterion is to minimize all spurious responses in the ambiguity function under the restraints of finite antenna aperture and signal bandwidth.**

The problem is completely solved for any set of consistent constraints. The interplay between aperture size, range and azimuth resolution, and spurious response level for the optimum design is exhibited, and antenna aperture illuminations, signal amplitude and phase modulation, and pulse repetition frequency are specified.

Introduction

THE PROBLEM OF designing the ambiguity function for a coherent side-looking synthetic-antenna range-azimuth radar (hereafter abbreviated "S/L radar") is investigated. A general familiarity with the concept and geometry of this type of radar [1] would be helpful. Very simply, the problem of designing the ambiguity function arises precisely because a function and its Fourier transform cannot both be identically zero outside a finite interval, or, roughly, such a pair has a minimum allowable "time-bandwidth product." The design is aggravated by inherent dependence between parameters which we would like to specify.

To elaborate somewhat, recall that the conventional S/L radar derives its range information by pulsing and its azimuth information from the thereby sampled azimuth modulation, hereafter often called the azimuth "signal." It is well known that if a signal is not sampled at a sufficiently high frequency then pieces of the alias spectra overlap ("fold") into the fundamental spectrum. The azimuth signal's extent and bandwidth, however, is determined by geometry and the far-field antenna pattern in azimuth, which, of necessity, is infinite in extent since it is derived from the finite aperture of the physical antenna. Constraints in practical operation make some aliasing inevitable. The sampling frequency also determines, along with the geometry, the unambiguous range interval; since that interval and the antenna's vertical aperture are both finite, some ambiguous range return is also inevitable.

Clearly, by adjusting the sampling frequency, the amount of one ambiguity can be reduced at the cost of increasing the other; hence, the sampling frequency is one parameter that must be specified by a design. For example, the sampling frequency could, in theory, be raised high enough to sample the highest Doppler an increase of two orders of magnitude, typically but at the cost of proportionately reducing the unambiguous range interval, generally below the minimum at which it is useful in practice. Hence, this does not represent a useful solution. The amount of ambiguous energy received depends on the size of the angular interval where the antenna pattern is desirably contained relative to the size of the antenna aperture. As the sampling frequency is varied, the optimum antenna patterns in both dimensions must also be varied.

Performance requirements and geometry, e.g., orbital applications, are assumed which require use of all of the resulting unambiguous range interval. Thus, it may be said we are concerned with the problem of design for operation at the ambiguous limit. In fact, the results are more general.

The critera to use in designing the aperture illuminations are supplied by the ambiguity function. Recall that the ambiguity function may be defined as the output of the maximum likelihood processor for a received signal from a point target [2]. The ambiguity function could also be defined as the output of a maximum signal-to-noise ratio (SNR) processor. In the presence of white normal noise both processors are a filter matched to the signal.

In conventional S/L radars the received signal is recorded and, hence, available for processing in a two-dimensional format. A two-dimensional complex filter [3] matched to the two-dimensional received signal from a point target is used for the processing. The output for a point target will be defined to be the ambiguity function of the S/L radar for our purposes. It can be shown that the same SNR is obtained as would be obtained with the maximum likelihood processor acting on the received ("one-dimensional") data.

Clearly, the properties of the ambiguity function are of final interest. We will see from these properties that a certain energy criterion is dictated for the design of the azimuth pattern and a certain average gain criterion is appropriate to the design of the range (or vertical) pattern. (Note that the more conventional criterion of antenna-beamwidth and sidelobe level is not of direct interest.)

Manuscript received December 10, 1964. The research reported in this paper was sponsored by the Research and Technology Division, U. S. Air Force, Wright-Patterson AFB, Ohio, under Contract AF 33(616)-8365.

Since the Doppler, or azimuth, information is derived by sampling, the range performance in the unambiguous range interval, as specified by the ambiguity function, is essentially independent of the Doppler, or azimuth, performance; hence, the design of the range cross-section of the ambiguity function in the unambiguous range interval reduces to a signal design problem. It will be seen that a certain average response criterion is appropriate to the design of the signal.

The Ambiguity Function

Let the radiated real signal be a coherent train of identical pulses

$$s(t) \equiv \mathrm{Re}\{e^{i\omega_c t}\sum F(t - nT_s)\} \tag{1}$$

where

$$F(t) \equiv |F(t)|\, e^{i\phi(t)}$$

is the complex modulation, $|F|$ and ϕ the real envelope and phase modulations, respectively. The received signal, scattered by a stationary point target at range delay τ_0 and azimuth position t_0 is approximately

$$r(t) = K_1 a_v(\tau_0) a_h(t - t_0)\,\mathrm{Re}\{e^{i\omega_c(t-\tau_0) - ik_a(t-t_0)^2/2} \sum F(t - nT_s - \tau_0)\},\ t \in T, \tag{2}$$

where the azimuth linear FM rate is

$$k_a = k_a(\tau_0) \equiv \frac{4\pi v^2}{\lambda R_0}, \tag{3}$$

$v \equiv$ vehicle velocity,
$\lambda \equiv$ RF wavelength,
$R_0 \equiv \nu\tau_0/2$,
$\nu \equiv$ velocity of light,
$T \equiv$ observation time

a_v and a_h are (real) factors due to the far-field, two-way voltage pattern a of the antenna and K_1 is a constant not of immediate interest—it depends upon antenna gains, propagation loss, scattering cross section, etc. We assume that

$$a(\gamma, \gamma_0; \beta, \beta_0) \approx b_h(\beta - \beta_0) b_v(\alpha - \alpha_0)$$

for angles of interest, where

$\gamma_0 \equiv$ center of beam pattern
$\gamma \equiv$ angle in vertical plane
$\beta_0 \equiv$ center of beam pattern
$\beta \equiv$ angle in horizontal plane.[1]

For small angles (see Fig. 1)

$$a_v(\tau_0) \approx b_v\left[\frac{\tan\gamma_0}{\tau_*}(\tau_* - \tau_0)\right],$$

where τ_* is the signal delay to a scatterer at the center of the vertical pattern, and

$$a_h(t - t_0) \approx b_h\left[\frac{2v}{\nu\tau_0}(t - t_0)\right].$$

It will be convenient to define a complex modulation associated with the azimuth Doppler

$$A_h(t; \tau_0) \equiv a_h(t) e^{-ik_a(\tau_0)t^2/2}.$$

Then the received signal may be written

$$r(t) = \mathrm{Re}\{a_v(\tau_0) A_h(t - t_0) e^{i\omega_c(t-\tau_0)} \sum F(t - nT_s - \tau_0)\}.$$

[1] It is sufficiently accurate to adopt the usual definition of these planes for a vehicle moving horizontally over a flat stationary earth, the antenna boresighted normal to the flight path ($\beta_0 \equiv 0$) with a depression angle γ_0 from the horizontal. Both planes contain the boresight line, the "vertical" plane being normal to the earth and the "horizontal" plane being normal to the vertical plane.

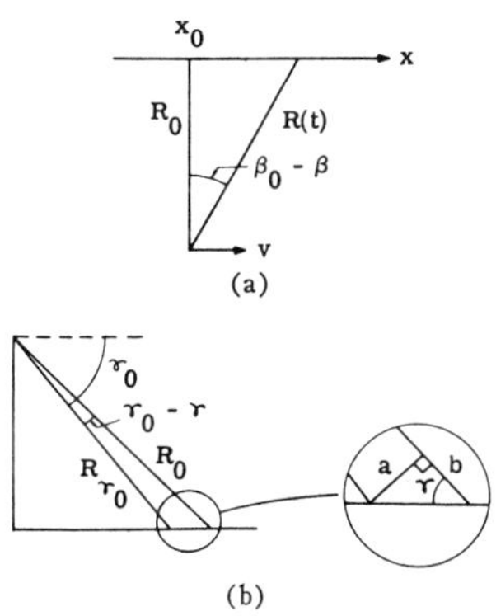

Fig. 1. Geometry of S/L radar. (a) Slant range (horizontal). (b) Vertical planes.

This signal can be written even more concisely as

$$r(t) = \mathrm{Re}\{K_0 S(t;\alpha_0)e^{i\omega_c t}\}$$

where $K_0 \equiv K_1 a_r(\tau_0)e^{-i\omega_c\tau_0}$ is the unknown complex amplitude and $\alpha_0 \equiv (\tau_0, t_0)$ is the 2-tuple of unknown true parameters that are to be estimated and where

$$S(t;\alpha_0) \equiv A_h(t - t_0;\tau_0)\sum F(t - nT_s - \tau_0),\ t\in T$$

is the complex modulation of the received signal.

In practice this received signal is coherently mixed with the output of a stable oscillator to give a signal on an offset frequency ω_0:

$$s(t;\alpha_0) = \mathrm{Re}\{K_0 S(t;\alpha_0)e^{i\omega_0 t}\}.$$

Further, as the pulses are received sequentially, they are recorded in a dimension orthogonal to delay, or time, which therefore corresponds to azimuth location. For simplicity a recording velocity of 1 m/s in range is assumed, thereby preserving delay itself as the coordinate, and v m/s in azimuth, thereby giving azimuth position itself as the coordinate. Then the resulting two-dimensional signal is

$$s(t,x;\alpha_0) = \begin{cases}\mathrm{Re}\{K_0 S(t,x;\alpha_0)e^{i\omega_0 t}\}, & \text{range offset case,}\\ \mathrm{Re}\{K_0 S(t,x;\alpha_0)e^{iu_0 x}\}, & \text{azimuth offset case,}\end{cases}$$

where $u_0 \equiv \omega_0/v$ is called the azimuth offset angular (spatial) frequency, and

$$S(t,x;\alpha_0) \equiv F(t - \tau_0)A_h\left(\frac{x - x_0}{v};\tau_0\right)\sum h(x - nX_s),\quad x\in X$$

where $X_s \equiv vT_s$ is the sampling spatial distance, $x_0 \equiv vt_0$ is the true azimuth position, and h is the recorder impulse response in the azimuth dimension. (The latter is carried along until a convenient point is reached to discard it, since the recorder responses can be approximately accounted for easily.) The two offset choices will be discussed below. The signal spatial duration is $X \equiv vT$.

It will be convenient to introduce the azimuth complex modulation

$$B(x;\tau_0) = |B(x)|\,e^{-ik(\tau_0)x^2/2}$$

where

$$|B(x)| \equiv a_h\left(\frac{x}{v}\right)$$

and

$$k(\tau_0) \equiv k_a(\tau_0)/v^2.$$

Then we have

$$S(t,x;\alpha_0) = F(t - \tau_0)B(x - x_0;\tau_0)\sum h(x - nX_s),\ x\in X.$$

This two-dimensional signal is processed by a complex matched filter. As usual, it is conceptually simpler to use the frequency domain to describe the filtering operation. It is extremely convenient that the complex modulations are of large time-bandwidth product, for there is then a very simple approximate relationship between the signal and its spectrum. In particular, let such a signal be

$$F(x) \equiv |F(x)|\,e^{ikx^2/2};$$

then its Fourier transform is

$$\tilde{F}(\omega) \approx \left|F\left(\frac{\omega}{k}\right)\right| e^{-i\omega^2/2k},$$

within uninteresting constants [7]. It is seen that as a target passes through the azimuth pattern, the latter is replicated in the frequency domain as the envelope of the spectrum. Time or azimuth position is related uniquely to spectral location; e.g., the spectrum of any large time-bandwidth product piece of the azimuth signal is trivially seen to be the corresponding piece of the spectrum of the azimuth signal.

The spectrum can now be quickly written down for the sampled complex azimuth signal

$$\sum B_n(u;x_0)e^{i(u-nu_s)^2/2k+i(u-nu_s)x_0}$$

$$B_n(u;x_0) \equiv \left|B\left(\frac{u - nu_s}{k}\right)\right| \mathrm{rect}\left(\frac{u - nu_s + x_0 k}{kX}\right)$$

where it was assumed that $\tilde{h}(nu_s) \approx 1$ for n interest and $u_s \equiv 2\pi/X_s$ is the sampling angular frequency. The two-dimensional spectrum of the signal is, therefore,

$$\tilde{s}(\omega,u;\alpha_0) = \begin{cases} K_0\tilde{S}(\omega-\omega_0, u;\alpha_0) + \overline{K}_0\tilde{\overline{S}}(-\omega-\omega_0, -u;\alpha_0), \\ \qquad\qquad \text{range offset,} \\ K_0\tilde{S}(\omega, u-u_0;\alpha_0) + \overline{K}_0\tilde{\overline{S}}(-\omega, -u-u_0;\alpha_0), \\ \qquad\qquad \text{azimuth offset,}\end{cases}$$

where

$$\tilde{S}(\omega,u;\alpha_0) = \tilde{F}(\omega)e^{i\omega\tau_0}\sum B_n(u;x_0)e^{i(u-nu_s)^2/2k+i(u-nu_s)x_0}.$$

The offset frequencies are chosen in the following way. Consider, first, range offset operation. If the bandwidth of F is Ω, then to avoid overlapping of spectra and the attendant clutter we want $\omega_0 \geq \Omega/2$. Recorder bandwidth limitations generally force us to the lower bound. In azimuth offset operation $\Delta \equiv kX$ is the bandwidth of the (largest) adequately sampled segment of B; then for the analogous reason we want $u_0 \geq \Delta/2$ and are forced to the lower bound.

The sampling angular frequency is different in the two cases. For range offset operation we need

$$u_s(\text{r.o.}) = \Delta(\text{rad}/m)$$

by the definition of Δ, in order that the adequately sampled pieces of the azimuth alias spectra do not overlap. In azimuth offset operation we must have

$$u_s(\text{a.o.}) = 2\Delta(\text{rad}/m)$$

in order to make "holes" of width Δ in the spectra of the sampled RF signal so that after coherent mixing the "positive" and "negative" spectra can interleave without overlapping. If the reader sketches the two-dimensional spectra before and after coherent mixing down to the offset, the operations involved are rendered rather transparent.

If the two-dimensional signal s were received along with an additive two-dimensional Gaussian noise process, with white spectral density over the frequencies of interest then the optimum, or matched, complex filter would be $\bar{S}(\omega, u; 0)$ on the appropriate offset. However, the noise present is actually much different: it was derived from a one-dimensional noise process, assumed white. It can be shown that in this case the filter reduces to

$$\bar{\tilde{F}}(\omega) B_0(u; 0) e^{-iu^2/2k}$$

on the appropriate offset frequency.

The squared magnitude of the complex output, i.e., the squared envelope of the optimally processed real signal, is given by

$$|q(\tau, \mu|^2 - |K_0|^2 \left| \int e^{i\omega\tau} |\tilde{F}(\omega)|^2 d\omega \right|^2 \cdot \left| \int e^{iu\mu} B_0(u; 0) e^{-iu^2/2k} \cdot \sum B_n(u; \mu) e^{i(u-nu_s)^2/2k - inu_s\mu} du \right|^2 \tag{4}$$

for range offset operation, where

$$\tau \equiv \tau_0 \quad \text{and} \quad \mu \equiv x_0.$$

The function $|q|^2$ is the ambiguity function of the S/L radar when operated as described, with a range offset. The azimuth linear FM rate k is a function of delay; except for this the ambiguity function factors with the two complex modulations determining the behavior in the τ and μ dimensions, respectively.

Recalling that B has infinite extent, $B_0(u; 0)$ passes only Δ-width segment centered about zero. Each term is matched in phase (but, except for $n=0$, not in amplitude) and hence will essentially compress to a pulse-like shape at an azimuth location determined by the linear phase delay:

$$\mu_n = n \frac{u_s}{k} = nX.$$

These are the locations of the azimuth ambiguities, $n \neq 0$.

Of special interest is the response of $|q|^2$ about zero:

$$|q(\tau, \mu)|^2 \approx |K_0|^2 \left| \int e^{i\tau\omega} |\tilde{F}(\omega)|^2 d\omega \right|^2 \cdot \left| \int e^{iu\mu} |B_0(u; 0)|^2 du \right|^2.$$

It will be sufficient for present purposes to note the peak response of each ambiguity—assumed to be at approximately μ_n:

$$|q(0, \mu_n)|^2 \approx |K_0|^2 \|F\|^4 \left| \int B_0(u; 0) B_n(u; \mu_n) du \right|^2.$$

For azimuth offset operation the S/L ambiguity function is somewhat more complicated because both the positive and negative μ spectra can be passed by the complex filter $B_0(\omega/k)$. We find the squared magnitude, or squared envelope, of the output signal to be

$$\begin{aligned} |q(\tau, \mu)|^2 = \Big| & K_0 \int e^{i\omega\tau} |\tilde{F}(\omega)|^2 d\omega \int e^{i\mu u} B_0(u; 0) e^{-i(u^2/2k)} \\ & \cdot \sum B_n(u; \mu) e^{i(u-nu_s)^2/2k - inu_s\mu} du \\ & + K_0 \int e^{i\omega\tau} [F(\omega)]^2 d\omega \int e^{i\mu u} B_0(u; 0) e^{-i(u^2/2k)} \\ & \cdot \sum B_n(-u - 2u_0; \mu) \\ & \cdot e^{-i(-u-2u_0-nu_s)^2/2k - i(nu_s+2u_0)\mu} du \Big|^2 \end{aligned} \tag{5}$$

where we remind ourselves that now $u_s = 2\Delta$.

It can be seen that the first integral has a behavior much like that for the range offset case: it gives compressed pulse-like outputs but at locations

$$\mu_n = n \frac{u_s(\text{a.o.})}{k} = n \frac{2\Delta}{k} = 2nX.$$

The second integral is different in its behavior: each term is completely mismatched in phase in the sense that the filter is the conjugate of the correct one. This results in even greater dispersion in *both* dimensions about locations

$$\frac{2u_0 + nu_s}{k} = (2n + 1)X.$$

Design Criteria Based on the Ambiguity Function

Phase-matched processing with range offset is first considered as it is somewhat simpler. The height of the ambiguous peaks in azimuth are observed to be proportional to the energy under the one-way voltage pattern in azimuth of the antenna in the angular sectors centered on nX/R, $n = \pm 1, 2, \ldots$, and of width X/R, which will be termed "ambiguous (azimuth) intervals." (See Fig. 2.) The main response is propor-

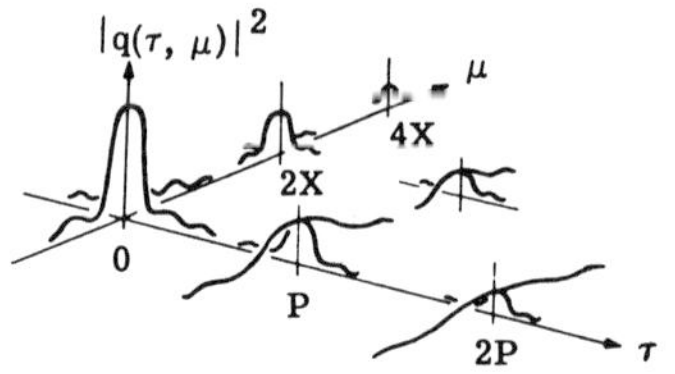

Fig. 2. Typical ambiguity function for a side-looking radar. Azimuth offset operation.

tional to the energy in the unambiguous interval $(-X/2R, +X/2R)$. Thus, maximization of the energy of the one-way voltage pattern in azimuth in the unambiguous interval is sought relative to the energy in any particular ambiguous interval. Since a target is as likely to be in any one of them and have a given response, we decide that: *The design criterion for the one-way voltage pattern in azimuth is to maximize the energy in the unambiguous interval.*

With azimuth offset operation we observe that all the odd-order ambiguous azimuth intervals disperse, i.e., their phase is mismatched to the processor, and only the even-order ambiguous azimuth intervals compress. Experience has shown that such dispersed energy only slightly deteriorates the output and, hence, one might be tempted to allow antenna pattern energy to fall in these odd intervals in order to better control the pattern in the even intervals. An alternate criterion appropriate to the azimuth pattern that one might then decide upon is: *Maximize the energy of the one-way voltage pattern in the unambiguous interval relative to the energy in the even ambiguous intervals.* In cases of practical interest it appears that this criterion leads to approximately the same patterns as the criterion first proposed; its further discussion is confined to Appendix A. Thus, the criterion first proposed is retained for the azimuth pattern.

From the ambiguity function it is also seen that the two-way voltage gain of the vertical pattern governs the response of targets lying in ambiguous range intervals. The gain assigned to a target at any location in the unambiguous interval should thus be maximized, relative to the gain assigned to a target at any of its ambiguous locations. One way is to make the average gain over the unambiguous interval large relative to the average gain over any ambiguous interval, i.e., make the energy in the former interval large relative to the energy in the latter interval, or simply, to make the energy in the unambiguous interval as large as possible.

In deciding upon such a criterion two things have been ignored. First, fairly uniform gain over the unambiguous range interval might be desired: the criterion will not be modified but comments will be made on this property of our solutions. Second, the S/L radar has the interesting inherent property of suppressing ambiguous range returns because their azimuth histories are not matched to a processor that is matched to the return from a target in the unambiguous range interval. This property implies that some relaxation in the requirement for the vertical pattern could be made. However, two things militate against this: 1) average gain is sacrificed in the unambiguous interval and 2) more ambiguous energy (dispersed, to be sure) is allowed into the output. The second reason experience proved to be not particularly important and so, it is decided that in fact: *The design criterion for the one-way voltage pattern in range is to maximize the energy in the unambiguous interval.*

It will be seen that the same criterion for both the horizontal and vertical patterns of the antenna have been arrived at: the problem is to choose a function that when truncated, then Fourier transformed, and then truncated again, has maximum energy.

For matched processing the peaks of the azimuth ambiguities are not so simply related to the patterns. However, the amplitude of the matched filter will not be far from uniform; the taper will be found to increase with increasing "freedom" in the design parameters but generally the approximation is valid for practical situations. Then again the ambiguous responses are proportional to the energy in intervals of the one-way, far-field voltage pattern and the criterion just decided upon for the azimuth pattern remains appropriate.

Thus far, criteria have been established for designing the antenna patterns by examining the range and azimuth ambiguities. But of the same order of importance is the sidelobe levels of the central, or unambiguous, response of the ambiguity function. The factoring of the ambiguity function means that the design of $|q(\tau, 0)|$ near the central response is the design of the complex signal modulation F. It also means that the behavior of $|q(0, \mu)|$ near the central response is determined by the far-field pattern of the azimuth antenna pattern in the unambiguous angular interval.

A reasonable and, as it develops, quite interesting criterion to choose for designing the complex modulation is: *Maximize the energy of the range cross-section* $|q(\tau, 0)|$ *of the ambiguity function in the range resolution interval.* We would like to choose the same criterion for the relevant piece of the azimuth pattern but a priori we would expect to have to compromise between the two criteria to be met by the azimuth pattern. We therefore do not modify our criterion for the azimuth pattern but will comment on this property of our solutions below.

SOLUTION OF THE ANTENNA DESIGN PROBLEM

From the foregoing discussion, we see that a mathematical statement of the design problem would be

$$\max_{E \in \mathfrak{L}^2} \nu,$$

where

$$\nu = \frac{\int_I b(u)du}{\int_{-\infty}^{\infty} b(u)du}.$$

In the horizontal pattern problem, $b = b_h$ and

$$I = I_h \equiv X/R_0; \tag{6}$$

in the vertical pattern problem, $b = b_v$ and

$$I_v \equiv (\nu \tan \gamma_0)/2R_0 f_s. \tag{7}$$

Of course, $b = |E|^2$ where the far-field, one-way voltage pattern of the antenna is, by the Kirchhoff-Huygens principle,

$$E(u) = \int_{-\Omega}^{\Omega} e^{-ikux} f(x)dx \tag{8}$$

where f is the voltage across the aperture $(-\Omega, \Omega)$ and $k \equiv 2\pi/\lambda$. Thus, our mathematical problem is to find a function $f \in \mathfrak{L}^2(-\infty, \infty)$ which, when "band-limited" to $(-\Omega, \Omega)$ and then "timelimited" to $(-I/2, I/2)$, has maximum energy. Thus, the bandlimited version of f is then the solution, i.e., the aperture illumination, that is sought.

By using (8) for $E(u)$ we see that

$$\int_{-\infty}^{\infty} |E(u)|^2 du = \frac{2\pi}{k} \int_{-\Omega}^{\Omega} |f(x)|^2 dx.$$

To calculate the numerator of ν we write

$$\int_I |E(u)|^2 du = \int_I du \int_{-\Omega}^{\Omega} dx f(x) \cdot e^{-ikux} \int_{-\Omega}^{\Omega} dy \bar{f}(y) e^{ikuy}$$

and find

$$\int_I |E(u)|^2 du = \int_{-\Omega}^{\Omega} dx f(x) \int_{-\Omega}^{\Omega} dy f(y) \frac{\sin \frac{kI}{2}(x-y)}{\frac{k}{2}(x-y)}.$$

Then

$$\nu = \frac{\int_{-\Omega}^{\Omega} dx f(x) \int_{-\Omega}^{\Omega} dy f(y) \frac{\sin \frac{kI}{2}(x-y)}{\pi(x-y)}}{\int_{-\Omega}^{\Omega} |f(x)|^2 dx}$$

or, if we define

$$S_0(\omega) \equiv f(\Omega\omega) \tag{9}$$

and

$$c \equiv kI\Omega/2, \tag{10}$$

we may write

$$\nu = \frac{\int_{-1}^{1} d\omega S_0(\omega) \int_{-1}^{1} dv S_0(v) \frac{\sin c(\omega - v)}{\pi(\omega - v)}}{\int_{-1}^{1} |S_0(\omega)|^2 d\omega}. \tag{11}$$

It is very interesting that the only parameter of concern is the "aperture-interval" product c, analogous to a "time-bandwidth" product.

The problem is written in the form

$$\max_{S_0 \in \mathfrak{L}^2} \frac{(AS_0, S_0)_1}{\|S_0\|_1^2}$$

where $(\cdot, \cdot)_1$ and $\|\cdot\|_1$ are the inner product and norm on $\mathfrak{L}^2\{[-1, 1], \mu\}$, μ Lebesgue measure, and the operator

$$Af(\omega) \equiv \int_{-1}^{1} \frac{\sin c(\omega - v)}{\pi(\omega - v)} f(v)dv.$$

Since A is linear, bounded, completely continuous, and symmetric, we know [4] the solution to the maximization problem exists and is equal to the largest positive characteristic value μ_0^+ of the integral equation

$$\mu S_0 = AS_0. \tag{12}$$

The associated characteristic function (s), S_{00}, is the solution to our problem (when bandlimited).

The solutions $\{S_{0n}\}$ to the integral equation (12) were recognized by Slepian [5] as the prolate spheroidal wave functions. Because of the paucity of the tabulations of the $\{S_{0n}\}$, (12) was solved numerically on an IBM 7090 digital computer, generating a tabulation useful here. The table of $\{\mu_0(c), S_{00}(c; x)\}$ is given in [6] for $c = 1/2, \cdots, 5(1/2)$ and $x = \pm 0, \cdots, 5(1/20)$. A summary of the results are shown in Figs. 3 and 4. Note that when c is very small the best that can be done is to choose uniform illumination, as would probably be expected. (In this case the solutions can be checked by an approximate solution of the integral equation.)

To calculate the far-field voltage pattern E of the antenna we must calculate

$$E(u) = \int_{-\Omega}^{\Omega} e^{-ikux} f(x)dx, \qquad f(x) = S_{00}\left(c; \frac{x}{\Omega}\right).$$

The prolate spheroidal wave functions have an amazing property: if they are bandlimited to $[-1, 1]$, the transform is a scaled version of the original function, i.e. [5],

$$2i^n R_{0n}^{(1)}(c; 1) S_{0n}(c; t) = \int_{-1}^{1} e^{-ic\omega t} S_{0n}(c; \omega) d\omega.$$

Using this relation we find

$$E(u) = 2\Omega \sqrt{\frac{\pi\mu_0(c)}{2c}}\, S_{00}\left(c; \frac{2u}{I}\right). \tag{13}$$

Thus Fig. 4, with a changed abscissa scale, also shows the far-field patterns (except for a multiplicative constant) over the primary interval. Sufficiently accurate continuations of S_{00} outside the interval $[-1, 1]$ were not available and hence were calculated on a digital computer. The tabulations are given in [6] and summarized in Fig. 5 in normalized form.

The azimuth cross section $|q(0, \mu)|$ is now calculated in the unambiguous interval about $\mu = 0$ to observe the resolution and sidelobe level of the central spike of $|q(\tau, \mu)|$ in the azimuth dimension. The far-field voltage pattern $E(u)$, as given by (13), determines the form of the azimuth signal; in spatial variables,

$$B(x) = \left| E\left(\frac{x}{R}\right) \right|^2 e^{ikx^2/2}.$$

The azimuth cross section of the ambiguity function is then, within a scale,

$$|q(\mu, 0)|^2 = \left| \int_{-1}^{+1} e^{i\pi x u} |S_{00}(c_h; u)|^2 du \right|^2.$$

This output signal is given in Figs. 6 and 7.

The relative heights (average relative height in the range dimension) of the major spurious responses of the ambiguity function $|q(c; \tau, \mu)|^2$ are extracted from Figs. 3 and 4; the results are summarized in Figs. 8, 9, and 10.

Curve (b) of Fig. 10 takes into account a typical value of the suppression of the range ambiguities that occur. Knowing the suppression, curve (b) is trivially obtained from curve (a). The amount of suppression is determined by the geometry and a fairly good estimate can be obtained of its effect in a routine calculation.

The price to be paid for the reduction of the spurious azimuth responses of the ambiguity function is a somewhat decreased resolution of $|q(c; 0, \mu)|$ as is summarized in Fig. 11, using the data of Figs. 6 and 7. Note that although Fig. 11 was plotted using a particular definition of resolution— the width between half-power points of $|q(0, \mu)|$—a similar plot for any other definition one may choose can be made from the data of Figs. 6 and 7. Note that azimuth sidelobe levels are very close to azimuth ambiguity levels, indicating a well-balanced design.

The price to be paid for the reduction of spurious range responses of the ambiguity function is reduced antenna gain at the edges of the unambiguous range interval (Fig. 3). One value of the inherent range ambiguity suppression of the S/L radar is that it reduces the value of c_v required to achieve a given level of spurious range responses, thereby giving more uniform gain over the unambiguous range interval. The uniformity of gain is acceptable; in fact, it is probably as well as we can do under the aperture length restraints encountered in practice. If one does have excess vertical aperture then it can be devoted to "squaring-up" the gain pattern—a problem in its own right.

Solution of the Signal Design Problem

With the choice of criterion made above for $|q(\tau, 0)|^2$ we see that again a mathematical statement of the design problem would be

$$\max_{q \in \mathcal{L}^2} \nu$$

where now

$$\nu \equiv \frac{\int_{I_s} |q(\tau, 0)|^2 d\tau}{\int_{-\infty}^{\infty} |q(\tau, 0)|^2 d\tau}.$$

I_s the delay resolution interval.

For phase-matched filtering

$$q(\tau, 0) = \int_{-\Omega_s}^{\Omega_s} e^{i\omega\tau} |\hat{F}(\omega)| \, d\omega$$

and for matched filtering

$$q(\tau, 0) = \int_{-\Omega_s}^{\Omega_s} e^{i\omega\tau} |\hat{F}(\omega)|^2 d\omega.$$

$2\Omega_s$ the bandwidth of the complex signal modulation.

As the variational problem is the same as that already solved for the antenna problem, we can write down the solution immediately:

$$\hat{F}_m(\omega) \equiv |\hat{F}(\omega)|$$

$$= \begin{cases} \operatorname{rect}\left(\frac{\omega}{2\Omega_s}\right) S_{00}\left(c_s; \frac{\omega}{\Omega_s}\right), & \text{phase matched,} \\ \operatorname{rect}\left(\frac{\omega}{2\Omega_s}\right) S_{00}^{1/2}\left(c_s; \frac{\omega}{\Omega_s}\right), & \text{matched,} \end{cases}$$

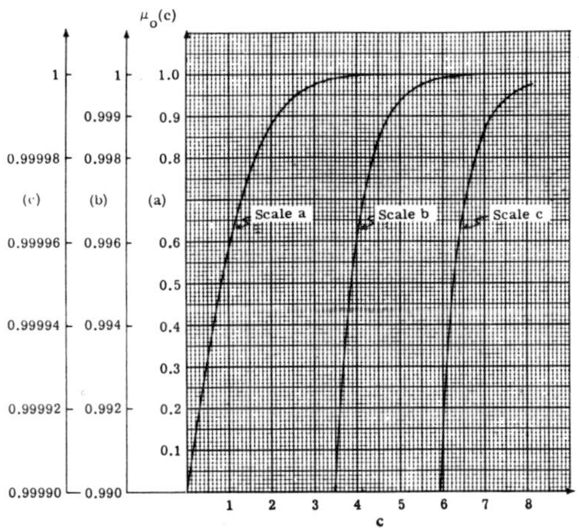

Fig. 3. The largest characteristic value $\mu_0(c)$. This is the fraction of total energy in the unambiguous intervals for the antenna patterns or in the resolution interval for the delay cross section of the ambiguity function.

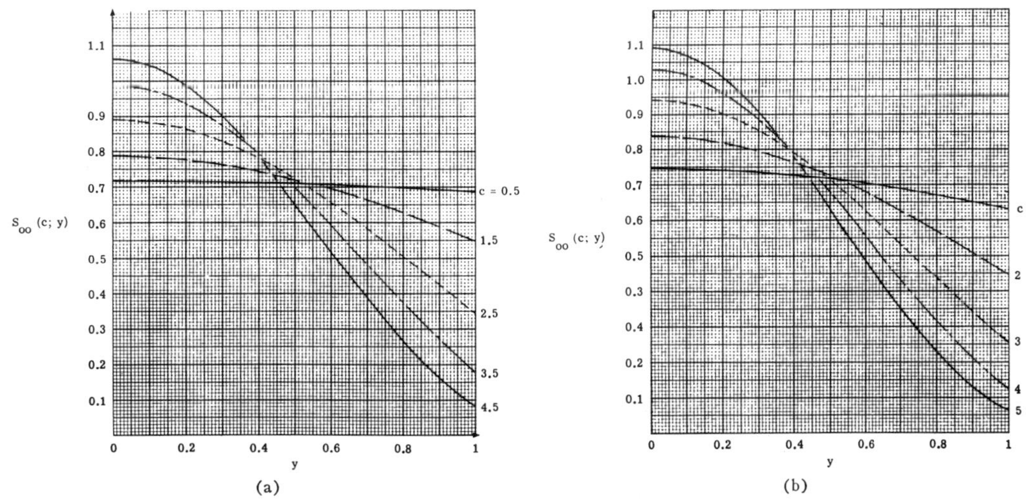

Fig. 4. The characteristic function $S_{00}(c;\ y)$, associated with the largest characteristic value. These are the optimum antenna aperture illumination functions (setting $y=x/\Omega_h$ or $y=x/\Omega_v$) or the optimum signal spectrum square magnitude (setting $y=\omega/\Omega_s$).

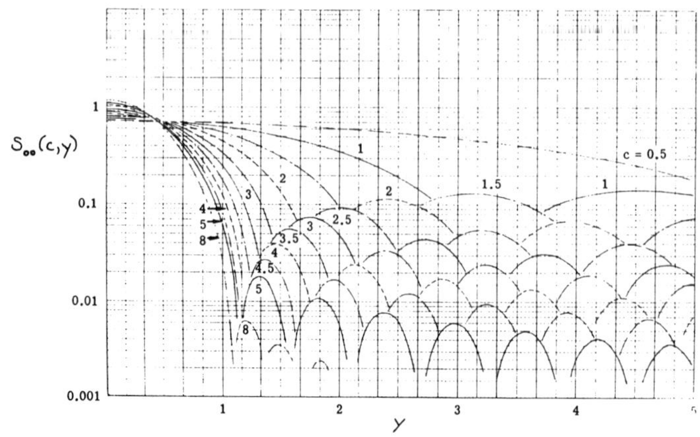

Fig. 5. Continuation of the characteristic function $S_{00}(c; y)$. These are the optimum antenna far-field one-way voltage patterns (setting $y=2\mu/I_h$ or $y=2\mu/I_c$) or the optimum delay cross section of the ambiguity function (setting $y=2\tau/I_s$).

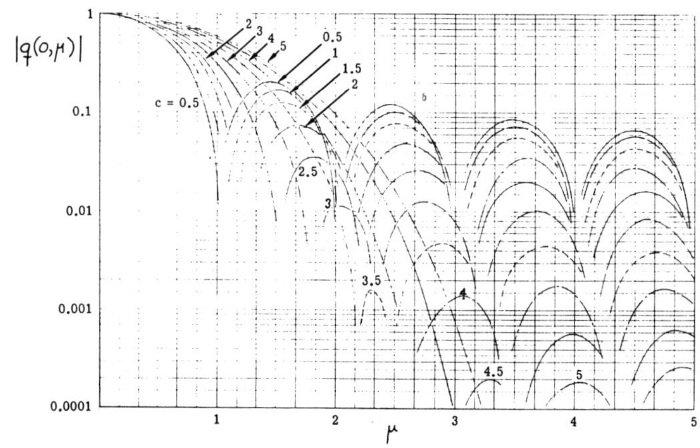

Fig. 6. The azimuth cross section of the ambiguity function for phase-matched processing.

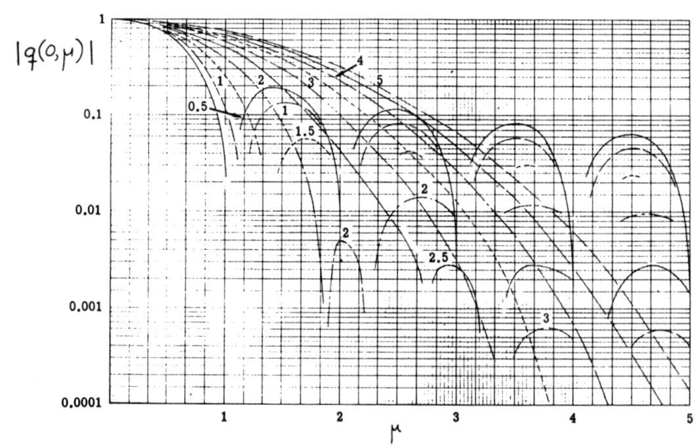

Fig. 7. The azimuth cross section of the ambiguity function for matched processing.

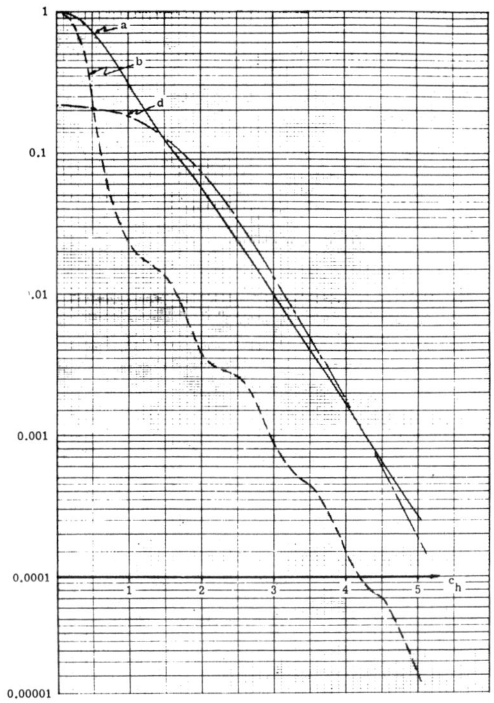

Fig. 8 (*left*). Important characteristics of azimuth cross section $|q(0, \mu)|$ for phase-matched processing. Curve (a): relative height of the first ambiguous peak, range offset. Curve (b): relative height of the first ambiguous peak, azimuth offset. Curve (d): relative height of the largest sidelobe.

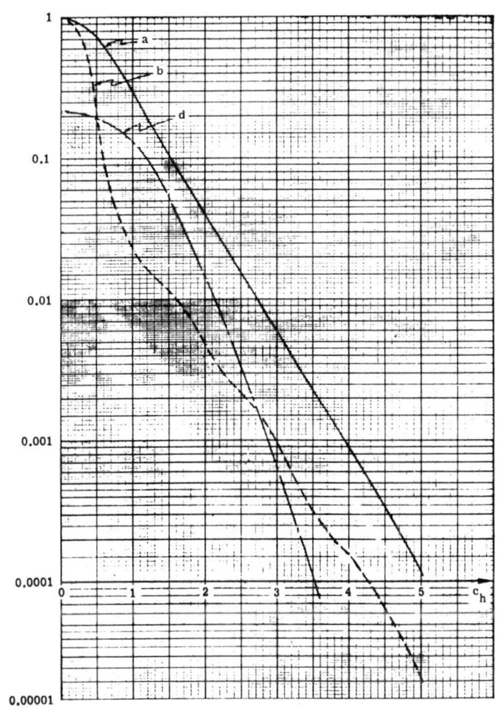

Fig. 9 (*below left*). Important characteristics of azimuth cross section $|q(0, \mu)|$ for matched processing. Curve (a): relative height of the first ambiguous peak, range offset. Curve (b): relative height of the first ambiguous peak, azimuth offset. Curve (d): relative height of the largest sidelobe.

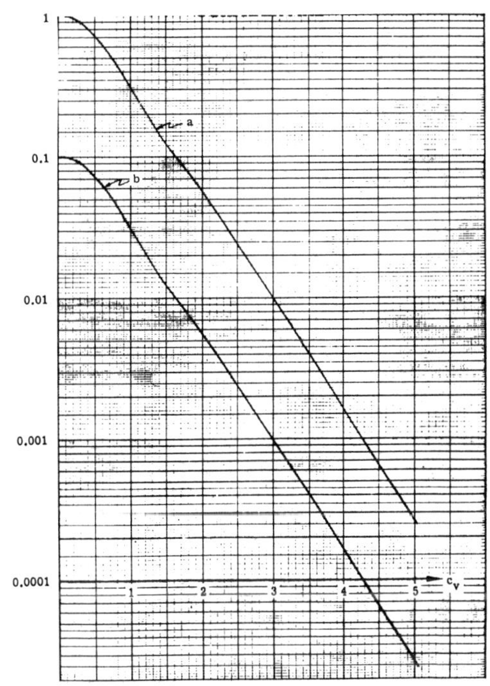

Fig. 10 (*below right*). Important characteristics of range cross-section $|q(\tau, 0)|$. Relative average height of largest ambiguous peak for inherent suppression of 0 dB [curve (a)] and 10 dB [curve (b)].

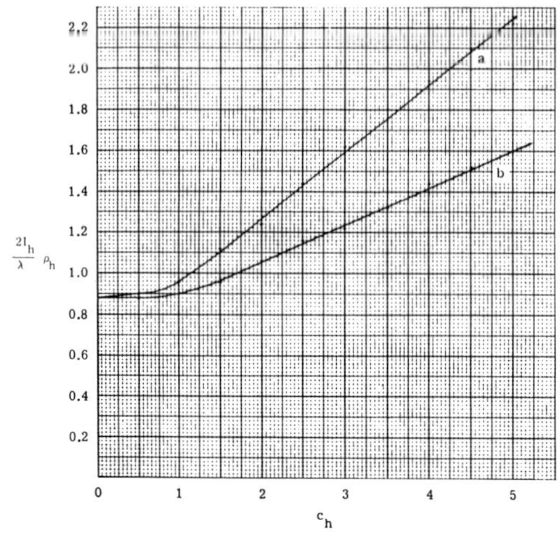

Fig. 11. Azimuth resolution ρ_h. Defined as the half-power width of main response of $|q(0, \mu)|$: (a) Matched processing; (b) Phase-matched processing.

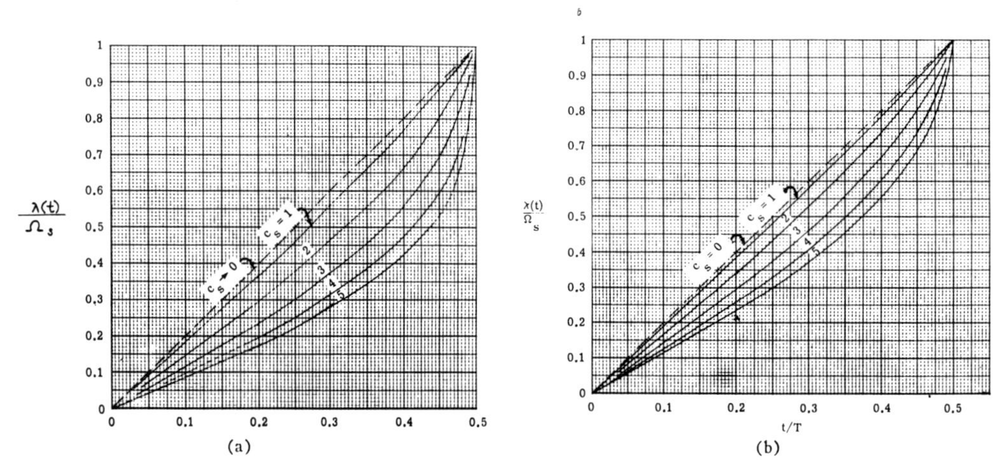

Fig. 12. The optimum signal frequency modulation. (a) Phase matched processing. (b) Matched processing. $\lambda(-t)s - \lambda(t)$.

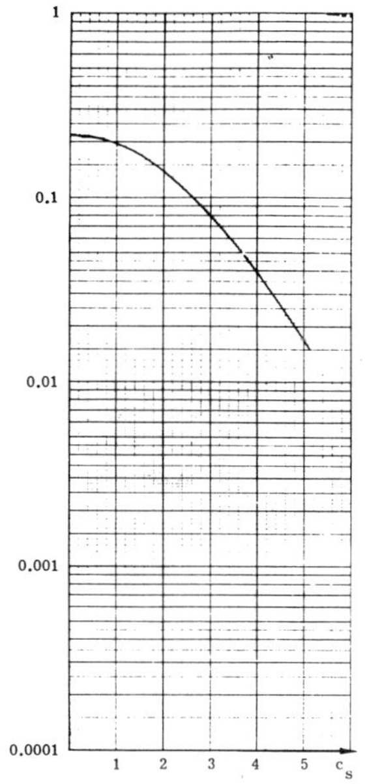

Fig. 13. Relative height of largest sidelobe of range cross section $|q(\tau, 0)|$.

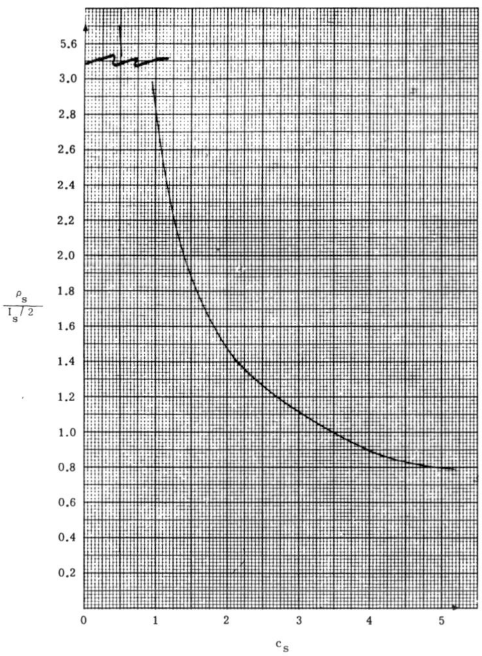

Fig. 14. Range resolution ρ_s. Defined as the half-power width of main response of $|q(\tau, 0)|$.

where

$$c_s \equiv \frac{\Omega_s I_s}{2} \cdot \tag{14}$$

The magnitude of the spectrum of the complex modulation is, therefore, specified. For various practical reasons we would also like to specify the magnitude of the complex modulation, $F_m \equiv |F|$. The problem of finding the phase modulation implied by these choices has been solved for signals of large time-bandwidth product [7], [8]—a condition we assume to be satisfied here. In the notation of [7] the frequency modulation (derivative of phase modulation) is given by

$$\lambda(c_s; t) = \tilde{G}^{-1}[2\pi G(t)]$$

where

$$G(t) \equiv \int^{t} F_m^2(s)ds, \qquad G(\infty) = 1,$$

and $\tilde{G}^{-1}$ is the inverse of the function

$$\tilde{G}(x) \equiv \int^{x} \tilde{F}_m^2(y)dy.$$

For example we take what is the most practical choice of time envelope:

$$F_m(t) = \frac{1}{\sqrt{T}} \operatorname{rect}\left(\frac{t}{T}\right).$$

The signal duration T is determined then by detectability (SNR) requirements, usually with a peak power restraint.

The frequency modulations for these envelopes are shown in Fig. 12 for phase matched and matched filtering. Such signals can be generated by using a filter of this impulse response; also, active generation is straightforward. One could, of course, generate actively or passively a linear FM signal (for example, since the technique is well established) and then add the above specified weighting in the processing. This is a nonoptimum procedure but for practical applications would probably be satisfactory.

The resolution and sidelobe levels are plotted in Figs. 13 and 14. Again an arbitrary choice of 3dB resolution was made and one may substitute his own. The usual trade-off between resolution and sidelobe level is evident.

THE EFFECTS OF NONLINEARITIES ON THE CHOICE OF THE AMBIGUITY FUNCTION

The essential effects of the nonlinearities encountered in normal operation of the S/L radar can be seen by considering the model shown in Fig. 15. It is standard operating procedure to rather severely limit the data $x(t)$; the limiting is usually done at RF (IF) with a "no-memory" nonlinearity. The noise can include any thermal sources, etc., present but more importantly it includes the returns from all other targets that overlap $x(t)$ in time: such an ensemble approaches a Gaussian process with increasing numbers of targets and dispersion ratios.

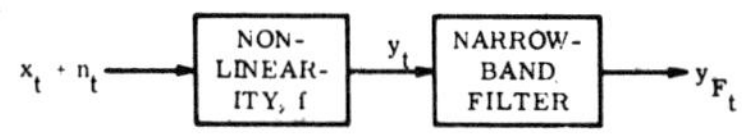

Fig. 15. Model of nonlinearities.

We wish to show that the azimuth modulation is, with large probability, preserved. Thus, suppose the extreme case of "infinite" limiting is considered by letting

$$f(x) = \operatorname{sgn} x.$$

In general

$$x(t) = A(t) \cos [\omega_0 t + \alpha(t) + \theta],$$

where θ is the azimuth phase modulation. A includes antenna weighting, in particular, the azimuth pattern weighting.

Using the characteristic function method we write

$$y(t) = \frac{i}{\pi} \int \frac{du}{u} \exp \{iuA(t) \cos [\omega_0 t + \alpha(t) + \theta] + n(t)\}$$

where the generalized Fourier transform of f is $2/iu$. The Jacbobi-Anger formula [9] is used to write

$$\exp \{iuA(t) \cos [\omega_0 t + \alpha(t) + \theta]\}$$

$$= \sum_{m=0}^{\infty} \epsilon_m i^m J_m[A(t)u] \cos [m\omega_0 t + m\alpha(t) + m\theta]$$

and observe that for the $A(t)$'s of the type we are interested in here, the new amplitude modulation $J_m[A(t)u]$ does not significantly change the bandwidth of the mth term. Then the narrow-band filter gives at its output a random process whose mean (the signal) is

$$E\{y_F(t)\} = \frac{i}{\pi} \int \frac{du}{u} \phi_n(u) J_1(A_t u) \cdot \cos [\omega_0 t + \alpha(t) + \theta]$$

where

$$\phi_n(u) = e^{-\sigma^2/2 u^2}$$

is the characteristic function of the noise whose variance is σ^2. This integral is readily evaluated by the formula of Weber and Sonine [9] to give

$$E\{y_F(t)\} = \frac{\sqrt{\pi}}{4\sigma} {}_1F_1\left(\frac{1}{2}; 2; -\frac{A_t^2}{4\sigma^2}\right)$$

$$\cdot A_t \cos [\omega_0 t + \alpha(t) + \theta],$$

${}_1F_1$ being the confluent hypergeometric function.

The special case when $A_t^2/4\sigma^2 \ll 1$ is of primary interest here; the case $A_t^2/4\sigma^2 \gg 1$ is worth observing. Using asymptotic forms of the confluent hypergeometric function we find

$$E\{y_F(t)\} \approx \begin{cases} \frac{\sqrt{\pi}}{4\sigma} A(t) \cos [\omega_0 t + \alpha(t) + \theta], & A_t^2/4\sigma^2 \ll 1, \\ \cos [\omega_0 t + \alpha(t) + \theta], & A_t^2/4\sigma^2 \gg 1. \end{cases}$$

Because any particular return is received imbedded in thousands of other returns being simultaneously received, it is highly probable that the case $A_t^2/4\sigma^2 \ll 1$ prevails.

Observe that the following conclusions can be based on this model; severe limiting of the radar sampling pulses followed by narrow-band filtering

1° Tends not to effect the azimuth phase history in either phase or amplitude.

2° Tends not to effect the range return in either phase or amplitude.

Curiously, had we decided to follow a single return through the system incorporating the nonlinearities the case $A_t^2/4\sigma^2 \gg 1$ would prevail and tell us that the envelope modulation would tend to be stripped off. This would lead one to the erroneous conclusion that rectangular envelopes would always be presented to the processor.

EXAMPLE

An example is discussed to illustrate a typical application of our results. Suppose we wish to design an S/L radar to be on board a vehicle with the mapped range interval starting at R_* nautical miles with a velocity of ν_* feet per second. The radar is to provide ρ_* foot resolution in both slant range and azimuth; all spurious responses of the ambiguity function $|q(\tau, \mu)|^2$ are to be $-\xi_*$ dB relative to the main response of the ambiguity function $|q(0, 0)|^2$. The RF wavelength is λ_* and the depression angle γ_*. Azimuth offset operation and matched filter processing are assumed.

First, from Fig. 9, curves (b) and (d) we find the minimum allowable c_{h*}. The horizontal aperture illumination function is given in Fig. 4. From Fig. 11 we find $2I_h\rho_*/\lambda_*$, which determines $I_h = I_{h*}$. Using (10) we find the horizontal aperture length $2\Omega_{h*}$.

Second, the sampling frequency f_{s*} is given by (6): $f_{s*}=4v_* I_{h*}/\lambda_*$.

Third, from Fig. 10 curve (b), say, we find the minimum allowable $c_v=c_{v*}$. The vertical aperture illumination function is given in Fig. 4. Then I_{v*} is found using (7). Using (10) we find the vertical aperture length $2\Omega_{v*}$.

Fourth, from Fig. 13 we find the minimum allowable $c_s=c_{s*}$. The frequency modulation, assuming a rectangular signal envelope, is found from Fig. 12. From Fig. 14 we find $2\rho_{s*}/I_s$, which determines $I_h=I_{h*}$. Using (14) we find the RF bandwidth $2\Omega_{s*}$.

Appendix A

Solution and Discussion of an Alternate Criterion

In the remarks summarizing the assumptions, it was noted that one might be led to the criterion: *Maximize the energy of the one-way voltage pattern in the unambiguous intervals relative to the energy in the certain ambiguous intervals.* To simplify the mathematical problem and hint at the general result, one such criterion is discussed: *Maximize the energy of the one-way voltage pattern in the unambiguous interval relative to the energy in the first ambiguous intervals.*

In a manner quite analogous to that used to solve the problem posed by the other criterion, we are led to the formulation:

$$\max_{E\in\mathcal{L}^2} \gamma$$

where

$$\gamma \equiv \frac{\displaystyle\int_{-I_h/2}^{I_h/2} \{A_h(u) - [A_h(u+I_h) + A_h(u-I_h)]\}du}{\displaystyle\int_{-\infty}^{\infty} A_h(u)du}.$$

The integral equation that results is

$$\mu_n f_n = Bf_n,$$

where

$$Bf_n(t) \equiv \int_{-1}^{1} \frac{\sin c(t-s)}{\pi(t-s)}[1-\cos 2c(t-s)]f_n(s)ds.$$

The largest positive characteristic value γ_0, the maximum of ν, and its associated characteristic function were found by numerical computation on a digital computer. The resulting aperture functions are shown in Fig. 16, along with the solutions to the other problem.

It is observed that over the range of c commonly used, the solutions are remarkably similar to the solutions of the other maximization problem. When c is small, it is quite clear that the solutions give patterns that have a null in the ambiguous interval.

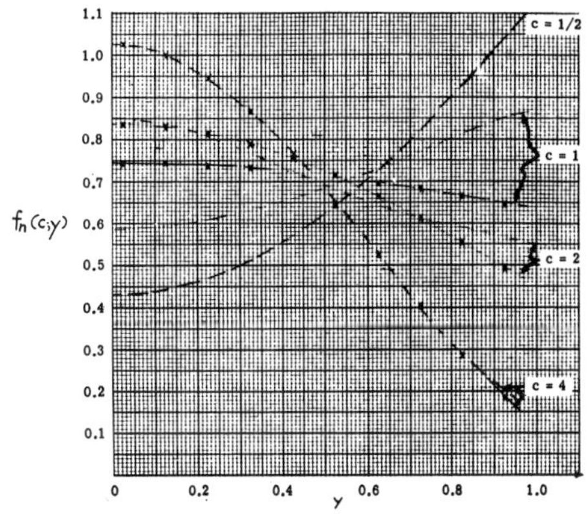

Fig. 16. Optimum aperture functions under an alternate criterion.

In the more general problem that takes into account the energy in the even ("compressible") ambiguous intervals, partial results indicate that the solutions are very close to the functions $\{S_{00}(c;\cdot)\}$.

Appendix B

An Approximation and Lower Bound on the Sampling Frequency

If there were no inherent suppression of range ambiguities the sampling frequency could be found very easily. This will be discussed because the result will then be useful as an approximation and a lower bound. From Fig. 5 we see that, with no range suppression, curve (a) essentially determines both c_v and c_h and their equality for a specified level of spurious responses. Note that this is equivalent to making the energies in the unambiguous intervals under the antenna patterns equal. From the equation

$$\hat{c} \equiv c_v = c_h,$$

or

$$I_v\Omega_v = I_h\Omega_h,$$

and using (12) and (13) we find the sampling frequency, range offset,

$$\hat{f}_s = \sqrt{\frac{\nu v \tan\theta}{R_s\lambda}\frac{D_v}{D_h}} \text{ cps},$$

and, in turn

$$\hat{c} = \frac{\pi}{2}\sqrt{\frac{\nu\tan\theta}{\lambda R_s v}D_v D_h}.$$

If azimuth offset is used, the first formula should be doubled and the second halved.

Acknowledgment

The author acknowledges the assistance of T. R. Crimmins and J. O. Barney in performing the numerical work.

References

[1] L. J. Cutrona, W. E. Vivian, E. N. Leith, and G. O. Hall, "A high-resolution combat-surveillance system," *IRE Trans. on Military Electronics*, vol. MIL-5, pp. 127–131, April 1961.

[2] E. J. Kelly, I. S. Reed, and W. L. Root, "The detection of radar echoes in noise," I and II, *J. Soc. Ind. Appl. Math.*, vol. 8, pp. 309–341, June 1960, and vol. 8, pp. 481–507, September 1960.

[3] L. J. Cutrona, E. N. Leith, C. J. Palermo, and L. J. Porcello, "Optical data processing and filtering systems," *IRE Trans. on Information Theory*, vol. IT-6, pp. 386–400, June 1960.

[4] F. Reisz and B. Sz.-Nagy, *Functional Analysis*, trans. from second French ed. by L. F. Boron, New York: Ungar, 1955.

[5] D. Slepian and H. O. Pollock, "Prolate spheroidal wave functions, Fourier analysis and uncertainty—I," *Bell Sys. Tech. J.*, vol. 40, pp. 43–63, January 1961.

[6] R. O. Harger and T. R. Crimmins, "A tabulation of prolate spheroidal angle functions of the first kind and their associated characteristic values," Rept. 4563-65-T, Inst. of Sci. and Tech., University of Michigan, Ann Arbor, March 1964.

[7] R. O. Harger, "A note on the realization of ambiguity functions," *IRE Trans. on Space Electronics and Telemetry*, vol. SET-9, pp. 127–130, December 1963.

[8] E. N. Fowle, "A method of designing FM pulse compression signals," *IRE Trans. on Information Theory*, vol. IT-10, pp. 61–67, January 1964.

[9] W. Magnus and F. Oberhettinger, *Special Functions of Mathematical Physics*. New York: Chelsea, 1949.

Performance Of A Synthetic-Aperture Mapping Radar System

J.A. Develet, Jr.
(Aerospace Corporation, El Segundo, California)

IEEE Transactions on Aerospace and Navigational Electronics, Vol. ANE-11, No. 3
(September 1964)
Pages 173-179

Summary—An ideal synthetic-aperture mapping radar system is postulated. The response of this system is then computed for 1) a point target, 2) locally homogeneous terrain, and 3) terrain of arbitrary reflectivity.

Relationships for achievable resolution, swath width, and final map SNR vs the independent parameters of vehicle altitude, terrain scattering coefficient, carrier wavelength, vehicle velocity, and average radiated power are developed.

The ideal system considered in this paper can be approximated in practice, for small real-antenna beamwidths, by using large time-bandwidth chirp waveforms and sin x/x illumination of the real-antenna aperture.

I. Introduction

PREVIOUS work [1,2] in the area of synthetic-aperture radar mapping systems is extended to an analysis of an ideal system. An ideal system is defined herein as one with square real-antenna beams, flat bandlimited passbands, and minimum radar pulse repetition frequency (PRF) to obtain distortion-free maps. The ideal system has no range or azimuth ambiguities, and it allows a precise definition of resolution.

This system demonstrates in a tractable fashion the interrelation among the various parameters without resort to extensive machine computation.

II. Analysis

A. Ideal System Configuration

Consider the geometry of Fig. 1(a), (b), and (c); the vehicle traveling with velocity v, ft/sec, illuminates a ground swath L×W ft². It is assumed that all the energy radiated by the vehicle is contained within this area and that the antenna pattern is tailored such that under conditions of homogeneous terrain every elemental area within L×W returns the same energy to the radar receiver.

The synthetic-aperture mapping radar system (Fig. 2) operates as follows. The transmitter impulses the terrain every 1/PRF sec. Reflections from the terrain within the illuminated area L×W (which contain very fine range resolution information) are observed by the radar receiver through a flat bandlimited aperture, B cps wide, centered at the carrier frequency ω_0. A bandwidth B cps allows an independent amplitude and phase measurement to exist every $1/B$ sec. One may write [Fig. 1(b)]

$$B = \frac{c}{2x_0 \cos \psi}, \qquad \text{cps} \tag{1}$$

where

B = bandwidth of the radar receiver, cps
c = velocity of light, 9.84×10^8, ft/sec
x_0 = distance referred to ground range coordinates of allowable independent reflectivity samples, ft
ψ = radar depression angle [Fig. 1(b)].

The output of the radar receiver is then synchronously detected by an oscillator at ω_0 rps. Next, the in-phase and quadrature outputs from the detector are stored, each PRF line registered side by side, on a suitable storage medium. These stored signals are finally read out in azimuth and processed by a matched filter (Fig. 2) to give fine azimuth resolution essential to a final high resolution map.

B. Geometrical Relationships

The path of an elemental area $\Delta x\Delta y$ on the ground is examined as an observer on the vehicle would see it move. The azimuth geometry of concern is shown in Fig. 1(c). For large values of range R, the change in range with respect to time may be written

$$R(t) = R + \frac{v^2t^2}{2R}, \qquad \text{ft} \tag{2}$$

where

R = perpendicular range to target from vehicle path, ft
v = vehicle velocity, ft/sec
t = time with $t=0$ referring to the closest approach of the elemental area, sec.

The highest Doppler frequency may be derived from (2) while noting the maximum value for $t=\pm L/2v$, sec

$$f_d = \frac{2\dot{R}(t)}{\lambda}, \qquad \text{cps}$$

$$= \frac{Lv}{\lambda R}, \qquad \text{cps} \tag{3}$$

Manuscript received January 21, 1964.

[1] L. J. Cutrona, *et al.*, "A high-resolution radar combat-surveillance system," IRE Trans. on Military Electronics, vol. MIL-5, pp. 127–131; April, 1961.

[2] IRE Trans. on Military Electronics, vol. MIL-6; April, 1962.

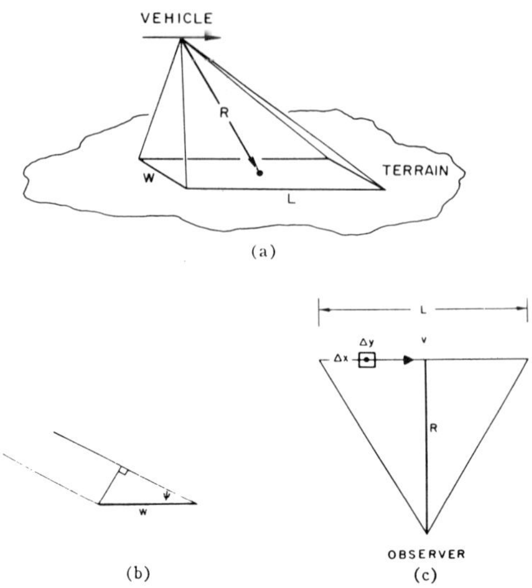

Fig. 1—Mapping system geometry. (a) Terrain illumination. (b) Depression angle geometry. (c) Azimuth geometry.

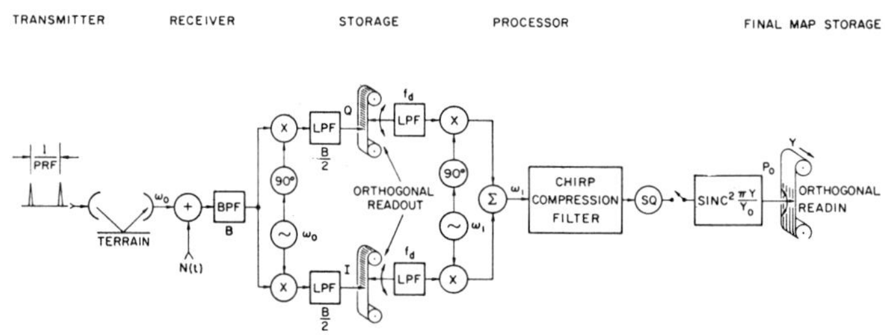

Fig. 2—Synthetic-aperture mapping radar system.

where

f_d = highest Doppler frequency, cps
$\lambda = c/f_0$, carrier wavelength, ft.

The sampling theorem (see Fano[3]) states that in order to properly reproduce the Doppler history from the elemental area $\Delta x \Delta y$ the radar must have a PRF at least equal to

$$\mathrm{PRF} = 2f_d = \frac{2Lv}{\lambda R}, \quad \text{pps} \tag{4}$$

Thus, with the radar PRF set at the value stipulated by (4), the PRF lines registered on the storage medium contain all the information necessary to recreate the Doppler histories of every elemental area within the illuminated area L×W.

If one refers to Fig. 1(b), it is clear that the PRF given by (4) cannot permit, without range-return overlap, a swath width W in excess of

$$W = \frac{c}{2(\mathrm{PRF}) \cos \psi}, \quad \text{ft} \tag{5}$$

where

W = maximum swath width, ft
PRF = minimum PRF given by (4), pps.

C. Radar Range Equation

The basic simple geometric formulations which will be needed subsequently in system response determination are presented in Section II, B. An important rela-

[3] R. M. Fano, "Transmission of Information," John Wiley and Sons, Inc., New York, N. Y., pp. 154–155; 1961.

tion which connects the required average transmitter power to final map SNR is derived in this section.

Consider the geometry of Fig. 1(a). One may apply the well-known radar equation to an elemental area on the ground.[4] Thus[5]

$$P_r = \frac{P_t G^2 \sigma \lambda^2}{(4\pi)^3 R^4}, \quad \text{w} \tag{6}$$

where

P_r = ensemble average peak received power in a band B cps, w
P_t = peak transmitter power radiated in a band B cps, w
G = real antenna, one-way power gain
σ = ensemble average radar cross section of the elemental area, $\Delta x \Delta y$, ft².

In Cosgriff, *et al.*,[6] the ensemble average radar cross section for terrain targets is given by

$$\sigma = \gamma(x, y)\Delta x \Delta y \sin \psi, \quad \text{ft}^2 \tag{7}$$

where

$\Delta x \Delta y$ = elemental area, ft²
$\gamma(x, y)$ = ensemble average scattering coefficient at the ground coordinates, x, y.

The definition of γ given[6] is extended to include point targets; *i.e.*, for a point target at x, y, γ goes to infinity such that

$$\sigma_p = \lim_{\Delta x \Delta y \to 0} \gamma_p(x, y)\Delta x \Delta y \sin \psi \tag{8}$$

where

σ_p = radar cross section of the point target, ft²
$\gamma_p(x, y)$ = scattering coefficient of the point target at x, y.

Assuming perfectly square beams shaped properly to yield equal ensemble average energy from all scatterers within the beam of the same scattering coefficient γ, the gain G in (6) is obtained as

$$G = \frac{4\pi R^2}{L \times W \sin \psi}. \tag{9}$$

Eq. (9) is the result of dividing 4π steradians by the beamwidth in steradians.

Since waveforms are considered here to have flat bandlimited spectra (bandlimited impulses), it can be found that by Fourier transformation and integration

$$\mathcal{E}_t = P_t\left(\frac{1}{B}\right), \quad \text{joules} \tag{10}$$

where

$\mathcal{E}_t$ = energy transmitted per pulse, joules
P_t = peak radiated transmitter power, w

The substitution of (10), (9), and (7) into (6) yields

$$P_r = \frac{\mathcal{E}_t B \gamma(x, y)\Delta x \Delta y \lambda^2}{4\pi (LW)^2 \sin \psi}, \quad \text{w.} \tag{11}$$

The received power does not depend on the range to the elemental area. This is a direct result of efficient real-beam design in which only the area to be mapped is illuminated.

D. Point-Target Response

The ensemble average peak received power P_r to be expected at the output of the radar receiver bandlimitation B (Fig. 2) to a target at ground coordinates x, y localized to $\Delta x \Delta y$ in the mapped area is derived in Section II, C. For all practical purposes, this return may be considered to result from a true point reflector at x, y or an elemental portion of terrain with scattering coefficient, $\gamma(x, y)$. Regardless of the origin of P_r, consider the following received signal voltage due to reflection from the elemental area, $\Delta x \Delta y$, at x, y.

$$v(t) = \sqrt{2P_r}\cos(\omega_0 t + \beta t^2 + \alpha) \cdot \sum_k \frac{\sin \pi B(t - t_k)}{\pi B(t - t_k)}, \quad \text{v} \tag{12}$$

where

β = phase constant, rad/sec²
α = arbitrary phase constant, rad
t_k = time of arrival of kth pulse, sec ($t = 0$ has been referred to the array center).

It can be shown from (2) that

$$\beta = \frac{-2\pi v^2}{\lambda R}, \quad \text{rad/sec}^2. \tag{13}$$

Since registration of the PRF lines on the storage medium (Fig. 2) is carefully controlled to place targets at the same range side by side,[7] orthogonal azimuth readout of the synchronously detected voltage of (12) yields for the in-phase and quadrature channels

$$I = \sqrt{2P_r}\cos\left[\frac{2\pi(y + \eta_y)^2}{\lambda R} - \alpha\right]\frac{\sin[\pi(\eta_x/x_0)]}{\pi(\eta_x/x_0)}, \tag{14}$$

$$Q = \sqrt{2P_r}\sin\left[\frac{2\pi(y + \eta_y)^2}{\lambda R} - \alpha\right]\frac{\sin[\pi(\eta_x/x_0)]}{\pi(\eta_x/x_0)} \tag{15}$$

[4] M. I. Skolnik, "Introduction to Radar Systems," McGraw-Hill Book Co., Inc., New York, N. Y., ch. 2; 1962.
[5] It is assumed here that the elemental area is much smaller than the resolution of the system.
[6] R. L. Cosgriff, *et al.*, "Terrain Scattering Properties for Sensor System Design," Engineering Experiment Station, The Ohio State University, Columbus, Ohio, Bull. No. 181, p. 4; May, 1960.

[7] Again, it is assumed $L/R \ll 1$.

where

η_x = range offset of elemental area from orthogonal azimuth scan, ft

η_y = azimuth offset of elemental area trajectory from orthogonal azimuth scan, ft

$y = vt$, ft

$x_0 = c/2B \cos\psi$, ft.

In (14) and (15), distance variables along the ground map are substituted instead of time variables. The in-phase and quadrature readouts are then combined (Fig. 2) to yield a single linear FM Doppler history at ω_1. Thus

$$v = \sqrt{2P_r}\cos\left[\omega_1\left(\frac{y}{v}\right) + \frac{2\pi(y+\eta_y)^2}{\lambda R} - \alpha\right] \cdot \frac{\sin[\pi(\eta_x/x_0)]}{\pi(\eta_x/x_0)}, \quad \text{v.} \tag{16}$$

Upon readout, the low-pass filters perform the perfect sinc interpolation necessary to recreate the band-limited waveform of (16).[8]

The voltage of (16) is now compressed in a matched filter. The operation of matched filtering can be shown to be multiplication by a reference waveform and averaging the product over the time interval L/v.[9] The reference waveform is

$$R = 2\exp\left\{j\left[\omega_1\left(\frac{y}{v}\right) + \frac{2\pi y^2}{\lambda R} + \phi\right]\right\}, \quad v. \tag{17}$$

After multiplication, the averaging of $v \times R$ from $-L/2$ to $+L/2$ yields

$$\underline{v}(\eta_x, \eta_y) = \sqrt{2P_r}\,\frac{\sin[\pi(\eta_x/x_0)]}{\pi(\eta_x/x_0)} \cdot\left[\frac{1}{L}\int_{-L/2}^{L/2}\cos\left(\frac{4\pi\eta_y y}{\lambda R} + \mu\right)dy - j\frac{1}{L}\int_{-L/2}^{L/2}\sin\left(\frac{4\pi\eta_y y}{\lambda R} + \mu\right)dy\right]. \tag{18}$$

Eq. (18) represents the in-phase and quadrature outputs from the compression filter (Fig. 2). The square of the envelope given by (18) followed by ensemble averaging yields the power output of the system:

$$P_0 = 2P_r\frac{\sin^2[\pi(\eta_x/x_0)]}{[\pi(\eta_x/x_0)]^2}\cdot\frac{\sin^2[\pi(\eta_y/y_0)]}{[\pi(\eta_y/y_0)]^2}, \quad \text{v}^2 \tag{19}$$

where $y_0 = \lambda R/2L$, ft.

It is usual for mapping radar systems to have equal azimuth and range resolution, thus

$$x_0 = y_0 = \frac{\lambda R}{2L}, \quad \text{ft.} \tag{20}$$

Eq. (19) represents the point target power response of an ideal synthetic aperture radar system. Note the $(\text{sinc}\times\text{sinc})^2$ dropoff of power as the target moves away from the range and azimuth under surveillance. This is expected, because all waveforms have been assumed bandlimited.

E. Local Homogeneous Terrain Response

The development of the ensemble average system power response for locally homogeneous terrain follows exactly the preceding development, except the power output becomes the summation of contributions from all the elemental areas. Thus utilizing (18), the system output voltage becomes

$$v = \sum_{\eta_x}\sum_{\eta_y}\underline{v}(\eta_x, \eta_y), \quad \text{v.} \tag{21}$$

The vector given by (21) is Rayleigh distributed. From (21), the system ensemble average power is

$$P_0 = \sum_{\eta_x}\sum_{\eta_y}\overline{v(\eta_x, \eta_y)\times v^*(\eta_x, \eta_y)}, \quad \text{v}^2. \tag{22}$$

Derivation of (22) makes use of the fact that only the direct products of the components in (21) are correlated.[10] One obtains from (19) and (22)

$$P_0 = \sum_{\eta_x}\sum_{\eta_y}2P_r\frac{\sin^2[\pi(\eta_x/x_0)]}{[\pi(\eta_x/x_0)]^2}\cdot\frac{\sin^2[\pi(\eta_y/y_0)]}{[\pi(\eta_y/y_0)]^2}. \tag{23}$$

If one substitutes (11) into (23), letting $x = \eta_x$ and $y = \eta_y$, and passes to the limit as $\Delta\eta_x\Delta\eta_y \to 0$,

$$P_0 = \frac{\mathcal{E}_t B\lambda^2}{2\pi(\text{LW})^2\sin\psi}\int_{-\infty}^{\infty}\int_{-\infty}^{\infty}\gamma(\eta_x, \eta_y)\frac{\sin^2[\pi(\eta_x/x_0)]}{[\pi(\eta_x/x_0)]^2}\cdot\frac{\sin^2[\pi(\eta_y/y_0)]}{[\pi(\eta_y/y_0)]^2}d\eta_x d\eta_y. \tag{24}$$

The infinite limits are substituted, since it is usual that $x_0 \ll L$ and $y_0 \ll W$.[11] For locally homogeneous terrain, (24) reduces to

$$P_0 = \frac{\mathcal{E}_t B\gamma\lambda^2}{2\pi(\text{LW})^2\sin\psi}\cdot\int_{-\infty}^{\infty}\int_{-\infty}^{\infty}\frac{\sin^2[\pi(\eta_x/x_0)]\sin^2[\pi(\eta_y/y_0)]}{[\pi(\eta_x/x_0)]^2[\pi(\eta_y/y_0)]^2}d\eta_x d\eta_y = \frac{\mathcal{E}_t B\gamma x_0 y_0\lambda^2}{2\pi(\text{LW})^2\sin\psi}, \quad \text{v}^2. \tag{25}$$

[8] It is assumed here that the linear FM Doppler history has large time-bandwidth product.

[9] C. W. Helstrom, "Statistical Theory of Signal Detection," Pergamon Press, Inc., New York, N. Y., p. 91; 1960.

[10] The phase μ in (18) is independent from elemental area to elemental area and is equally distributed from 0 to 2π.

[11] Edge effects are neglected here.

For the usual situation of $x_0 = y_0$,

$$P_0 = \frac{\mathcal{E}_t B \gamma x_0^2 \lambda^2}{2\pi (\mathrm{LW})^2 \sin \psi}, \quad \mathrm{v}^2. \tag{26}$$

Eq. (26) represents the system power output due to locally homogeneous terrain of constant scattering coefficient γ; P_0 is the ensemble average power of the Rayleigh distributed voltage vector at the matched filter output (Fig. 2).

F. Response to Terrain of Arbitrary Reflectivity

For terrain of arbitrary reflectivity, $\gamma(\eta_x, \eta_y)$ cannot be removed from the integral in (24). In this situation, for an arbitrary scan located at x, y, (24) becomes

$$P_0(x, y) = \frac{\mathcal{E}_t B \lambda^2}{2\pi (\mathrm{LW})^2 \sin \psi} \int_{-\infty}^{\infty} \int_{-\infty}^{\infty} \gamma(x + \eta_x, y + \eta_y) \cdot \frac{\sin^2 [\pi(\eta_x / x_0)]}{[\pi(\eta_x / x_0)]^2} \cdot \frac{\sin^2 [\pi(\eta_y / y_0)]}{[\pi(\eta_y / y_0)]^2} d\eta_x d\eta_y. \tag{27}$$

It is shown in (27) that the ensemble average power output at any point x, y in the map is a weighted sum of all the incremental components.

Consider the case of an abrupt change in reflectivity from $\gamma = 0$ to $\gamma = \gamma$ along the line $y = 0$. The system ensemble average power output, $P_0(x, y)$, is now computed. Since γ is independent of η_x, one may first integrate (27) with respect to η_x then with respect to η_y using the appropriate value for γ. Thus

$$P_0(y) = \frac{\mathcal{E}_t B x_0 y_0 \gamma \lambda^2}{2\pi (\mathrm{LW})^2 \sin \psi} \left(\frac{1}{\pi} \int_{-\infty}^{\pi y / y_0} \frac{\sin^2 z}{z^2} dz \right), \quad \mathrm{v}^2. \tag{28}$$

The system response at the edge of a sudden reflectivity change is shown in Fig. 3(a).

Fig. 3(b) presents the system response at the edge of a sudden reflectivity change for the situation of a system point-target response which is Gaussian instead of sinc. Since a system with a Gaussian point-target response is not the subject of this report, this curve is presented only as a matter of interest.

In a similar manner, (27) may be used to find the ensemble average power response to an arbitrary scattering coefficient function $\gamma(x, y)$.

It must be remembered that in all the foregoing, P_0 is the average power of a Rayleigh distributed vector that is allowed to have an independent value every x_0 ft in range and y_0 ft in azimuth.

G. Noise Performance

Independent additive noise at the input to the system can be treated separately from the signal. The noise power at the coordinates x, y competing with the signal is determined in this section.

Assume that the additive noise at the radar receiver input has the following one-sided, power-spectral density referred to the antenna terminals prior to any system losses.

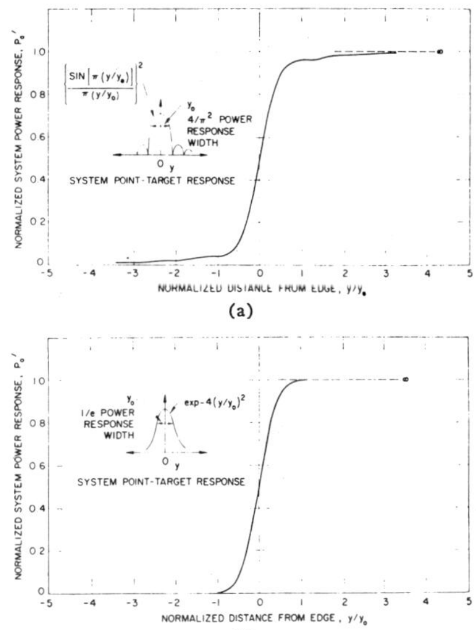

Fig. 3—System ensemble average power response to an edge. (a) Sinc point-target power response. (b) Gaussian point-target response.

$$\Phi = \mathrm{KT_{eq}}, \quad \mathrm{v}^2/\mathrm{cps} \tag{29}$$

where

K = Boltzmann's constant, 1.38×10^{-23} joules/°K

T_{eq} = equivalent receiver temperature referred to the antenna terminals, °K.

The mean-square value of the noise is ΦB and is not affected by synchronous demodulation; *i.e.*, both the in-phase and quadrature storage media contain noise voltage of mean-square ΦB. Upon readout, however, the highest frequency contained in the noise changes from $B/2$ to f_d. Thus, at the output of the orthogonal readout device, low-pass white noise is bandlimited to f_d, cps, with mean-square ΦB. The power-spectral density at the output of the low-pass filters becomes

$$\Phi' = \Phi \frac{B}{f_d}, \quad \mathrm{v}^2/\mathrm{cps}. \tag{30}$$

At the output of the translation device, the spectral density is halved by translation to the intermediate frequency, ω_1 (Fig. 2). Thus, the new spectral density becomes

$$\Phi'' = \Phi \frac{B}{2 f_d}, \quad \mathrm{v}^2/\mathrm{cps}. \tag{31}$$

A matched filter, such as that in Fig. 2, has a noise bandwidth equal to the reciprocal of the integration time, L/v, sec (see Develet,[12]). Thus, the matched filter noise output power becomes

$$N_0 = \frac{\Phi B v}{2 L f_d}, \quad \text{v}^2. \tag{32}$$

There are two situations of importance that must be treated: first, the point target SNR and second, the terrain SNR. Because of the envelope detector, a nonlinear element, superposition does not hold beyond the matched filter output (Fig. 2).

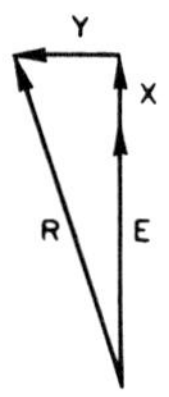

Fig. 4—Point-target matched filter output.

1) *Point-Target Signal-to-Noise Ratio:* At high SNR, the vector diagram of Fig. 4 represents the voltage at the output of the matched filter for the response to a point target. The ensemble average output power of the envelope detector at high SNR is

$$\begin{aligned}\overline{R^2} &= \overline{E^2} + \overline{X^2} + 2\overline{EX} \\ &= \overline{E^2} + \overline{X^2}, \quad \text{v}^2.\end{aligned} \tag{33}$$

Since $\overline{X^2} = N_0$, the SNR (power) is given by

$$\left(\frac{S}{N}\right)_{ff} = \frac{\overline{E^2}}{\overline{X^2}} = \frac{\overline{E^2}}{N_0}. \tag{34}$$

For peak SNR, consider the target maximum signal output to noise ratio ($\eta_x = \eta_y = 0$). The substitution of (19) and (32) into (34) yields

$$\left(\frac{S}{N}\right)_{ff} = \frac{2 P_r 2\, L f_d}{\Phi B v}. \tag{35}$$

From (3) and (10),

$$\left(\frac{S}{N}\right)_{ff} = \frac{2\mathcal{E}_r}{\Phi}\left(\frac{L}{x_0}\right). \tag{36}$$

Noting from (4) and (20) that L/x_0 is equal to the number of pulses received from the target, one obtains

$$\left(\frac{S}{N}\right)_{ff} = \frac{2\mathcal{E}}{\Phi}. \tag{37}$$

Eq. (37) is a familiar result: at high SNR, the system SNR to a point target is twice the total received energy from the target on its pass through the real-antenna beam divided by the one-sided, radar-receiver, power-spectral density.

At arbitrary SNR, the voltage response to a point target has a modified Rayleigh probability distribution function (PDF) with $P^2 = 2\mathcal{E}$ and $\psi_0 = \Phi$. (see Rice[13]).

2) *Terrain Signal-to-Noise Ratio:* For homogeneous terrain, the ratio of (26) to twice (32) yields the desired SNR.[14]

$$\begin{aligned}\left(\frac{S}{N}\right)_{ff} &= \frac{\mathcal{E}_t B \gamma x_0^2 \lambda^2}{2\pi (LW)^2 \sin\psi}\left(\frac{L f_d}{\Phi B v}\right) \\ &= \frac{P_{av} \gamma \lambda x_0^2}{2\pi H (\text{PRF}) \Phi W^2}\end{aligned} \tag{38}$$

where

P_{av} = average radiated power, $\mathcal{E}_t$ PRF, v²
H = vehicle altitude, $R \sin\psi$ (Fig. 1), ft.

Rearranging (38), one may solve for the required average radiated power vs the independent system parameters

$$P_{av} = \frac{2\pi (\text{PRF}) K T_{eq} W^2 H}{\gamma x_0^2 \lambda}\left(\frac{S}{N}\right)_{ff}. \tag{39}$$

If one calculates the energy returned from a patch of ground x_0^2 in area, one finds that

$$\left(\frac{S}{N}\right)_{ff} = \frac{\mathcal{E}}{\Phi} \tag{40}$$

where $\mathcal{E}$ = total energy returned from a square x_0^2 in area on the ground in its pass through the real-antenna beam, joules.

Eq. (40) holds for all SNR and is 3 db poorer than the point target response at high SNR. Of course, since the PDF's are different, it is hard to relate performance to SNR directly.

The PDF for terrain return combined with the additive, white, Gaussian noise of the radar receiving system is Rayleigh with an independent value allowable every x_0 ft in range and y_0 ft in azimuth.

III. Conclusions and Formula Summary

Specific interrelationships for an ideal synthetic-aperture mapping radar system are placed on a quantitative basis. The most important relations along with symbol definitions are summarized in Table I.

[12] J. A. Develet, Jr., "A note on demodulation of PCM/PM signals with switching-type phase detectors," IRE Trans. on Space Electronics and Telemetry, vol. SET-8, p. 41; March, 1962.

[13] S. O. Rice, "Mathematical analysis of random noise," in "Selected Papers on Noise and Stochastic Processes," Nelson Wax, Ed., Dover Publications, Inc., New York, N. Y., pp. 238–241; 1954.

[14] The noise power output of an envelope detector is $\overline{X^2} + \overline{Y^2} = 2N_0$ or twice the input noise power N_0.

TABLE I

FORMULAS FOR THE IDEAL MINIMUM PRF HIGH-RESOLUTION RADAR SYSTEM*

Formula	Definitions
$PRF = \frac{v}{x_0}$	PRF = pulse repetition frequency, pps v = vehicle velocity, ft/sec x_0 = separation in range or azimuth, at the target, of allowable independent reflectivity samples on the final map (resolution), ft
$W = \frac{kc}{2\ PRF \cos\psi}$	W = swath width, ft k = radar receiver duty cycle, $0 \leq k \leq 1$ c = velocity of light, 9.84×10^8 ft/sec ψ = depression angle [Fig. 1(b)]
$B_v = kB_r$ $B_r = \frac{c}{2x_0 \cos\psi}$	B_r = radio frequency square bandwidth (Fig. 2), cps B_v = square baseband width necessary for real-time transmission of radar data, cps
$L = \frac{\lambda R}{2x_0}$	L = length of real beam at the target [Fig. 1(c)], ft λ = carrier wavelength, ft R = range to target [Fig. 1(c)], ft
$\left(\frac{S}{N}\right)_{ff} = \frac{\mathcal{E}}{\Phi}$ (locally homogeneous terrain) $\left(\frac{S}{N}\right)_{ff} = \frac{2\mathcal{E}}{\Phi}$ (point target high SNR)	$\left(\frac{S}{N}\right)_{ff}$ = ensemble average signal to noise power ratio on final film $\mathcal{E}$ = ensemble average total energy received in a bandwidth B cps from the resolution element x_0^2 in area on its pass through the real beam, L (referred to receiver antenna terminals), joules Φ = one-sided radar receiver spectral density (referred to receiver antenna terminals), w/cps
$P_{av} = \frac{2\pi(PRF)KT_{eq}W^2(S/N)_{ff}H}{\lambda\gamma x_0^2}$ (locally homogeneous terrain) $P_{av} = \frac{\pi(PRF)KT_{eq}W^2(S/N)_{ff}H \sin\psi}{\lambda\sigma}$ (point target high SNR)	P_{av} = average radiated power in the band B, w γ = normalized scattering coefficient[6] K = Boltzmann's constant, 1.38×10^{-23} joules/°K T_{eq} = equivalent radar receiver temperature referred to antenna terminals before any system loss, °K H = vehicle altitude, ft σ = radar cross section of point target in direction of radar, ft²

* For purposes of this table, range and azimuth resolution are assumed equal.

Synthetic Aperture Radar System Design For Random Field Classification

R.O. Harger
(University of Maryland, College Park)

IEEE Transactions on Aerospace and Electronic Systems, Vol. AES-9, No. 5 (September 1973)
Pages 732-740

Abstract

The optimum design of synthetic aperture radar (SAR) systems intended to classify randomly reflecting areas, such as agricultural fields, characterized by a reflectivity density spectral density is studied. Assuming areas of known shape and location, the binary case, and a certain Gaussian signal field property, and ignoring interfield interference, the problem solution is given. The optimum processor includes conventional matched filter processing, but is nonlinear; a coherent optical system realization is outlined. The performance is approximated using a χ^2 assumption and bounded by the Cernov bound. A fundamental design problem involves the system bandwidth analogously, in a special case, as in diversity communication systems; a solution is given based on the Cernov bound. A set of summary design curves is given and exemplified by a satellite SAR system design.

Also discussed is the measurement of reflectivity spectral density amplitude with imaging sidelooking (synthetic or "brute-force") radars and the maximum likelihood estimator's accuracy and realization with a coherent optical system. It is also shown that a CW modulation is useable if the random reflectivity is, effectively, isotropic. Finally, the reflectivity density spectral density amplitude, when constant over the spatial bandpass of the measuring system, is related to the scattering cross-section density commonly measured.

Introduction

The application of synthetic aperture radar (SAR) to remote sensing uses can raise interesting and basic system design questions. Such an application is the classification of random fields, e.g., agricultural fields. Heretofore, SAR systems have generally beeen designed to produce, with desired fidelity, images of selected terrain areas. Minimum mean-square error and resolution (system impulse response design) criteria are nearly universally used, and the resulting optimum processor is the matched filter and related variants [1, ch. 4]. System bandwidths (and, hence, the along-track dimension of the physical antenna aperture) are fixed usually by a resolution specification, and the average power by a signal-to-noise specification for either point scatters of specified scattering cross section or terrian type with specified scattering cross-section density.

Opposedly, a reasonable criterion for a classification role would be to minimize probability of classification error $P(\mathcal{E})$: For fields described by random reflectivities, it is clear from known results in communication theory that the optimum processor will not be a matched filter (though it may include such an operation) and that, e.g., the choice of average power and bandwidth will likely be interdependent, as in a "diversity" communication problem.

We therefore will pose and answer, under certain assumptions, the following set of SAR system design problems: What is the structure of the minimum probability of error processor? What is its performance in terms of SAR system parameters? How should said parameters be optimally chosen? Generally, the implications of this problem set have not been realized (see, e.g., [2], especially the report of Panel 6: "Sensors and Data Systems"). The work reported here is an extension of earlier work [3] of the author.

For concreteness, we consider the problem of classifying agricultural fields, though generally the results can have broader application and, certainly, still broader implication. A casual examination of an aerial photograph (see, e.g., Fig. 43 of [4]) emphasizes what is commonly known: agricultural regions are typically predominantly covered by rectangular areas, each consisting of one crop of a set of possible types which, hopefully, differ in some aspect of their radar reflectivity characteristics. For example, if their reflectivity is characterized by a "reflectivity density" that is a sample function of a homogeneous (over each field) random field, then the level and shape of the spectral density characterize (at least partially) the field types. The level is, however, known to not differ greatly for the usual agricultural types to be classified; that is, the scattering cross-section density differs by less than 10 dB over agricultural type (see, e.g., Fig. 22a of [4]). It seems likely (the measurements have not been made to our knowledge) that the spectral shape will be flat over spatial frequencies selected by typical radar systems. (So-called multispectral radars may encounter different, constant levels in their several spectral bands.) The relatively small differences in reflectivity imply that an optimum, or near optimum, system design for classification is worthy of study, especially since some earlier work [5] has been discouraging.

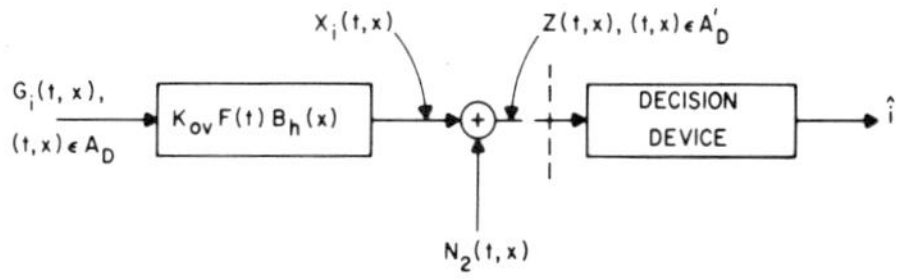

Fig. 1. SAR system model for classification of a random field.

The Model

Assuming that the SAR system is operated in the usual way [1], the data available for processing is a two-dimensional dispersion of the terrain scene called the "signal film". The relation between the reflectivity density $G(t, x)$ characterizing the terrain and the signal $X(t, x)$ recorded on the signal film is a convolution, as indicated in Fig. 1. $F(t)$ is the periodic pulse modulation, with t being the (unscaled) range direction; $B_h(x)$ is the azimuth modulation due to the far-field antenna pattern (amplitude) and Doppler due to relative motion (phase), with x being the (unscaled, orthogonal) along-track direction; and, to correctly account for power, $K_{0v} \equiv (ik/4\pi R_0)^2/v$, where v is the velocity of the vehicle carrying the SAR radar set, R_0 is the mean range to the mapping interval, and $k \equiv 2\pi/\lambda$, λ being the RF (radiated) wavelength.

The field to be classified is assumed to be a known region A_D of known location. Its reflectivity $G_i(t, x)$ belongs to a class, $i = 1, \cdots, M$ of M possible reflectivity densities. For simplicity, we consider $M = 2$; the error approximations and bounds for the $M = 2$ case can be used to find $P(\mathcal{E})$ bounds for $M > 2$ using the "union bound" [6]. Also, we neglect interfering effects with any adjacent field. This is an interesting extension we are presently considering; as we shall see here, the "resolution" (in the conventional sense) that typically results makes such an approximation useful.

In order to render the analysis tractable while still retaining the random nature of the reflectivity, we assume $X_i(t, x)$, $i = 1, 2$, is a sample function of a (complex) zero-mean, Gaussian random field. Because at every $(t, x) \in A_D'$ the X_i is the sum of (typically) a very great number (e.g., thousands) of returns from individual resolution areas, and since they, being the sum of reflections from a great number of small (general order of a wavelength), presumably independent scatterers, are approximately independent, an appeal to a central limit theorem makes this assumption quite reasonable. Of course, if $G_i(t, x)$, $(t, x) \in A_D$ is itself a sample function of a Gaussian field, then $X_i(t, x)$, being a linear transformation of $G_i(t, x)$, is also. We assume the spectral density of G_i and the spectrum of FB_h is constant over the band where FB_h is not zero; the fields characterized by G_1 and G_2 must be distinguished by the difference in the spectral density amplitudes $S_{X_i}(0, 0)$, $i = 1, 2$.

In order to account for unavoidable thermal noise, film grain noise, and other additive noises, the random function $N_2(t, x)$ is added to $X_i(t, x)$ to give the signal film data $Z(t, x)$ available for processing. It is assumed $N_2(t, x)$ is a sample function of a zero-mean (complex), Gaussian random field, of spectral density N_{02} over the spectrum of FB_h. If N_2 accounts solely for RF thermal noise of spectral density amplitude $N_0/2$, then $N_{02} = N_0/vT_s$ (see Appendix 4.A of [1]). In sum,

$$Z(t, x) = G_i{*}(FB_h)(t, x) + N_2(t, x), \quad (t, x) \in A_D'. \quad (1)$$

Optimum Processor

Under the above assumptions the decision problem belongs to the so-called "Gaussian signal in Gaussian noise" class of problems; see, e.g., [7] for a recent exposition. Minimum probability of error decision rule is: choose H_2 (field type characterized by G_2 is present) if $Q(z) > l$; otherwise choose H_1. Here, l is a number (threshold) depending on a priori information, such as the probability of H_i occurring a priori, and the random variable

$$Q(z) \equiv \frac{1}{N_{02}} \int_{A_D'} \bar{Z}(\mathbf{r}') \int_{A_D'} Z(\mathbf{r}'')[q_2(\mathbf{r}', \mathbf{r}'') - q_1(\mathbf{r}', \mathbf{r}'')]\, d\mathbf{r}' d\mathbf{r}'' \quad (2)$$

where $\mathbf{r} \equiv (t, x)$, A_D' is the aperture (extent) of $G_i{*}(FB_h)$, and q_i is the solution of the integral equation

$$N_{02}q_i(\mathbf{r}, \mathbf{r}') + \int_{A_D'} R_{X_i}(\mathbf{r}, \mathbf{r}'')q_i(\mathbf{r}'', \mathbf{r}')\, d\mathbf{r}'' = R_{X_i}(\mathbf{r}, \mathbf{r}'), \quad \mathbf{r}, \mathbf{r}' \in A_D' \quad (3)$$

where R_{X_i} is the covariance function of $\{X_i\}$, which, in general, can not be solved explicity for q_i. Equation (3) can be solved in two cases of interest to us here: the threshold case of weak signal and the large time-band width product (TBP) case.

The optimum processing characterized by (2) and (3) is complicated by the nonhomogeneity of $\{X_{t,x}\}$ after passage through FB_h; there will be "edge effects" for (t, x) near the boundary of A_D. These should be negligible if the TBP is large, as it generally will be. Further, we intuitively feel that the "first step" of the optimum processing should reasonably be a matched filtering, or "compression", to help reduce adjacent field intereference. After such compression, we could then use a processing aperture A_D and have fields approximately homogeneous over A_D. We can, in fact, show that, in the threshold case, a filtering matched to FB_h is the first step of optimum processing.

In the threshold case[1] it is known [7] that

$$q_i(\mathbf{r}, \mathbf{r}') \approx \frac{1}{N_{02}} R_{X_i}(\mathbf{r}, \mathbf{r}'). \tag{4}$$

As

$$R_{X_i}(t_1, x_1; t_2, x_2) = \int_{A_D'}\int dt'dx'F(t_1 - t')B_h(x_1 - x') \int_{A_D'}\int dt''dx''\bar{F}(t_2 - t'') \cdot \bar{B}_h(x_2 - x'') \cdot R_{G_i}(t' - t'', x' - x'') \tag{5}$$

for (t_1, x_1) and $(t_2, x_2) \in A_D'$ and is zero otherwise, where R_{G_i} is the covariance function of $\{G_i\}$, it is easily seen that (2) may be written

$$Q(z) \approx \frac{1}{(N_{02})^2}\int_{A_D} \bar{Z}_C(\mathbf{r}') \int_{A_D} Z_C(\mathbf{r}'') [R_{G_2}(\mathbf{r}', \mathbf{r}'') - R_{G_1}(\mathbf{r}', \mathbf{r}'')] \tag{6}$$

where

$$Z_C(t_1 x) \equiv \int_{A_D'}\int dt'dx'Z(t', x')\bar{F}(t' - t)\bar{B}_h(x' - x) \tag{7}$$

is a matched filtering, or compression, of the data $Z(t, x)$.

Thus, after this first step of matched filter processing, the data to be further processed is $Z_C(t, x)$, $(t, x) \in A_D$. But, the spectral density of $\{Z_C\}$ may equivalently be taken to be the spectral density of $\{Z\}$, because all spectra are either constant or zero and both signal and noise spectra are treated the same by the matched filtering.

When the TBP $\gg$ 1, as generally obtains here, $\{Z_C\}$ is approximately homogeneous over A_D, and we can solve (3) by the Fourier transform. We find

$$\tilde{q}_2(\omega, u) - \tilde{q}_1(\omega, u) \approx \frac{S_{X_2}(\omega, u)}{N_{20} + S_{X_2}(\omega, u)} - \frac{S_{X_1}(\omega, u)}{N_{20} + S_{X_1}(\omega, u)} \tag{8}$$

where $\tilde{q}_i$ and S_{X_i} are the Fourier transforms of q_i and R_{X_i}, respectively. Also, we can write, setting $\tilde{q} \equiv \tilde{q}_2 - \tilde{q}_1$,

$$Q(Z) \approx \frac{1}{4\pi^2}\iint |\tilde{Z}_C(\omega, u)|^2 q(\omega, u) \tag{9}$$

where $\tilde{Z}_C$ is the Fourier transform of $\tilde{Z}_C$ extended as zero outside of A_D. Of course,

$$S_{X_i}(\omega, u) = |\tilde{F}(\omega)\tilde{B}_h(u)|^2 S_{G_i}(\omega, u). \tag{10}$$

[1] We do not define this condition precisely here, as we have no further interest in it (see, e.g., [7] for the condition).

In addition to the forms (2) and (9), there are other interesting forms for $Q(Z)$. For example, applying the well-known fact that q_i as given by (3), is that filter that gives as output the minimum variance estimate $\hat{X}_i$ of X_i given the input Z, (2) can be written in the "estimator-correlator" form

$$Q(Z) = \frac{1}{N_0}\int_{A_D} \bar{Z}(\mathbf{r}) [\hat{X}_2(\mathbf{r})\, d\mathbf{r} - \hat{X}_1(\mathbf{r})\, d\mathbf{r}],$$

an obvious extension of the matched filter processor.

An even more interesting variant of (8) is [8], [9] the "mixed-integrator"

$$Q(Z) = \int_{A_D} \left| \int_{A_D} Z_C(\mathbf{r}')h(\mathbf{r} - \mathbf{r}')\, d\mathbf{r}' \right|^2 d\mathbf{r} \tag{11}$$

where the weighting function h is such that $|\tilde{h}|^2 = \tilde{q}$ and it can be shown to have an extent roughly the coherent area of the random field $\{X_i\}$. Thus, (11) performs coherent integration—the interior integrals—over all possible coherent areas, and then incoherently sums—the exterior integral—all such contributions.

Physical Realization

The form (9) suggests an optimum processor realization employing a coherent optical system [8], [9] that is compatible with existing SAR processors [1]. It is based on the well-known fact that such an optical system can compute a Fourier transform. The compression or matched filtering "first step" (or a close variant) is what is done by the usual SAR processor. We now need only to properly arrange a lens at the "usual" SAR processor output (see figure 2) to obtain $\tilde{Z}_C$. Passing through a transmission function $\sqrt{\tilde{q}}$ and integrating the square modulus with a photoemissive device gives $Q(Z)$.

Performance

The calculation of the $P(\mathcal{E})$ performance is, in general, difficult, but with the assumptions already made above we can establish (see, e.g., [9]) that $Q(Z)$ is approximately a chi-square random variable of order $2K$, and, hence, that

$$P(\mathcal{E}) = \frac{1}{2}[1 - Q(\chi_1{}^2 | 2K)] + \frac{1}{2}Q(\chi_2{}^2 | 2K) \tag{12}$$

where

$$\chi_1{}^2 \equiv 2K\frac{1 + \beta_2}{\beta_1 - \beta_2} \ln\left(\frac{1 + \beta_1}{1 + \beta_2}\right)$$

$$\chi_2{}^2 \equiv 2K\frac{1 + \beta_1}{\beta_1 - \beta_2} \ln\left(\frac{1 + \beta_1}{1 + \beta_2}\right),$$

$\beta_i \equiv S_{X_i}(0, 0)/N_{20}$, we have assumed $\beta_1 > \beta_2$, and $Q(\cdot\, | 2K)$ is the distribution function of a chi-square random variable of unit parameter and order $2K$.

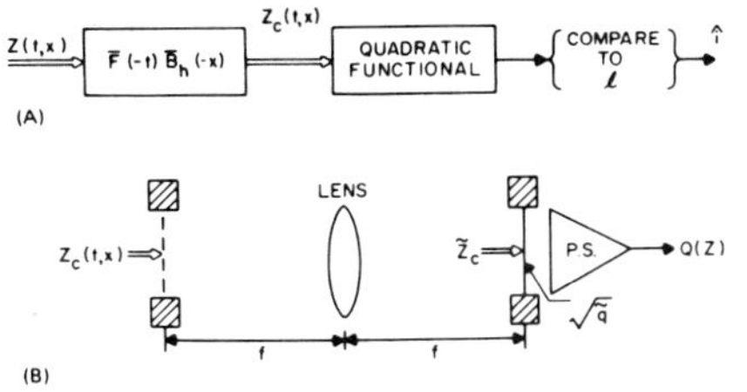

Fig. 2. (A) Optimum decision device (processor) for classification of random fields given SAR system data. (B) Realization of quadratic functional calculation by a coherent optical system. $\sqrt{q}$ is a transmission function and P.S. a photoemissive surface. ⇒ denotes a light field and → an electrical signal, e.g., a voltage.

Since $S_{X_i} = |K_{0\nu}\tilde{F}\tilde{B}_h|^2 S_{G_i}$, we can express β_i in terms of the SAR system parameters:

$$\beta_i = \frac{A^2}{4\pi\lambda R_0{}^3}\left[\frac{4\pi^2\nu}{\Delta\Omega}\;\frac{\pi S_{G_i}(0,0)}{\lambda^2}\right]\frac{P_{av}}{(N_0/2)}\;\frac{Q}{\nu D_h} \tag{13}$$

where A is the physical antenna (unit efficiency assumed) area, ν is the velocity of light, $\Delta/2\pi$(Hz) and $\Omega/2\pi$(Hz) are the range and azimuth modulation bandwidths, respectively, P_{av} is the average radiated power, D_h is the along-track (azimuth) dimension of the physical antenna, and Q is a number of the order of unity depending only on the shape of the physical antenna aperture illumination (see [1, sec. 3.6]: for uniform illumination $Q = 1/6$). We can interpret this form as follows:

1) β_i is the signal-to-noise mean power ratio "per complex sample".
2) $\pi S_{G_i}(0,0)/\lambda^2$ is the scattering cross-section density[2].
3) $4\pi^2\nu/\Delta\Omega$ is the SAR system resolution (m^2).
4) Therefore, $(4\pi^2\nu/\Delta\Omega)(\pi S_{G_i}(0,0)/\lambda^2)$ is the scattering cross section of the SAR system resolution area.

With this last interpretation, the form (8) generalizes nicely the signal-to-noise ratio formula for a point scattering object [1, sec. 3.6].

The TBP parameter K is, equivalently, the number of resolution elements in A_D. As a resolution element has size $4\pi^2/\Delta\Omega(\text{s}\cdot\text{m})^{-1}$ and $D_h/2 = 2\pi/\Omega$ (a famous relation for SAR systems), we have

$$K = \frac{2}{\nu}\;\frac{\Delta_f A_D}{D_h} \tag{14}$$

where $\Delta_f = \Delta/2\pi$ and where A_D is in meters^2 and D_h is in meters.

It will be convenient to write

$$\beta_i = a\frac{\Gamma_i P_{av}}{\Delta_f}, \tag{13a}$$

[2] A discussion of terrain reflectivity measurements follows in a later section.

explicitly separating out parameters $\Gamma_i \equiv S_{G_i}(0,0)$, which characterizes the field type, and P_{av} and Δ_f, important SAR system design parameters.

A fundamental design problem is now relatively intuitively clear without having yet plotted the $P(\mathcal{E})$ (12): by increasing Δ_f, with other parameters fixed, we increase the number of (independent, as we can show) samples K, but at the cost of decreasing signal-to-noise ratio β_i (there may well exist a best Δ_f).

In Fig. 3 we plot $P(\mathcal{E})$ versus $A_1 \equiv \beta_1 K$. If E_{S1} is the total received energy when the field has reflectivity density Γ_1 (a number we restrain as it typically will be in practice), then

$$A_1 \equiv \beta_1 K = \frac{E_{S1}}{2N_{02}} \tag{15}$$

is the overall received signal-to-noise ratio. Also set

$$\alpha \equiv \frac{\beta_2}{\beta_1}. \tag{16}$$

As $\beta_1 > \beta_2$, $0 \leqslant \alpha < 1$. These curves show a "cross-over", indicating that there is a best K for any particular A_1; that is, there will be a best signal-to-noise ratio per complex sample (or diversity path) $\lambda \equiv A_1/K = \beta_1$, and it will depend also upon $\alpha \equiv \beta_2/\beta_1$.

Cernov Bound

In order to determine the best K for each A_1 more accurately and analytically, we use the so-called Cernov bound [10], [7] on $P(\mathcal{E})$. One can show

$$P(\mathcal{E}) \leqslant \exp\{-A_1 F(\lambda;\alpha)\} \tag{17}$$

where

$$F(\lambda;\alpha) \equiv \frac{1}{\lambda}\ln\left\{\frac{[1+(1+\alpha)\lambda/2]^2}{(1+\lambda)(1+\alpha\lambda)}\right\} \tag{18}$$

where $\lambda \equiv A_1 K$ is the signal-to-noise ratio per complex sample, as before.

The function $F(\lambda;\alpha)$ of (18) is plotted as a function of λ, with α as a parameter, in Fig. 4. Clearly, there is a best λ for each α. For $\alpha = 0$, $\lambda_{opt} = 3.07$, a well-known number in diversity communication where the alternative hypothesis is "noise only". A brief tabulation of this result is:

α	λ_{opt}
0	3.07
1/4	2
1/2	1.4

A plot of the Cernov bound (17) on $P(\mathcal{E})$ versus λ, with the pair (α, A) as parameter, is shown in Fig. 5, again clearly showing a best choice of λ. The Cernov bound with optimum choice of λ will be denoted P_{eo}(CB).

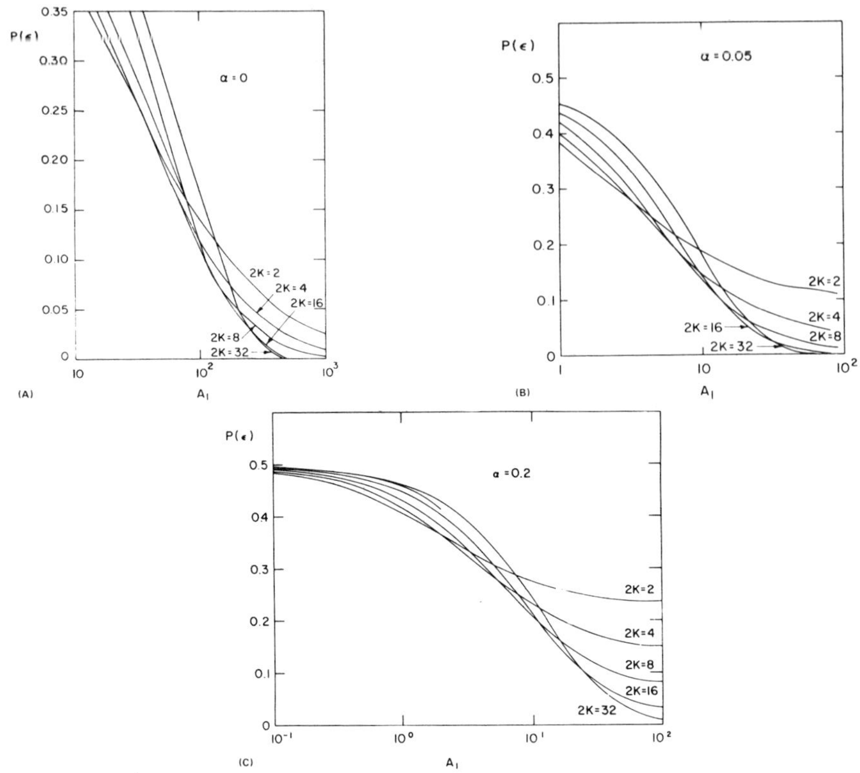

Fig. 3. Missclassification probability $P(\mathcal{E})$ versus received signal-to-noise ratio A_1 with number of independent samples $2K$ as parameter for $\alpha = 0.0, 0.5, 0.2$.

Optimum SAR Design

For a given pair $(\Gamma_{1*}, \Gamma_{2*})$ of field types that are to be distinguished, $\alpha_* \equiv \beta_2/\beta_1 = \Gamma_{2*}/\Gamma_{1*}$ is fixed. We have seen that there is fixed a best signal-to-noise ratio "per diversity path" λ_*; e.g., using the Cernov bound, Fig. 4 gives λ_* for specified α_*. Next, for a specified probability of misclassification $P_*(\mathcal{E})$, the overall signal-to-noise ratio A_{1*} is determined, e.g., by the Cernov bound via Fig. 5. Fig. 6 conveniently summarizes this. Finally, the optimum number of diversity paths $K_* = A_{1*}/\lambda_*$.

We can combine Figs. 4, 5, and 6 to reduce this selection to examination of a single plot–Fig. 7. With abscissa A_{1*} and ordinate K_*, a family of loci parameterized by α can be plotted [e.g., by using Fig. 6 and the tabulation of $\lambda_*(\alpha)$]. Along each locus the achieved P_{eo}(CB) can be indicated and like values connected by another set of loci, as shown. Thus, for a specified $(\alpha, P_{eo}(\text{CB}))$, the intersection of these loci determines, by inspection, (A_{1*}, K_*).

Having now solved the "diversity problem", the remaining SAR system parameters are chosen subject to the following conditions:

$$\lambda_* = a \frac{\Gamma_{i*} P_{\text{av}}}{\Delta_f} \tag{19}$$

and

$$K_* = \frac{2}{\nu} \frac{\Delta_f A_D}{D_h}. \tag{20}$$

Generally, the solution $(P_{\text{av}}, \Delta_f, D_h)$ will not be unique, of course. Other considerations can then be involved to get a final, unique solution, as shown in the following example.

Example

1) Suppose the following SAR system parameters are given: mean slant range $R_0 = 200$ nautical miles; RF frequency $f_0 = 10^{10}$Hz (X band); vehicle (satellite) velocity $\nu = 2\,500$ ft/s; mean depression angle $\gamma_0 = \pi/4$; range offset processing mode [1]; and, as the radar views the

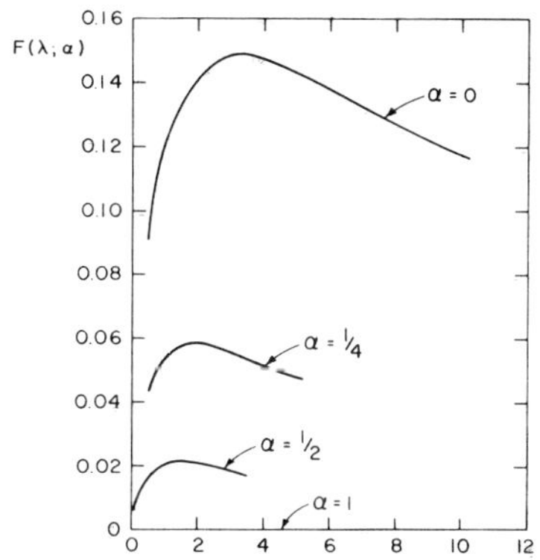

Fig. 4. Cernov bound exponent $F(\lambda; \alpha)$ versus λ for $\alpha = 0, 1/4, 1/2$.

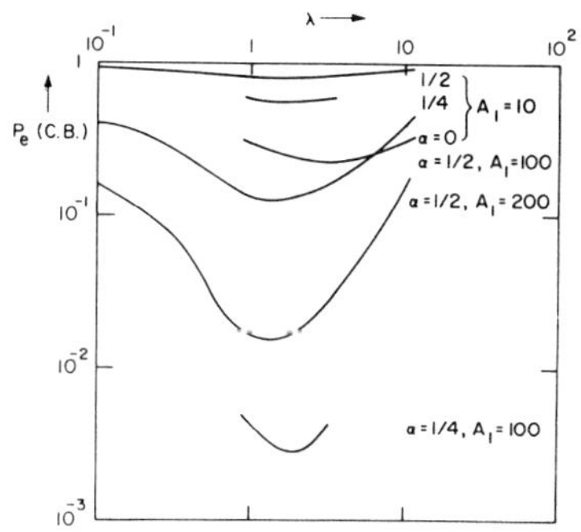

Fig. 5. Cernov bound P_e(CB) on missclassification probability versus signal-to-noise ratio per complex sample λ.

Fig. 6. Cernov bound P_{eo}(CB) with optimum diversity parameter λ_{opt} versus signal-to-noise ratio A_1.

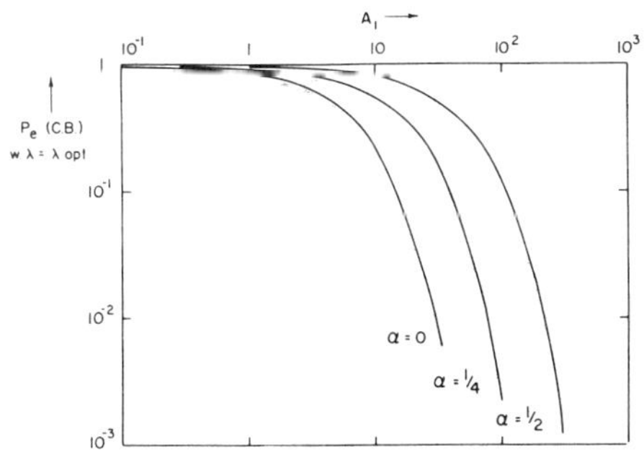

Fig. 7. Summary of optimum diversity (signal-to-noise ratio per independent, complex sample) problem.

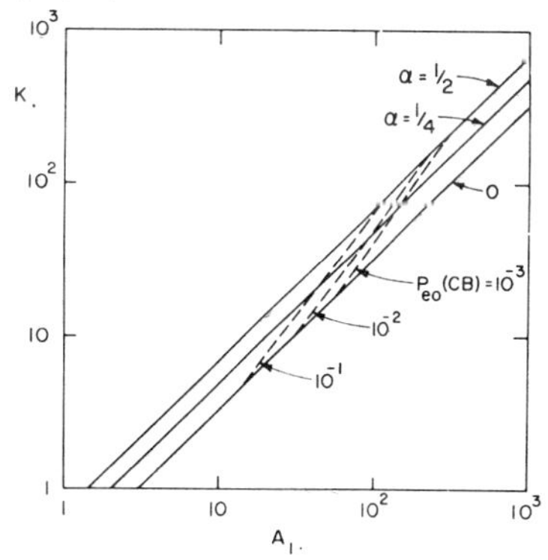

terrain, giving an effective noise temperature about 300°K, $N_0/2 = k_B T_s = 1.4 \times 10^{-21}$. Then,[3]

$$a \approx 10^9 .$$

2) Suppose the two field types to be distinguished differ by 3 dB in their reflectivity density spectral density amplitudes: $\alpha = 1/2$. Suppose also that the classification error probability is to be not greater than $10^{-2} = P(\mathcal{E})$. Using Fig. 7, $A_{1*} = 220$ and $K_* = 160$ are sufficient to ensure at least $10^{-2} = P(\mathcal{E})$.

3) Suppose the reflectivity spectral density amplitude $\Gamma_1 = -20$ dB. Then (19) and (20) give

$$\frac{P_{av}}{\Delta_f} = 1.5 \times 10^{-7} \tag{21}$$

[3] We assume operation at the ambiguous limit, as would generally be desirable in a satellite SAR system [11].

and

$$\frac{\Delta_f A_D}{D_h} = 2.4 \times 10^{10}. \tag{22}$$

Equations (21) and (22) embody the parameter specifications and the optimum diversity solution.

4) Equations (21) and (22) have no unique solution, of course. To restrict the allowable solutions we take into account other desired system parameters. For example, recall [1] that the unambiguous slant range interval

$$\Delta R_s = \frac{\nu D_h}{4\nu} \approx 5 D_h \text{ nautical miles}$$

in the present example. Generally, we want ΔR_s at least 50 nautical miles, but cannot reasonably mount an antenna with D_h greater than 10 meters, so we assume $D_h = 10$ meters. Now (22) becomes

$$\Delta_f A_D = 2.5 \times 10^{11} \tag{22a}$$

5) Finally, suppose the smallest field to be classified is 10^4 m^2, about 2.5 acres. Then we find $P_{av} \approx 3.4$ watts and $\Delta_f \approx 24$ MHz, very reasonable parameters.

On Terrain Reflectivity Measurements

We have assumed that, though we did not know which field type was present in a prescribed area, we did have the a priori probabilities and knew the spectral densities of the random reflectivity densities. Such "a priori" probabilities are presumably accurately known—e.g., from government records. We discuss here a second problem: measurement of the spectral density of a known field type in order to learn this a priori information. This is an example of establishing relevant "ground truth", in the vernacular of this area.

The specific problem we examine here is estimating the amplitude of the field's spectral density, assuming its shape is white (constant) over the system spatial bandpass. We give here the analytic result, an optical system that computes the amplitude from the SAR data film, and give the formula for the best accuracy that is possible—and can be approached with the described system—in terms of the SAR system parameters.

We also discuss the question of using systems other than an SAR system to gather this information. We first observe that a "brute force" side-looking radar's data is directly useable with no change required in the processor, assuming, of course, that the same azimuth bandwidth (and "shape") is achieved.

We next observe that a side-looking CW radar can also be used, provided the random field is isotropic in the direction of the vehicle velocity vector—which, in practice, essentially means "provided the random field is isotropic". (We remark that cultural milieu are typically *not* isotropic.)

Results of this type are important to guide the gathering of a priori information ("ground truth") to enable good SAR system design in this application.

Two-Dimensional Sensors

We consider the questions outlined above first for SAR systems, and assume we have the data film for an isotropic, homogeneous random field $G(t, x)$ with spectral density $S_G(\omega, u)$ and zero mean. To make the following calculations tractable we assume $\{G(t, x)\}$ is a complex Gaussian field. The model for the SAR system from object to data film is shown in Fig. 1, as discussed before.

With this model we measure "ground truth" directly by using the SAR system to measure the unknown parameters of the spectral density S_G, and we are asking: "What is the best processor to use?" Note that such a measuring device avoids complicated questions that crop up when other types of sensors are used; in a sense, we "measure the answer" directly.

It should not be a surprise, therefore, that the "best processor" is intimately related to the "best classifier" that we discussed earlier. In fact, one such best processor computes exactly the quadratic functional Q discussed earlier. Here we shall give a somewhat simpler, but still very similar, processor.

We omit the details here, but we can show [12] that the maximum likelihood estimator $\tilde{A}_0$ of the unknown amplitude A_0 of the reflectivity spectral density $S_G(\omega, u) = A_0 S_0(\omega, u)$ is

$$A_0 \approx \frac{1}{T_f X_f \Delta\Omega} \int_\Delta \int_\Omega \frac{|\tilde{Z}(\omega, u)|^2}{S_0(\omega, u)\psi(\omega, u)} \, d\omega du$$

where

$$\psi(\omega, u) = |K_{0v}|^2 \, |\tilde{F}(\omega)|^2 \, |\tilde{B}_h(u)|^2,$$

all as defined earlier, and $T_f X_f$ is the area of the homogeneous field sample. Here we have assumed that the measurement has a "good" signal-to-noise ratio in that $A_0 S_0(\omega, u)\psi(\omega, u) \gg N_0$ over the bandpass of the system.

This estimate, $\tilde{A}_0$, can, in principle, easily be made by a coherent optical system, as shown in Fig. 8. The Fourier transform $\tilde{Z}_f$ of the data over its extent is easily done with a lens. (Again, we have neglected the slight complication due to offset frequencies.)

With the same approximations we can show [12] that an unbiased estimator of A_0 cannot have an error variance smaller than

$$\frac{2A_0^{\,2}}{T_f X_f \Delta_f \Omega_f}.$$

In fact, the maximum likelihood estimator $\tilde{A}_0$ approaches this asymptote as $T_f X_f \Delta_f \Omega_f$, the "space-bandwidth product" of the data, becomes large.

We make the important observation that only $|\tilde{B}_h|^2$ enters these expressions. This implies [1] that either SAR or "brute force" side-looking radar data can be used here with, other things equal, no change necessary.

One-Dimensional (CW) Sensors

A much simpler sensor is a side-looker with no signal modulation ($F = 1$); i.e., a CW radar. The signal returned by terrain scattering is [1, p. 64, eq. 3.9]

$$Y(vt) = K_0 \int dx \int d\rho \, B_v(\rho) B_h(x - vt; \rho) G(x, \rho)$$

plus the usual thermal noise $N(t)$. As G is Gaussian, so is Y, and so we can measure its spectral density and estimate the latter's parameters exactly, as described above, except that everything is "one dimensional". But the question is: "How is S_Y related to S_G?"

It is straightforward to calculate S_Y. But first we note that G enters only as

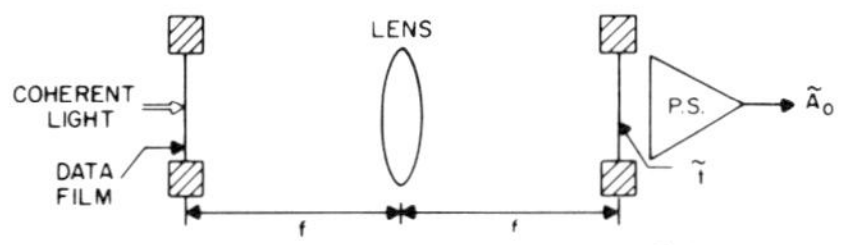

Fig. 8. Coherent optical system realization of the calculation of maximum likelihood estimator $\tilde{A}_0$ of reflectivity density spectral density amplitude of a random field, given SAR data. The transmission function $\tilde{t} \equiv (T_f X_f \Delta\Omega)^{-1}\psi$.

$$\Gamma(x) = \int d\rho B_v(\rho) G(x, \rho).$$

If $\{\Gamma_x\}$ is a wide-sense stationary (WSS) process, with spectral density S_Γ, then $\{Y_t\}$ is WSS with

$$S_Y(v\omega) = \frac{|K_0|^4}{v} S_\Gamma(\omega) |B_h(\omega)|^2$$

where v is the vehicle velocity. We recall that the x direction is the direction of the vehicle velocity vector.

The fact that $\{\Gamma_x\}$ is WSS means that its covariance function

$$R_\Gamma(x_1, x_2) = \int d\rho B_v(\rho_1) \int d\rho_2 B_v(\rho_2) E\{G(x_1, \rho_1) G(x_2, \rho_2)\}$$

is a function only of $(x_1 - x_2)$. For this to be true it is necessary (for arbitrary B_v) and sufficient that

$$E\{G(x_1, \rho_1) G(x_2, \rho_2)\} = R_G(x_1, \rho_1; x_2, \rho_2)$$

be a function of x_1 and x_2 only as $(x_1 - x_2)$ for any ρ_1, ρ_2. We could call this "isotropy in the x direction", but, clearly, in practice this will generally happen only if G is, in fact, isotropic.

On the Use of Terrain Handbooks

The scattering cross section Γ we have used above essentially is that measured by Cosgriff *et al.* [13], at least when the terrain is "incoherent"—or, more precisely, when its spectral density is "white" over the effective spatial bandwidth of the measuring apparatus. We now show this.

Cosgriff *et al.* [13] use a CW signal and measure returned power. For a random field this would be done often "independently" and the results averaged. From [1, Eq. 3.9], for example, we find the (ensemble) average power returned is

$$P_r = P_{av} \left(\frac{k}{4\pi R_0}\right)^4 \frac{1}{4\pi^2} \int du \int dv |B_v(v)|^2 \cdot |B_h(u)|^2 S_G(u, v).$$

Now suppose that over the bandwidth of $B_v B_h$, $S_G(u, v) \approx S_G(0, 0)$. Then

$$P_r \approx P_{av} \left(\frac{k}{4\pi R_0}\right)^4 \frac{S_G(0,0)}{4\pi^2} \int du |B_h(u)|^2 \int du |B_v(v)|^2.$$

The integrals can be evaluated as in [1, ch. 3]. We find

$$P_r = \frac{A^2 P_{av}}{4\pi\lambda^2 R_0^4} \left\{ \left(\frac{R_0^2 \lambda^2}{A}\right) \frac{\pi S_G(0,0)}{\lambda^2} \right\} \frac{Q_1}{16\pi^2}$$

where Q_1 is a factor depending on the illumination function of the antenna aperture. Presumably this is uniform in the set-up of Cosgriff *et al.* Then one can show $Q_1/16\pi^2 \approx 1/2$; arbitrarily replacing this 1/2 by unity we have

$$P_r \approx \frac{A^2 P_{av}}{4\pi\lambda^2 R_0^4} \left\{ \left(\frac{R_0^2 \lambda^2}{A}\right) \frac{\pi S_G(0,0)}{\lambda^2} \right\}.$$

Comparing this last with equation 2.1 of Cosgriff *et al.* [13], which defines their σ, the scattering cross section

$$\sigma \rightarrow \left(\frac{R_0^2 \lambda^2}{A}\right) \frac{\pi S_G(0,0)}{\lambda^2}.$$

Further, they define the scattering cross-section density

$$\gamma = \frac{\sigma}{\perp \text{area}},$$

where $\perp$ area is a nominal area, normal to the antenna boresight, of the nominal illuminated terrain area. If we define said nominal area using the 3-dB beamwidth of the antenna, then we get, within a factor nearly unity,

$$\perp \text{area} = R_0^2 \frac{\lambda^2}{A}.$$

Thus we have the correspondence

$$\gamma \rightarrow \frac{\pi S_G(0,0)}{\lambda^2} = \Gamma.$$

We can thus use the tabulated values of γ in [13] as the values of Γ in the system design of the SAR.

Acknowledgment

Mr. M.A. Tanney computed Fig. 3.

References

[1] R. O. Harger, *Synthetic Aperture Radar Systems: Theory and Design.* New York: Academic Press, 1970.
[2] "Useful applications of earth oriented satellites," report of summer studies by the National Research Council, National Academy of Sciences, 1969.
[3] R. O. Harger, "Study of the feasibility of using orbiting imaging radars for surveys of earth resources," NASA Goddard Space Flight Center, Greenbelt, Md., Final Rept., NASA Contract NAS 5-21065, July 15, 1970.
[4] R. K. Moore, "Radar return from the ground," University of Kansas, Lawrence, Bull. Engrg. 59, 1969.
[5] R. M. Haralick and G. L. Kelly, "Pattern recognition with measurement space and spatial clustering for multiple images," *Proc. IEEE,* vol. 57, pp. 654-665, April 1969.
[6] J. M. Wozencraft and I. M. Jacobs, *Principles of Communication Engineering.* New York: Wiley, 1965.
[7] H. L. VanTrees, *Detection, Estimation, and Modulation Theory; Part III: Radar/Sonar Signal Processing and Gaussian Signals in Noise.* New York: Wiley, 1971.
[8] R. O. Harger, "On processing images propagated through the atmosphere," *IEEE Trans. Aerospace and Electronic Systems,* vol. AES-3, pp. 819-828, September 1967.
[9] —, "Maximum likelihood and optimized coherent heterodyne receivers for strongly scattered Gaussian fields," *Optica Acta,* vol. 16, pp. 745-760, December 1969.
[10] R. G. Gallager, *Information Theory and Reliable Communication.* New York: Wiley, 1968.
[11] R. O. Harger, "An optimum design of ambiguity function, antenna pattern, and signal for side-looking radars," *IEEE Trans. Military Electronics,* vol. MIL-9, pp. 264-278, 1965.
[12] —, "Maximum likelihood estimation of spectral density parameters of Gaussian processes," University of Michigan, Ann Arbor, Institute of Science and Technology Rept., November 1966.
[13] R. C. Cosgriff, W. H. Peaks, and R. C. Taylor, "Terrain scattering properties for sensor system design (Terrain handbook II)," Ohio State University Rept., Columbus, May 1960.

Robert O. Harger (S'53—M'55) was born in Flint, Mich. on September 15, 1932. He received the B.S.E. degree in mathematics and the M.S.E. and Ph.D. degrees in electrical engineering, from the University of Michigan, Ann Arbor, in 1955, 1959, and 1961, respectively.

From 1955 to 1958 he was a Line Officer in the U.S. Navy. He worked at Lincoln Laboratory, Lexington, Mass., in 1958, and the Cornell Aeronautical Laboratory, Buffalo, N.Y., in 1959. From 1960 to 1968 he was with the University of Michigan Department of Electrical Engineering and the Institute of Science and Technology (Radar and Optics Laboratory). He is presently with the University of Maryland, College Park. He is the author of *Synthetic Aperture Radar Systems* (Academic Press, 1970).

Dr. Harger is a member of Sigma Xi, Phi Kappa Phi, Tau Beta Pi, Eta Kappa Nu, and Phi Eta Sigma.

Statistics Of Focused And Defocused Radar Maps

B.P. Hildebrand
(Battelle Pacific Northwest Laboratories, Richland, Washington)

IEEE Transactions on Aerospace and Electronic Systems, Vol. AES-10, No. 5 (September 1974)
Pages 615-621

Abstract

Coherent high-resolution synthetic-aperture radar systems achieve their range resolution by pulse compression and azimuth resolution by compression of naturally generated FM coding due to Doppler shifts as the aircraft flies by the target. If the data is left unprocessed, it is, in effect, a defocused map of the terrain. As such, it should exhibit less dynamic range than if the data is compressed. This paper describes an experimental study to verify the above conjecture. The results of this study indicate that if dynamic range of the data link is a problem, the radar data should be transmitted in its unprocessed form. This might very well be the case for planetary mapping by means of satellites.

Introduction

This paper describes a theoretical and experimental study of the effect of azimuth defocusing in a high-resolution radar system. The investigation includes the consideration of dynamic range, statistics of terrain, and optimization of the radar data for most efficient transmission.

The fundamental basis on which the conclusions are drawn is the first-order probability density function of the amplitude of simulated radar returns. The simulation was done optically, starting from a high-quality radar map. The motivation for the study stemmed from the fact that a radar map normally has a large dynamic range. The hypothesis is that a defocused map should have less dynamic range. Hence, the data should be transmitted in this form to avoid saturation of the transmission link and attendant information loss.

Manuscript received March 1, 1973.

This work was done while the author was at the Institute of Science and Technology, University of Michigan, Ann Arbor.

A by-product of the study was the derivation of statistics of typical terrain as seen by the radar. From these statistics it was also possible to show that the probability density function for a radar map approximates the one with the highest entropy. The theoretical and experimental density functions turn out to be similar.

Two high quality radar maps were used as the model for the terrain. One was a map made over a city and the other over farmland. The negatives were inserted in the coherent optical system shown in Fig. 1.

In the radar system, the signal returns are recorded on film, which is an intensity sensing element. As a result, the phase associated with the terrain is lost. In order to simulate the terrain from the finished map, a piece of ground glass which introduces a random phase on the scene is placed as shown in Fig. 1. The ground glass is such as to produce a flat phase spectrum across the spatial bandwidth of the terrain. The two spherical lenses form a simple imaging system. The cylindrical lens performs the function of defocusing in the azimuth direction. It is placed in the Fourier transform plane of the object so as to provide defocusing without magnification. Cylinders of different focal lengths can be used to simulate various amounts of dispersion.

The light distribution can now be recorded at the image plane. It can be scanned and recorded directly on some electronic storage medium or photographed and then scanned. The latter method was used in this experiment. The negatives made at the image plane were scanned off in a flying spot scanner and the output stored on magnetic tape. The tape was then sampled and quantized and the resulting signal recorded on digital tape. The digital tape was then analyzed on an IBM 7090 computer.

There are two physical constraints placed on the intensity of the light impinging on the image plane. The first is that the intensity is always a positive quantity. The second is that the total energy on the image plane remains invariant irrespective of the focal length of the cylindrical lens.

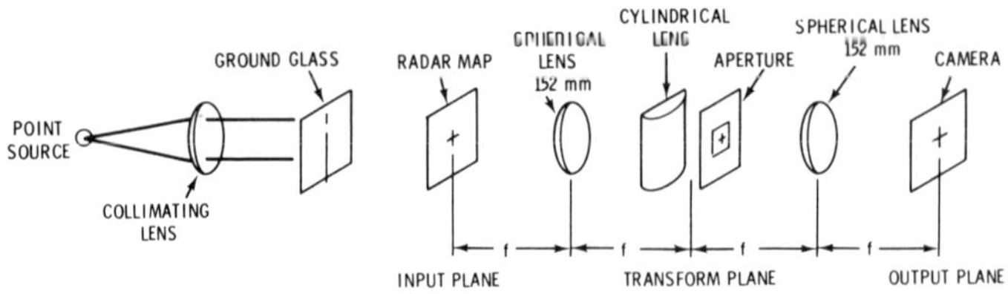

Fig. 1. Optical simulation system.

Fig. 2. Experimental probability density curves for a radar map of urban terrain.

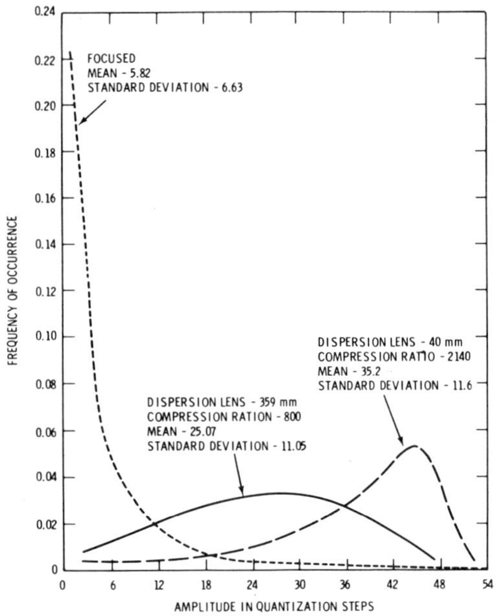

Fig. 3. Experimental probability density curves for a radar map of rural terrain.

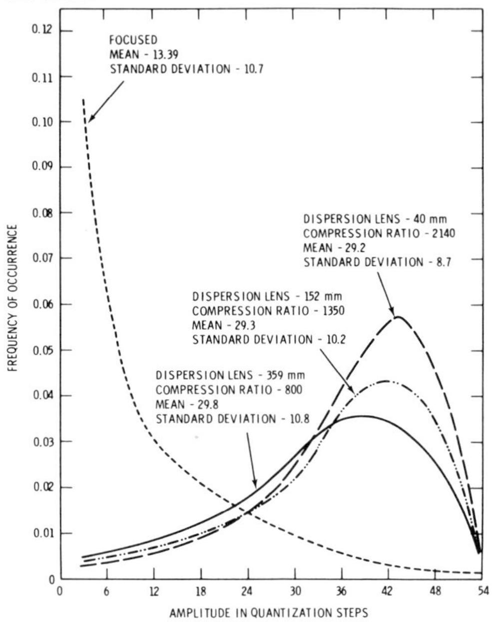

Since the average illumination on the image plane does not change for different cylinders, the mean of the electronic signal should be the same for all dispersions. However, during the recording processes, both optical and electronic biases may be introduced, such as a change in the source brightness, etc. Hence, the mean is changed. This was taken care of in the computer by normalizing the amplitude variable with respect to the mean.

The computer was programmed to obtain the mean and variance of each negative and the probability density function of the amplitude. An additional statistic obtained was the probability density function of the derivative of the amplitude variation. The details of the experiment and computer program are given in Appendix I.

Results

The results of the experiment are shown in Figs. 2 through 7. Figs. 2 and 3 are plots of the probability density function of the two pictures with respect to the non-normalized amplitude variable. Here it can be seen that the means are considerably different for different dispersions. This was due to brightness changes in the light source and amplifier gain adjustments to take advantage of the full dynamic range of the recorder.

Figs. 4 and 5 are the same as Figs. 2 and 3, except that the amplitude variable has been normalized to the mean. That is, the means of these density functions are now equal.

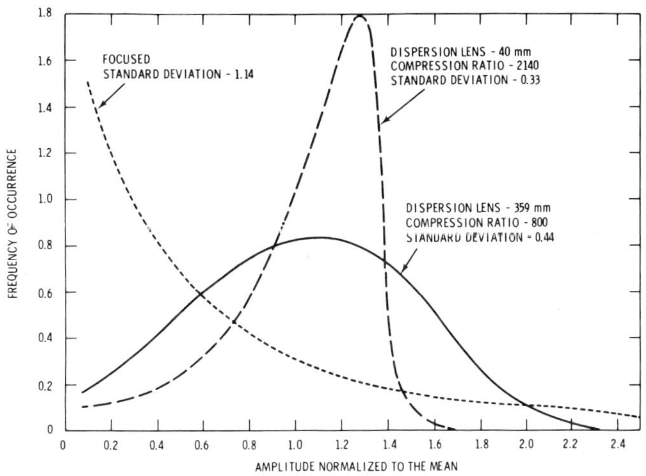

Fig. 4. Normalized probability density curves for a radar map of urban terrain.

Fig. 5. Normalized probability density curves for a radar map of rural terrain.

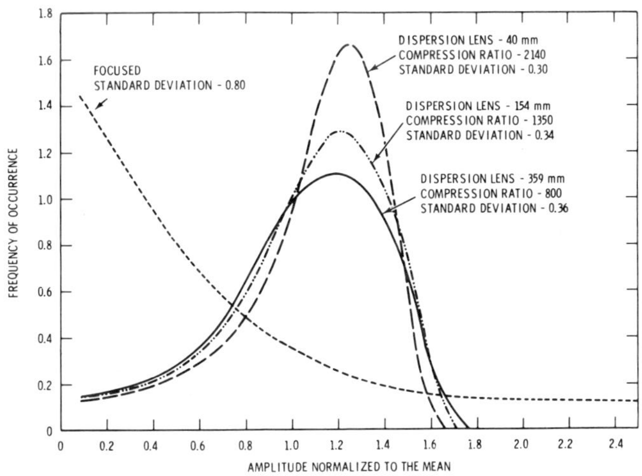

Hence, valid conclusions may be drawn on the effects of defocusing.

Figs. 6 and 7 are plots of the probability density function of the derivative versus amplitude and normalized amplitude, respectively.

Interpretation of the Results

We will restrict our attention to Figs. 4, 5, and 7 since these represent the density functions with respect to the normalized amplitude.

One of the first things to notice in Figs. 4 and 5 is that the range of the argument decreases with defocusing. The focused map exhibits a much larger amplitude variation than the defocused version. This is certainly an expected result.

Another important observation is that the variance of the amplitude decreases with defocusing. These two observations lead one to conclude that the total amplitude variation, or dynamic range, of the radar map is decreased considerably by defocusing in azimuth.

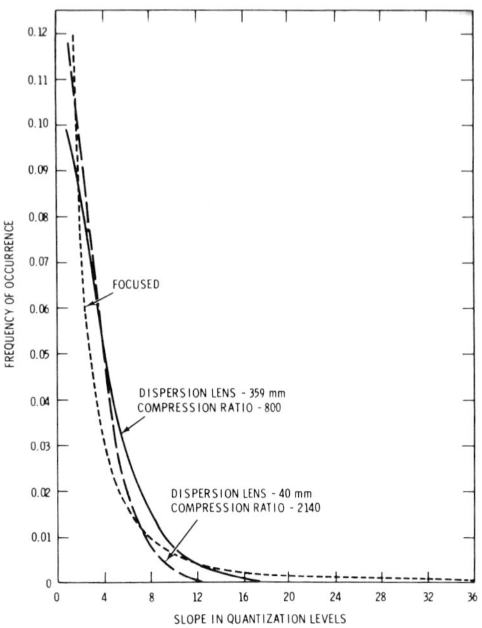

Fig. 6. Experimental probability density curves for the rate of change of amplitude for a radar map of urban terrain.

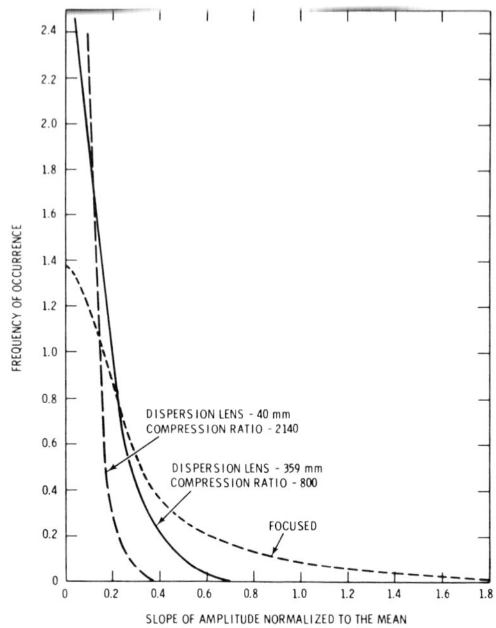

Fig. 7. Normalized probability density curves for the rate of change of amplitude for a radar map of urban terrain.

The above conclusion is reinforced by examing Fig. 7, which is a plot of the probability density function of the rate of change of amplitude. This figure shows quite clearly that a high rate of change in amplitude is much less likely to occur in the defocused map.

The statistics of radar terrain as shown in the focused curve in Figs. 4 and 5 exhibit a smooth exponential-like curve from low-energy to high-energy signals. The curve for the maps made over farmland is slightly less peaked than the map made over a city. This is to be expected, since there is more ground painting at intermediate energy levels than in the city returns.

The question of efficiency of transmission of the radar data is not easy to answer. It is shown in Appendix II that for the class of random variables bounded between zero and some upper limit and having a specified mean, the density function leading to the greatest entropy is a truncated exponential:

$$f(x) = \alpha e^{\lambda_1 - x}, \quad 0 \leqslant x \leqslant M$$

$$= 0, \quad \text{elsewhere}$$

where

$$\alpha = (1/M) - \lambda_1 \, [(M - \bar{x})/M]$$

and

$$\lambda_1 = (e^{\lambda_1 M} - 1)/[e^{\lambda_1 M}(M - \bar{x}) + \bar{x}] .$$

These equations, when solved, show that, for $\bar{x} > M/2$, the density function is a rising exponential, for $\bar{x} < M/2$, it is a falling exponential, and for $\bar{x} = M/2$, it is uniform. The closer $\bar{x}$ is to one of the limits, the steeper is the density curve. Fig. 8 shows density functions where $\bar{x} = 1$ and $M = 10$, $\bar{x} = 1$ and $M = 1.7$, and $\bar{x} = 1$ and $M = 2$.

It is interesting to note that curve *a* in Fig. 8 resembles the density function for the focused case in Fig. 5. In this case, $\bar{x} = 1$ and $M = 10$, as in the theoretical curve plotted above. Thus, it appears that the amplitude of a focused radar scene has a probability density function which approaches fairly closely the distribution with maximum entropy.

As the map is defocused, the relationship between the mean and the upper bound moves over to a ratio of $\bar{x}/M >$ 1/2 and, hence, the optimum density function changes to a rising exponential. The experimental curves likewise change to a rising function which resembles an exponential. Curve *b* of Fig. 8 should be compared to the 359-mm dispersion curve of Fig. 4, curve *c* to the 40-mm dispersion curve of Fig. 5, and curve *d* to the 40-mm dispersion curve of Fig. 4.

These observations show that the experimental density curves are near optimum with respect to entropy. The defocused curves, however, have the additional characteristics of smaller dynamic range and smaller

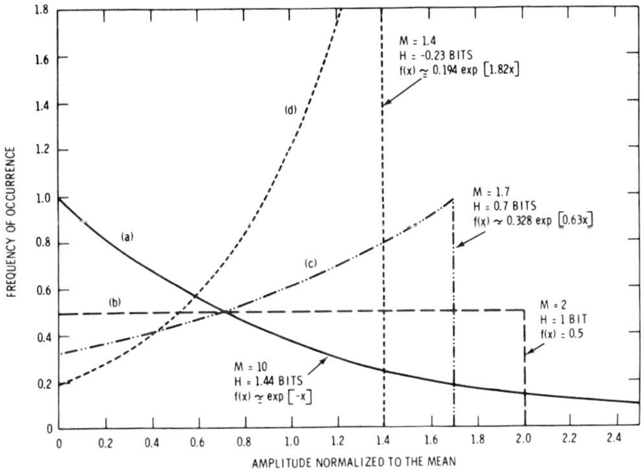

Fig. 8. Maximum entropy probability density functions for a positive random variable with a mean of unity and bounded above by *M*.

variance. Based on these observations alone, it appears that some advantage may be gained by transmitting the radar data in a defocused state. That is, the data link requires less dynamic range.

Conclusions

The one-dimensional probability density function of the amplitude of a radar scene has a shape which resembles a truncated falling exponential. The dynamic range of the map is rather large. As the scene is defocused, the dynamic range and the variance decrease. The shape of the density curve changes from a falling exponential to a rising exponential as the ratio of the mean to the maximum amplitude changes from less than 1/2 to greater than 1/2. The steepness of the curve varies as a function of the same ratio.

A theoretical study showed that this type of a density function has maximum entropy. Thus, the defocused scene has an optimum probability density function, decreased dynamic range, and a smaller variance than the focused scene.

Appendix I

Description of the Experiment

Two high-quality radar maps were used as the test objects. They were inserted in the optical system shown in Fig. 1. The light source was a CW laser operating at the 6328-Å wavelength and the spherical imaging lenses were 152-mm Baltars. The cylindrical dispersing lenses had focal lengths of 359, 152, and 40 mm. These focal lengths correspond to signals having compression ratios of 800, 1350, and 2140, respectively. The image was recorded on photographic film at exposures designed to utilize the linear portion of the film response.

The developed film was then scanned by a flying spot scanner capable of resolving at least 15 lines per mm. The film drive speed and repetition rate of the scanner were such as to obtain approximately 30 scans per negative. The scan rate was adjusted to give an electronic spatial frequency ratio of 100 to 1.

The output of the flying spot scanner was recorded on magnetic instrumentation tape by means of an Ampex FR1100 tape recorder operating at 30 inches per second in the FM mode. The analog tape was then sampled at a 15 kHz rate, quantized to 64 levels, and recorded on digital tape. This was accomplished on a Raytheon designed analog-to-digital converter owned and operated by the Cooley Electronics Laboratory of The University of Michigan. The digital tape was then inserted into an IBM 7090 computer for analysis.

The computer was programmed to do the following:

1) Plot the probability density function of the amplitude for each run.
2) Find the mean and variance of each run.
3) Plot the probability density function of the rate of change of the amplitude.
4) Repeat 1, 2, and 3 with respect to the amplitude normalized to the mean.

Appendix II

Maximum Entropy Distribution

The purpose of this appendix is to find distribution functions which provide maximum entropy. The distribution functions of interest here have rather specialized conditions. They are bounded, and the values of one or more of the various moments are designated. The problem is then essentially one of maximizing the entropy while employing these specific conditions.

$$H = \int_{-\infty}^{\infty} p(x) \log p(x)\, dx \tag{1}$$

where $p(x)$ is the probability density function which we wish to find, and the log may be to any base we wish. Here we shall use natural logarithms.

Incorporating the first condition, that $p(x)$ is nonzero, between O and M, we may write (1) as

$$\int_0^M p(x) d(x) \ln p(x)\, dx. \tag{2}$$

The moments of the distribution are denoted as x, x^2, x^3, $\cdots$, some of which are held invariant as specified. Also, the total probability equals unity, a condition which always holds for probability density functions, i.e.,

$$\int_0^M p(x) d(x) = 1.$$

The technique used to maximize (2), while using these constraints, is the Lagrange method of undetermined multipliers.[1] A state function for H is

$$\begin{aligned} H = &-\int_0^M p(x) \ln p(x)\, dx + \lambda_0 \left\{\int_0^M p(x)\, dx - 1\right\} \\ &+ \lambda_1 \left\{\int_0^M p(x) x dx - \overline{x}\right\} \\ &+ \lambda_2 \left\{\int_0^M p(x) x^2\, dx - \overline{x^2}\right\} \\ &+ \lambda_3 \left\{\int_0^M p(x) x^3\, dx - \overline{x^3}\right\} + \cdots. \end{aligned} \tag{3}$$

Since H is to be an extremum,

$$\begin{aligned} \delta H = \delta \Big[&\int_0^M p(x) \left\{-\ln p(x) + \lambda_0 + \lambda_1 x + \lambda_2 x^2 + \lambda_3 x^3 + \cdots\right\} dx \\ &- \lambda_0 - \lambda_1 \overline{x} - \lambda_2 \overline{x^2} - \lambda_3 \overline{x^3} - \cdots\Big] \\ = \delta \Big[&\int_0^M I[p(x), x]\, dx - \lambda_0 - \lambda_1 \overline{x} - \lambda_2 \overline{x^2} \\ &- \lambda_3 \overline{x^3} - \cdots\Big] = 0 \end{aligned} \tag{4}$$

where

$$I[p(x), x] = p(x)\left\{-\ln p(x) + \lambda_0 + \lambda_1 x + \lambda_2 x^2 + \lambda_3 x^3 + \cdots\right\}.$$

However, the variation is independent of x. Therefore,

$$\begin{aligned} H &= \int_0^M I[p(x), x]\, dx \\ &= \int_0^M \left\{ (\partial I[p(x), x])/[\partial p(x)] \right\} \delta p(x) dx \\ &= 0. \end{aligned} \tag{5}$$

When H is an extremum, it is independent of small variations of $p(x)$, subject to the constraints which we have considered. This indicates that H is independent of $\delta p(x)$ and that $\partial I[p(x), x])/[\partial p(x)] = 0$. After differentiation, the following condition is established:

$$p(x) \left\{-1/[p(x)]\right\} - \ln p(x) + \lambda_0 + \lambda_1 x + \lambda_2 x^2 + \lambda_3 x^3 + \cdots = 0 \tag{6}$$

$$\begin{aligned} p(x) &= \exp(-1 + \lambda_0 + \lambda_1 x + \lambda_2 x^2 + \lambda_3 x^3 + \cdots) \\ &= \alpha \exp(\lambda_1 x + \lambda_2 x^2 + \cdots) \end{aligned} \tag{7}$$

where

$$\alpha = \exp(-1 + \lambda_0).$$

This is the variational form for $p(x)$ which gives maxima and minima for the integral H. We must now solve for the constants λ_0, λ_1, λ_2, λ_3, etc. in terms of the moments of the distribution $\overline{x}, \overline{x^2}, \overline{x^3}$, etc.:

$$\int_0^M p(x) x^n\, dx = \overline{x^n}. \tag{8}$$

There are $n + 1$ such equations which may be used to solve for the $n + 1$ constants $\lambda_1, \lambda_2, \cdots, \lambda_n$ and α. It should be noted that in solving for the constants of (7), several local maxima and minima distributions may result. However, it is only the absolute maximum distribution which is of interest here. As an example, consider the following case.

The Special Case of Fixed Mean

Here $\overline{x}$ is given, and

$$\lambda_2 = \lambda_3 = \lambda_4 = \cdots = 0.$$

Equation (7) becomes

$$p(x) = \alpha \exp[\lambda_1 x]. \tag{9}$$

Two equations result from (8):

[1] Margenau and Murphy, *The Mathematics of Physics and Chemistry*, 1943, ch. 6.

$$\int_0^M \alpha \exp [\lambda_1 x]\, dx = 1 \quad (10)$$

$$\int_0^M \alpha \exp [\lambda_1 x]\, x dx = \bar{x}. \quad (11)$$

Equation (10) normalies the distribution to have unity area. Using (10) and (11) results in a transcendental equation for

$$\lambda_1 = [\exp (\lambda_1 M - 1] / [(M - \bar{x}) \exp (\lambda_1 M) + \bar{x}] . \quad (12)$$

Also, a linear equation relating α to λ_1 may be derived, and is

$$\alpha = (1/M) - \lambda_1 [(M - \bar{x})/M] . \quad (13)$$

Given the point of truncation of the distribution M and the first moment $\bar{x}$, α and λ_1 may be found from (12) and (13).

As an example, consider the case for which $\bar{x} = 1$ and $M = 1.7$. This is the case which is plotted in Fig. 8. Two extreme distribution functions result. One of these is a delta function at $x = \bar{x}$. This obviously yields a minimum entropy distribution, since

$$H = - \int_0^M p(x) \ln p(x) dx$$

$$= - \infty.$$

However, a second solution is

$$p(x) = 0.328 \exp (0.63x).$$

The resulting entropy for this distribution is $H = 0.486$, and is the maximum valued distribution of interest.

Acknowledgment

The author wishes to acknowledge the help of Dr. K.A. Haines in both the experimental and theoretical aspects of this work.

B. Percy Hildebrand received his B.A.Sc. and M.A.Sc. degrees in Electrical Engineering in 1954 and 1956 respectively at the University of British Columbia and the Ph.D. degree in 1967 from the University of Michigan. During the years 1956 to 1958 he worked for Canadair Ltd. on missile guidance systems. After immigrating to the U.S. he held assignments at the Hallamore Electronics, Co., and Beckman Systems Division where he participated in the design of automatic test equipment. He then spent a year and a half in the Undersea Warfare Department of the Hughes Aircraft Co. in Fullerton, California, where he worked in the area of sonar signal processing. In 1961 he joined the Radar and Optics Laboratory of the University of Michigan where he was involved in imaging radar systems, optical data processing and holography under Professor Emmett N. Leith. After completing his Ph.D. in 1967 he accepted the position of Senior Research Scientist at Battelle-Northwest in Richland, Washington. Here, he has contributed significantly to the development of acoustical holography, resulting in a number of patents for scanning systems. In addition to his research he has taught graduate courses in control theory, optical processing and systems theory at the Joint Center for Graduate Study at Richland. Dr. Hildebrand is the author of more than 30 journal articles and co-author of the first comprehensive book on acoustical holography.

Aperture Size And Ambiguity Constraints For A Synthetic Aperture Radar

R.W. Bayma and P.A. McInnes
(Bayma — Environmental Research Institute of Michigan, Ann Arbor; McInnes — Univeristy of Sheffield, England)

The Record of the IEEE 1975 International Radar Conference (21-23 April 1975)
Pages 499-504

ABSTRACT

Aperture size and PRF constraints for a terrain mapping SAR are reviewed. Ambiguous Doppler responses are plotted as a function of PRF and normalized antenna beamwidth. A design procedure for selecting aperture size and PRF such that the ambiguous responses are below an acceptable level is presented.

INTRODUCTION

In order to map a planetary surface with a Synthetic Aperture Radar (SAR), the antenna pattern must adequately illuminate the required swath, and the radar pulse-repetition frequency (PRF) must adequately sample the along-track Doppler frequencies. However, a high PRF may generate range ambiguities. This paper presents some practical aperture size and PRF constraints for a terrain mapping radar. In order to simplify the presentation, it is assumed that the mapped swath is small compared to the slant range. This permits interleaving of transmit and receive pulses so that the maximum PRF is determined by the two-way propagation delay over the illuminated swath. It also allows the illumination beamwidths to be sufficiently narrow so that small angle approximations are valid.

MAPPING REQUIREMENTS

Consider the geometry shown in Figure 1. Let W be the required cross-track swath, ψ the grazing angle and R the average slant range. Then the required elevation beamwidth is given by

$$\beta_W = \frac{W \sin \psi}{R} \qquad (1)$$

Let L be the required along-track illumination interval. Then the required azimuth beamwidth is

$$\beta_L = \frac{L}{R} \qquad (2)$$

The dimension L is equal to the synthetic aperture length L_A plus the length of the patch processed during a synthetic aperture interval. In most references [1,2], a continuous strip mapping mode is considered for which the mapped interval is the desired azimuth resolution ρ_A. The required synthetic aperture length is approximately

$$L_A = \frac{R\lambda}{2\rho_A} \qquad (3)$$

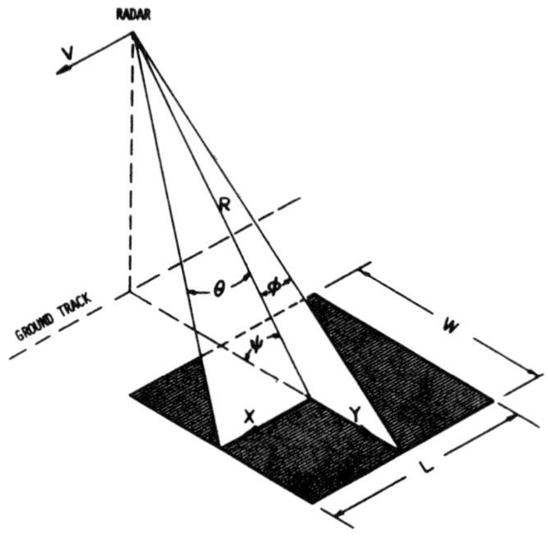

FIGURE 1. SYNTHETIC APERTURE RADAR GEOMETRY

DELAY-DOPPLER REQUIREMENTS

Let θ and ϕ be azimuth and elevation angles measured from boresight. Then the received Doppler frequency is given by

$$f = \frac{2V}{\lambda} \theta \quad \text{Hz} \qquad (4)$$

where V is the vehicle velocity and λ is the transmitted wavelength.

The two-way cross-track delay from boresight is related to elevation angle by

$$\tau = \frac{2R}{c \tan \psi} \phi \qquad (5)$$

where c is the propagation velocity.

Using Eqs. (2) and (4), the Doppler bandwidth required for processing is

$$B_{min} = \frac{2V}{\lambda} \cdot \beta_L = \frac{2VL}{R\lambda} \quad \text{Hz} \qquad (6)$$

Similarly, from Eqs. (1) and (5) or from Figure 1, the required delay interval over the swath is

$$T_{min} = \frac{2R}{c \tan \psi} \cdot \beta_W = \frac{2W \cos \psi}{c} \qquad (7)$$

Then to prevent aliasing of the azimuth Doppler and range delay signals, we require that the PRF satisfy the constraints:

$$B_{min} < PRF < 1/T_{min} \qquad (8)$$

If we now substitute Eqs. (6) and (7) for B_{min} and T_{min}, we arrive at the constraint

$$\frac{2LW \cos \psi}{R\lambda} < \frac{c}{2V} \qquad (9)$$

But the minimum value of L is L_A. Substituting Eq. (3) for L, we get the well known result (Ref. [2], pg. 29)

$$\frac{W \cos \psi}{\rho_A} < \frac{c}{2V} \qquad (10)$$

i.e., the ratio of unambiguous slant range interval to along-track resolution is bounded by one-half the ratio of propagation velocity to vehicle velocity.

However, the bounds of Inequalities (8) - (10), cannot be made arbitrarily tight for a terrain mapping radar unless the antenna directivity is zero outside the mapped swath. This would require an infinite aperture, and hence is not realizable. In this paper we present a practical design procedure and give results for an SAR using a uniformly illuminated or cosine weighted rectangular aperture.

AMBIGUITY SPECTRUM

Let $A(\theta,\phi)$ be the far field amplitude pattern of the radar antenna. Then if we assume that the slant range and grazing angle are essentially constant, the signal energy from a resolution element of area $\Delta x \Delta y$ is proportional to

$$S_{XY}(x,y) = |A^2(\theta(x),\phi(y))|^2 |\sigma_o(x,y)| \Delta x \Delta y \qquad (11)$$

where x and y are ground coordinates relative to boresight and $\sigma_o(x,y)$ is the complex terrain reflectivity coefficient.

To illustrate the effect of PRF sampling ambiguities, it is necessary to relate θ and ϕ to Doppler and delay. Then the signal energy is proportional to

$$S(f,\tau) = G(f,\tau)\sigma(f,\tau) \qquad (12)$$

where we set

$$\sigma(f,\tau) = \sigma_o(x(f),y(\tau))\Delta x \Delta y \qquad (13)$$

and we use Eqs. (4) and (5) to define the two-way directivity in delay-Doppler coordinates as

$$G(f,\tau) = \left|A^2\left(\frac{\lambda}{2V} f, \frac{c \tan \psi}{2R} \tau\right)\right|^2 \qquad (14)$$

Equation (12) defines the response for a single pulse. If we assume that the response from non-overlapping clutter patches add noncoherently, the sampled return is given by

$$S^*(f,\tau) = \sum_{m,n=-\infty}^{\infty} S\left(f + m\,\mathrm{PRF}, \tau + \frac{n}{\mathrm{PRF}}\right) \qquad (15)$$

The response is periodic in f and τ with periods equal to PRF and 1/PRF respectively. The response of interest is centered at zero delay and zero Doppler. Then we can rewrite Eq. (15) as

$$S^*(f,\tau) = S(f,\tau) + S_A(f,\tau) \qquad (16)$$

where $S(f,\tau)$ is the desired response, and

$$S_A(f,\tau) = \sum_{\substack{m,n=-\infty \\ m\neq n\neq 0}}^{\infty} S\left(f + m\,\mathrm{PRF}, \tau + \frac{n}{\mathrm{PRF}}\right) \qquad (17)$$

is the ambiguous delay-Doppler response. For an ideal antenna pattern,

$$S(f,\tau) = 0, \ |f| > \frac{\mathrm{PRF}}{2}, \ |\tau| > \frac{1}{2\cdot\mathrm{PRF}} \qquad (18)$$

and hence the ambiguous response is zero, and $S^*(f,\tau) = S(f,\tau)$.

Figure 2 illustrates the ambiguous Doppler regions for the case of a realistic azimuth illumination pattern. Doppler frequencies displaced by integer multiples of the PRF are completely ambiguous. By substituting delay for Doppler, and pulse repetition period (PRP) for PRF, the same figure illustrates the range ambiguity structure of an SAR. Terrain intervals with slant-range delay displaced by integer multiples of the PRP are completely ambiguous. From Figure 2, we note that Doppler ambiguities can be reduced by increasing the PRF. However, from Eq. (15), we observe that this will simultaneously increase the amplitude of the range ambiguities. A reasonable solution to this design problem is to choose the PRF and a minimum antenna aperture such that at each point in the mapped delay-Doppler space, the total ambiguous response is below a fixed level.

Uniformly Illuminated Antenna Aperture

In order to obtain quantitative results, we consider the case of a uniformly illuminated rectangular aperture. Then the normalized two-way far-field directivity in delay-Doppler coordinates is

$$G(f,\tau) = \mathrm{sinc}^4(f/B)\mathrm{sinc}^4(\tau/T) \qquad (19)$$

where $\mathrm{sinc}(x) = \sin \pi x/\pi x$, and

$$B = \frac{2V}{\lambda} \cdot \frac{\lambda}{d_A} = \frac{2V}{d_A} \qquad (20)$$

and

$$T = \frac{2R}{c \tan \psi} \cdot \frac{\lambda}{d_R} \qquad (21)$$

are the Doppler frequency and delay of targets located at the first azimuth and elevation nulls of

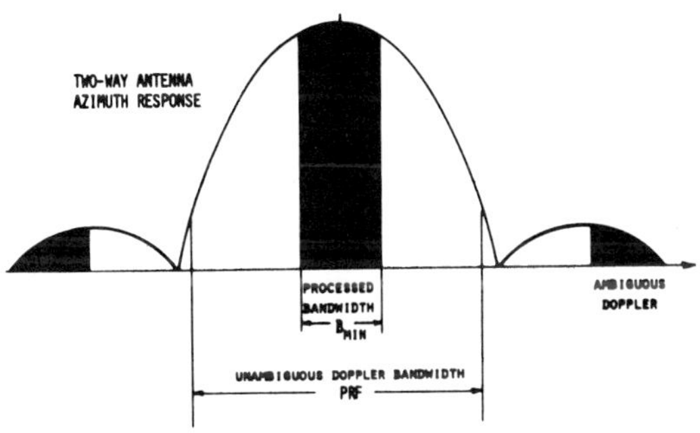

FIGURE 2. ILLUSTRATION OF DOPPLER SAMPLING AMBIGUITIES.

the antenna pattern, and d_A and d_R are azimuth and elevation aperture dimensions respectively. In this paper, we illustrate the ambiguity spectrum for the cases in which the PRF is set to 6 dB, 16 dB or 26 dB bandwidths which are given respectively by

$$PRF = B_{6dB} = 0.885\ B \tag{22}$$

$$PRF = B_{16dB} = 1.356\ B \tag{23}$$

or

$$PRF = B_{26dB} = 1.615\ B \tag{24}$$

and the corresponding delay response is down 6 dB, 16 dB or 26 dB at $\pm 1/(2 \cdot PRF)$. This requires the elevation aperture d_R to be chosen so that the delay parameter T satisfies the equation

$$0.885\ T = T_{6dB} = \frac{1}{PRF} = PRP \tag{25}$$

$$1.356\ T = T_{16dB} = \frac{1}{PRF} = PRP \tag{26}$$

or

$$1.615\ T = T_{26dB} = \frac{1}{PRF} = PRP \tag{27}$$

where the PRF is specified by Eqs. (22), (23), or (24). The mapped swath will then be determined by the 6 dB elevation beamwidth. Note then, that as the PRF is increased, the unambiguous delay interval and mapped swath must decrease.

Figures 3 and 4 illustrate the mid-range and far-range ambiguity spectrum as a function of normalized Doppler frequency for a PRF equal to the 6 dB azimuth bandwidth. For case (a) in which the PRP just covers the mapped swath, the delay response is only down 6 dB at the edge of the unambiguous interval, and the ambiguity levels exceed the mainlobe response. This, of course, is expected, since in this case the ambiguous delay regions are well illuminated by the antenna.

If the elevation beamwidth is reduced so that the delay response is down 16 dB at the edges of the interpulse period, then both the mid- and far-range ambiguity levels are at least 20 dB down in the region of zero Doppler. This satisfies the range ambiguity requirement, but we note that only a small fraction of the total Doppler bandwidth is free of excessive azimuth ambiguities.

By increasing the PRF to the 16 dB bandwidth we obtain the responses shown in Figures 5 and 7. In this case, the signal-to-ambiguity level over the 6 dB delay and Doppler intervals varies from 21 dB to 23 dB at mid-range and from 18 to 21 dB at near or far range if the PRP is at least as large as the 16 dB delay interval.

Responses for the PRF equal to the 26 dB Doppler bandwidth are shown in Figures 7 and 8. Here we note that increasing the PRF does not significantly improve the signal-to-ambiguity level in the regions of interest.

If we set

$$B_{min} = B_{6dB} \tag{28}$$

and

$$T_{min} = T_{6dB} \tag{29}$$

and use the PRF constraint

$$B_{16dB} < PRF < \frac{1}{T_{16dB}} \tag{30}$$

then for a uniformly illuminated aperture, we have the numerical result

$$1.53\ B_{min} \leq PRF \leq \frac{1}{1.53\ T_{min}} \tag{31}$$

where the factor 1.53 is the ratio of 16 dB to 6 dB beamwidth, and B_{min} and T_{min} are determined from Eqs. (6) and (7) and mapping requirements.

Using the factors given in Eq. (29), the practical mapping constraint becomes

$$\frac{W \cos \psi}{\rho_A} < \frac{c}{2(1.53)^2 V} = \frac{c}{4.7V} \tag{32}$$

i.e., the mapping capability of an aperture limited SAR is less than half that of an ideal system.

<u>Cosine Weighted Antenna Aperture</u>

If cosine weighted antenna apertures are used the illumination function becomes

$$f(x,y) = \cos(\pi x/d_A) \cdot \cos(\pi y/d_R) \tag{33}$$

and the normalized two-way directivity is

$$G(f,\tau) = \left[\frac{\cos(\pi f/B)}{1-(2f/B)^2} \cdot \frac{\cos(\pi\tau/T)}{1-(2\tau/T)^2}\right]^4 \tag{34}$$

where the normalizing constants B and T refer to the nulls of a uniform aperture and are still given by Eqs. (20) and (21). The responses for PRP and PRF equal to the 6 dB, 16 dB and 26 dB delay and Doppler intervals are shown in Figures 9 through 14. Then if we set

$$B_{min} = B_{6dB} = 1.186\ B \tag{35}$$

$$T_{min} = T_{6dB} = 1.186\ T \tag{36}$$

and

$$1.856\ B < PRF < \frac{1}{1.856T} \tag{37}$$

the signal-to ambiguity level varies from 30 dB to 40 dB at midrange and from 24 dB to 29 dB at far range. The ratio of 16 dB to 6 dB intervals is 1.56, hence the essential results are the same as the uniform case, except that larger antenna apertures must be used for the same swath requirements and the same PRF.

CONCLUSION

A practical procedure for estimating the antenna aperture size and PRF requirements for a terrain mapping SAR is to size the antenna such that the required swath and processed Doppler bandwidth are within the 6 dB two-way antenna pattern, and then select the PRF such that the unambiguous delay and Doppler intervals are within the 16 dB response.

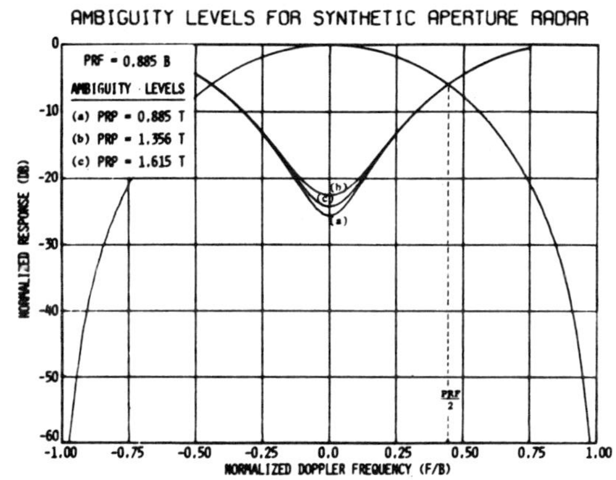

FIGURE 3. MID-RANGE AZIMUTH RESPONSE WITH UNIFORM ANTENNA ILLUMINATION

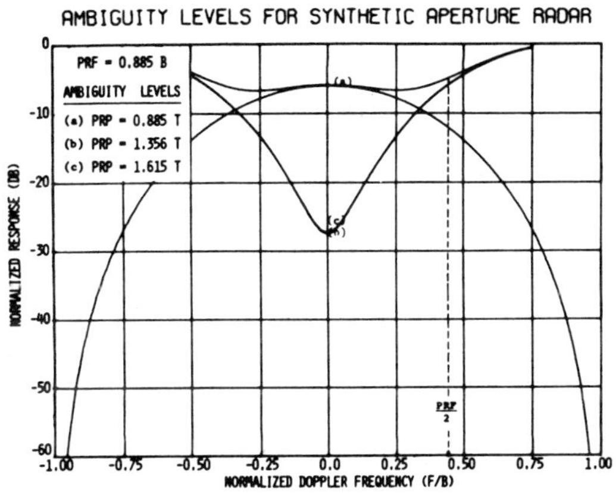

FIGURE 4. FAR-RANGE AZIMUTH RESPONSE WITH UNIFORM ANTENNA ILLUMINATION

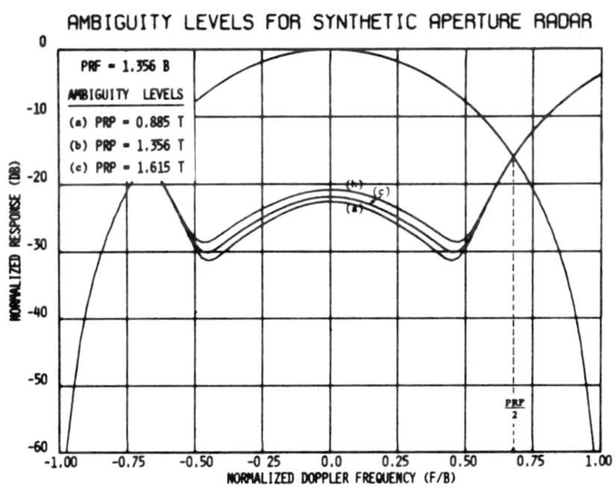

FIGURE 5. MID-RANGE AZIMUTH RESPONSE WITH UNIFORM ANTENNA ILLUMINATION

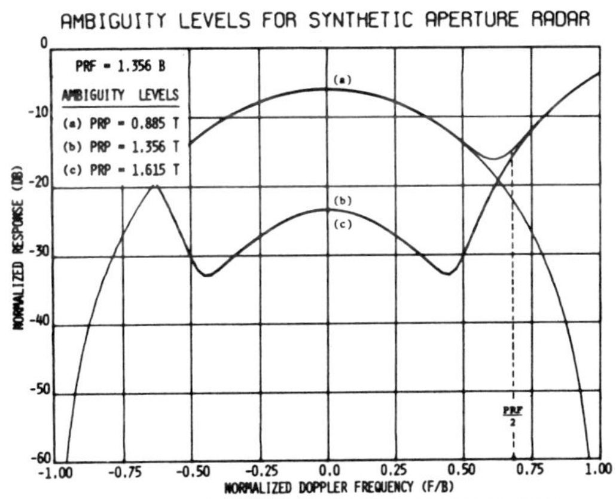

FIGURE 6. FAR-RANGE AZIMUTH RESPONSE WITH UNIFORM ANTENNA ILLUMINATION

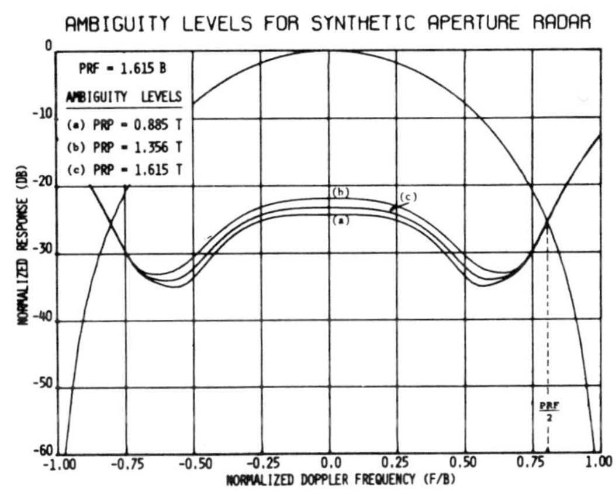

FIGURE 7. MID-RANGE AZIMUTH RESPONSE WITH UNIFORM ANTENNA ILLUMINATION

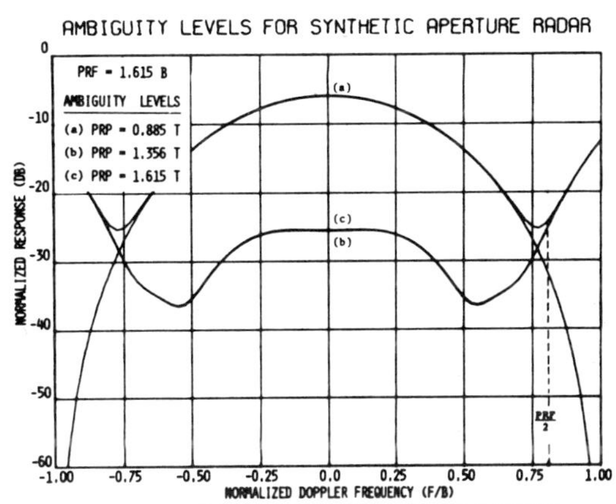

FIGURE 8. FAR-RANGE AZIMUTH RESPONSE WITH UNIFORM ANTENNA ILLUMINATION

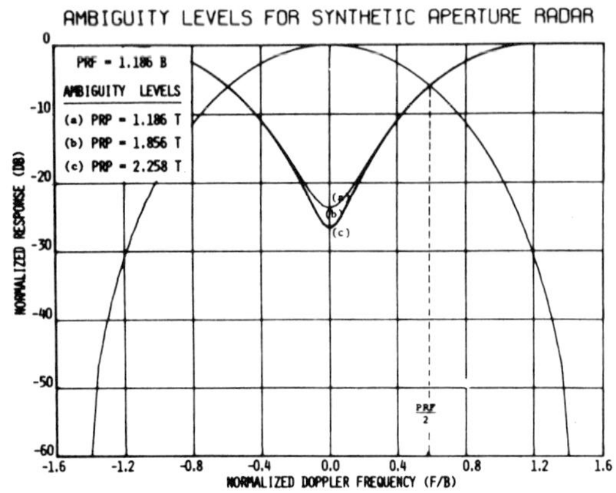

FIGURE 9. MID-RANGE AZIMUTH RESPONSE WITH COSINE ANTENNA ILLUMINATION

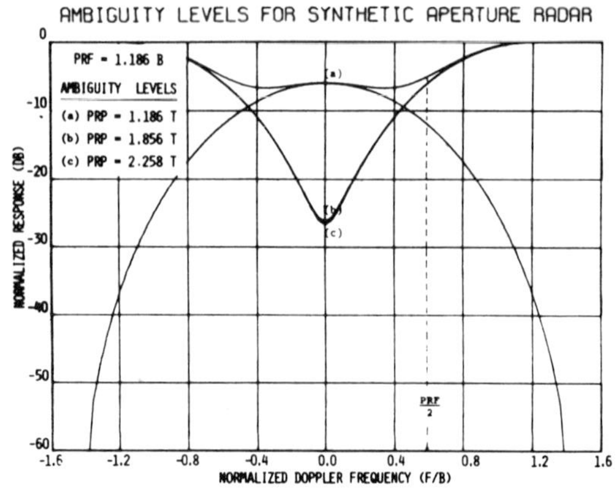

FIGURE 10. FAR-RANGE AZIMUTH RESPONSE WITH COSINE ANTENNA ILLUMINATION

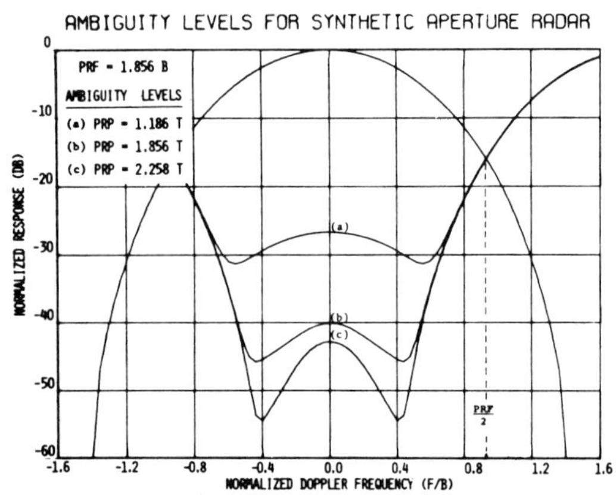

FIGURE 11. MID-RANGE AZIMUTH RESPONSE WITH COSINE ANTENNA ILLUMINATION

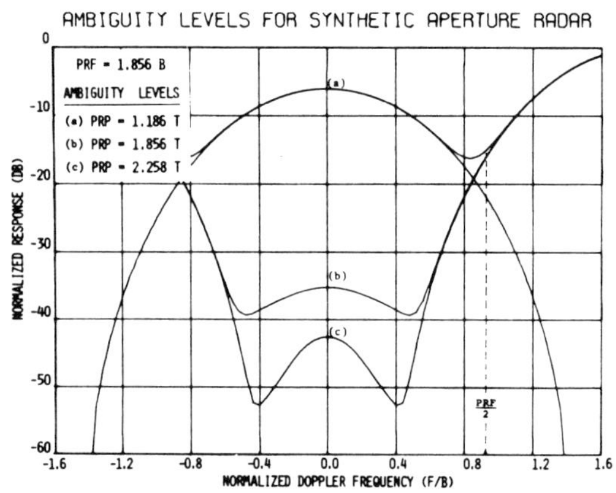

FIGURE 12. FAR-RANGE AZIMUTH RESPONSE WITH COSINE ANTENNA ILLUMINATION

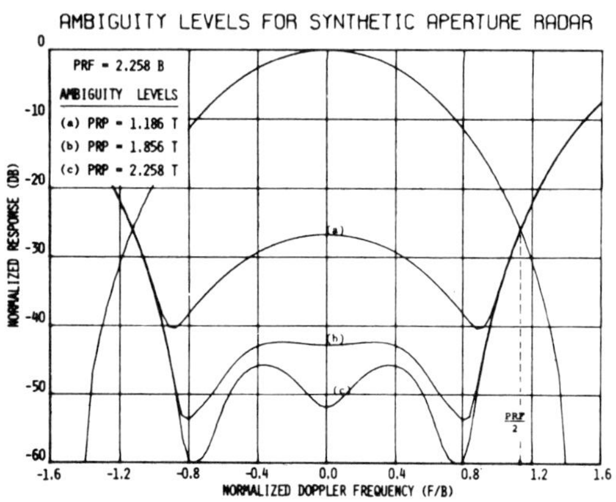

FIGURE 13. MID-RANGE AZIMUTH RESPONSE WITH COSINE ANTENNA ILLUMINATION

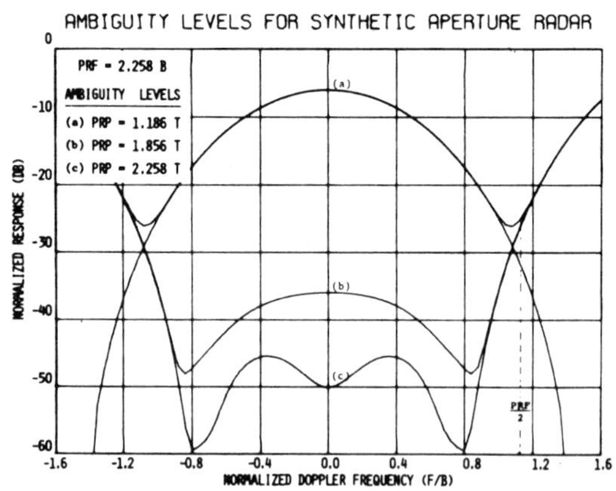

FIGURE 14. FAR-RANGE AZIMUTH RESPONSE WITH COSINE ANTENNA ILLUMINATION

For typical antenna patterns, the unambiguous intervals will be approximately 1.5 times the mapping requirements, and the signal-to-ambiguity levels will be approximately 20 dB over the mapped swath.

ACKNOWLEDGEMENTS

This work was originally performed by ERIM for Martin-Marietta Aerospace, Denver Division, during the period October 1972 through August 1973 under support from NASA Ames Research Center, Contract NAS2-7204. During this period, P. A. McInnes was a visiting ESRO/NASA Fellow at ERIM and the Electrical and Computer Engineering Department of The University of Michigan.

REFERENCES

1. William M. Brown, Synthetic Aperture Radar, IEEE Transactions, AES-3, pp. 217-229, 1967.
2. Robert O. Harger, Synthetic Aperture Radar Systems, Academic Press, 1970.

Synthetic Aperture Imaging Radar And Moving Targets

R.K. Raney
(Institute of Science and Technology, University of Michigan, Ann Arbor)

IEEE Transactions on Aerospace and Electronic Systems, Vol. AES-7, No. 3 (May 1971)
Pages 499-505

Abstract

This paper considers the effects of slowly moving targets as they appear in the output of an airborne coherent side-looking synthetic aperture imaging radar. The image of a moving reflector is described, and two approaches to airborne moving target indication (AMTI) are summarized. It is shown that the effects of target movement are decreased as the radar scan rate is increased, and are increased as the (Doppler processed) compression ratio is increased.

Symbols

$a_c(a_r)$ target acceleration in the cross-range (range) direction
$a(x)$ effective amplitude weighting of the azimuth signal
B_a Doppler bandwidth of the azimuth signal [see (8)]
$\beta \cong \lambda/d$ real antenna beamwidth (radians)
d horizontal aperture dimension of the real radar antenna
D phase center separation of dual antenna array
δt time delay between reception of dual observation signal pair
ΔR range image smear due to excessive radar/target motion [see (13)]
$\Delta \rho$ azimuth image defocus due to radial target acceleration and cross-range target velocity [see (18)]
Δv_r velocity resolution of two differentially moving targets in same spatial resolution cell
ΔX azimuth target shift due to radial target velocity [see (19)]
$\epsilon_{\dot{c}} = v_c/V$ cross-range velocity motion coefficient
$\epsilon_{\ddot{r}} = \dfrac{a_r R_0}{V^2}$ radial acceleration motion coefficient
$\epsilon_{\dot{r}} = v_r/V$ Doppler (radial velocity) motion coefficient
$f_D = 2\dfrac{v_r}{\lambda}$ Doppler frequency shift (approximate)
$k = 2\pi/\lambda$ wave number
K azimuth compression ratio of a fully processed synthetic aperture imaging radar [see (6)]
λ radar wavelength
μ scale factor $\leqslant 1$, useful in Doppler shift AMTI studies
$m_\Delta(x)$ difference signal obtained by dual antenna delay-and-subtract processing [see (24)]
$m(x)$ complex azimuth signal from a moving target [see (12)]
$\tilde{m}, \tilde{s}$ Fourier transform of m and s, respectively
R_0 reference range at $t = 0$
$R(t)$ radar/target slant range variation as a function of time
$\rho_r(\rho_a)$ ultimate resolution of a stationary target in the range (azimuth) direction
$s(x)$ complex azimuth signal from a nominally stationary target [see (11)]
T time interval during which a given stationary target is illuminated by a sequence of pulses from the radar [see (7)]
TBP_a azimuth time-bandwidth product [see (9)]
V aircraft velocity along the line of flight, and, therefore, the along-track or azimuth scan rate
$v_c(v_r)$ target velocity in the cross-range (range) direction
$x = Vt$ scaled along-track spatial coordinate.

Manuscript received June 12, 1970; revised November 15, 1970.

This work was performed by the Radar and Optics Laboratory of the University of Michigan's Willow Run Laboratories under Contract F33615-C-69-1015, sponsored by the Air Force Avionics Laboratory.

I. Introduction

Several unclassified papers have appeared in the literature describing synthetic aperture side-looking imaging radars (SAR) [1]-[5]. The response and sensitivity of such a radar to a slowly moving scatterer has not been considered. It is the purpose of this paper to review this question. The reader is assumed to be familiar with the basic principles of synthetic aperture coherent imaging radars.

Target motions degrade the radar target image, although in special circumstances striking improvement in image quality may be achieved if *a priori* knowledge of the target behavior is available [6], [7]. Rapidly moving scatterers are discussed in the context of a pulsed Doppler radar by Kelly and Wishner [8]. In this paper we are concerned with the decrease in accuracy and resolution of the image of a rectilinearly slowly moving target, and with the possibility of achieving moving target detection.

Target/Radar Range Variation

The classical airborne moving target indication (AMTI) viewpoint is that the Doppler shift of a moving target may be observable directly in each return signal. Since the range bandwidth of an imaging radar is too large to allow such an observation, target-induced phase perturbations are evidenced only in comparing a sequence of return signals in the cross-range or Doppler dimension. This is equivalent to observing a slowly varying change in the carrier phase on a pulse-to-pulse basis. Consequently, the resulting moving target modulation appears primarily in the cross range dimension in scanning high resolution radars. Simultaneously, a SAR uses the cross-range Doppler structure to achieve improved azimuth resolution. The target-induced phase perturbations interact with the SAR signal processing, leading to the particular image properties discussed in this paper.

The cross-range phase structure of the coherent signal from a moving target depends on the (slowly varying) slant range $R(t)$ from the radar to the target in question (Fig. 1). The conventional SAR signal structure, determined by the ground track velocity of the scanning radar, is modified by the target motions. For the geometry of a side-looking scanning radar, in terms of the target velocities and accelerations (v_r, a_r; v_c, a_c) (projected onto the slant range vector) in the range and cross-range directions, respectively, the range variation with time may be written

$$R^2(t) = \left[Vt - v_c t - (1/2)a_c t^2\right]^2 + \left[R_0 - v_r t - (1/2)a_r t^2\right]^2 \qquad (1)$$

where V is the scan rate (sensor vehicle speed) and R_0 is the reference range at $t = 0$. A simple expression may be derived for $R(t)$ by expansion of (1) in a Maclaurin series. Retaining terms through the quadratic, over the time T of

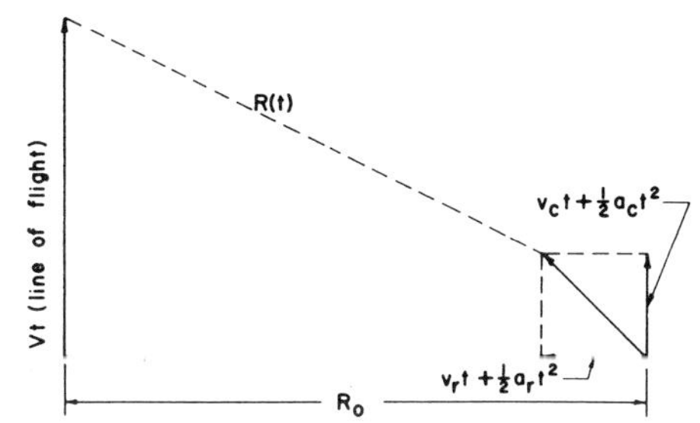

Fig. 1. Moving target/radar geometry in the slant range plane.

target illumination, the range variation is

$$R(t) \cong R_0 - v_r t + [(V - v_c)^2 - R_0 a_r] \frac{t^2}{2R_0}, \qquad t \in T.$$

It is convenient to normalize by the vehicle scan rate V, and convert to the scaled spatial coordinate $x = Vt$. Then

$$R(x) \cong R_0 - \epsilon_{\dot r} x + [(1 - \epsilon_{\dot c})^2 - \epsilon_{\ddot r}] \frac{x^2}{2R_0}, \quad x \in VT. \quad (2)$$

The normalized motion coefficients are

$$\epsilon_{\dot r} = \frac{v_r}{V} \text{ (Doppler or radial velocity)} \qquad (3)$$

$$\epsilon_{\ddot r} = \frac{a_r R_0}{V^2} \text{ (radial acceleration)} \qquad (4)$$

$$\epsilon_{\dot c} = \frac{v_c}{V} \text{ (cross-range velocity).} \qquad (5)$$

These coefficients in combination with the basic parameters of the side-looking radar may be used to describe the effects of target motion on the resulting output image.

Radar Resolution Parameters

Before describing the effect of target motion on the image, we need to summarize the conventional response. The imaging radar we are considering achieves resolution ρ_r in range and ρ_a in azimuth by known techniques (1) – (5). Normally for such a SAR, the azimuth resolution ρ_a is considerably smaller than the unprocessed resolution $R\beta$ that would be achieved by the real antenna beamwidth β at range R without coherent Doppler processing.[1] One measure of the resolution improvement is the azimuth compression ratio K (the ratio of pre-processed resolution $R\beta = R(\lambda/d)$ to post-processed resolution $\rho_a = d/2$, where d is the horizontal antenna aperture dimension). This SAR compres-

[1] The real beamwidth β is assumed sufficiently small such that $\beta \cong \sin \beta$. If this constraint is not satisfied, then $\sin \beta$ should be used in the sequel where appropriate.

sion ratio is

$$K = \frac{\lambda R}{2\rho_a^2} \ . \tag{6}$$

For processing over the full beamwidth, K is equal to the time-bandwidth product TBP_a of the azimuth channel of the radar. To show this, we calculate the time interval T that a given resolution cell is in the real antenna of beamwidth β,

$$T = \frac{R\beta}{V} \text{ seconds} \ . \tag{7}$$

Likewise, the Doppler bandwidth B_a of the azimuth signal from a nominally stationary target is

$$B_a = \frac{2}{\lambda} \beta V \text{ Hz.} \tag{8}$$

The azimuth time-bandwidth product of such a radar is thus

$$TBP_a = \frac{2\beta^2 R}{\lambda} \ . \tag{9}$$

If the ultimate azimuth resolution is achieved, then

$$\beta = \frac{\lambda}{2\rho_a} \ ,$$

which, when substituted into (9), shows that TBP_a equals K. In terms of these parameters, the time during which the target is illuminated is

$$T = \frac{R\lambda}{2V\rho_a} \ . \tag{10}$$

II. Image Perturbations

The resolution improvement predicted by (6) is instrumented by constructing a filter matched to the structure of the received signal [4], [5], which is determined by the modulation on the ranging pulse (assumed to be a simple short CW burst in this paper) and the Doppler modulation across the set of signals received from each target. The azimuth phase history of the signal from an individual scatterer as a function of x is given by $2kR(x)$, where the wave number $k = 2\pi/\lambda$, as usual. If the scatterer is stationary, then the motion coefficients are zero, and the low-pass equivalent signal in the Doppler dimension from a stationary scatterer is, from (2),

$$s(x) = a(x) \exp j \frac{k}{R_0} x^2 \tag{11}$$

where $a(x)$ represents the azimuth weighting of the real antenna, the target reflectivity, and the constant but unknown phase $2kR_0$. The signal would appear in a two-dimensional memory format as in Fig. 2(A), for which the corresponding processed image would be as in Fig. 2(B).

This performance is that expected of the matched processor. In the event that the target perturbs the signal,

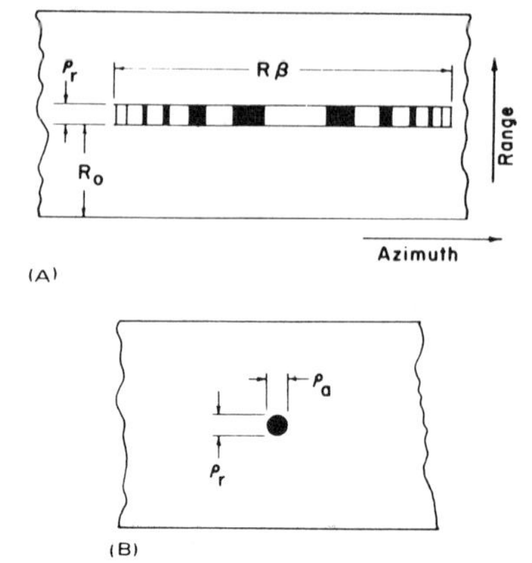

Fig. 2. (A) Data format of stationary target signal. (B) Image of stationary target.

such results are not obtained since the processor, in general, is not matched to the perturbed signal. In addition, if cross-range target velocity is present, then the resultant foreshortening of the along-track coordinate scaling factor is reflected in the antenna azimuth weighting. The received signal from a moving target is, therefore, from (2),

$$m(x) \cong a[(1 - \epsilon_{\dot c})x] \exp(-j2k\epsilon_{\dot r}x) \cdot \left\{ \exp + jk \frac{x^2}{R_0} [(1 - \epsilon_{\dot c})^2 - \epsilon_{\ddot r}] \right\} . \tag{12}$$

We expect that, compared to an equivalent stationary scatterer, the image corresponding to (12) will be smeared in range and shifted in azimuth due to $\epsilon_{\dot r}$, and defocused due to $\epsilon_{\dot c}$ and $\epsilon_{\ddot r}$. These effects, which are discussed below, are sketched in Fig. 3.

The detailed effects of target motion should be studied using a modified ambiguity function to investigate the mismatch of $s(x)$ and $m(x)$, represented by (12). The intermediate results presented here bear strong resemblance to problems encountered in studying swept sonar pulses [9]. The conclusions reached below are somewhat more specialized than the question of "Doppler tolerance" that is of concern in sonar, and in general wideband signal theory.

Radial Velocity and Range Smear

One consequence of fine range resolution is that excessive relative radial target movement, in general, causes the target to translate through more than one range resolution cell during the radar observation time T, a phenomenon known as "range walk". The radial target total displacement ΔR is, from (2),

$$\Delta R = VT \left| \epsilon_{\dot r} - [(1 - \epsilon_{\dot c})^2 - \epsilon_{\ddot r}] \frac{VT}{2R_0} \right| . \tag{13}$$

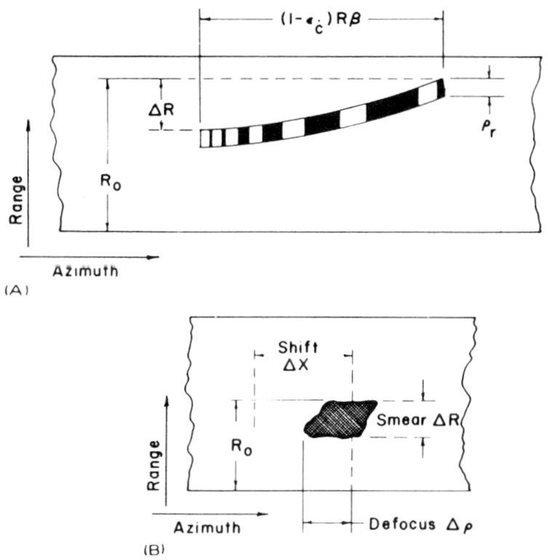

Fig. 3. (A) Moving target signal. (B) Image of a moving target if processed as for a stationary target.

In the absence of target motion, ΔR is due solely to the radar-induced range variation and is called "range curvature", forming a lower bound on the attainable resolution ρ_r that can be achieved without compensation for the effect. The lower bound is formed by letting $\epsilon_{\dot r} = \epsilon_{\dot c} = \epsilon_{\ddot r} = 0$ in (13), which leads to the minimum resolution limited by range curvature for stationary targets

$$\frac{\lambda K}{4} < \rho_r . \tag{14}$$

The principal range walk component due to target motion in (13) is due to the radial velocity component $\epsilon_{\dot r}$. In practice, additional quadratic perturbations due to target motion as influenced by the coefficients $\epsilon_{\dot c}$ and $\epsilon_{\ddot r}$ in (13) are dominated by the range curvature effect, since

$$|\epsilon_{\dot c}| \ll 1$$

$$|\epsilon_{\ddot r}| \ll 1.$$

We therefore have the constraint on range resolution of a radially moving target,

$$|\epsilon_{\dot r} VT| < \rho_r ,$$

which, by (3), may be written

$$v_r T < \rho_r . \tag{15}$$

This result states clearly that the target cannot move through more than one range resolution cell ρ_r during the observation time T if the image is not to be smeared in range. Through (10), this bound may be written

$$\left|\frac{v_r}{V}\right| \frac{\lambda R}{2\rho_a} < \rho_r .$$

If the resolution cell is symmetrical, $\rho_a = \rho_r = \rho$, then the bound may be expressed in terms of the azimuth compression ratio (6) and the target Doppler coefficient, leading to the *fundamental constraint* in the dimensionless and normalized form,

$$|\epsilon_{\dot r}| < 1/K \tag{16}$$

if the image of a moving target is not to be smeared in the range dimension due to range walk. The severity of this constraint is decreased as the radar scan rate V is increased. Likewise, for more sophisticated data processing, such that the azimuth compression ratio K is increased, the constraint is tightened. Note that K increases linearly with range R for a given radar.

It is possible to adjust the (two-dimensional) matched filter so that the image is not smeared—although at the expense of inducing a smear on all other target images. The moving target data format [Fig. 3(A)], in effect, is sheared through the angle $\epsilon_{\dot r}$ radians due to radial target velocity. A corresponding shear of the filter would eliminate the smear of the image. In an optical processor, a rotation of a cylindrical lens could realize this adjustment.

Radial Acceleration, Tangential Velocity, and Azimuth Defocus

The coefficient of the x^2 term in (12) represents a linear frequency modulated perturbation of the nominal azimuth signal $s(x)$. For the cognizenti, this term obviously corresponds to a change in the optical focal length of $m(x)$ compared to $s(x)$, but does not otherwise disrupt the signal structure [3]. From this point of view, an upper bound on $\epsilon_{\dot c}$ and $\epsilon_{\ddot r}$ is determined by the depth of focus of the processed signal. An equivalent criterion based on the principle of stationary phase requires that the phase perturbation not exceed π radians over the observation interval T if the image is not to be defocused. This implies the constraint, from (12),

$$\frac{k}{R}\left|(1 - \epsilon_{\dot c})^2 - \epsilon_{\ddot r} - 1\right| (VT)^2 < \pi .$$

For typical cross-range target velocities such that $|\epsilon_c| \ll 1$, this equation simplifies to the normalized form

$$\left|2\epsilon_{\dot c} + \epsilon_{\ddot r}\right| < 1/K . \tag{17}$$

For a defocus $\Delta\rho$ of zero to several resolution elements, the defocused resolution is given by

$$\Delta\rho = \rho[1 + K^2(2\epsilon_c + \epsilon_r)^2]^{1/2} . \tag{18}$$

The constraint of (17) must be satisfied if the defocusing due to target motion is not to exceed the size of the resolution cell. We see that as the azimuth processing becomes more demanding, corresponding to increasing the compression ratio K, allowable target motions are reduced (expressed by the coefficients $\epsilon_{\dot c}$ and $\epsilon_{\ddot r}$). Furthermore, for a given radar and given range, (4), (5), and (18) show that image sensitivity to radial target acceleration or tangential velocity is decreased as the scan rate V is increased.

The factor of two on $\epsilon_{\ddot{c}}$ in (17) is of interest. The defocus component, corresponding to a mismatched quadratic term, accounts for $(1)\epsilon_{\dot{c}}$, and the actual cross-range displacement of the target accounts for an additional $(1)\epsilon_{\dot{c}}$. Thus, (17) includes some *smear* in addition to *defocus*. It is clear, however, that these two phenomena are ambiguously combined, together with the effects of radial target acceleration, producing the total apparent azimuth defocus of the image.

Radial Velocity and Accuracy

A synthetic aperture side-looking radar maps all targets at an azimuth angle corresponding to a reference Doppler frequency that is nominally equivalent to the center of the side-looking antenna pattern. However, if a target is moving radially with respect to the aircraft, then the change in angle at which the reference Doppler frequency occurs is $\epsilon_{\dot{r}} = v_r/V$ (for small angles). This corresponds to an azimuth image shift and consequent mapping error [Fig. 3(B)] of $\Delta X = Rv_r/V$, strongly suggesting that for accurate location of moving targets, a relatively high-speed surveillance vehicle should be used. Strictly speaking, if a large error ΔX occurs such that the azimuth error angle $\epsilon_{\dot{r}}$ exceeds the real beamwidth β, then the image may be reshifted by an integral number of beamwidths to satisfy the *a priori* condition that the target was, in fact, in the beam when it was illuminated by the radar. Thus, *the azimuth mapping relative accuracy (image shift) of a target moving radially with velocity* v_r *is*

$$\Delta X = R\epsilon_{\dot{r}} \text{ modulo } R\beta . \qquad (19)$$

The corresponding Doppler shift f_D is often sufficiently large so that the azimuth bandwidth of the system is exceeded, causing the target image to be obliterated. Thus, the system Doppler bandwidth of an AMTI imaging radar must be several times larger than B_a of (8).

It should be observed that if the proper target position is known, then the shift may be measured, and the radial speed of the target at the time of radar observation may be estimated. This is a technique suitable only for specialized instances. The velocity resolution Δv_r of such an estimate is

$$\Delta v_r = V \frac{\rho}{R}.$$

Two targets in the same resolution element ρ with different radial velocities may be resolved if their velocities differ by Δv_r where $\Delta v_r > \rho(V/R)$, which follows directly from the velocity resolution limit. There is a singular ambiguity in this result, however. Two targets at the same range element that are separated in azimuth by ΔX may be superposed and unresolvable in the output image if their velocities differ by $\Delta v_r = V(\Delta X/R)$.

III. Detection

The signal corresponding to a moving scatterer (12) as it arrives in a synthetic aperture imaging radar suggests several possibilities for detecting moving targets. We are assuming that the targets of interest are moving relatively slowly with respect to the scattering environment, so that (12) represents the predominant target characteristics. The first-order perturbation term is, of course, the Doppler frequency shift, resulting from radial target velocity. Second-order perturbations are produced by cross-range target velocity or radial target acceleration. Another second-order effect introduced above is image smearing due to target traversal of range resolution cells (range walk) during radar observation.

Detection and Estimation Based on Second-Order Terms

Consider the second-order effects. For small defocusing or smearing errors, the most noticeable result is an apparent reduction in output SNR of 3 dB for each mismatch of K^{-1} [see (3) – (5), (15), (17), and (18)]. Since these reductions result from true mismatches, some recovery is possible by retuning the processing filter for optimum output for each moving target in question. This requires two adjustments, one for azimuth focus $(2\epsilon_{\dot{c}} + \epsilon_{\ddot{r}})$ and one for range walk $(\epsilon_{\dot{r}})$. Therefore, at the price of increasing the dimensionality of the processing filter by two, it is theoretically possible to obtain a parameter estimation (with resolution K^{-1}) on $\epsilon_{\dot{r}}$ and $(2\epsilon_{\dot{c}} + \epsilon_{\ddot{r}})$ of the moving target. If this is desirable, then one designs the radar to accentuate these effects, increasing the azimuth compression ratio K and decreasing the scan rate V of the radar. The effectiveness of this approach depends on several implicit and explicit assumptions. First, it depends directly on the signal-to-clutter ratio (SCR). [SCR is defined for such applications as the ratio of the peak of the processed moving target image (optimally focused) to that of a typical competing processed stationary target image (optimally focused).] Even for a small SCR, it can be shown that a relatively large $\epsilon_{\dot{r}}$ or $(2\epsilon_{\dot{c}} + \epsilon_{\ddot{r}})$ would be required for detection using such a technique. Second, the focus parameter is one-to-one ambiguous in $\epsilon_{\dot{c}}$ and $\epsilon_{\ddot{r}}$, reflecting the cross-range target velocity and radial target acceleration, respectively, so that any particular estimations based on this approach also would be ambiguous. Third, these results assume a coherently integrable point scatterer. If the target is distributed (or is to be distinguished from a distributed stationary target), or if the coherence length of the signature is less than T (11), then these results must be considered optimistic. Clearly, the relevance of these three qualifications on the direct use of the second-order effects in moving target detection must be judged as they pertain to a given imaging radar problem.

The balance of this discussion summarizes AMTI based on appropriate Doppler processing. Two methods are discussed below, frequency detection and phase detection, respectively.

Frequency Detection

The basic limitation of a frequency detection AMTI technique is that the moving target Doppler shift must be greater than the clutter Doppler spectrum width in order to be detected. Consider the Fourier transform (of the Doppler modulation and shift terms) of (12), in terms of spatial frequency ω,

$$\tilde{m}(\omega) = \tilde{s}(\omega - 2k\epsilon_{\dot{r}}) .$$

The Doppler frequency $f_D = 2(v_r/\lambda)$ [related to the spatial Doppler coefficient $\epsilon_{\dot{r}}$ by $f_D = (2V/\lambda)\epsilon_{\dot{r}}$] represents the mean frequency shift of the moving target spectrum with respect to the clutter spectrum. The clutter spectrum bandwidth B_a [see (8)] for a side-looking radar with an antenna of nominal beamwidth β is $B_a = (2V/\lambda)\beta$. Clearly, for $f_D > B_a$ we require $\epsilon_{\dot{r}} > \beta$, which is the classic constraint. This lower limit may be somewhat improved by employing special antenna patterns and data processing. The effective bound is then $\epsilon_{\dot{r}} > \mu\beta$, where μ is some number less than one, but in practice is usually close to unity.

It follows from the above that *for a moving target to be detected by a single antenna (Doppler shift) side-looking AMTI radar, the radial target speed v_r must satisfy*

$$v_r > \frac{\mu V \lambda}{2\rho_a} . \tag{20}$$

The parameter μ also serves as the threshold, to be selected according to false-alarm probabilities and detection probability specifications. Clearly, for a given synthetic resolution, the minimum detectable velocity becomes larger as either a) the surveillance aircraft speed is increased, or b) a longer wavelength radar is used. In addition, if higher resolutions are sought, then the minimum v_r is increased as c) the aperture d is decreased to achieve the required resolution. Note that this constraint is not range-dependent.

The Doppler shift lower bound for target detectability and the range walk upper bound may be combined to define a useful region of optimally detectable radial velocity. From (16) and (20), we have

$$\mu\beta < \frac{v_r}{V} < \frac{1}{\beta}\frac{1}{(R/\rho)} , \tag{21}$$

which indicates a detectability window of normalized radial velocities (v_r/V) on the $(R/\rho,\ v_r/V)$ plane, as shown in Fig. 4. The lower bound is soft, as it depends on the beam-edge gradient, sidelobes, SCR, and detection probabilities, a familiar situation. The upper bound is soft, too, as discussed above. The effect of the pulse repetition frequency (prf) is to create aliases of the clutter band about multiples of the prf, as shown. If the Doppler bandwidth of the system is narrower than the prf, then the upper limit of v_r/V (before ambiguities) would be constrained by the system bandwidth, and not the aliased clutter spectrum, which is shown in the figure.

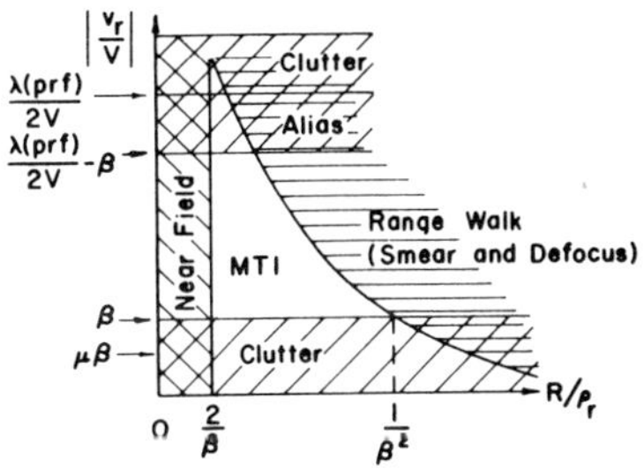

Fig. 4. MTI detectability window.

Phase Detection

An alternative technique is available that materially reduces the lower target velocity constraint by measuring the change in signal phase due to a moving target rather than the Doppler frequency shift. Two mutually coherent observations of the target signal are made, at times which are separated by the (small) amount δt. The resulting signals will be identical, except for a difference $v_r\delta t$ in the nominal reference range R. Note that the first signal must be delayed and stored in memory until the second signal is available for comparison. These two signals may be written (again neglecting the second-order effects)

$$t_1\text{, delayed: } m_1(x) = s(x)\exp\left[-j2k\left(\frac{v_r}{V}x + R_0\right)\right] \tag{22}$$

$$t_2 : m_2(x) = s(x)\exp\left[-j2k\left(\frac{v_r}{V}x + R_0 + v_r\delta t\right)\right] \tag{23}$$

where the form of (22) includes the required delay. Subtracting (23) from (22), the difference signal $m_\Delta(x)$ is

$$m_\Delta(x) = m_1(x)\left[1 - e^{-j2kv_r\delta t}\right] , \tag{24}$$

which, in terms of the magnitude, is

$$|m\Delta(x)| = 2\,|m_1(x)|\,|\sin(kv_r\delta t)| \tag{25}$$

where constant phase terms have been cancelled, as usual. The subtraction operation may be shown to be the optimum method if the clutter is highly correlated for the two samples, which is normally the case in applications. For a radar employing a dual side-looking antenna pattern,

$$V\delta t = D$$

where D is the phase center separation of the dual pattern.

The potential for moving target detection theoretically predicted by this result may be compared to the Doppler shift case above. If we approximate $\sin\theta \cong \theta$, valid for $\theta \lesssim \pi/5$, then for very slowly moving targets the moving target difference signal is amplitude weighted by

$$\left|\frac{2\pi}{\lambda}\frac{v_r}{V}D\right| .$$

If we further assume a detection threshold established by receiver noise, for example, of η radians, $\eta < \pi/5$, then we shall achieve detection if

$$\left|\frac{2\pi}{\lambda}\frac{v_r}{V}D\right| > \eta$$

or

$$\left|\epsilon_{\dot{r}}\right| > \frac{\eta}{Dk} . \tag{26}$$

In practice, η depends on the SCR as well as the noise level. Our point here is to provide some comparison of the phase detection technique with the Doppler shift case (20), (21) for which the normalized threshold was $\mu\beta$. We see that for the phase detection method, targets with small radial velocity v_r may be detected for large along-track aircraft velocity V by suitable selection of antenna spacing D, independent of the radar compression ratio, or beamwidth.

The moving target signal (difference signal) is weighted by $|\sin\ [(2\pi/\lambda)\ (v_r/V)D]|$. This result is consistent with the ground-based delay-and-subtract AMTI discussed in many references (e.g., Urkowitz [10]), and displays the usual phenomenon of blind speeds. Problems associated with prf aliasing, offset frequencies, ambiguities, and other radar performance characteristics are not sufficiently different in the dual antenna pattern imaging radar case compared to the single antenna imaging radar case to warrant separate discussion here [2], [4].

IV. Summary

Use of the phase detection method allows one to achieve AMTI against slowly moving targets with large aircraft velocities V, therefore minimizing degradation due to moving target image defocus (18) and smear (16). The trade-off suggests that large scan rate V should be used, together with an imaging radar with modest beam compression ratio K (6).

In the case of conventional SAR imagery, the output indication of a moving target is highly dependent on the target parameters. This dependence is reduced with higher aircraft speed V and lower compression ratios K. The image of a moving target may be blurred and attenuated. Blur is due to azimuth defocus, caused by radial target acceleration and cross-range target velocity (18), and due to range smear, caused by relative radial velocity (16). Attenuation is due to the concomitant reduction in peak intensity caused by image blur, together with Doppler frequency shift beyond the azimuth passband (8) of the radar. Finally, the image of a moving target is shifted in the cross-range direction (19) by an amount proportionate to its radial velocity.

References

[1] L.J. Cutrona, W.E. Vivian, E.N. Leith, and G.O. Hall, "A high resolution radar combat-surveillance system," *IRE Trans. Military Electronics,* vol. MIL-6, pp. 119-133, April 1961.

[2] W.M. Brown, "Synthetic aperture radar," *IEEE Trans. Aerospace and Electronic Systems,* vol. AES-3, pp. 217-229, March 1967.

[3] L.J. Cutrona, E.N. Leith, L.J. Porcello, and W.E. Vivian, "On the application of coherent optical processing techniques to synthetic aperture radar," *Proc. IEEE,* vol. 57, pp. 1026-1032, August 1966.

[4] R.O. Harger, "An optimum design of ambiguity function, antenna pattern, and signal for side-looking radars," *IEEE Trans. Military Electronics,* vol. MIL-9, pp. 264-278, July-October 1965.

[5] ——, *Synthetic Aperture Radar Systems: Theory and Design.* New York: Academic Press, 1970.

[6] A.W. Rihaczek, "Radar resolution of moving targets," *IEEE Trans. Information Theory,* vol. IT-13, pp. 51-56, January 1967.

[7] W.M. Brown and R.J. Fredricks, "Range Doppler imaging with motion through resolution cells," *IEEE Trans. Aerospace and Electronic Systems,* vol. AES-5, pp. 98-105, January 1969.

[8] E.J. Kelly and R.P. Wishner, "Matched-filter theory for high-velocity accelerating targets," *IEEE Trans. Military Electronics,* vol. MIL-9, pp. 56-59, January 1965.

[9] B. Harris and S.A. Kramer, "Asymptotic evaluation of the ambiguity function of high-gain FM matched filter sonar systems," *Proc. IEEE,* vol. 56, pp. 2149-2157, December 1968.

[10] H. Urkowitz, "The effect of antenna pattern on the performance of dual-antenna radar airborne moving target indicators," *IEEE Trans. Aerospace and Naval Electronics,* vol. ANE-11, pp. 218-223, December 1964.

R. Keith Raney (S'61–M'62) was born in Auburn, N.Y., on December 26, 1937. He received the B.A. degree in physics from Harvard University, Cambridge, Mass., in 1960, the M.S. degree in electrical engineering from Purdue University, La Fayette, Ind., in 1962, and the Ph.D. degree in computer information and control engineering from the University of Michigan, Ann Arbor, in 1968.

While at Purdue, he was a Research Assistant and Instructor in electromagnetic fields and probability theory. He joined Bendix Systems Division, Ann Arbor, in 1962, where he participated in the design and analysis of feedback control systems, including numerical machine tools, tracking systems, and satellite stabilization techniques. He joined the Institute of Science and Technology of the University of Michigan in 1963. Since that time, he has been active in theoretical investigation of MTI radar, synthetic aperture imaging radar, and optical data processing techniques, having published several papers in these areas. He is presently a Research Engineer with the Radar and Optics Division of the Willow Run Laboratory of the University of Michigan, and leader of the Radar Section. He has been active in teaching courses in remote sensing applications of radar, and as co-chairman of the intensive summer course on imaging radar.

Dr. Raney is a member of Eta Kappa Nu, Sigma Xi, and the Optical Society of America.

Hard Limiting In Synthetic Aperture Signal Processing

B.D. Steinberg
(University of Pennsylvania, Philadelphia)

IEEE Transactions on Aerospace and Electronic Systems, Vol. AES-11, No. 4 (July 1975)
Pages 556-561

Abstract

In synthetic aperture radar a large linear phased array is formed from the rapid movement of a single element through each position in the array. Storage and coherent combining of the successive radar echoes are central to the array-forming process. Optical processing is the most common technique because of the efficiency with which Fourier transformation may be accomplished with simple optics. Real-time operation, however, requires all-electronic processing, which is difficult to accomplish because of the huge quantity of data to be manipulated. Dynamic range compression by hard limiting may ease the problem by reducing the number of bits per frame. The effects of hard limiting are analyzed in this paper.

It is shown that large targets simultaneously illuminated by the radar antenna will produce image targets or ghosts displaced in angle. Statistically homogeneous clutter will "linearize" the hard-limited receiver and suppress the ghosts without loss in contrast, as does thermal noise if it is larger than the target echoes. Pulse compression reduces the probability of images from prominent targets. Judicious choice of the pulse-compression waveform is a powerful tool for destroying coherent buildup of images from all large targets not in the same range resolution cell. Linear FM, the most common choice, unfortunately does not exhibit this desirable property.

Introduction

The earliest sophisticated signal-processing array is the airborne synthetic aperture [1]. Its most important application has been in terrain mapping by radar. Synthetic aperture radar forms a large linear phased array from the rapid movement of a single element, the radar antenna, through each position in the array. Storage and coherent combining of the successive radar echoes are central to the array-forming process. Of the many early processing techniques developed for this purpose, which includes cathode ray tube storage [2], range-gated analog filters, and acoustic delay-line processors [3], optical processing of the signals has been the most common technique [4]. Its success is due to the efficiency with which Fourier transformation may be accomplished with simple optics.

Optical processing is not without limitations, one of which is the difficulty of achieving real-time operation. For this reason interest exists in all-electronic digital processors. Here the problem is the large quantity of data which needs to be manipulated at one time (the number of range sweeps forming the synthetic aperture × the pulse-compression ratio × the number of bits per radar sample video). Dynamic range compression of the input signal to the processor is a practical approach toward reducing this number.

The extreme of dynamic range compression is reduction of the input signal to one bit, i.e., infinite peak clipping or hard limiting. Work in the last decade (e.g., [5], [6]) on phased arrays with nonlinear processing has shown the following: 1) Intermodulation of signals from more than one target creates false targets or "ghosts" in angle. 2) If the target echoes are at different frequencies, the frequencies of their angular images are displaced by multiples of the difference frequency. This phenomenon causes a displacement in apparent target range when a linear FM waveform is used for pulse compression. 3) Receiver noise can suppress images when the noise at each antenna element exceeds the signal strength and the noise is independent from element to element. 4) Receiver noise reduces signal-to-noise ratio (SNR) only 1 dB, a result consistent with earlier theory [7].

Clutter rather than noise constitutes the background in synthetic aperture radar. Clutter is the collection of return echoes from scatterers illuminated by the radar antenna. The "target" signal is the return from the smaller set of scatterers within the main lobe of the synthetic aperture. In statistically homogeneous terrain the average clutter-to-target ratio at the input to the receiver is the angular compression ratio, i.e., beamwidth ratio of the radar antenna to the synthetic aperture. If range compression, i.e., pulse-compression, also is used, the ratio is further increased by the range compression factor. This fact, and the fact that the first-order amplitude statistics of clutter are similar to noise, leads to the conclusion that clutter should "linearize" a hard-limiting receiver, much as does noise, even though the clutter is correlated from pulse to pulse. Thus, terrain maps should not exhibit ghosts. Further, it is shown that

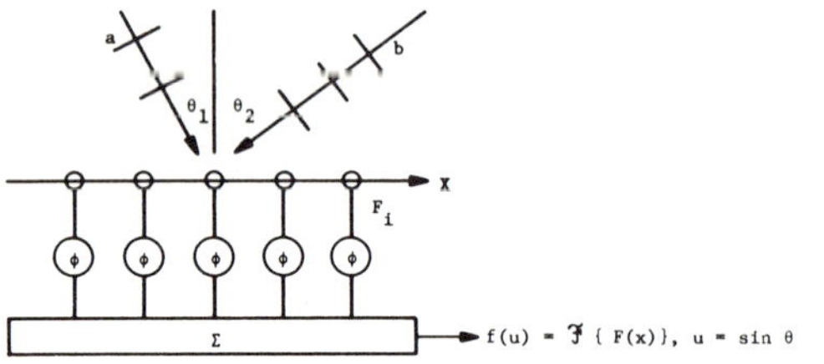

Fig. 1. Conventional antenna array.

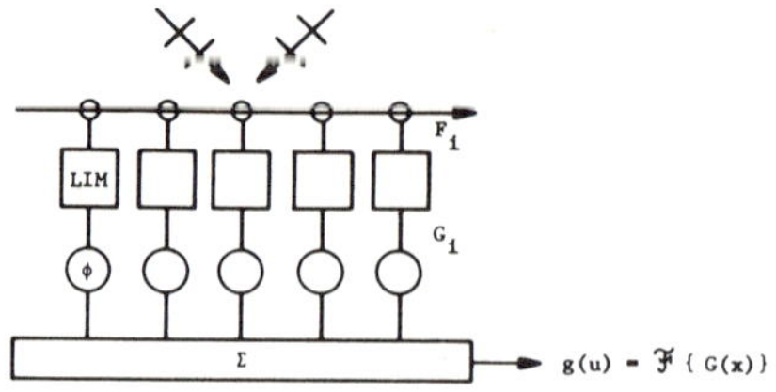

Fig. 2. Hard-limited array.

the contrast in the map between target and clutter remains virtually unaffected by hard limiting.

The second of the four points listed above provides another hint toward ghost suppression. With linear FM pulse compression, ghosts are displaced in range from the targets by an amount related to the beat frequency between the targets. Only the linear FM waveform yields images which integrate efficiently in the signal processor. Proper waveform design will suppress images except from prominent targets in the same range resolution cell.

Model

In a synthetic aperture system, the radar is moved at constant velocity V. The radar transmits pulses every τ seconds. The distance between transmissions is $d = V\tau$. N receptions are stored and summed, thereby forming an array of length $L = Nd$. Because of the near identity of operations to a conventional antenna array, it is possible to model the synthetic aperture as in Fig. 1. Shown is a set of discrete elements, phase shifters, and a summer. Plane waves from direction θ_i, having strengths a_i (each proportional to the E field), arise from targets. Receiver noise n is added at each element. The noise is independent from element to element, which accurately models the synthetic aperture situation.[1] Clutter also is received at each element, although it is not independent from element to element. Following the addition of the noise and clutter, the signal on each elementary channel is hard limited (Fig. 2). The set of limited signals enters the beam former. For simplicity, equal aperture weights are chosen; therefore, the implicit form of the beam pattern is $\sin x/x$. The particular shape is immaterial for the purpose of the analysis.

An additional simplifying assumption which has no effect upon the one-bit analysis places the target at infinity. The beam former introduces a set of phase shifts which varies linearly across the aperture. The phase progression may be written $\phi(x) = kxu$, where $k = 2\pi/\lambda$, x = position in the aperture, $u = \sin\theta$, and θ = angle from array normal. Letting the received signal at the ith element be F_i, the excitation across the array is $F(x) = \Sigma F_i \delta(x - x_i)$, and the beamformer output in the absence of the limiters (Fig. 1) is[2]

$$f(u) = \sum_{i=1}^{N} F_i e^{jkx_i u} = \mathfrak{F}\{F(x)\}. \tag{1}$$

Hard limiting converts $F(x)$ into $G(x) = F(x)/|F(x)|$. The beam former driven by the limiters is unchanged. Its output is

$$g(u) = \Sigma G_i e^{jkx_i u} = \mathfrak{F}\{G(x)\}. \tag{2}$$

Analysis

$$F(x) = \Sigma a_i e^{jkxu_i} + n(x)e^{j\phi(x)} \tag{3}$$

represents several plane waves plus noise or clutter having amplitude $n(x)$ and phase $\phi(x)$. After limiting,

$$G(x) = F(x)/[n^2(x) + \Sigma a_i^2 + 2\Sigma\Sigma a_i a_j \cos kx(u_i - u_j) + 2n(x)\,\Sigma a_i \cos(kxu_i - \phi(x))]^{1/2}. \tag{4}$$

The third term in the denominator gives rise to discrete ghosts because it is periodic in the difference angle. The fourth term is not periodic because of the randomness in the phase variation of noise of clutter across the aperture. Hence, discrete images do not arise from it. Whether or not isolated false targets appear in the display is determined by which of these dominate. A few examples illustrate the important relationships.

Consider two targets of amplitudes a and b, each very much larger than noise. Let $a \gg b$. Then,

$$\begin{aligned} G_1(x) &= F(x)/a[1 + b^2/a^2 + 2b/a \cos kx(u_1 - u_2)]^{1/2} \\ &\cong (1/a)\,F(x)\,[1 + b^2/4a^2 - b/a \cos kx(u_1 - u_2)] \end{aligned} \tag{5}$$

$$\begin{aligned} g_1(u) &= \mathfrak{F}\{G_1(x)\} \cong (1/a)\ \{F(x)\} * \mathfrak{F}\{1 + b^2/4a^2 - b/a \cos kx(u_1 - u_2)\} \\ &= (1/a)f(u) * [(1 + b^2/4a^2)\delta(u) - (b/2a)\delta(u - u_1 + u_2) - (b/2a)\delta(u + u_1 - u_2)] \end{aligned} \tag{6}$$

[1] The time between noise samples is τ, which typically is 1000 times longer than the reciprocal of the noise bandwidth of the receiver.

[2] $\mathfrak{F}\{\ \}$ means Fourier transform.

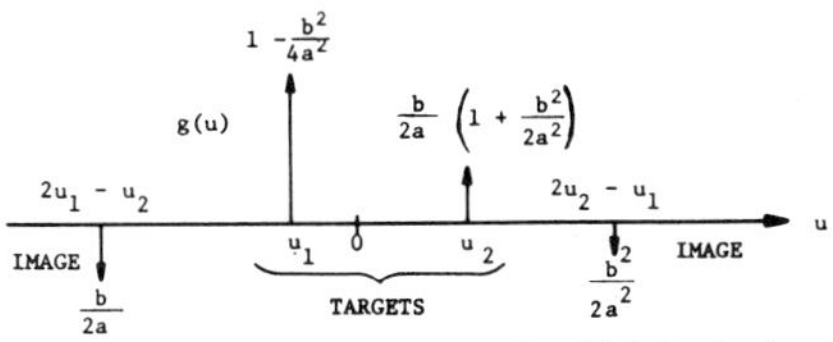

Fig. +. Angular spectrum of two targets ($a \gg b$) following hard limiting.

Fig. 4. Three targets without noise.

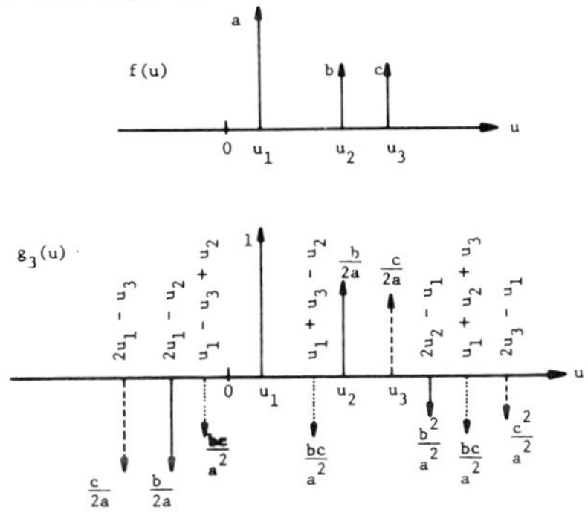

where $*$ means convolution and $\delta(u-U)$ is an impulse at $u=U$. Since $f(u)=a\delta(u-u_1)+b\delta(u-u_2)$,

$$g_1(u) \cong (1-b^2/4a^2)\delta(u-u_1) + b/2a(1+b^2/2a^2)\delta(u-u_2) - (b/2a)\delta(u+u_2-2u_1) - (b^2/2a^2) \cdot \delta(u+u_1-2u_2), \tag{7}$$

as shown in Fig. 3. Plane waves appear at the proper angles u_1 and u_2. The relative amplitude of the weaker wave is reduced approximately by a factor of two, or 6 dB. An image of the weaker wave appears with approximately equal strength. An image of the stronger wave also appears, but with significantly reduced amplitude.

The effect of increasing b is found by adding additional terms to the binomial expansion (5). Second- and higher-harmonic distortion terms arise which alter the amplitudes of the components of the angular spectrum of Fig. 3 and add new spectral lines. The line spacing is $|u_1-u_2|$ and the amplitudes have been tabulated [8].

When another small target is added, the first-order contributions are

$$G_2(x) \cong (1/a)F(x)[1-b^2/4a^2-c^2/4a^2 - b/a\cos kx(u_1-u_2) - c/a\cos kx(u_1-u_3)]. \tag{8}$$

The two smaller targets pass through the system, as did the small target in the earlier case, each losing 6 dB in strength relative to the larger and each generating images spaced from themselves and the large target by the angular separations between them. In addition to these seven terms, three additional spectral components appear at angles $u_1+u_2-u_3$, $u_1-u_2+u_3$, and $-u_1+u_2+u_3$. The angular locations are found from the convolution

$$\sum_{i=1}^{3}\delta(u-u_i) * [\delta(u)+\delta(u-u_1+u_2)+\delta(u-u_2+u_1) + \delta(u-u_1+u_3)+\delta(u-u_3+u_1)]. \tag{9}$$

Fig. 4 illustrates this case. The dashed lines represent the third target and the images associated with it. The dotted lines represent the three additional intermodulation components. It is evident that in the absence of noise even so simple a set of sources as one large and two small targets leads to a profusion of discrete ghosts.

When a significant amount of noise is added so that it dominates the signal in the receiver, the two-target case becomes

$$F(x) = ae^{jkxu_1} + be^{jkxu_2} + n(x)e^{j\phi(x)} \tag{10}$$

$$\begin{aligned} |F(x)| &= [(n+a\cos\psi_1+b\cos\psi_2)^2 + (a\sin\psi_1+b\sin\psi_2)]^{1/2} \\ &\cong n(1+a/n\cos\psi_1+b/n\cos\psi_2 + (a^2+b^2)/2n^2+ab/n^2\cos(\psi_1-\psi_2) \end{aligned} \tag{11}$$

where

$$\psi_1 = kxu_1-\phi(x), \quad \psi_2 = kxu_2-\phi(x) \text{ and } \psi_1-\psi_2 = kx(u_1-u_2). \tag{12}$$

$$\begin{aligned} G_3 = F/|F| &\cong e^{j\phi}[1+(a/n)e^{j\psi_1}+(b/n)e^{j\psi_2}] \cdot [1-(a^2+b^2)/2n^2 - a/n\cos\psi_1 - b/n\cos\psi_2 - ab/n^2\cos(\psi_1-\psi_2)] \\ &\cong (e^{j\phi}+(a/n)e^{jkxu_1}+(b/n)e^{jkxu_2}) \cdot [1-(a^2+b^2)/2n^2 - a/n\cos\psi_1 - b/n\cos\psi_2 - ab/n^2\cos(\psi_1-\psi_2)]. \end{aligned} \tag{13}$$

The targets appear correctly in position. In addition, the targets mix with each other and produce first-order images at angles $u=2u_2-u_1$ and $u=2u_1-u_2$. At the former angle, the output is the sum over (13) phase weighted by $\exp[-jkx(2u_2-u_1)]$:

$$g_3(2u_2-u_1) \cong \sum^{N} \exp[-jkx(2u_2-u_1)]\, G(x). \tag{14}$$

The image term arises from the product of the last terms in the two factors of (13):

$$\Sigma \exp\,[-jkx(2u_2 - u_1)(b/n)e^{jkxu_2}(ab/n^2)\cos kx(u_1 - u_2)$$

$$= \overset{N}{\Sigma}\,(ab^2/n^3)\cos^2 kx(u_1 - u_2), \qquad (15)$$

which is smaller than the signal by the order of b^2/n^2. Since the target is itself smaller than noise, these images and the weaker higher-order images may be ignored. In effect, the noise or clutter linearizes the hard-limiting process. This important observation was first made in [5].

Noise

Since images may be ignored when the targets are not larger than the noise, a single target suffices for analysis of the effects of hard limiting upon SNR:

$$F(x) = ae^{jkxu_1} + n(x)e^{j\phi(x)}. \qquad (16)$$

When the beam of the phased array is steered to $u = u_1$, which is equivalent to the aircraft motion of the synthetic aperture radar moving the target broadside to the flight path, the output voltage is the sum of N samples of (16) phase weighted by e^{-jkxu_1}:

$$f(u_1) = \overset{N}{\Sigma}\, a + \Sigma n_i \exp j(\phi_i - kx_iu_1). \qquad (17)$$

The number of samples N is the number of elements in the phased array or the number of pulses summed in the synthetic aperture. The first term is the signal voltage Na; its power is $(Na)^2$. The second term is noise or clutter. Its mean is zero because the argument $\phi(x) - kxu_1$ is a random variable uniform over 2π. When the noise samples are independent, as in the case of receiver noise, the expected value of the noise power is $E(\Sigma n_i^2) = N\overline{n^2}$. The output SNR in the absence of limiting, therefore, is

$$r_f = (Na)^2/N\overline{n^2} = Na^2/\overline{n^2} \qquad (18)$$

where $a^2/\overline{n^2}$ is the input SNR to the synthetic aperture receiver or to the individual elements of the phased array.

After limiting $G(x) = F(x)/|F(x)|$, [3]

$$|F| = (a^2 + n^2 + 2an\cos\psi)^{1/2}$$

$$\cong n(1 + a/n\cos\psi), \qquad n \gg a$$

$$G_4 \cong F/n\,(1 - a/n\cos\psi) = e^{jkxu_1}\,(a/n + e^{-j\psi})$$

$$\cdot\,(1 - a/n\cos\psi). \qquad (19)$$

[3] Being a random variable, occasionally $n < a$. The analysis pertains to that part of the aperture for which $n > a$. It may be shown that the minimum input SNR for which the analysis is justified is the order of unity.

Steering the beam to $u - u_1$, the array output (dropping the second-order term $a^2/n^2\cos\psi$) is

$$g_4(u_1) = \overset{N}{\Sigma}\, e^{-j\psi_i} + a\overset{N}{\Sigma}\,(1 - \cos^2\psi_i - j\cos\psi_i\sin\psi_i)/n_i. \qquad (20)$$

The first term is noise having zero mean and power N. The second term is signal. Its mean is

$$aN\,(\overline{\sin^2\psi/n}) = aN\,\overline{\sin^2\psi}(\overline{1/n}) = (aN/2)(\overline{1/n}) \qquad (21)$$

since ψ and n are independent. The post-limiting SNR is

$$r_g = [(aN/2)(\overline{1/n})]^2/N = (a^2N/4)\,[2/(1/n)^2. \qquad (22)$$

The effect of limiting upon the SNR, therefore, is

$$r_g/r_f = (a^2N/4)(1/n)^2/(Na^2/\overline{n^2}) = (\overline{n^2}/4)(1/n)^2 \qquad (23)$$

which may be calculated from the pdf of n. Since receiver noise is a two-dimensional Gaussian process, the envelope has a Rayleigh pdf $w(n) = (n/\sigma^2)\exp(-n^2/2\sigma^2)$. Thus,

$$\overline{n^2} = \int_0^\infty n^2\,w(n)\,dn = 2\sigma^2 \qquad (24)$$

$$(\overline{1/n}) = \int_0^\infty (1/n)\,w(n)\,dn = \sqrt{\pi/2}(1/\sigma) \qquad (25)$$

and

$$r_g/r_f = 1/4\,(2\sigma^2)(\pi/2\sigma^2) = \pi/4 \cong -1\text{ dB}. \qquad (26)$$

Therefore, small targets in thermal noise appear at their proper locations and are each suppressed by hard limiting only 1 dB on the average. The intermodulation spectrum has no specific structure and no images of the targets result from the mixing.

Clutter

The background signal in synthetic aperture radar normally is backscatter, commonly called clutter, from land or from the surface of the sea, rather than receiver noise. The effect of hard limiting upon the ratio of target-to-clutter (TCR) as they appear together in the map of the terrain is explored next.

Clutter, normally being the sum of echoes from a large number of independent backscatterers, is well modelled as a two-dimensional, zero-mean Gaussian process. However, unlike noise, the clutter correlation from pulse to pulse is not zero. The effect of clutter correlation is to modify the clutter power and, therefore, the TCR. The clutter term is represented by the second sum in (17). Changing the argument to ψ and letting $\nu_i = n_i\exp j\psi_i$, the clutter power is

$$|\overset{N}{\Sigma}\,\nu_i|^2 = \overset{N}{\Sigma}\,\nu_i\nu_i^* + \sum_{i\neq j}^{N}\sum^{N}\nu_i\nu_j^*$$

$$= \overset{N}{\Sigma}\,n_i^2 + \sum_{k=1}^{N-1}\sum_{i=1}^{N-k}\nu_{i+k}\nu_{i-k} \qquad j = i + k. \qquad (27)$$

Its average value is

$$E|\sum^{N}_{i} v_i|^2 = N\overline{n^2} + 2 \sum^{N-1}_{k} (N-k) R(k)$$

$$= N\overline{n^2}\,[1 + 2 \sum^{N-1}_{k} (1 - k/N)\,\rho(k)] \qquad (28)$$

where $R(k)$ is the autocorrelation function of the clutter echoes, $\rho(k) = R(k)/R(0)$, and $R(0) \equiv \overline{n^2}$. The bracketed factor is the increase due to nonzero correlation. After limiting, the normalized correlation function [9]

$$\rho_L = 2/\pi \sin^{-1} \rho. \qquad (29)$$

The effect of correlation is a modification of (23) by the factor

$$[1 + 2 \sum_{k} (1 + 2 \sum_{k} (1 - k/N)\rho(k)/[1 + 4/\pi \sum_{k} (1 - k/N)$$

$$\sin^{-1} \rho(k)]. \qquad (30)$$

The ratio of the post-limiting to the prelimiting TCR is the product of (23) and (30). Since $\rho \leqslant 1$, $\rho_L \leqslant \rho$ and $(30) \geqslant 1$, which leads to the interesting conclusion that the correlation of the clutter improves the post-limiting TCR. Evaluation of (30) for common SAR parameters shows the effect to be about 1 dB, which compensates almost exactly for the signal suppression (26) calculated earlier. Thus, the contrast in a map of statistically homogeneous terrain is hardly affected.

Pulse Compression

Strong targets produce intermodulation products which rise above the general clutter level and appear on the display as false targets. Pulse compression, although introduced to provide fine range resolution, has a further virtue in this situation. The pulse-compression transmitted wave spreads the energy from a target echo over a number of range bins equal to the compression ratio. At the input to the receiver, its energy density is reduced by the same factor. If the clutter is reasonably homogeneous statistically, the input TCR to the receiver is reduced by this factor. When such a reduction changes prominent targets into "weak" targets, the linearizing effect of noise and clutter destroys the images.

Choice of proper pulse-compression waveform further aids in image rejection. The images from targets at u_1 and u_2 appear at $u_{nm} = nu_1 + mu_2$ where n and m are integers. If the instantaneous frequencies of the echoes are not the same (as in linear FM pulse-compression radar when the targets are at different ranges), the frequencies of the images follow the image angles, i.e., $f_{2,-1} = 2f_1 - f_2$. Since range is related to frequency in linear FM radar, the image is also displaced in range.

To demonstrate, consider a linear FM wave of duration T and two targets at R_1 and R_2. The transmitted wave may be represented by $\cos(\omega_0 t + \alpha t^2)$, $0 < t < T$. The echoes are proportional to

$$\cos[\omega_0(t - t_1) + \alpha(t - t_1)^2] + \cos[\omega_0(t - t_2) + \alpha(t - t_2)^2].$$

The two components are of duration T and begin at $t = t_1$ and $t = t_2$, respectively. Provided that $t_2 - t_1 < T$, the returns overlap in time, coexist in the receiver, and mix or beat in the limiter. The beat $\cos[\omega_0(t_2 - t_1) + 2\alpha t(t_2 - t_1) + \alpha(t_1^2 - t_2^2)]$ has a constant difference frequency. The first and last terms in the argument are constants. The middle term varies with time. Its derivative is the angular frequency $2\alpha(t_2 - t_1)$, which is a constant. This constant frequency signal itself mixes with each of the original waveforms to form the first-order images. These arise from the upper sideband from the mixing of the first signal and the difference tone and from the lower sideband of the other mixing process.

The upper sideband has the form

$$\cos[\omega_0(t - 2t_1 + t_2) + \alpha(t - 2t_1 + t_2)^2 - 2\alpha(t_1 - t_2)^2].$$

This waveform appears as a legitimate echo arising from a range corresponding to time of arrival $2t_1 - t_2$, a frequency slope of α, and an additional phase term which has no significance. It is a linear FM wave having the same slope as the echoes from the legitimate targets. Hence, it will pass through the pulse-compression filter and will appear to arise at a different range from either target; its apparent distance from them is equal to the range separation between the targets. The lower sideband resulting from the other mixing process appears with equal and opposite displacement both in frequency and range. The frequency versus time relations are shown in Fig. 5.

Generalizing, let

$\omega(t)$ = frequency versus time of the transmitted wave, $0 < t < T$

$\omega_1(t) = \omega(t - t_1)$ = frequency versus time of the first received wave, $t_1 < t < t_1 + T$

$\omega_2(t) = \omega(t - t_2)$ = frequency versus time of the second received wave, $t_2 < t < t_2 + T$

$2\omega_1 - \omega_2$ = frequency versus time of one image $t_2 < t < t_1 + T$

$2\omega_2 - \omega_1$ = frequency versus time of the second image, $t_2 < t < t_1 + T$.

Higher-order images follow the same pattern. However, it is only those images whose frequency versus time patterns have the same character as $\omega(t)$ for which there are significant responses from the pulse-compression filter. Ideal waveform design is one for which none of the images satisfy this condition. A less restrictive waveform design is one in which none of the lower-order images have this property; the higher-order images rapidly diminish in amplitude with the order irrespective of the waveform.

The linear FM waveform, which is the common choice for SAR, is a particularly bad choice because a low (second) order mixing product falls in the passband and has the proper waveform. Few other waveforms, if any, have this undesirable property.

Two second-order phenomena are noted. The first is that the image strength is proportional to its lifetime, which is the coexistence time of the target signals which formed it. The image energy is proportional to $t_2 - t_k < T$. The second phenomenon also is due to the shortened duration; the normal output of the pulse-compression filter is the convolution of the impulse responses of the pulse stretching network in the transmitter and the pulse-compression filter in the receiver. The image waveform, being shorter in duration than that of a legitimate target, will not compress as well in the pulse-compression filter. In the limit as $t_2 - t_1 \rightarrow 1$, the input to the pulse-compression filter approaches a delta function. When this happens the output waveform approximates the impluse response of the pulse-compression filter, the duration of which is approximately T.

Conclusions

Hard limiting at the input to the synthetic aperture processor reduces the data handling requirements imposed upon it. Target echoes overlapping in time mix in the limiter and produce false targets or ghosts in the display. Terrain maps without prominent targets will be faithfully produced, however, because mixer products from terrain echoes are suppressed due to an effective linearizing of the receiver by the clutter. Pulse compression reduces the number of false targets produced by the mixing of large targets. Proper waveform design eliminates coherent buildup of ghosts from all prominent targets except those in the same range cell. Linear FM, the most common pulse-compression waveform, unfortunately does not exhibit this desirable property.

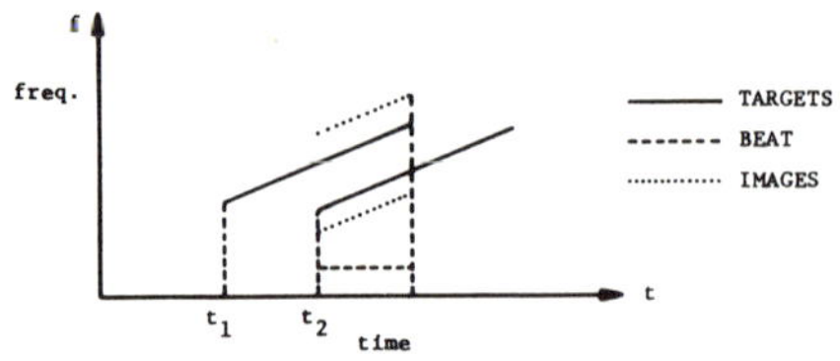

Fig. 5. Frequency-time relations in linear FM.

References

[1] R.O. Harger, *Synthetic Aperture Radar Systems, Theory and Design.* New York: Academic Press, 1970.

[2] C.W. Sherwin, "Further considerations on high resolution radar," *Proc. Fourth Symp. on Mapping and Charting by Radar* (St. Louis, Mo.), April 1954.

[3] B.D. Steinberg, "REDAP–A new method in achieving high resolution with airborne radar," *Proc. Fourth Symp. on Mapping and Charting by Radar* (St. Louis, Mo.), April 1954.

[4] L.J. Cutrona, "A high-resolution coherent radar system for airborne battlefield surveillance," *Airborne Battlefield Surveillance Radar Symp. Rec.* (University of Michigan, Ann Arbor), February 1955.

[5] E.D. Banta, "Far field properties of wide band planar arrays with nonlinear processing," *1961 IRE Conv. Rec.*, pt. 1, pp. 95-100.

[6] J.J. Kovaly and S.N. Miller, "A superlimiting phased-array receiving system in a two-source environment," *IEEE Trans. Aerospace and Electronic Systems,* vol. AES-3, pp. 518-526, May 1967.

[7] W.B. Davenport, Jr., "Signal-to-noise ratios in band-pass limiters," *J. Appl. Phys.*, vol. 24, pp. 720-727, June 1953.

[8] H.H. Woerrlein, "Capture and spurious target generation due to hard limiting in large time-band width product radars," NRL Naval Research Lab., Washington, D.C., NRL Rept. 7001, December 1969.

[9] J.L. Lawson and G.E. Uhlenbeck, *Threshold Signals,* M.I.T. Radiation Lab. Ser., vol. 24. New York: McGraw-Hill, 1950, ch. 3,p.58.

Bernard D. Steinberg (S'48–A'50–SM'64–F'66) was born in Brooklyn, N.Y., in 1924. He received the B.S. and M.S. degrees in electrical engineering from the Massachusetts Institute of Technology, Cambridge, in 1949, and the Ph.D. degree from the University of Pennsylvania, Philadelphia, in 1971.

He worked in the Research Division of Philco through the middle 1950's, specializing in radar backscatter and radar signal processing. He was one of the founders of General Atronics Corporation in Philadelphia in 1956 and served as its Vice President and Technical Director for 15 years. His work there was in signal processing techniques and their applications to radar, HF communications, hydroacoustics, and seismology. His most recent work is in self-adaptive signal processors, particularly in large antenna arrays. Currently, he is a Professor in the Moore School of Electrical Engineering at the University of Pennsylvania, Philadelphia, and Director of its Valley Forge Research Center.

Dr. Steinberg is a member of a U.S. Commission VI of the International Scientific Radio Union.

CHAPTER IV
Effect of Random Phase Errors on Synthetic Arrays

As a microwave signal is propagated through the troposphere, the temporal and spatial fluctuations of its index of refraction introduce random phase fluctuations to the RF energy. For a synthetic aperture, these fluctuations impart random phase errors across the aperture which are the fundamental limitation to the synthetic aperture in attaining its theoretical azimuth resolution. This chapter includes those theoretical considerations of random phase fluctuations and how the performance limitations can be described in these terms.

For small rms differences in normal distributed random phase error across the array, C.A. Greene and R.T. Moller derived the statistical measurement of synthetic aperture performance. Resultant performance includes the measure of expected beam broadening, rms beam canting, and expected loss in peak gain. However, their derivations show that the most sensitive performance parameter to residual uncompensated phase errors is the ratio of main lobe to integrated sidelobe power.

Following, L.A. Rondinelli and G.W. Zeoli reviewed the sources of data on index of refraction fluctuations and on phase and phase difference measurements to predict SAR performance. They predict that, for X-band arrays, synthetic apertures of 1000 feet at 30 to 60 nautical mile range are achievable, except for the reduction in mainlobe to sidelobe power ratio.

In the third paper, J.A. Develet, Jr. shows his method of measuring synthetic aperture performance in the presence of random phase errors. His performance measurement parameter is the mean envelope power at the system output. His assumptions are Gaussian for the real beam power, the phase error, and the random phase error correlation function. For the influence of random phase errors on the edge response, the reader is directed to Develet (1964). The ability of a synthetic aperture radar to map regions of sudden change in reflectivity, such as rivers, ravines, and roads, is influenced by the degree of random uncompensated phase errors.

L.J. Porcello combines the Tatarski formulation of phase error structure with Brown and Riordan's analysis of phase errors which have a power-law structure function to determine the best attainable SAR resolution in the presence of tropospheric phase errors.

These four papers demonstrate the fundamentals for determining SAR performance when its RF signal traverses the troposphere with random variations in index of refraction. As such, the reader can use these results in future similar SAR design requirements. For additional studies on the limits of antenna resolution due to atmospheric turbulence, the reader is referred to Ruina and Angulo (1963).

The Effect Of Normally Distributed Random Phase Errors On Synthetic Array Gain Patterns*

C.A. Greene and R.T. Moller
(Aerospace Group, Hughes Aircraft Company, Culver City, California)
IRE Transactions on Military Electronics, Vol. MIL-6, No. 2 (April 1962)
Pages 130-139

*Received by the PGMIL 2 December 1961

Summary—**In the practical formation of synthetic array patterns the individual signals which are integrated to form the array will sustain uncompensated phase shifts. If these uncompensated phase shifts are deterministic, the evaluation of their effect on the synthetic array radiation pattern is straightforward. On the other hand, if the uncompensated phase shifts are random, such as would be produced by propagation anomalies, a statistical study must be made to determine statistical measures of the effects on the radiation pattern.**

Part I describes an analytical study of the effects of normally distributed random phase errors on synthetic array performance. The statistical measures of performance derived include expected beam broadening, rms beam canting, and expected loss in peak gain. The analysis is limited to the case of small rms difference in phase error across the array.

Part II describes a Monte Carlo simulation study which removed the restriction of small rms phase error difference. In addition, the simulated radiation patterns generated in the simulation study allowed the computation of the statistics of the ratio of main lobe to integrated sidelobe power. This performance parameter is, in general, the most sensitive of all the performance measures to residual uncompensated phase errors.

Introduction

PHASE ERRORS occurring in the formation of synthetic antenna arrays[1] result in perturbations of the gain patterns which depend upon the detailed structure of the phase variations. For random errors certain statistical parameters of the perturbed gain pattern can be calculated. Random phase errors can be caused by tropospheric or ionospheric propagation effects or by random aircraft motion. To simplify the analysis, it is assumed that the phase errors can be represented as a stationary random process with a joint normal density function. It is not known exactly how well this model fits the physical processes mentioned above but it is thought to fit well enough so that the resulting synthetic array performance measures provide good estimates of the effects of those phenomena.

It is the purpose of this paper to derive appropriate statistical measures relating the effects of random phase shifts along a synthetic array to the degradation of the synthetic array pattern. The discussion will be presented in two parts. Part I presents the results of an analytical study in which it was found necessary to assume that the random phase shifts along the array were small in order that the analysis be tractable. No reasonable way has been found to extend the results to large phase errors. In order to obtain results for large values of the random phase shifts, the IBM 7090 computer was used to simulate the problem. The results of the simulation study are presented in Part II.

[1] R. C. Heimiller, this issue, pp. 122–129.

Before beginning Part I, a qualitative description of the effects of random phase errors on synthetic array performance will be given together with a brief discussion of the statistical measures of performance degradation that will be used in this paper.

Elements of Beam Degradations

A moderate amount of random uncompensated phase errors along a synthetic array will cause the synthetic pattern [denoted by $G(\theta)$] to exhibit a fluctuating main lobe having a broader beam, less peak gain and an axis which points in a slightly different direction from that of the unperturbed synthetic beam [denoted by $G_0(\theta)$]. The sidelobe structure will also be observed to fluctuate. The amount of beam spreading and wandering will, of course, vary as the phase errors across the array vary. It is possible that the fluctuation in the phase errors across an array length can be so severe that the main lobe is destroyed.

The statistics which will be used to describe the perturbed beam are the following:

1) Expected Pattern $\triangleq \langle G(\theta) \rangle$
2) RMS Beam Canting $\triangleq \sqrt{\langle \epsilon^2 \rangle}$
3) Expected Beam Spreading $\triangleq \langle \delta \rangle$
4) Expected Peak Gain Reduction $\triangleq \langle r \rangle$
5) RMS Azimuth Beam Rotation as a Function of the Forward Motion of the Radar Platform $\triangleq \sqrt{\langle (\Delta\epsilon)^2 \rangle}$
6) Expected Ratio of Highest Sidelobe to Main-Lobe Gain $\langle s \rangle$
7) Expected Ratio of Main-Lobe toTotal Sidelobe Power Where $\langle\,\rangle$ Indicates Taking the Ensemble Average.

In explaining what is meant by these statistics it is perhaps simplest to discuss items 2) through 4) before item 1). As mentioned above, the synthetic radiation pattern fluctuates as successive beams are formed with a randomly varying phase error structure. The axis of the synthetic main lobe wanders about its error-free direction. The rms beam-canting statistic is the rms value of the beam axis wander. The expected beam spreading is the expected increase in the beamwidth of the fluctuating main lobe, the beamwidth being defined at some constant gain level relative to the instantaneous peak gain. The expected peak gain reduction is the difference between the peak gain of the unperturbed pattern and the expected value of the peak gain of the perturbed pattern.

If the expected value of the synthetic array gain in each given direction is computed, the expected pattern of 1) above is obtained. Because in many cases the magnitude of the wander of the synthetic main lobe axis is large compared to the beam broadening, the expected pattern generally exhibits a peak gain which is lower than the expected peak gain, and a beamwidth which is wider than the expected beamwidth. The axis of the expected pattern is colinear with the axis of the unperturbed beam. The utility of the expected pattern is largely related to its providing bounds on the expected peak gain and beamwidth.

Statistic 5) provides a measure of the expected amount of rotation of the main lobe as the synthetic aperture moves ahead by one resolution distance. This property in addition to the expected beam broadening would determine the actual effective resolution of the synthetic array.

The last two statistics are self-explanatory. They are mentioned only in Part II.

I. Analytical Study

General Theory

The error-free two-way gain pattern $G_0(\theta)$ for a focused synthetic array[1] pointing broadside to the direction of motion is given by[2]

$$G_0(\theta) = \left| \sum_{n=-N/2}^{N/2} W_n \exp j(an \sin\theta) \right|^2, \tag{1}$$

where W_n is the amplitude weighting of the individual summands, a is a constant and θ is the angle measured from broadside. For the evaluation of the main lobe and the first few sidelobes ($\sin\theta \approx \theta$), consider the integral approximation to (1)

$$G_0(\theta) = \left| \int_{-N/2}^{N/2} W(z) \exp j(az\theta) dz \right|^2, \tag{2}$$

where z is a dummy variable of integration related to the distance along the array by

$$z = \frac{N}{L} l, \tag{3}$$

L is the array length and l is the distance along the array measured from the center. For uniform weighting $W_n = 1$, (2) reduces to the familiar pattern for small θ,

$$G_0(\theta) = \left[\frac{\sin a\theta N/2}{a\theta/2} \right]^2.$$

When the phase of each returned signal is randomly shifted by an amount $\phi(z)$, the resulting gain pattern for uniform weighting $W_n = 1$ takes the form

$$G(\theta) = \left| \int_{-N/2}^{N/2} \exp(j\phi(z)) \exp(jaz\theta) dz \right|^2. \tag{4}$$

This is the basic expression for $G(\theta)$ which will be used to derive the statistical description of the perturbed pattern.

[2] There is no loss of generality in assuming the synthetic array mainlobe points broadside.

Expected Pattern, $\langle G(\theta) \rangle$

If the random phase error function $\phi(z)$ is assumed to be a Gaussian variable, (4) becomes

$$\langle G(\theta) \rangle = \int_{-N/2}^{N/2} \int_{-N/2}^{N/2} \exp\left(-\frac{\sigma_{x-y}^2}{2} \right) \exp ja\theta(x-y) dx dy, \tag{5}$$

where

$$\sigma_{x-y}^2 \triangleq \langle (\phi(x) - \phi(y))^2 \rangle.$$

Eq. (5) follows from (4) by noting that

$$\langle \exp j(\phi(x) - \phi(y)) \rangle = \exp\left[-\sigma^2(1 - \rho(x-y)) \right] = \exp\left(-\frac{\sigma_{x-y}^2}{2} \right),$$

if $\phi(x)$ is a Gaussian variable, where $\sigma^2 = \langle \phi^2 \rangle$ is the mean-square phase shift at any point of the array and $\rho(x-y)$ is the correlation coefficient of the phase error at the two points x and y. Mathematical manipulations of (5) lead to the form

$$\langle G(\theta) \rangle = 2N^2 \int_0^1 (1-u) \exp\left(-\frac{\sigma^2_{Nu}}{2} \right) \cos a\theta N u \, du \tag{6}$$

which reduces to

$$\langle G(\theta) \rangle = G_0(\theta) - N^2 \int_0^1 (1-u) \sigma^2_{Nu} \cos a\theta N u \, du, \tag{7}$$

for small σ_{Nu}, where σ_{Nu} is the rms difference in phase error between two points separated by a distance $l = (L/N)z = Lu$. Note that $\langle G(\theta) \rangle$ is a "beam" pointing in the same direction as $G_0(\theta)$ as is to be expected since $\langle G(\theta) \rangle$ is an average over all possible perturbed beams. These beams point to each side of the peak of $G_0(\theta)$ with equal probability. Obviously, $\langle G(\theta) \rangle$ contains no information related to statistical properties (2) and (5); however, some measure of statistical properties (3) and (4) can be obtained. Figs. 2 and 3 show the beam spreading and gain reduction of $\langle G(\theta) \rangle$ as a function of the rms phase shift across the total array length, where $\rho(x-y)$ is assumed linear but greater than zero across the array.

As stated earlier, Figs. 2 and 3 provide upper bounds to the properties $\langle \delta \rangle$ and $\langle r \rangle$. For example, the peak gain of the expected pattern is $\langle G(0) \rangle$, and

$$G(0) \le G(\theta) \big|_{\theta = \text{direction of instantaneous beam axis}}.$$

RMS Beam Canting

Let ϵ equal the angle from broadside of the axis of the main lobe of $G(\theta)$. At the peak of the beam

$$G'(\epsilon) = G'(0) + \epsilon G''(0) + \frac{\epsilon^2 G'''(0)}{2} + \cdots = 0.$$

For ϵ sufficiently small (this implies the relative random phase shifts must likewise be small),

$$\epsilon \approx -\frac{G'(0)}{G''(0)}. \tag{8}$$

Taking expectations

$$\langle \epsilon^2 \rangle \approx \left\langle \left(\frac{G'(0)}{G''(0)} \right)^2 \right\rangle$$

$$\approx \frac{\langle G'(0)^2 \rangle}{[G_0''(0)]^2}.$$

Expected Beam Spreading

It can be shown, using the approach of the previous material, that the expected increase in main lobe beamwidth is given approximately by

$$\langle \delta \rangle \approx \frac{2}{|G_0'(\alpha)|} \left[\langle G(\alpha) \rangle - k \langle G(0) \rangle + \frac{k}{2} \frac{\langle G'(0)^2 \rangle}{G_0''(0)} - \langle G''(\alpha) \rangle \right] \tag{9}$$

where α is the angular deviation from the peak of the unperturbed beam such that

$$G_0(\alpha) = kG_0(0), \qquad 0 \leq k \leq 1$$

and where k and α will be chosen to give, say, the -3 db points of the unperturbed synthetic beam.

Expected Gain Reduction

Defining

$$r \triangleq \frac{G(\epsilon)}{G_0(0)},$$

$$r \approx \frac{G(0) + \epsilon G'(0) + \frac{\epsilon^2}{2} G''(0)}{G_0(0)}$$

$$\approx \frac{1}{G_0(0)} \left\{ G(0) - \frac{G'(0)^2}{2G''(0)} \right\}.$$

Taking expectations

$$\langle r \rangle \approx \frac{1}{G_0(0)} \left\{ \langle G(0) \rangle - \frac{\langle G'(0)^2 \rangle}{2G_0''(0)} \right\}. \tag{10}$$

RMS Rotation with Forward Motion of Radar Platform

Define $G_1(\theta)$ and $G_2(\theta)$ to be the synthetic patterns when the radar platform positions are separated by a distance l (l is assumed much less than the total array length L). From (8)

$$\epsilon_1 = -\frac{G_1'(0)}{G_1''(0)}$$

$$\epsilon_2 = -\frac{G_2'(0)}{G_2''(0)}.$$

Define

$$\Delta\epsilon \triangleq \epsilon_1 - \epsilon_2$$

and then

$$\langle (\Delta\epsilon)^2 \rangle \approx 2\langle \epsilon^2 \rangle - 2 \frac{\langle G_1'(0) G_2'(0) \rangle}{[G_0''(0)]^2}$$

$$\sqrt{\langle (\Delta\epsilon)^2 \rangle} \approx \sqrt{2\langle \epsilon^2 \rangle - 2 \frac{\langle G_1'(0) G_2'(0) \rangle}{[G_0''(0)]^2}}. \tag{10}$$

Application of Results

Three specific correlation functions will be considered in application of the formulas developed above: linear, quadratic, exponential. The particular examples are chosen since they lead to closed form expressions for the degradations and at the same time represent quite well the range of forms that a correlation function may take.

1) Linear Correlation Function: The linear correlation function has the form

$$\rho(u) = 1 - Nc|u|. \qquad 0 \leq |u| \leq 1.$$

The constant c is determined from the relationship

$$\sigma^2_N = 2\sigma^2[1 - \rho(1)]$$

to give

$$\rho(u) = 1 - \frac{\sigma^2_N}{2\sigma^2} |u|. \tag{11}$$

2) Quadratic Correlation Function: The quadratic correlation function has the form

$$\rho(u) = 1 - \frac{\sigma^2_N}{2\sigma^2} u^2. \qquad 0 \leq |u| \leq 1. \tag{12}$$

3) Exponential Correlation Function: The exponential correlation function has the form

$$\rho(u) = e^{-Nc|u|} \qquad 0 \leq |u| \leq 1$$

which will be approximated by

$$\rho(u) \approx 1 - Nc|u| + N^2c^2u^2/2.$$

Solving for cN in terms of σ^2 and σ^2_N

$$cN = 1 - \sqrt{1 - \frac{\sigma^2_N}{\sigma^2}}. \qquad \sigma_N \leq \sigma.$$

In the special case $\sigma^2 = \sigma^2_N$

$$\rho(u) = 1 - |u| + \frac{u^2}{2}$$

$$= 2\rho(u)_{\text{Linear}} - \rho(u)_{\text{Quadratic}}. \tag{13}$$

The last step follows from (11) and (12). In general, the synthetic array performance for the case of the exponential correlation function depends both on the variance of the phase error difference across the array *and* the value of the correlation function for a separation equal to the array length. This dependence will become clear in the later discussion of the simulation results. For the analytical study only the special case of (13) was treated, *i.e.*, the case where the correlation function drops to 0.5 across the array. For this case the beam degradation for the exponential correlation function can be approximated by combining the degradations associated with the linear and quadratic correlation function.

Now the various statistical synthetic array performance parameters can be computed. Each of the approximation formulas for these statistical parameters derived earlier requires the evaluation of the value, or the expected value, of the perturbed or unperturbed gain pattern, or one of its derivatives, at $\theta=0$ or α. Obtaining the values for the unperturbed beam is straightforward. The derivation of the expected values of the perturbed gain and its derivatives follows along the same lines as used in the development of the expected pattern, especially (5). These expectation values turn out to be functions of the synthetic array parameters, a and N, and the rms difference in phase error between two points along the array. The rms phase error difference between two points depends only on the rms phase error σ and the correlation function $\langle(\phi_1-\phi_2)^2\rangle = 2\sigma^2[1=\rho(u)]$. Therefore it is a straightforward but sometimes laborious process to compute the approximate statistical synthetic array performance parameters. The analytical results of these calculations are shown in Figs. 1–4.

Discussion of Results

Figs. 1–4 are plots of the rms beam canting, expected beam spreading, expected peak gain reduction and rms beam rotation as a function of the rms difference in phase error across the array (end-to-end) σ_N, where the equations of headings 1) through 5) are used and the correlation functions are given by (11), (12), and (13). The expected canting and spreading, and the rms rotation are given in units of the -4 db beamwidth of the unperturbed synthetic beam. The peak gain degradation is given in decibels relative to the peak gain of the unperturbed beam.

An approximate analysis has been made to determine the region of validity of the theory as limited by the approximations made regarding the size of σ_N and/or ϵ. This approximate analysis suggests that the results presented in Figs. 1–4 are in question for an rms phase difference across the array in excess of about one radian, regardless of the form of the correlation function.

It is clear from Figs. 1–3 that the beam spreading, beam pointing error, and peak gain loss are quite small over the region of validity of the theory, $\sigma_N<1$.

The rotation of the synthetic array main lobe as the radar moves along can produce map distortion and degraded resolution. Under severe conditions some of the surface to be mapped may be entirely missed. It is interesting to compare the amount of beam rotation during the time the center of the array moves a resolution distance to the beamwidth. For the case of the linear correlation function,

$$\frac{\sqrt{\overline{(\Delta\epsilon)^2}}}{\text{Unperturbed} - \text{4-db Beamwidth}} = \frac{3\sigma_N}{\pi}\frac{l}{L} = \frac{3}{2\pi}\left(\frac{\sigma_N D^2}{\lambda R}\right)$$

for $l=\lambda R/2L=D=-4$-db resolution distance.

Figs. 2 and 3 show that there is no expected beam spreading or expected gain reduction for the quadratic correlation function, case 2). At first glance this seems surprising. However, in order that a random process have a completely quadratic correlation function (at least across the array length), the random process must be approximately linear for an array length distance. But phase functions of this type would be expected to yield beams with exceedingly small amounts of spreading and gain reduction since a linear phase error produces only beam canting.

II. Simulation

The statistical synthetic array performance parameters 1) through 7) discussed in the Introduction were computed using a simulation technique commonly referred to as the Monte Carlo technique. The purpose was to investigate the performance under conditions of large phase errors for which the theory is invalid. Each beam degradation statistic was computed from a large number of perturbed synthetic beams, each of which was generated using Gaussian phase errors having the assumed statistics. The random phase errors were generated by simulating on the computer the passage of white Gaussian noise through a filter having the proper transfer function to yield a random variable with a prescribed covariance function. The correlation function, $\rho(u)$, was assumed to be of the exponential form $\rho(u)=e^{-\xi|u|}$.

In performing the simulation, sums of randomly phase-shifted terms of the form given in (4) were used to represent $G(\theta)$. The number of terms in the sums was reduced by first presumming groups of terms. The members of each subsum were assumed to suffer the same random phase error. The number of subsums chosen depended on σ_N across the array.

Fig. 5 is a plot of the three normalized exponential correlation functions assumed in the simulation. For $\xi=0.286$ the decorrelation of phase error across the array is quite small while for $\xi=4.0$ there is essentially complete decorrelation of the phase errors across the array length.

Figs. 6 and 7 are two typical runs made in the simulation. Fig. 6 shows the results when the rms phase shift across the array is one radian. Fig. 7 shows a typical pattern for $\sigma_N=2$ radians. In the latter case the variation among the sample patterns is much greater than in the former. The patterns in both figures have been plotted to the third null of the unperturbed pattern. Most significant features of the perturbed beam are evident. Since the total power in the synthetic array pattern over one period of the pattern is approximately constant, the ratio of main-lobe to integrated sidelobe power (over one period) can be computed from just the main-lobe curve.

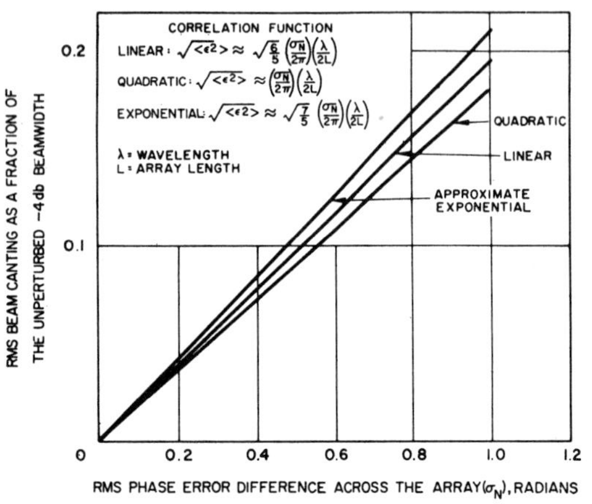

Fig. 1—RMS beam canting as a function of the rms difference in phase error across the array.

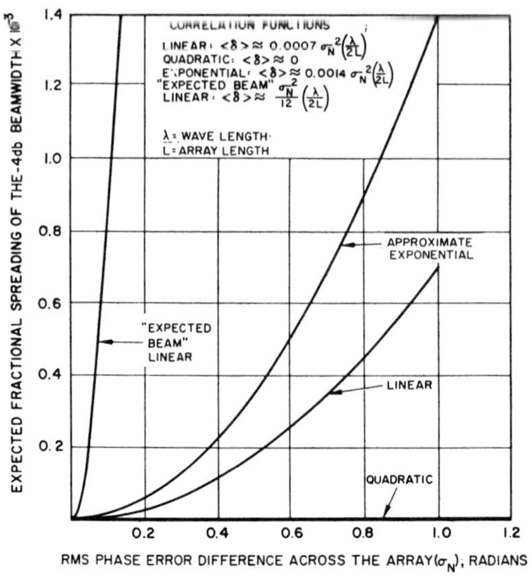

Fig. 2—Expected beam spreading as a function of the rms difference in phase error across the array.

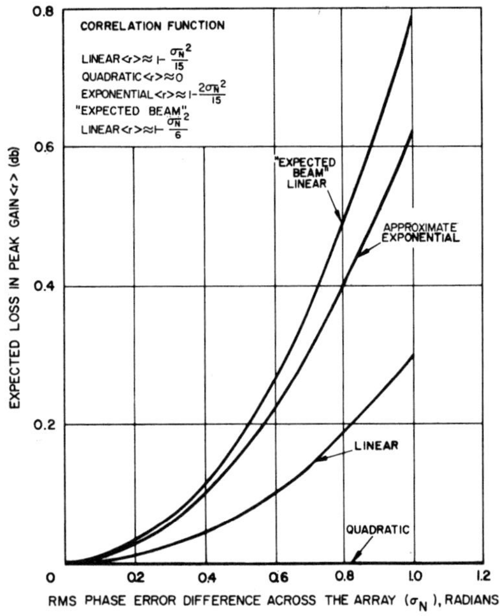

Fig. 3—Expected reduction in peak gain as a function of the rms difference in phase error across the array.

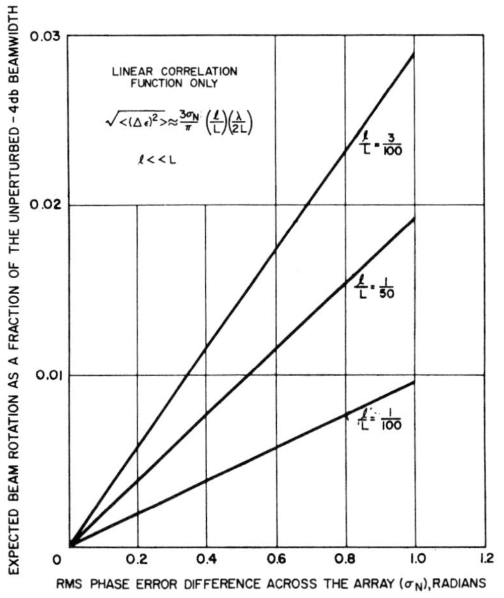

Fig. 4—Expected beam rotation during the time the radar platform moves a distance l as a function of the rms difference in phase error across the array.

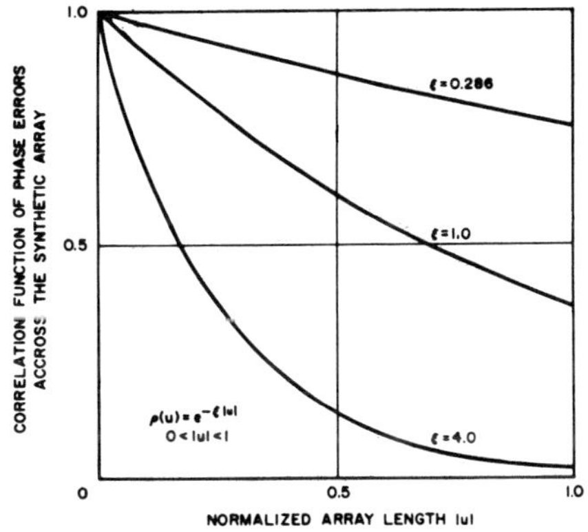

Fig. 5—Plots of the exponential correlation function of the phase errors across the synthetic array for the three conditions of array length relative to phase correlation distance assumed in the Monte Carlo simulation.

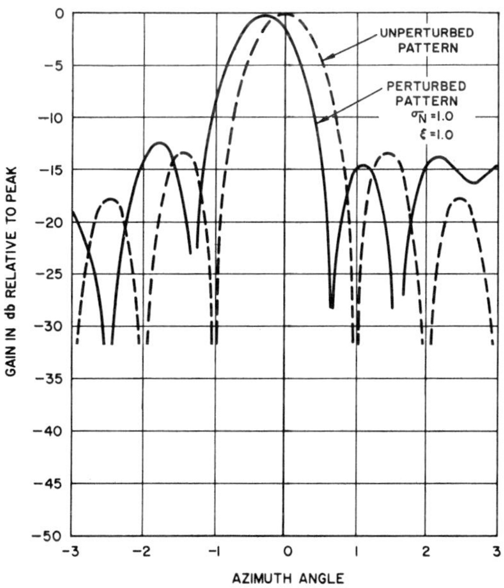

Fig. 6—Typical perturbed synthetic array pattern.

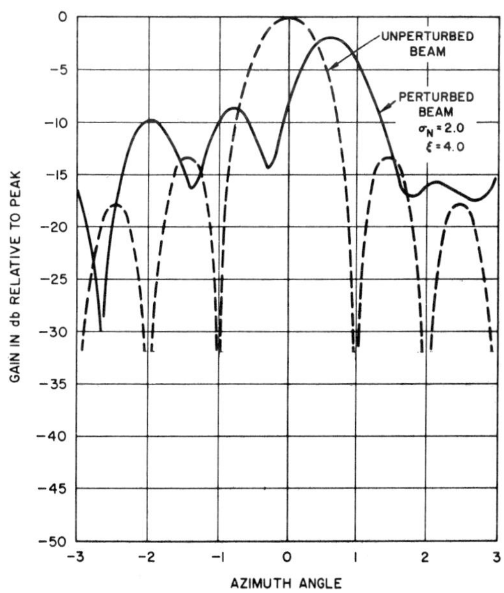

Fig. 7—Typical perturbed synthetic array pattern.

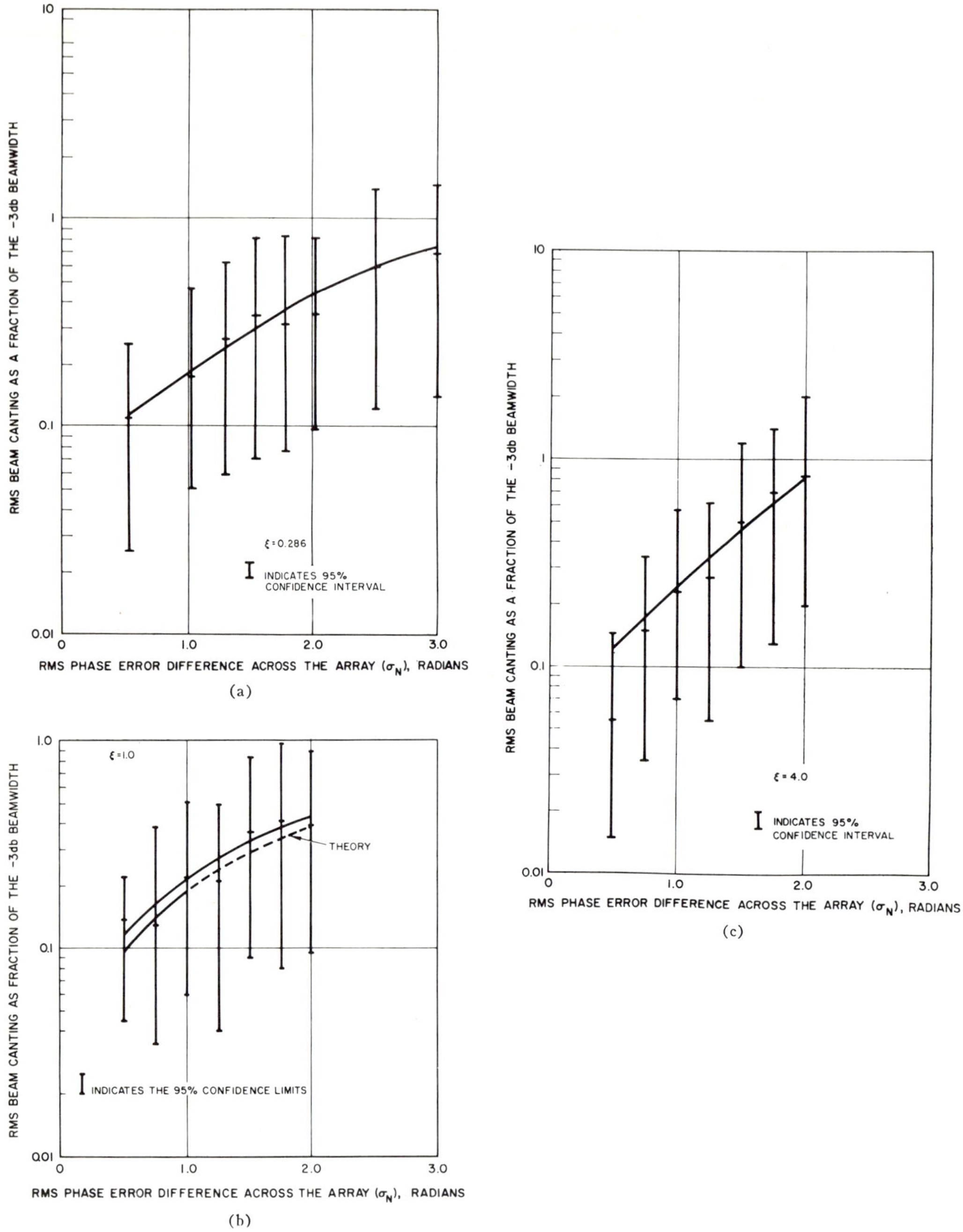

Fig. 8—RMS beam canting as a function of the rms difference in phase error across the array—Monte Carlo simulation results.

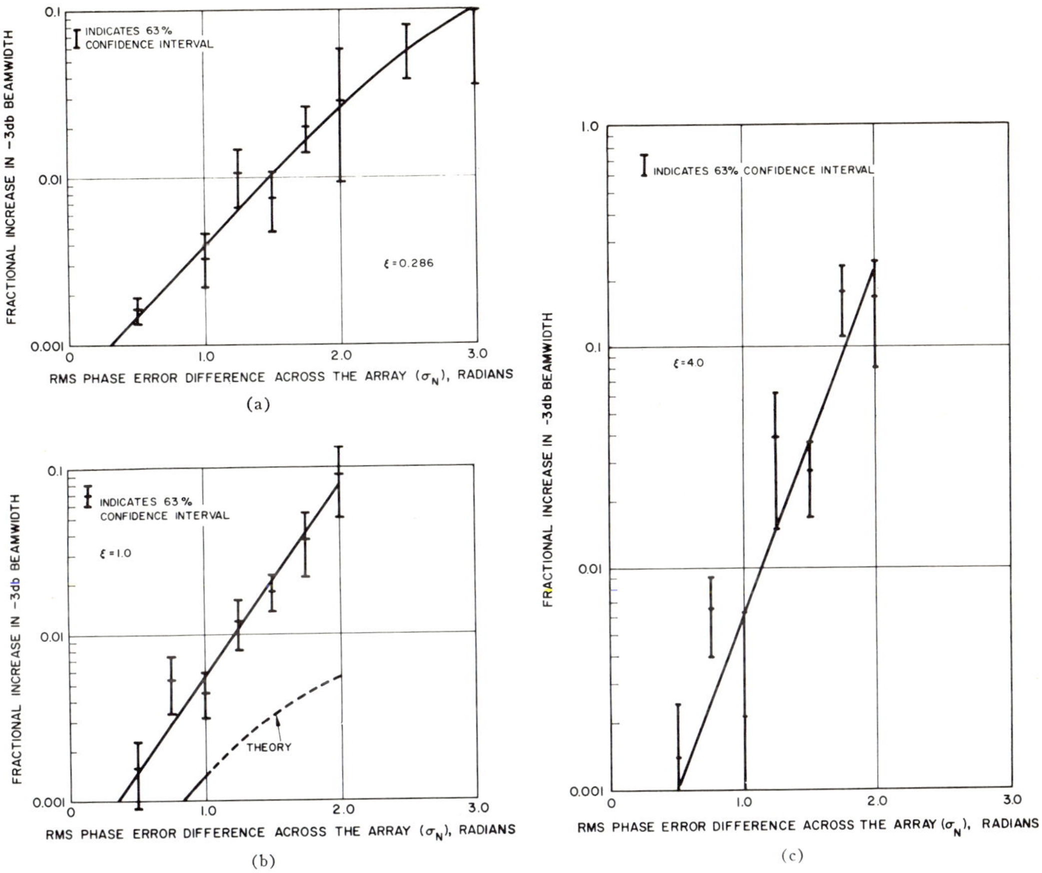

Fig. 9—Expected fractional increase in −3-db beamwidth as a function of the rms difference in phase error across the array—Monte Carlo simulation results.

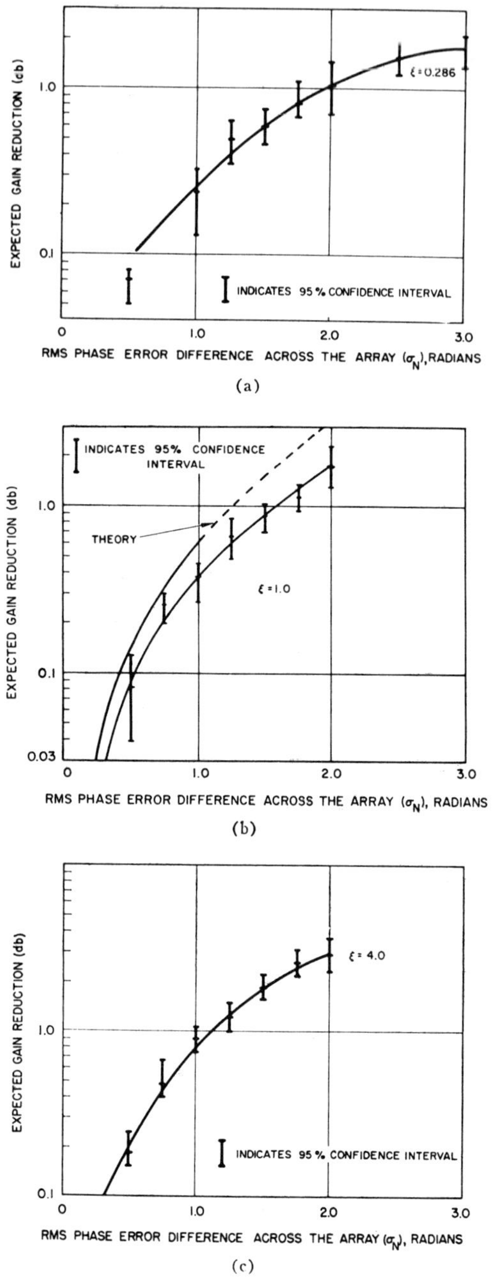

Fig. 10—Expected reduction in peak gain as a function of the rms difference in phase error across the array—Monte Carlo simulation results.

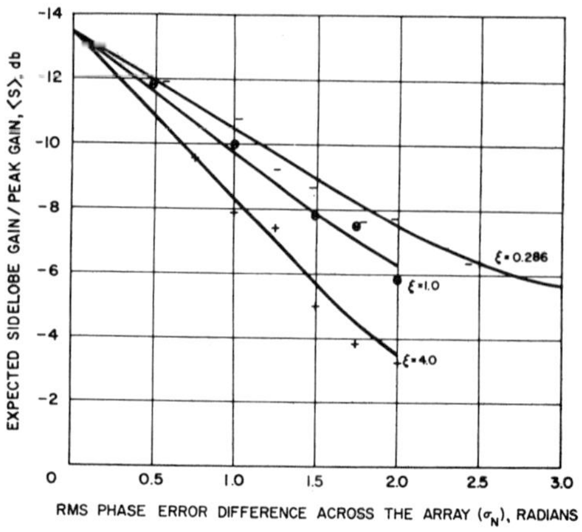

Fig. 11—Expected highest gain/peak gain vs rms phase shift σ_N.

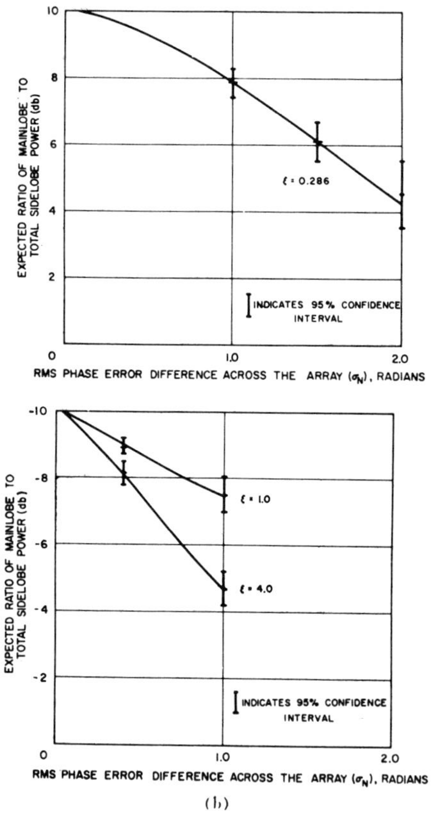

Fig. 12—Expected ratio of main lobe to total sidelobe as a function of the rms difference in phase error across the array—Monte Carlo simulation results.

The graphs of the results of the simulation study, Figs. 8–12, include a confidence interval about each point (except for Fig. 12).

Fig. 8 contains plots of the rms beam canting as a function of the rms difference in phase error across the array for the three assumed values of the phase error correlation distance relative to the array length. A comparison of the theory of Part I with the results of the simulation is included in the figure for the case of $\xi = 1.0$.

Fig. 9 shows the expected beam spreading as a fraction of the −3 db beamwidth for $\xi = 0.286$, 1 and 4. As the decorrelation of the random phase errors across the array increases there appears to be only a slight increase in the expected beam spreading.

Fig. 10 shows the expected gain reduction as a function of the rms difference in phase error across the array. The correspondence with theory of the $\xi = 1$ case is evident. Fig. 11 shows the expected ratio of the highest sidelobe level to the peak gain.

Thus far the computer program has not been set up to determine automatically the ratio of main-lobe power to total integrated sidelobe power (over one period of the pattern). However, an approximate computation using the data for the smaller values of σ_N for which the main lobe is well defined has been made and the results appear in Fig. 12. The approximation assumes a standard form for the main-lobe shape, the peak values and widths being determined from the simulated patterns. The main-lobe power, obtained from integrating under the main lobe, is subtracted from the known and constant total power within one period of the radiation pattern to yield the sidelobe power.[3]

Conclusions

The theory and simulation results agree quite well, and together provide a reasonably complete picture of the effects of random phase errors on synthetic array performance. The beam broadening, beam-pointing error, and peak gain loss are quite small for moderate amounts of uncompensated phase errors. The ratio of main-lobe to sidelobe power appears to be the synthetic array performance factor most sensitive to phase errors.

The simulation study needs to be extended to determine distributions as well as expected values of the performance parameters. Study of various amplitude weighting patterns is needed to determine a way to reduce the sensitivity of the main-lobe-to-sidelobe power ratio to phase errors. A more careful study is needed to obtain a significant measure of effective resolution and map distortion as affected by random variation of the direction of the beam axis.

Acknowledgment

The authors gratefully acknowledge the many helpful suggestions and criticisms offered by R. C. Heimiller and S. A. Meltzer, and the programming of the Monte Carlo Simulation for the 7090 IBM computer by E. S. Levitan and Shirley Meyers.

[3] The integral under the perturbed two-way synthetic array radiation pattern over an angular interval corresponding to one period of the error-free pattern is a constant if $W_n = 1$ [Eq. (1)] and if $\sin\theta \cong \theta$.

Evaluation Of The Effect Of Random Tropospheric Propagation Phase Errors On Synthetic Array Performance

L.A. Rondinelli and G.W. Zeoli
(Culver City, California)

Eighth Annual Radar Symposium Record, University of Michigan (June 1962)
Pages 235-256

ABSTRACT

A fundamental limitation on the performance of a synthetic array radar is imposed by the inability to compensate for phase errors produced by random propagation phenomena. For example, spatial and temporal fluctuations in the tropospheric index of refraction produce fluctuations in the r-f phase of a signal propagated over a fixed path near the earth's surface.

The results of a study are reported whose object was the prediction of the effects upon synthetic array radar performance of random phase errors produced by inhomogeneities in the tropospheric index of refraction. Two approaches were taken to the determination of the propagation phase errors that can be expected. One involves the determination of a statistical description of the fluctuations in the tropospheric index of refraction and the subsequent use of this information in a theoretical analysis leading to a prediction of the corresponding propagation phase errors. The second involves the use of direct measurements of the random phase fluctuations on a signal propagated over a single path and the fluctuations in a phase difference between two signals propagated over two paths having a common terminal at the transmitter.

Sources of data on index fluctuations and on phase and phase difference measurements are reviewed. The resulting predicted phase errors to be expected in the synthetic array radar application are compared and synthetic array radar performance computed.

I. INTRODUCTION

The synthetic array radar principle has attracted wide interest in recent years because it offers the possibility of achieving extremely high resolution radar mapping with real antenna apertures very much smaller than would be required for conventional non-coherent radar mapping of the same resolution. In order to successfully accomplish the formation of a synthetic aperture radiation pattern, however, it is necessary to maintain a high degree of phase coherence in the echo from a scattering element as the real radar aperture traverses the synthetic aperture. This phase coherence is disturbed by, among other things, variations in the propagation delay to and from the scattering element resulting from other than the usual geometrical considerations. This atmospherically produced phase decorrelation, unlike that due to non-linear motion of the real radar aperture phase center, cannot be compensated and the corresponding performance degradation must be tolerated.

If the radar propagation path is confined to the troposphere, as will be assumed throughout this paper, the random variations in the propagation delay referred to are caused by temporal and spatial variations in the density, temperature and relative proportions of the constituent gasses and other particles in the atmosphere. The resulting fluctuations in atmospheric index of refraction can be related to the variations of propagation delay.

The contribution of air gas components, other than water vapor, to the total index of refraction of the atmosphere is much greater than that from water vapor. However, the water vapor content generally varies considerably as a function of time and space and so plays the dominant role in determining the small scale index variations which are of interest here. Liquid water droplets and suspended solid particles contribute in lesser degree to index variations, but liquid water produces attenuation, and therefore amplitude variations, which may be significant at the higher microwave frequencies.

There are two approaches available for the evaluation of the effect of tropospheric inhomogeneities upon the propagation delay to and from a scatterer as the propagation path rotates about the scatterer. One approach is to use a theory, such as that of Muchmore and Wheelon[1], which relates phase disturbances to an appropriate statistical description of the index variations. Application of the theory requires the evaluation of the spatial correlation function of the index variations which are assumed to constitute a stationary random process. Microwave refractometer data are available to provide a basis for evaluating the index correlation function.

A second approach to the tropospheric propagation problem is directly empirical and provides data immediately applicable to synthetic array radar. It involves measuring the fluctuations in relative RF phase over two paths having a common terminal at the transmitter. This is precisely the synthetic array geometry. Programs to make such measurements have been carried out by the Central Radio Propagation Laboratory of the National Bureau of Standards and by the Electrical Engineering Research Laboratory of the University of Texas.[2, 3, 4]

The present paper will review available experimental data pertinent to these two approaches, demonstrate a reasonable degree of agreement between the two approaches, and present estimates of synthetic array radar performance predicted by the data. The analysis required to convert phase disturbances of the otherwise coherent synthetic array signals to degradations of synthetic array performance is discussed in a recent paper by Moller and Greene.[5]

II. THEORETICAL CONSIDERATIONS

This section will review briefly the theory of focused synthetic array performance and of tropospheric electromagnetic propagation as appropriate to the quantitative prediction of the effect of propagation phase errors on focused synthetic array performance.

SYNTHETIC ARRAY THEORY

As is well known, one form of the expression for the two-way power pattern of a focused synthetic array is given by

$$P(\theta) = \left| \sum_{n=1}^{N} A_n \exp j\,(\beta\, n \sin\theta + \alpha_n) \right|^2 \tag{1}$$

where

A_n = amplitude weighting factors

N = number of elements in the synthetic array

$\beta = \dfrac{4\pi VT}{\lambda}$

VT = spacing between elements of the synthetic array

λ = radar wavelength

α_n = any residual uncompensated phase error in the signal component from the nth element.

In the ideal case of no residual phase errors the pattern of Equation (1) has a peak value at $\theta = 0$

$$P(0) = \left| \sum_{n=1}^{N} A_n \right|^2 \tag{2}$$

and a -4db beamwidth of

$$[\theta]_{-4db} = \sin^{-1} \frac{2\pi}{\beta N} \cong \frac{\lambda}{2L} \tag{3}$$

where $L = NVT$ = synthetic array length.

In a realistic situation the presence of uncompensated residual phase errors will cause the synthetic beam to deform. The direction of the maximum of $P(\theta)$ for any one beam formation will, in general be different from $\theta = 0$ (with unity probability) and will wander about $\theta = 0$ as successive synthetic beams are formed. The width of the synthetic beam will also fluctuate from the value given in Equation (3). While it is more probable that the width of any given beam formation will be greater than that given by Equation (3), it may also be narrower. In addition, phase errors tend to reduce the maximum value of $P(\theta)$ from that given in Equation (2) and to increase the sidelobe level correspondingly.

The quantitative results of a detailed analysis of the effect of normally distributed stationary random phase errors on the focused synthetic array performance characteristics of peak gain, beamwidth, sidelobe level, and direction of the beam axis are given in Reference 5. Analytical results have been obtained for the effect on performance of small phase errors, and a Monte Carlo study provides some information regarding the effect of large phase errors.

Under the assumption of normally distributed stationary random phase errors only the mean square difference of the phase errors along the array is required in the analysis (see Appendix A). The analytical study for the case of small phase errors has included several forms for the correlation coefficient of the phase errors: linear, quadratic, exponential. The Monte Carlo study for the case of large phase errors has so far been limited (due to time) to the exponential correlation coefficient which, incidentally, includes the case of the linear correlation coefficient for array lengths short compared to the phase correlation distance.

As will be seen later, most experimental evidence indicates that the correlation distance of tropospheric-produced phase errors is large relative to synthetic array lengths of current interest. Also, the data appears to be in reasonable agreement with the assumption of a linear correlation coefficient over array lengths ranging from a hundred to several thousand feet. That is,

$$\left\langle \left[\phi(x) - \phi(x+D)\right]^2 \right\rangle = 2\sigma_\phi^2 \left[1 - \rho(D)\right] \simeq 2\sigma_\phi^2 K^2 D$$

σ_ϕ^2 = phase error variance

D = distance along the array

K^2 = constant of proportionality (4)

Therefore, only the results for a linear correlation coefficient are needed for this discussion and are reproduced from Reference 5 in Figures 1 through 4. These figures show the expected beam broadening as a fraction of the -3db beamwidth, the rms deviation of the synthetic beam axis from its average, zero phase error, value as a fraction of the -3db beamwidth, the expected reduction in peak gain, and the expected ratio of mainlobe to sidelobe power when illuminating a uniform diffuse surface, all plotted as a function of the rms difference in phase error across the array. The results of both the analytical and Monte Carlo studies are shown. All of the results assume uniform amplitude weighting, $A_n = 1$, and neglect weighting due to the real aperture radiation pattern.

A word of explanation in interpreting the latter performance characteristic of the expected ratio of mainlobe to sidelobe power is in order. This ratio is of particular interest in the discussion of the effects of phase errors on synthetic array performance because a relatively minor deterioration of the main synthetic beam can produce a large relative increase in sidelobe power. However, this ratio and therefore the effects of phase errors on it is not unique but to some degree is at the control of the designer. For example, the synthetic array radiation pattern of Equation (1) is periodic in θ, of period $\frac{\lambda}{2VT}$, if the A_n are assumed to be independent of θ and to represent only amplitude weighting employed by the processor, for instance to control the magnitude of the principal sidelobes. (The complete synthetic array power pattern would include an element by element weighting due to the real aperture radiation pattern and so would be aperiodic.) Thus, large synthetic beams will appear at $\theta = \frac{n\lambda}{2VT}$ and introduce large amounts of sidelobe power (or angle ambiguity) unless steps are taken to suppress them. It is common to attempt to suppress the angular ambiguity that would arise if more than one synthetic array mainlobe was contained within the illumination of the real aperture pattern by causing the first pair of spurious mainlobes to fall at or beyond the first pair of nulls of the real pattern. To achieve this suppression of angular ambiguity it may be necessary to increase the PRF to a value which introduces range ambiguities. Signal coding and processing techniques must then be sought to control the range ambiguity. The point here is that the total spurious

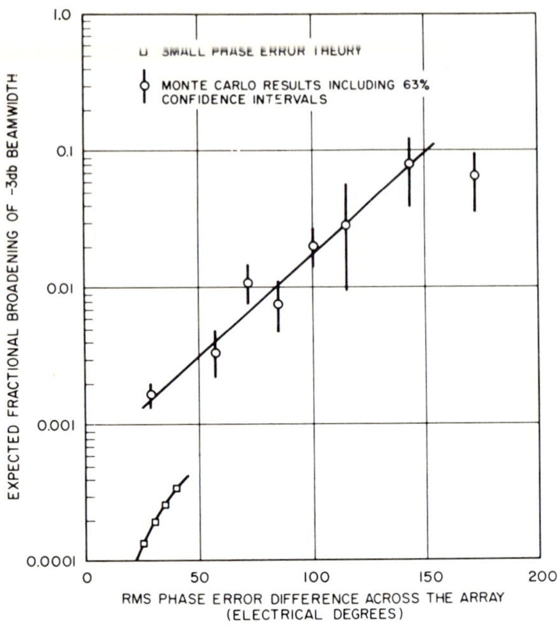

FIGURE 1. Expected fractional broadening of -3 beamwidth, assuming a linear covariance function.

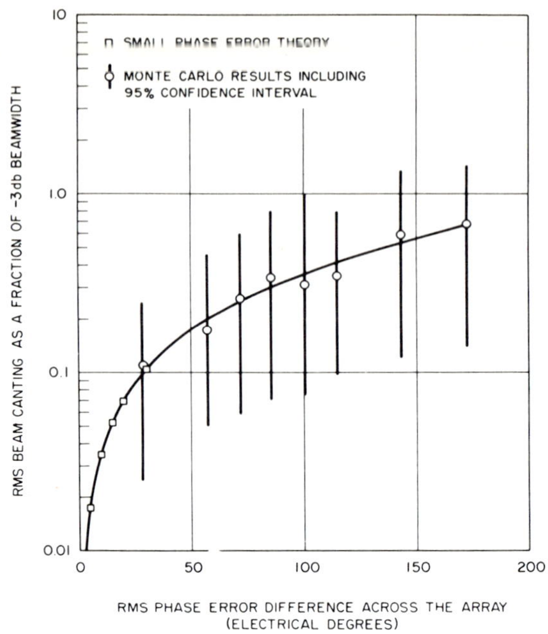

FIGURE 2. Rms beam canting as a fraction of the -3db beamwidth, assuming a linear phase error covariance function.

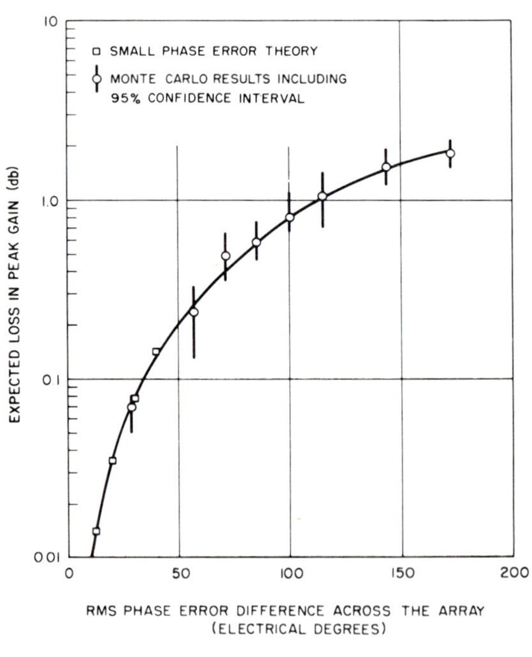

FIGURE 3. Expected loss in peak gain of the perturbed beam relative to the error-free peak gain, assuming a linear phase error covariance function.

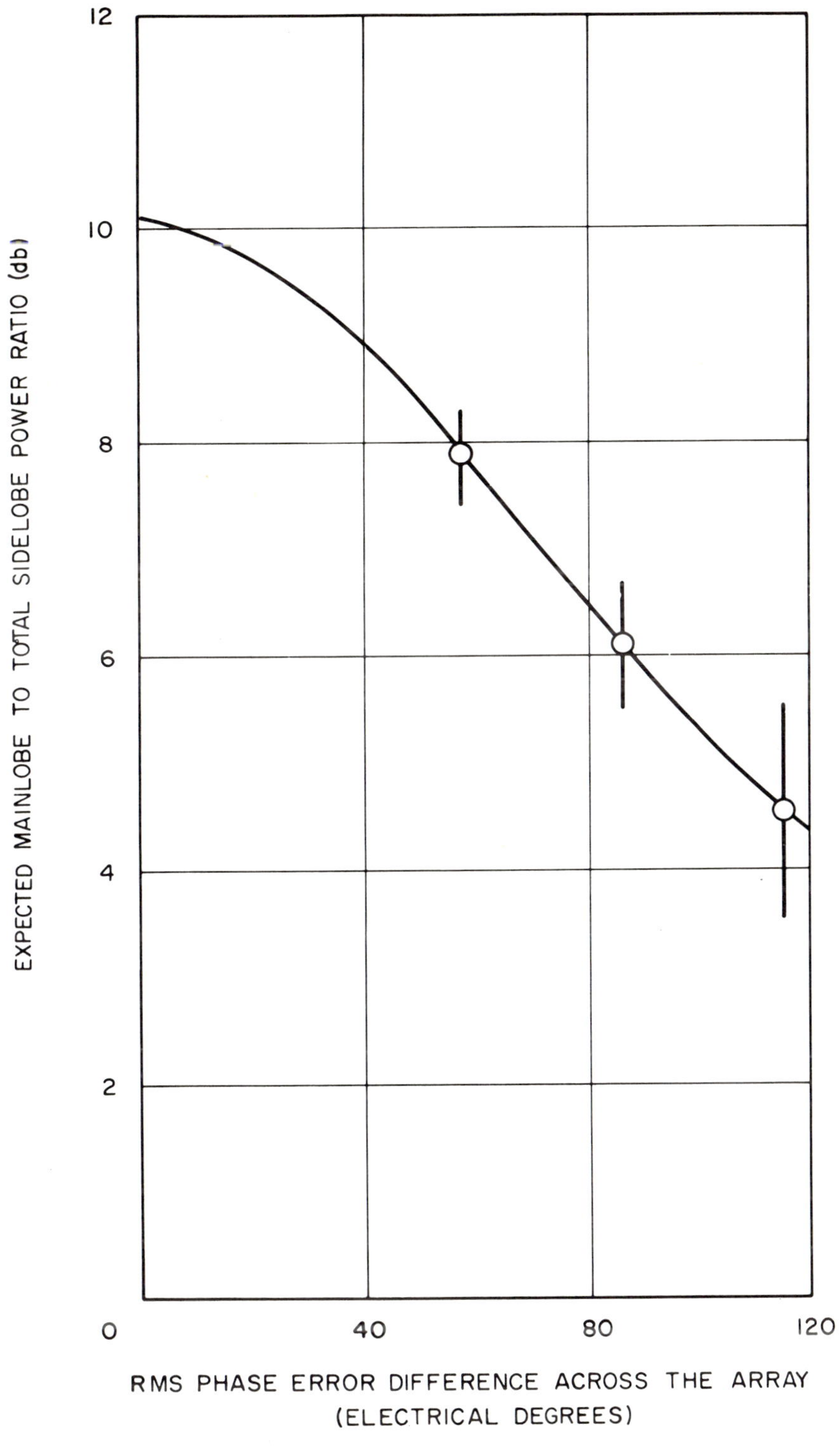

FIGURE 4. Expected mainlobe to total sidelobe power ratio, assuming a linear phase error covariance function (Monte Carlo results, showing 95 percent confidence intervals).

power from the synthetic array sidelobes, and perhaps from ambiguous range intervals, that competes with the principal mainlobe power is not unique. Therefore, it will be necessary to be somewhat arbitrary in choosing a measure which will convey an indication of the dependence of the expected ratio of mainlobe to spurious power on phase errors. The measure of the true ratio of mainlobe to sidelobe power which has been chosen is the ratio of the power in the mainlobe of Equation (1) to that in the sidelobes in the angular region $|\theta| < \frac{\lambda}{4VT}$ (one-half the distance to the first pair of spurious synthetic mainlobes), assuming uniform weighting $A_n = 1$. For this assumption of uniform weighting, the expected total power from the synthetic array pattern over this angular interval is a constant for each array formation regardless of the phase errors, and the ratio of the mainlobe to sidelobe power is approximately 10 db in the absence of phase errors.

This ratio provides a good estimate of the ratio of mainlobe to complete sidelobe power which would result in illuminating more-or-less uniform ground if the first pair of spurious synthetic mainlobes were located at or slightly beyond the first pair of nulls of the real aperture radiation pattern, and if the contribution from ambiguous range intervals is negligible. Incidentally, if a region of higher backscattering coefficient than the general background is illuminated (and of angular extent $\ll \frac{\lambda}{2VT}$), the ratio of mainlobe to sidelobe power would be correspondingly higher.

TROPOSPHERIC PROPAGATION THEORY

For the purpose of investigating the various aspects of the stochastic variation of signals propagating through the troposphere, the propagation medium is characterized by a dielectric constant which is both spatially and temporally varying about a mean value $\bar{\epsilon}$.

$$\epsilon(\underline{r}, t) = \bar{\epsilon} + \Delta\epsilon(\underline{r}, t) \tag{5}$$

$\bar{\epsilon}$ is usually a function of altitude alone. $\Delta\epsilon(\underline{r}, t)$ is usually assumed to be a zero mean stationary process which is both spatially homogeneous and isotropic. The assumption of spatial homogeneity implies that the statistical mean square value of $\Delta\epsilon(\underline{r}, t)$ is independent of position. The assumption of isotropy implies that the spatial correlation function of the index of refraction fluctuations at two points separated by a distance $\underline{R}$ depends only upon the magnitude of the distance $\underline{R}$ and not upon its vector direction. Using these assumptions, the first two moments of $\Delta\epsilon(\underline{r}, t)$ are

$$\langle \Delta\epsilon(\underline{r}, t) \rangle = 0 \tag{6}$$

$$\langle \Delta\epsilon(\underline{r}, t)\, \Delta\epsilon(\underline{r} + \underline{R}, t) \rangle = \langle \Delta\epsilon^2 \rangle\, C(|\underline{R}|) \tag{7}$$

where $C(\underline{R})$ is the spatial correlation function of the irregularities in the dielectric constant. Strictly speaking, the brackets in Equations (6) and (7) refer to ensembled averages but may be replaced equivalently by space or time averages for the assumed statistically homogeneous and stationary process $\Delta\epsilon(\underline{r}, t)$.

An evaluation of the fluctuation of signals propagating through a tropospheric medium characterized by Equations (5) through (7) may be obtained either by the Born single scattering approximation to the solution of the inhomogeneous wave equation or by use of geometrical optics.[6] In synthetic array applications the pattern deterioration produced by small amplitude variations in the received signals can usually be neglected in comparison to the deleterious effects of phase fluctuations. If one were interested in obtaining an accurate evaluation of amplitude fluctuations, it would be necessary to use the single scattering theory; however, for the purpose of determining only phase fluctuations along line-of-sight propagation paths, it is found that geometrical optics yields a good approximation to the more exact Born single scattering theory.[6]

GEOMETRICAL OPTICS

As a wave travels a distance L through the inhomogeneous atmospheric medium, geometrical optics predicts that the phase of the arriving signal will fluctuate due to the random variations in the velocity of the wave as it propagates along its ray path. The predicted phase fluctuation is given by[6]

$$\alpha = \frac{\pi}{\lambda} \int_0^L \frac{\Delta\epsilon}{\epsilon_0} (r, t)\, dr \qquad (8)$$

From Equation (8) it is seen that if $\Delta\epsilon(\underline{r}, t)$ is assumed to be a stationary Gaussian random process, then the phase fluctuation $\alpha(t)$ will also be a stationary Gaussian process. For such a process, by neglecting the effect of amplitude errors, the degradation of the synthetic array pattern can be related to the statistics of the difference in phase errors along the array (see Appendix A). The following statistical quantities are useful in describing degradation effects:

$$\left\langle [\alpha(x) - \alpha(x + D)]^2 \right\rangle = 2 \left[\langle \alpha^2 \rangle - \langle \alpha(x)\, \alpha(x + D) \rangle \right]$$

$$= 2 \langle \alpha^2 \rangle [1 - \rho(D)] \qquad 0 \le D \le \begin{matrix}\text{Array} \\ \text{Length} \\ D\ \text{Max.}\end{matrix} \qquad (9)$$

where $\langle \alpha^2 \rangle$ is the mean square phase error at any point along the array and $\rho(D)$ is the statistical correlation coefficient of the phase fluctuation at two points along the synthetic array separated by a distance D. In Appendix A it is shown that all statistics of the phase differences (considered as a random process) can be obtained from a knowledge of the product given in Equation (9), and it is not necessary to know $\langle \alpha^2 \rangle$ and $\rho(D)$ separately.

Both the mean square phase perturbation $\langle \alpha^2 \rangle$ and the correlation $\langle \alpha(x)\, \alpha(x + D) \rangle$ of the phase perturbations induced along two tropospheric paths having a common terminal at a transmitting source (see Figure 5) may be evaluated from Equations (7) and (8) provided the correlation function of the index of refraction is known.

Muchmore, Wheelon, and Levin[1, 7] have evaluated both the mean square phase perturbation and the phase correlation function at spaced receivers for the following models of index of refraction correlation function: Modified Bessel, Exponential, Gaussian, and Cauchy. Table I lists these correlation models along with the predicted mean square phase errors for each.

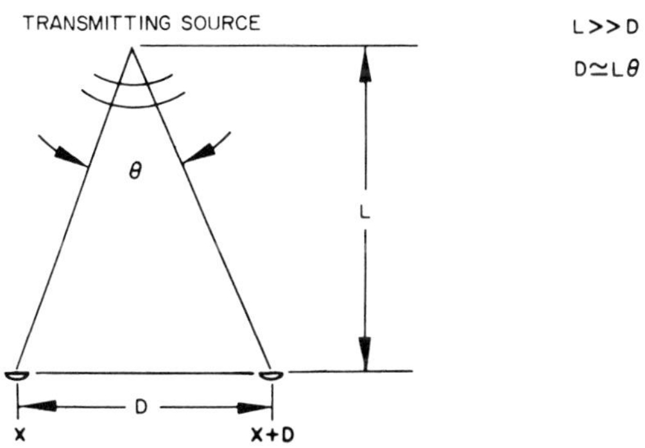

FIGURE 5. Geometry for space correlation function of phase fluctuations.

	BESSEL	EXPONENTIAL	GAUSSIAN	CAUCHY
SPACE CORRELATION OF REFRACTIVE INDEX C(R)	$\frac{R}{\ell_o} K_1\left(\frac{R}{\ell_o}\right)$	e^{-R/ℓ_o}	$e^{-(R/\ell_o)^2}$	$\left[1+\left(\frac{R}{\ell_o}\right)^2\right]^{-2}$
MEAN SQUARE PHASE ERROR $<\alpha^2>$	$\Pi^3 \frac{<\Delta\epsilon^2>}{\epsilon_o^2} \frac{L\ell_o}{\lambda^2}$	$2\Pi^2 \frac{<\Delta\epsilon^2>}{\epsilon_o^2} \frac{L\ell_o}{\lambda^2}$	$\Pi^{5/2} \frac{<\Delta\epsilon^2>}{\epsilon_o^2} \frac{L\ell_o}{\lambda^2}$	$\frac{\Pi^3}{8} \frac{<\Delta\epsilon^2>}{\epsilon_o^2} \frac{L\ell_o}{\lambda^2}$

TABLE I.

In Table I the parameter l_o in the index of refraction correlation models is referred to as the scale length of the turbulent irregularities. Its value may be estimated from refractometer measurement of index of refraction fluctuations as will be discussed later in this paper.

Figure 6 shows curves of the phase correlation coefficient for receivers spaced a distance D for each of the four models of the space correlation function of the index of refraction. In the defining equations of Figure 6 $K_n(z)$ is the modified Bessel function of order n, $L_n(z)$ denotes the modified Struve function and

$$\mathrm{erf}(z) = \frac{2}{\sqrt{\pi}} \int_0^Z e^{-x^2} dx.$$

In applying the theoretical results contained in Table I and Figure 6 to the synthetic array problem, two further assumptions are required. First of all, in a synthetic array the phase errors of interest are the two way phase fluctuations experienced by a wave as it travels to and from a scatterer. Since the two way traversal time of the transmitted signals is only of the order of milliseconds, the geometrical optics paths to and from a scatterer are essentially identical and hence the two way phase error is twice the value predicted by Equation (8). Thus, for synthetic array applications the two way mean square phase errors for a scatterer at range L would be four times the values in Table 1.

Furthermore, the phase correlation curves of Figure 6 were derived for the case of two receivers spaced a distance D apart simultaneously receiving signals from a common transmitter. In the synthetic array application $\alpha(x)$ and $\alpha(x + D)$ stand for the phase errors in the received signals at two points along the path traversed by the synthetic array vehicle. Thus $\alpha(x)$ and $\alpha(x + D)$ are phase errors produced by tropospheric irregularities at different times. However, since typical synthetic array integration periods are only of the order of seconds, experimental evidence indicates that it is valid to assume that the turbulent structure of the medium remains essentially unchanged in this time and hence, that the curves of Figure 6 are applicable.

III. EXPERIMENTAL DATA

This section will review the experimental data available for the evaluation of the mean square difference in phase error (Equation (9)) along a synthetic array aperture due to tropospheric irregularities by the two approaches described in the introduction. First, measurements of the statistical properties of the fluctuations in index of refraction of the troposphere will be described and the variance of the phase error difference that they imply will be derived.

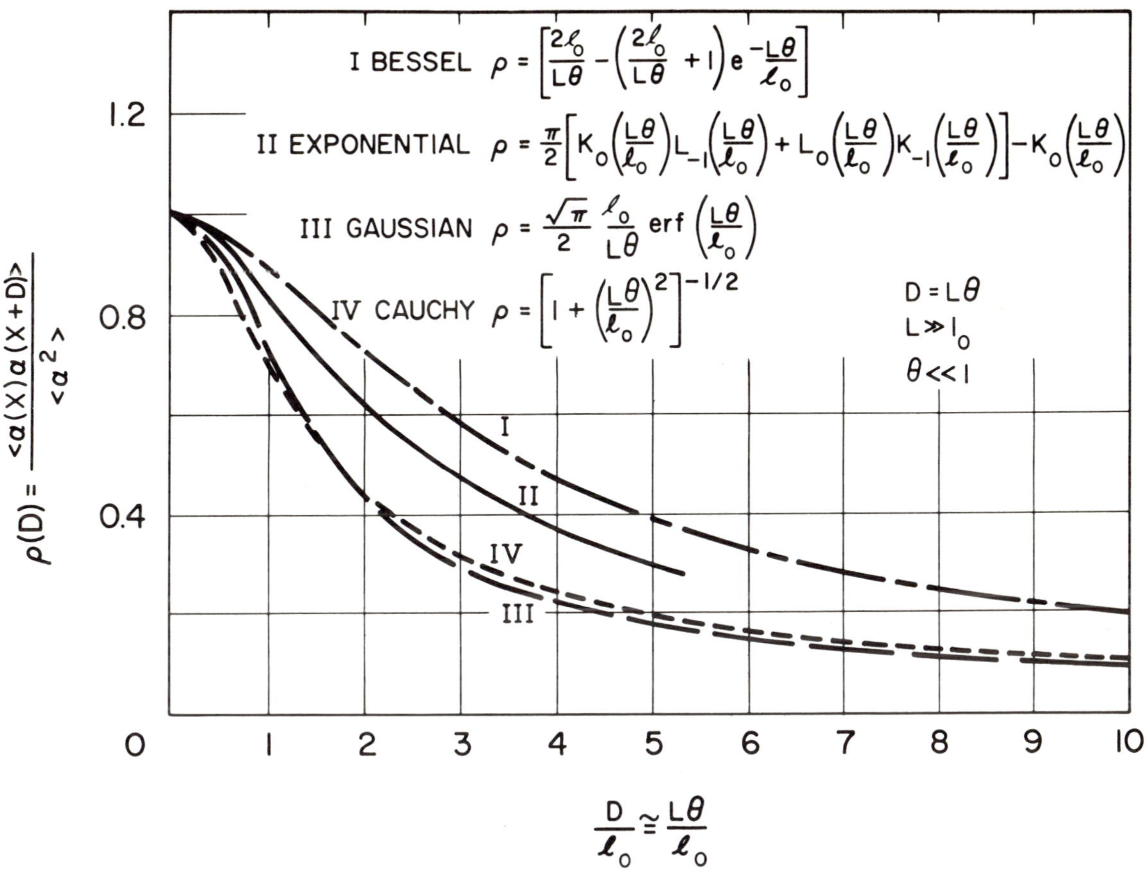

FIGURE 6. Phase correlation coefficients for rotated lines of sight.

Second, direct measurements of the phase difference between the signals received at two spaced receivers originating at a single distant transmitter will be reviewed and compared with the results of the first approach.

REFRACTIVE INDEX FLUCTUATION DATA

In describing atmospheric index fluctuation measurements, the following parameters are often used interchangeably:

$$n = \sqrt{\frac{\epsilon}{\epsilon_o}} \cong \sqrt{\frac{\bar{\epsilon}}{\epsilon_o}} + \frac{1}{2} \frac{\Delta\epsilon}{\sqrt{\epsilon_o \bar{\epsilon}}} \cong n + \Delta n$$

$$N = (n - 1)\, 10^6 = \bar{N} + \Delta N \qquad (10)$$

Hence

$$\frac{\langle \Delta\epsilon^2 \rangle}{\epsilon_o \epsilon} = 4 \langle \Delta N^2 \rangle \; 10^{-12} \simeq \frac{\langle \Delta\epsilon^2 \rangle}{\epsilon_o^2} \qquad (11)$$

where n = index of refraction
ϵ = air dielectric constant
$\bar{\epsilon}$ = mean value of air dielectric constant
ϵ_o = dielectric constant of vacuum
N = refractivity

Two of the more extensive studies of refractive index fluctuations in the troposphere have been conducted by the National Bureau of Standards[8] and the University of Texas E. E. R. L.[9] In the NBS study, refractometers were located approximately 20 feet above the local terrain at several altitudes between 5,000 and 15,000 feet. Refractivity fluctuations were analyzed in samples of 5 minutes duration. The variance of the refractivity was obtained for each sample and used as an individual piece of data from which a cumulative distribution was obtained. Figures 7, 8, and 9, taken from Reference 8, present the cumulative distributions of the five-minute variances of refractivity at four locations in Colorado.

Since the refractometers used in this study were positioned at fixed locations, the spatial correlation function of the refractive index was not experimentally determined. However, index correlation scale lengths were determined by the following procedure: Of the four correlation models listed in Table I, the modified Bessel model appears to predict results which are in best agreement with experimental scatter propagation measurements[10]. Accordingly, this correlation model was assumed to be valid. Then the variance of the phase fluctuations was measured for one way propagation paths for samples of 5 minutes duration. By substituting the results of these phase measurements and the measured variance of the refractivity into the equation for phase variance given in Table I for the modified Bessel correlation function, an estimate of the scale length l_o was obtained. By repeating this procedure for each five-minute sample, a cumulative distribution of the scale lengths was obtained. Several cumulative distribution curves obtained from Reference 8 are plotted in Figures 10 through 13.

In the University of Texas study[9] refractivity fluctuations were measured by means of an airborne microwave refractometer. Measurements were made from surface levels to altitude of approximately 12,000 feet at each of three locations in the United States: (1) Ohio, (2) west coast of Washington, and (3) west coast of central California. From individual samples of

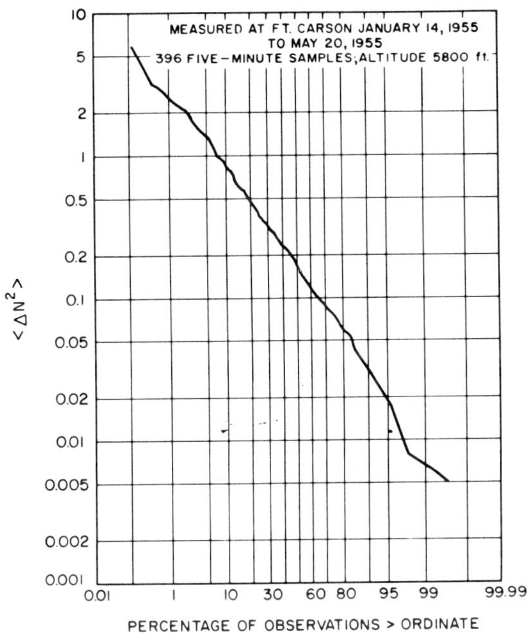

FIGURE 7. Cumulative distribution of the five-minute variances of refractivity.

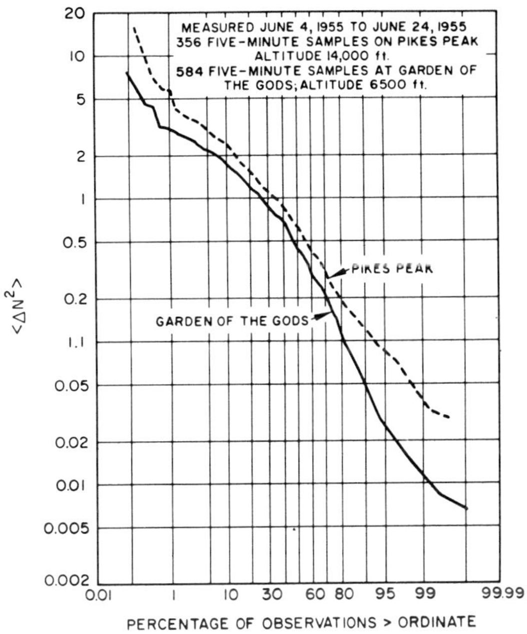

FIGURE 8. Cumulative distribution of the five-minute variances of refractivity.

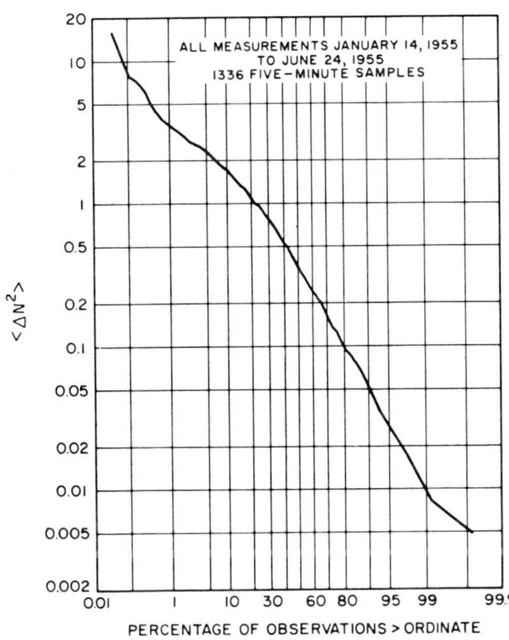

FIGURE 9. Cumulative distribution of the five-minute variances of refractivity.

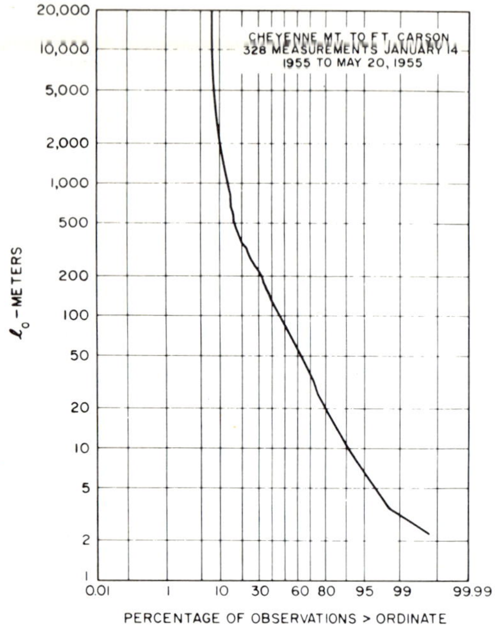

FIGURE 10. Cumulative distribution of five-minute sample values of ℓ_o.

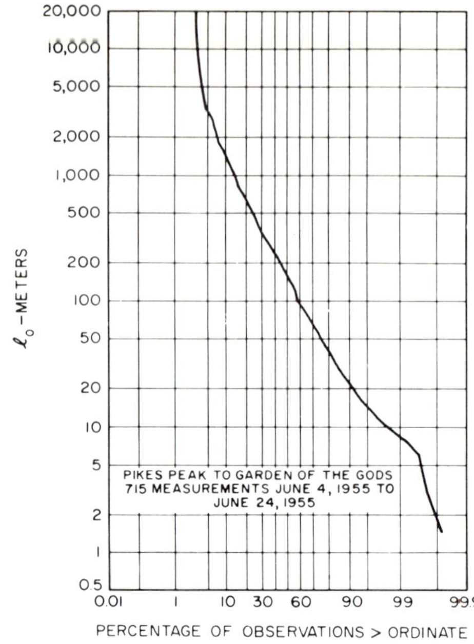

FIGURE 11. Cumulative distribution of five-minute sample values of ℓ_o.

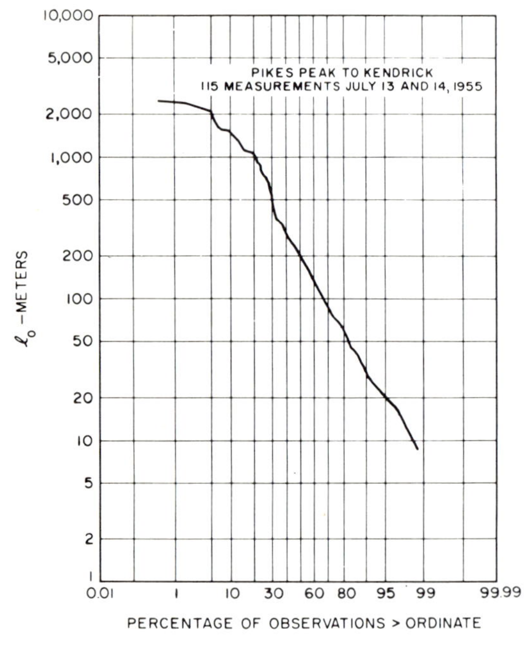

FIGURE 12. Cumulative distribution of five-minute sample values of ℓ_o.

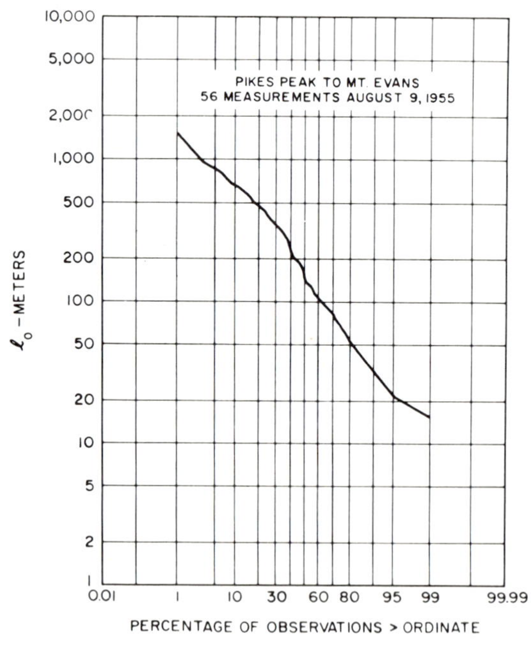

FIGURE 13. Cumulative distribution of five-minute sample values of ℓ_o.

refractometer measurements of approximately one-half minute duration the sample standard deviation of the refractivity $N = (n - 1) 10^6$ was determined. In addition, for each sample refractometer recording the correlation coefficient of the refractive index was estimated. From each sample correlation coefficient the abscissa of the 0.5 point was taken as the definition of the scale length of the fluctuations. (The modified Bessel correlation coefficient equals 0.5 at a distance 1.24 times the modified Bessel scale length l_o.) By taking repeated samples at each altitude increment, cumulative distributions of refractivity standard deviation and scale length were obtained. Although the study shows that there is a general tendency for the "intensity" of fluctuations (i.e., refractivity standard deviation) to decrease with altitude, it is not uncommon to obtain a peak in "intensity" at altitudes where a high moisture layer exists. Scale length measurements were not found to indicate any significant altitude trends. Figures 14 and 15 obtained from Reference 9 present the cumulative distribution of refractivity sample standard deviation and scale length respectively for each of the three locations combining the data for all altitudes.

In comparison with Figures 10 through 13, the absence of any large scale lengths in Figure 15 is partially accounted for by the fact that refractivity fluctuations having scale lengths of 600 feet or more were filtered out in analyzing the data. It should also be noted in comparing the scale length data that the difference in definition would tend to make the NBS scale lengths about 20 percent shorter than the corresponding University of Texas scale lengths.

Since the mean value of the refractive index should be expected to exhibit long term variations due to diurnal variations in temperature, air pressure and partial water vapor pressure, it is not surprising that the measured variance in refractivity exhibits considerable change from sample to sample. In fact, in analyzing samples of short duration it is reasonable to expect (and it has been observed) that the sample variance of refractivity increases as the sample duration is increased. The measured characteristic scale length l_o is also observed to increase as the length of the samples increases. Insufficient data analysis has been performed to determine what effect the relative rate of increase in measured refractivity variance and scale length with sample length has upon the derived phase difference variance.

The refractivity variance data of both studies are in good agreement. The NBS data gives a median refractivity standard deviation of about 0.6 N unit, whereas the University of Texas data gives about 0.4 N unit (Ohio over-land data).

The same cannot be said for the agreement of scale length data. The University of Texas scale length is defined to be the distance corresponding to the 0.5 point of the index correlation coefficient. The modified Bessel function correlation coefficient equals 0.5 at $R/l_o = 1.24$. Thus the NBS median scale length of about 150 meters is roughly four times the 120 foot value that would have been derived from the University of Texas over-land (Ohio) data had a consistent definition of scale length being used.

It should be remembered that the NBS scale lengths were derived from a theoretical expression for single path phase variance, assuming a modified Bessel correlation coefficient (see Table I), into which were substituted empirical refractivity variance data obtained at fixed points on the ground and single path phase variance data. On the other hand, the University of Texas data were taken by a microwave refractometer aboard an aircraft and both refractivity variance and index scale length were derived from the refractivity data. However, the University of Texas data were processed in such a fashion as to filter out large scale lengths, although presumably passing scales lengths up to 600 feet. At first sight, the University of Texas data appears to provide more nearly what is needed for the synthetic array application: namely, the nearly instantaneous spatial index correlation function at altitudes appreciable above the local

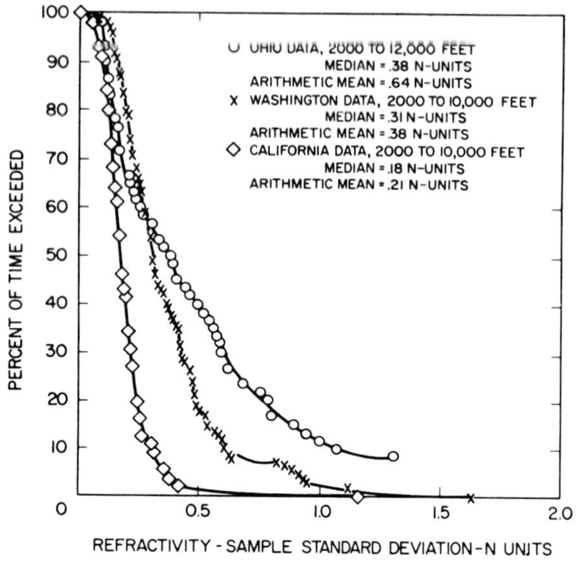

FIGURE 14. Cumulative distributions of sample standard deviations of refractivity.

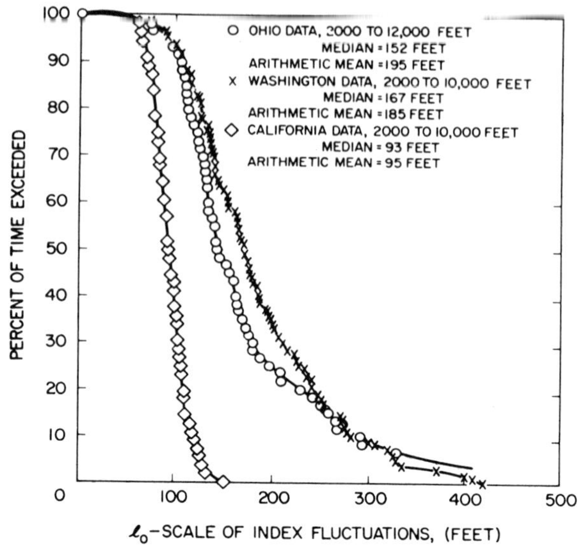

FIGURE 15. Cumulative distributions of index scale lengths.

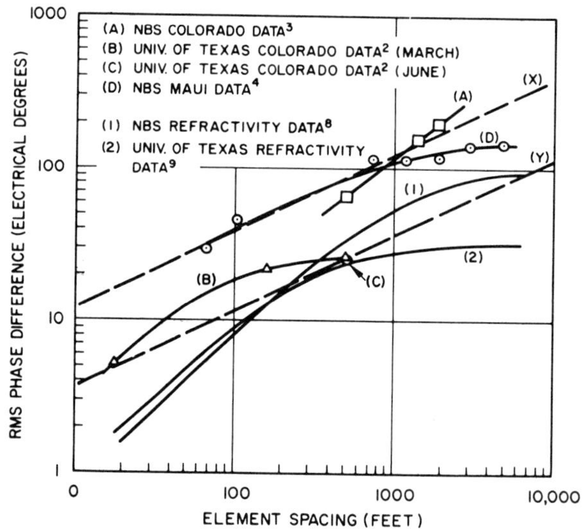

FIGURE 16. Experimentally derived two-way rms phase error difference versus spacing along the array for a range of 30 n.m. and a frequency of 10,000 mc.

ground level. However, in addition to the scale length filtering, questions are raised about the distortion in blob refractivity statistics caused during ascending or decending flights by passage through strong inversion layers.

The NBS data is perhaps less appropriate to the present application. The refractivity variance data were taken close to the ground at fixed locations. It is therefore necessary to assume that the index fluctuations experienced at a group of ground stations represents the instantaneous index fluctuations that would occur along a propagation path most of which would be appreciably above the ground level. Also, since the scale lengths are derived from propagation phase variance data, it is not proper to consider the NBS index fluctuation data as providing an estimate of phase difference variance which is independent of the direct phase difference measurements described in the next section.

Recognizing these limitations of the refractivity data, curves (1) and (2) of Figure 16 show the rms phase difference curves derived respectively from the NBS data (rms $\Delta N = 0.6$, $l_o = 150$ meters) and the University of Texas over-land (Ohio) data (rms $\Delta N = 0.4$, $l_o = 120$ feet), assuming the modified Bessel index correlation coefficient and using Equation (9), $\langle \alpha^2 \rangle$ from Table I, and Equation 1 of Figure 6.

RMS PHASE DIFFERENCE AT SPACED RECEIVERS

As is shown in Appendix A, if the phase errors across the synthetic array are assumed to constitute a stationary Gaussian process, then all statistics of the power gain pattern may be obtained from a knowledge of the rms phase difference for all "element" spacings less than or equal to the synthetic array length. The use of this statistic allows the prediction of synthetic array pattern degradation without the need of assuming any particular form for the index of refraction spatial correlation function. Unfortunately the measurement accuracy required in obtaining phase difference measurements dictates that the transmitting antenna be rigidly fixed, thereby necessitating that it be relatively close to the local terrain. In order to simulate the condition in which the transmitted waves originate in the upper troposphere and propagate downward, all the pertinent measurement programs have consisted of a transmitting antenna located at the top of a mountain and spaced receiving antennas at a lower altitude.

Both the National Bureau of Standards[3, 4] and the University of Texas[2] have carried out programs which have included the measurement of rms phase difference for spaced receivers. The various measurement programs have employed frequencies from 1,000 to 10,000 MC, ranges from 3 to 60 miles, antenna spacings from 2 to 5,000 feet and declination angles from 7 to 9 degrees. At each antenna spacing the phase difference was measured for time intervals ranging from 10 minutes to an hour and the rms phase difference evaluated. Several such samples were usually evaluated for each antenna spacing and an average or median value of rms phase difference established.

More specifically, the major measurement programs have been the following:

1. National Bureau of Standards[3], Colorado sites, May - July, 1955, freq. - 1046 mc, ranges of 3-1/2, 10 and 60 statute miles, receiver antenna spacings of 500, 1420, 1920 feet.

2. University of Texas[2], Colorado sites, March 1955, freq. - 9350 mc, range - 3-1/2 statute miles, receiver antenna spacings - 18, 162, and 500 feet.

3. University of Texas[2], Colorado sites, June 1955, freq. - 9350 mc, range - 10 statute miles, receiver antenna spacing - 500 feet (and perhaps others).

4. National Bureau of Standards[4], Maui, Hawaii site, Fall 1956, freq. - 9414 mc, range - 15-1/2 statute miles, receiver antenna spacings - 2, 67, 109, 733, 1181, 1914, 3003,

and 4847 feet.

The first program gave the single path rms phase fluctuation and phase difference correlation coefficient for a path range of 10 statute miles. These results have been scaled to 10,000 mc and a two-way range of 30 nautical miles (as a common base for comparison of all the data, and for convenience of application in the next section), and appear in Figure 16 as curve (A). The same reference (Figure 10 of Reference 3), demonstrates experimentally the linear frequency dependence predicted in Table I for rms phase fluctuations. The median data for 10 and 60 statute miles (Figure 9 of Reference 3) also indicates that the rms phase fluctuations vary approximately as the square root of range, in this interval, and it is this relationship that was used in the scaling of curve (A).

The rms phase difference data of the second program were taken at a range of 3-1/2 miles. The median data from Reference 2, Figure 3, were first scaled to 10 miles using the ratio of median values obtained for these two ranges in the first program, Figure 9 of Reference 3. The square root relationship was used to extend the results to 30 nautical miles. These data appear as curve (B).

The third program involved measurements made a few months after completion of the second at a range of ten miles and an antenna separation of 500 feet. The results are reported briefly in the concluding paragraphs of Reference 2 and appear as point (C) in Figure 16.

The fourth program, located in Hawaii under atmospheric moisture conditions probably more severe than would usually be encountered in the continental United States, provides a great deal of consistent data for many antenna spacings. These results, scaled to the same base conditions, appear as curve (D).

The agreement among the various sets of data is reasonably good. The general behavior of the curves is reasonably consistent with an assumption that the rms phase error difference varies as the square root of the spacing along the array. Dashed curves x and y are two such square root curves. Curve x might be thought to represent relatively poor weather conditions since it agrees well with the Maui data. Curve y might correspondingly be considered to represent relatively average to good U.S. weather conditions.

Admittedly, these latter statements are difficult to defend rigorously. What is needed is a great deal more data taken under a wide range of weather conditions, locale, and times of the day and year. Since the conditions for phase difference measurements, principally a convenient mountain, and the complexity of the experiment make them more difficult to obtain than airborne refractivity data, the latter type of experimental program probably offers the greatest hope of achieving better understanding and a better mathematical description of tropospheric propagation phase fluctuations.

IV. SYNTHETIC ARRAY PERFORMANCE

This section makes use of the experimental phase error difference data of Figure 16 and the synthetic array performance theory of Reference 5 and Figures 1 through 4 to make an estimate of the effect of tropospheric propagation induced phase errors on synthetic array performance.

From Figure 16 it is seen that the phase difference data do not appear to be in substantial disagreement with the assumption that the rms phase error difference between two points is proportional to the square root of their separation. Two such relationships which roughly bound the data are shown dashed in as curves x and y. With reference to Equation (4), these two curves differ in the ratio of $K\sigma_{\phi}$.

Using these two curves, x and y, and the synthetic array performance curves of Figures 1 through 4, the effect of tropospheric produced phase errors on synthetic array performance can be estimated. The results are presented in Figure 17. Here are plotted the expected peak gain loss, the rms beam canting, and the expected mainlobe to sidelobe power ratio as a function of array length or unperturbed resolution for a frequency of 10, 000 mc and for ranges of 30 nautical miles (solid curves) and 60 nautical miles (dashed curves).

It should be remembered that each pair of curves for each range roughly bracket the experimental median data for a wide range of conditions (Colorado and Maui). Therefore, while there will be individual performances outside each pair of curves, it might be expected that typical average performance would fall somewhere between.

Since index fluctuations are generally confined to altitudes below about 15, 000 feet, the performance estimates for a given array length and range also apply to array lengths and ranges several times larger if the radar altitude is correspondingly several times 15, 000 feet.

V. CONCLUSIONS

Experimental index of refraction and phase difference data have been examined and the median values thereof have been found to yield reasonably consistent estimates of median phase error difference statistics. However, the individual data are found to vary widely and therefore it is to be expected that synthetic array performance will likewise vary widely. For X-band arrays, median performance degradation does not appear to be severe for array lengths up to 1, 000 feet or so at ranges of 30 to 60 miles (and longer for high altitude radars), except for the reduction in mainlobe to sidelobe power. Since this performance factor can to some degree be controlled by the designer, and since the actual numbers for this factor in Figure 17 are only indicative, the conclusion is that long-range high resolution radar performance likely will not be limited by tropospheric propagation considerations provided adequate care is taken in controlling the sidelobe structure.

ACKNOWLEDGEMENT

The authors wish to acknowledge that the initial work on this problem at Hughes Aircraft Company was performed by R. C. Heimiller.

REFERENCES

1. Muchmore, R. B., and A. D. Wheelon, "Line of Sight Propagation Phenomena - I. Ray Treatment," Proc. IRE, October 1955, p. 1437.

2. Deam, A. P., and B. M. Fannin, "Phase Difference Variations in 9, 350 Megacycle Radio Signals Arriving at Spaced Antennas," Proc. IRE, 43, October 1955, p. 1402

3. Herbstreit, J. W., and M. C. Thompson, "Measurements of the Phase of Radio Waves Received over Transmission Paths with Electrical Lengths Varying as a Result of Atmospheric Turbulence," Proc. IRE, 43, October 1955, p. 1391

4 "An Experimental Study of Phase Variations in Line of Sight Microwave Transmissions," N. B. S. Report No. 6CB105.

5. Moller, R. T., and C. A. Greene, "The Effect of Normally Distributed Random Phase Errors on Synthetic Array Gain Patterns," IRE Mil. Elec. April 1962, pp 104 - 117.

6. Wheelon, A. D., "Radio Wave Scattering by Tropospheric Irregularities," Journal Res. N. B. S., Vol. 63D, No. 2, September-October 1959, p. 205.

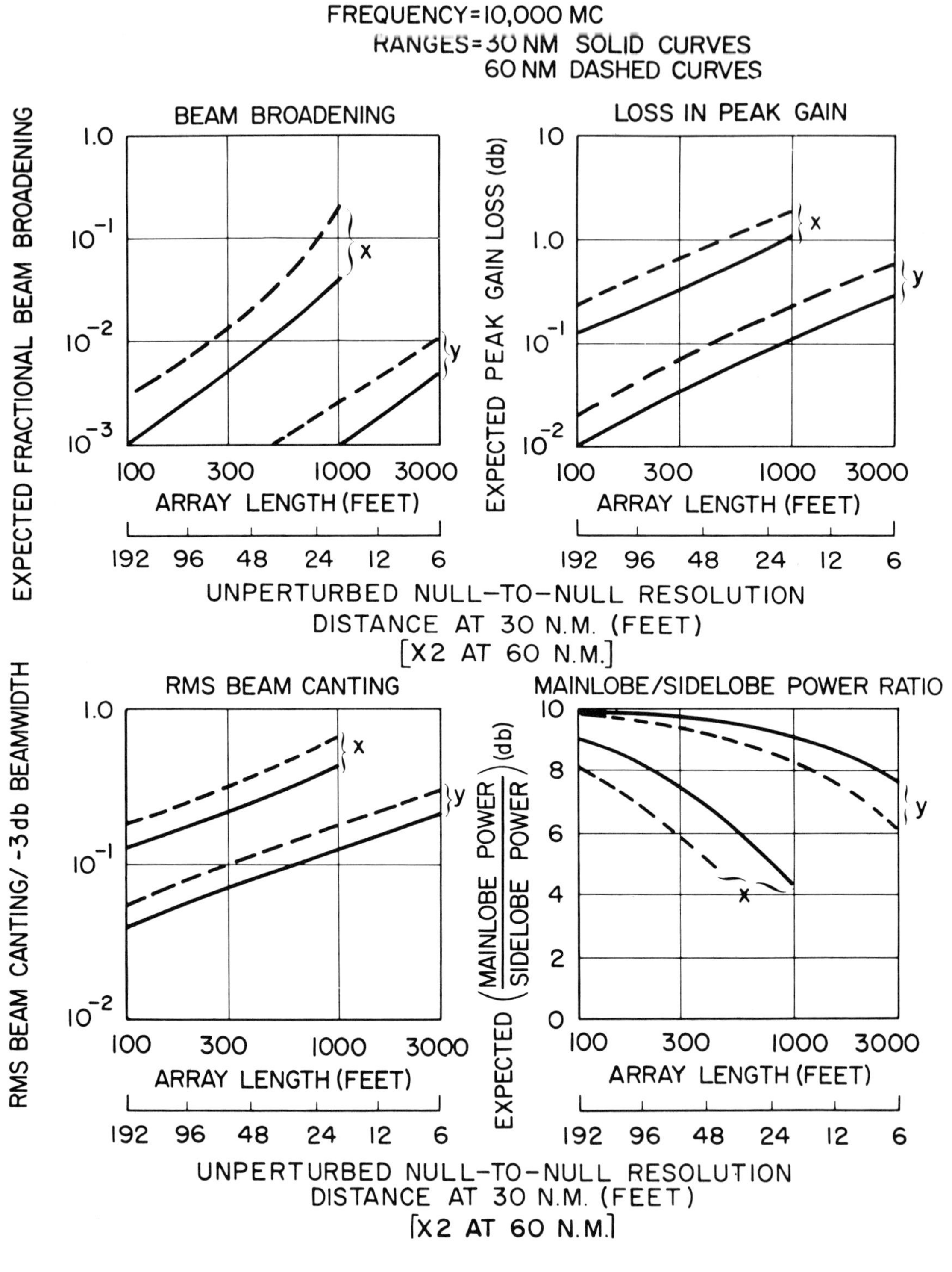

FIGURE 17. Focused synthetic array performance as limited by tropospheric propagation considerations.

7. Levin, E., R. B. Muchmore and A. D. Wheelon, "Aperture to Medium Coupling on Line of Sight Paths", Fresnel Scattering, IRE Trans. PGAP, April 1959, p. 142.

8. Private communication from K. A. Norton, National Bureau of Standards, Boulder, Colorado.

9. Von Rosenberg, C. E., et. al.; "Atomspheric Refractive Index Fluctuations as Recorded by an Airborne Microwave Refractometer", University of Texas E. E. R. L. Report No. 6-01.

10. Private communication from Messrs. K. A. Notron, J. W. Herbstreit, and M. C. Thompson Jr., National Bureau of Standards, Boulder, Colorado.

APPENDIX A

The power gain pattern of a broadside focused synthetic array is given by

$$G(\theta) = \left| \sum_{n=1}^{N} A_n e^{i(n\beta \sin\theta + \alpha_n)} \right|^2 \quad \text{(A-1)}$$

where the α_n are the random phase errors in the received signal at the successive "elements" of the synthetic array.

$$G(\theta) = \sum_{n=1}^{N} \sum_{s=1}^{N} A_n A_s^* e^{i\beta \sin\theta (n-s)} e^{i(\alpha_n - \alpha_s)} \quad \text{(A-2)}$$

Thus the phase errors enter the power pattern expression in terms of the phase differences between various "elements" of the array. The phase errors $\alpha_1, \alpha_2, \ldots \alpha_N$ are assumed to form an N dimensional Gaussian process with joint p. d. f.

$$p(\alpha_1, \alpha_2, \ldots \alpha_N) = \frac{\exp\left[-\frac{1}{2|\Lambda|} \sum_{n=1}^{N} \sum_{m=1}^{N} |\Lambda|_{nm} (\alpha_n - \bar{\alpha}_n)(\alpha_m - \bar{\alpha}_m) \right]}{(2\pi)^{N/2} |\Lambda|^{1/2}}$$

where Λ is the covariance matrix $\Lambda = \begin{bmatrix} \lambda_{11} & \lambda_{12} & \cdots & \lambda_{1N} \\ \lambda_{21} & & & \\ \lambda_{N1} & & & \lambda_{NN} \end{bmatrix}$

and $\lambda_{nm} = E\left[(\alpha_n - \bar{\alpha}_n)(\alpha_m - \bar{\alpha}_m) \right]$, also $|\Lambda|_{nm}$ is the cofactor of the element λ_{nm} in the determinant $|\Lambda|$ of the covariance matrix.

$= \sigma_n \sigma_m \rho_{nm}$

First let it be assumed that the phase error process is characterized by a zero mean and a constant variance σ^2. That is,

$$\overline{\alpha}_n = \overline{\alpha}_m = 0 \qquad \text{all } m, n$$

$$\sigma_n = \sigma_m = \sigma \qquad \text{all } m, n$$

Thus $\overline{\alpha_n^2} = \overline{\alpha_m^2} = \sigma^2$ all m, n

For these assumptions

$$\lambda_{nm} = E\left[(\alpha_n - \overline{\alpha}_n)(\alpha_m - \overline{\alpha}_m)\right]$$

$$= E\left[(\alpha_n \alpha_m\right] = \sigma^2 \rho_{nm}$$

All statistics of the N dimensional Gaussian phase process may then be determined from a knowledge of all the λ_{nm}. But for all terms $n = m$ $\lambda_{nn} = \sigma^2$ since $\rho_{nn} = 1$. Thus it is necessary to know separately σ^2 and ρ_{nm} all n, m. However, in the power gain pattern of a synthetic array one is interested in the statistics of the phase error difference process rather than the phase error process itself. It will now be shown that for the phase error difference process, all statistical quantities can be determined from a knowledge of the product $\sigma^2(1 - \rho_{nm})$ without knowing σ^2 and ρ_{nm} separately.

Lets define

$$A_{12} = \alpha_1 - \alpha_2 \qquad A_{13} = \alpha_1 - \alpha_3 \tag{A-3}$$

or in general $A_{ij} = \alpha_i - \alpha_j$ $\quad i \neq j \quad i = 1, \ldots N$

$j = 1, \ldots N$

Then since each of the A's is a linear transformation of the α's, the A's will be jointly Gaussian. Furthermore we know that the joint p. d. f. of the A's (and hence any statistic of the A's) is obtainable provided the covariance matrix of the A's is known. A typical nondiagonal element in this covariance matrix is

$$\overline{(A_{ij} - \overline{A}_{ij})(A_{kl} - \overline{A}_{kl})} \qquad \overline{A_{ij} A_{kl}}$$

since $\overline{A}_{ij} = \overline{\alpha_i - \alpha_j} = 0$

$$\overline{A}_{kl} = \overline{\alpha_k - \alpha_l} = 0$$

But $\overline{A_{ij} A_{kl}} = \overline{(a_i - a_j)(a_k - a_l)} = \overline{a_i a_k} + \overline{a_j a_l} - \overline{a_i a_l} - \overline{a_j a_k}$

But $\overline{a_i a_k} = \sigma^2 \rho_{ik}$

$\overline{a_j a_l} = \sigma^2 \rho_{jl}$ etc.

Hence

$$\overline{A_{ij} A_{kl}} = \sigma^2 \left[\rho_{ik} + \rho_{jl} - \rho_{il} - \rho_{jk}\right]$$

$$= \sigma^2 \left[(1 - \rho_{il}) + (1 - \rho_{jk}) - (1 - \rho_{ik}) - (1 - \rho_{jl})\right]$$

$$\overline{A_{ij} A_{kl}} = \sigma^2 (1 - \rho_{il}) + \sigma^2 (1 - \rho_{jk}) - \sigma^2 (1 - \rho_{ik}) - \sigma^2 (1 - \rho_{jl}) \tag{A-4}$$

In addition we must determine the diagonal terms in the covariance matrix of the A's. A typical diagonal element is

$$\sigma^2_{A_{ij}} = \overline{A_{ij}^2} = \overline{(a_i - a_j)^2} = \overline{a_i^2} + \overline{a_j^2} - \overline{2a_i a_j}$$

$$= \sigma^2 + \sigma^2 - 2\sigma^2 \rho_{ij}$$

$$\overline{(a_i - a_j)^2} = 2\sigma^2 (1 - \rho_{ij}) \qquad i = 1, 2 \ldots N \quad j = 1, 2, \ldots N \tag{A-5}$$

From Equations (A-4) and (A-5) we see that all statistics of the phase difference process may be determined from a knowledge of the quantity $\sigma^2 (1 - \rho_{ij})$ for all $1 \leq i \leq N$, $1 \leq j \leq N$

From Equation (A-5) we see that the required product $\sigma^2 (1 - \rho_{ij})$ is directly obtainable from the statistical mean square of the phase difference of the ith and j^{th} elements of the synthetic array. Thus all statistics of the synthetic array power gain pattern may be obtained from a knowledge of the statistical mean square phase difference for all element spacings less than or equal to the array length.

The Influence Of Random Phase Errors On The Angular Resolution Of Synthetic Aperture Radar Systems

J.A. Develet, Jr.
(Aerospace Corporation, El Segundo, California)

IEEE Transactions on Aerospace and Navigational Electronics, Vol. ANE-11, No. 1 (March 1964)
Pages 58-65

Summary—The influence of random phase errors on the angular resolution of a focused synthetic aperture radar system is treated. The principal measure of performance has been taken as the mean envelope power at the system output. This system output power is evaluated exactly, although not in closed form, based on the following assumptions: 1) the real beam pattern is Gaussian; 2) the random phase error is essentially a geometry-independent ergodic process with a Gaussian amplitude distribution and zero mean; and 3) the random phase error has a Gaussian correlation function.

The curves presented in this paper can be used to estimate expected system power response, expected system resolution and effective aperture length beyond which, in the presence of phase error, little gain in resolution is expected.

It was found that multiple sources of error with different correlation intervals make explicit solution of the integral equation for system power response practically impossible. In this situation, a reasonable approach is to evaluate the system power response separately for each error. If one of the errors is clearly dominant, it may be regarded as bounding achievable performance.

I. Introduction

Previous analyses regarding the effects of random phase errors on focused synthetic aperture radar systems[1] are extended for purposes of radar resolution determination and system design.

Greene and Moller[1] present an approximate (Monte Carlo) analysis of the expected system response for various conditions of mean-square phase error and phase-error correlation length for an arbitrarily assumed exponential correlation function. The range of parameters in this analysis, for the most part, is restricted to investigation of system response in the vicinity of the 3-db beamwidth.

This author maintains that the 3-db beamwidth has questionable significance in resolution of a radar system. When expected beamwidth is used to estimate resolution,[2] then beamwidth at the limits of system dynamic range (20 to 30 db) is the significant parameter. In other words, when receiving a strong target at maximum system input, the main question is how close a target of minimum system input can be placed and its presence still be resolved. This latter definition of resolution is obviously the most stringent because, as will be shown later, the 20- to 30-db beamwidth expands very rapidly with phase errors, while the 3-db beamwidth is comparatively insensitive to these errors.

The high sensitivity of the 20- to 30-db beamwidth to phase errors allows fairly accurate estimates of system resolution from the expected power output alone. Considering the accuracy of phase-error data, use of statistical decision theory is unnecessary for all practical purposes. This is fortunate because an exact analysis of the present highly nonlinear problem using statistical decision theory would be a formidable undertaking, to say the least.

An omission in previous analyses of synthetic aperture radar systems has been the determination of array length for a specified resolution in the presence of random phase errors. In the author's estimation, this quantity is the heart of system design. It is unreasonable to build a processor, storage facility, etc., compatible with a 4000-foot array when (due to irreducible random phase errors) nearly the same resultant resolution is achieved with a 2000-foot array.

Considering the foregoing, it is worthwhile to reattack this problem from a new perspective. It is not anticipated that the approach in this paper is the final answer, but the results obtained from it are considered to be of greater utility in system design.

Certain assumptions are made as to the character of the system in order to obtain exact mathematical solutions. The parameter calculated is the expected system average power output. Exact curves are presented to allow the system engineer to judge for himself the resultant resolution. Since the more logical 20- to 30-db resolution is highly sensitive to phase errors, the results obtained using a simple and somewhat arbitrary resolution estimate based on power response alone are not far from those that would be calculated using an exact statistical decision theory approach.

II. Analysis

A vast amount of literature is available on synthetic aperture radar systems. No duplication or repetition of these works will be attempted in this paper. It is assumed that the reader is familiar with the contents of the April, 1962, issue of the IRE Transactions on Military Electronics.

[1] C. A. Greene and R. T. Moller, "The effect of normally distributed random phase errors on synthetic array gain patterns," IRE Trans. on Military Electronics, vol. MIL-6, pp. 130–139; April, 1962.

[2] This is a reasonable measure due to its analytical tractability.

This paper deals only with the fully coherent focused system. A number of assumptions are made to simplify analytical study of the system.

1) Sampling rate of the radar is sufficiently high that no angular ambiguities appear in the data and also high enough that received signal spectrum splatter due to very short-term random phase fluctuations can be accommodated. This assumption allows filtering of the coherently detected received pulses to yield a continuous waveform Doppler history of a target. Continuous waveforms are much easier to treat mathematically and their use results in no loss of generality.
2) Range ambiguities for particular system geometries are suppressed by proper signal design. No further consideration of ambiguities will be treated.
3) Linearity of the radar system is assumed so that system output to any arbitrary field of targets can be obtained by voltage summation of the individual target responses.
4) A single dominant target is being viewed in each resolution cell.[3] Many targets of approximately the same return strength within one resolution cell may be approximately handled by assuming that the system output is a Rayleigh distribution with some mean power level. A small number of targets of approximately equal strength in one resolution cell can cause "breakup" of the return due to cancellation and reinforcement. This is one of the reasons for the different appearance of optical and radar maps.

Consider now Fig. 1, which depicts the system geometry while responding to a *single* point target. It is assumed that an observer at 0 fixed relative to the real antenna watches the target, T, go by with velocity, v. Time is referenced to zero as indicated in the figure. The observer and his data processor are "matched" to the geometry of Fig. 1. Any change in the true situation from that of Fig. 1 due to improper target (vehicle) motion, clock (oscillator) instabilities or propagation-medium scintillation must be compensated for by either a change in data processing or by acceptance of the change as random error, with the resulting system degradation. This paper is concerned only with *uncompensated* phase errors and their effects on *angular* resolution.

Let the radar at time, t, transmit the signal

$$V_t = E_t \cos \omega t \tag{1}$$

where

E_t = transmitted signal amplitude, volts
ω = carrier radian frequency, radian/sec.

[3] If a dominant target exists and is surrounded by much smaller targets all within one resolution cell, the effect of these smaller targets may be approximately characterized as random phase variations on the return of the dominant target.

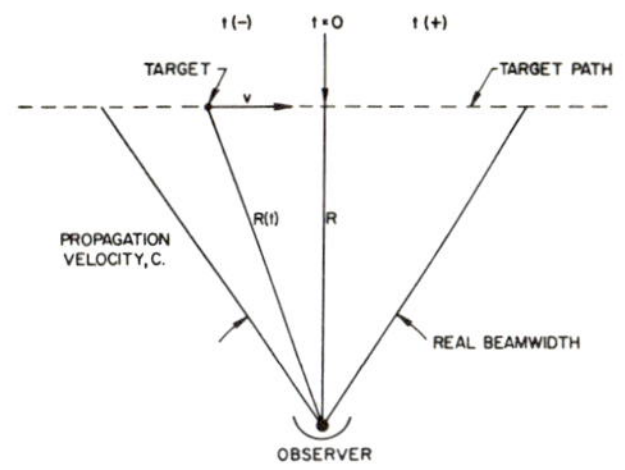

Fig. 1—System geometry.

Assuming Gaussian real antenna beam patterns, it is a simple matter to show that the return signal from a point target is given to first order by

$$V_r = E_r \exp(-2kt^2) \cos [\omega t + \alpha + \beta t^2 + \phi_i(t)] \tag{2}$$

where

E_r = received signal amplitude, volts
k = beam factor, $\sec^{-2}$
ω = carrier radian frequency, radian/sec
α = fixed phase shifts which may exist to and from target, *e.g.*, 2 ωR/c plus propagation and equipment biases
β = coefficient for change in phase vs time due to quadratic range change (Fig. 1), $\sec^{-2}$
$\phi_i(t)$ = ith ensemble member of an ergodic random Gaussian phase process of zero mean and Gaussian correlation function, radian.

It is not necessary to know the value of α as it will cancel in later calculations. The terms β and k may be computed from simple geometric considerations. The results are

$$\beta = -\frac{2\pi v^2}{\lambda R}, \quad \sec^{-2} \tag{3}$$

$$k = 2\left(\frac{v}{L}\right)^2, \quad \sec^{-2} \tag{4}$$

where

L = length at the target to the $1/e(-4.34$ db) points of the real antenna one-way power pattern, feet
v = vehicle velocity, feet/sec
λ = carrier wavelength, feet
R = range to target at $t=0$ (see Fig. 1), feet.

For convenience, β and k are used in the following analysis to minimize symbology.

The synthetic aperture radar system correlates the received signal, given by (2), with a delayed in-phase and quadrature replica and presents an average power output equal to the sum of the squares of the in-phase and quadrature correlator responses. Fig. 2 depicts this process. The limits of integration are shown to be infinite in Fig. 2. In a real system, a finite integration would be accomplished. Since mathematical tractability

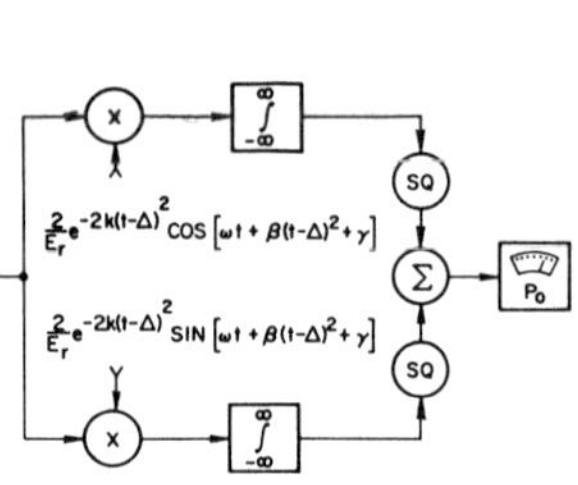

Fig. 2—System transfer function.

is sought and, further, since, with the assumed Gaussian weighting, there is little difference in the correlator output between reasonably large finite and infinite limits, the analysis to follow uses infinite limits.

Having dispensed with all the assumptions and system descriptions, injection of the voltage given by (2) into the system of Fig. 2 with the reference functions of (5) and (6) gives the following average power output:

$$P_0 = \overline{\int_{-\infty}^{\infty}\int_{-\infty}^{\infty}\left[\exp\left\{-2k\left[x^2+y^2+\left(\frac{x-\Delta}{\eta}\right)^2+\left(\frac{y-\Delta}{\eta}\right)^2\right]\right\}\right] \times \cos\left[2\Delta\beta(x-y)+\phi_i(x)-\phi_i(y)\right]dxdy}. \quad (7)$$

As depicted, the correlators are "matched" filters for the signal of (2) and, as such, maximize the signal-to-noise ratio in the output for a time shift $\Delta=0$. However, there are occasions when all the data received from a target on its pass through the real beam cannot and should not be processed. These situations occur when 1) processing time is precious and resolutions corresponding to real beam length L are not needed, 2) lack of phase coherence across length L yields a point of diminishing returns and processing more data gives little increase in system resolution and 3) vehicle constraints have forced the use of a smaller real antenna than desirable from a resolution point of view.

As a consequence of the above, the reference functions chosen for analysis are generalized to the following:

$$\frac{2}{E_r}\exp\left[-2k\left(\frac{t-\Delta}{\eta}\right)^2\right]\cos\left[\omega t+\beta(t-\Delta)^2+\gamma\right] \quad (5)$$

$$\frac{2}{E_r}\exp\left[-2k\left(\frac{t-\Delta}{\eta}\right)^2\right]\sin\left[\omega t+\beta(t-\Delta)^2+\gamma\right] \quad (6)$$

where $\eta=$ (by definition) the fraction of the data utilized in processing,

$$0\leq\eta\leq 1.$$

As depicted in (5) and (6), η may be greater than 1 and not necessarily restricted to the range $0\leq\eta\leq 1$. However, in the presence of noise the signal-to-noise ratio at the correlator outputs will drop off for $\eta>1$.[4] Since for every $\eta>1$ there is a $\eta<1$ with the same signal-to-noise ratio, it is obvious from an engineering point of view that the range of η resulting in the least data processing will always be used; *ergo*, $0\leq\eta\leq 1$.

[4] This is because only a matched filter, $\eta=1$, can yield the largest signal-to-noise ratio for $\Delta=0$.

The bar over (7) is the ensemble average over ϕ_i. Since the average of a sum is the sum of the averages and, further, since $\overline{\sin[\phi_i(x)-\phi_i(y)]}\equiv 0$ and $\overline{\phi_i(x)-\phi_i(y)}\equiv 0$, expansion of the cosine function in (7) yields

$$P_0 = \int_{-\infty}^{\infty}\int_{-\infty}^{\infty}\left[\exp\left\{-2k\left[x^2+y^2+\left(\frac{x-\Delta}{\eta}\right)^2+\left(\frac{y-\Delta}{\eta}\right)^2\right]\right\}\right] \times \cos\left[2\Delta\beta(x-y)\right]\overline{\cos\left[\phi_i(x)-\phi_i(y)\right]}dxdy. \quad (8)$$

Since $\phi_i(t)$ has been assumed to be Gaussian, it is recognized[5] that

$$\overline{\cos\left[\phi_i(x)-\phi_i(y)\right]} = \exp\left\{\frac{-\overline{\left[\phi_i(x)-\phi_i(y)\right]^2}}{2}\right\}. \quad (9)$$

Simplifying (9) we get

$$\overline{\cos\left[\phi_i(x)-\phi_i(y)\right]} = \exp\left\{-\left[R(0)-R(x-y)\right]\right\} \quad (10)$$

where

$R(x-y)=$ correlation function of the random phase perturbations, radian^2.

Let us assume now that $R(x-y)$ is Gaussian in nature; thus

$$R(x-y) = \sigma^2\exp\left[-\frac{(x-y)^2}{\tau_c^2}\right], \quad \text{radian}^2 \quad (11)$$

[5] J. A. Develet, Jr., "A threshold criterion for phase-lock demodulation," Proc. IEEE, vol. 51, pp. 349–356; February, 1963. Correction: Proc. IEEE, vol. 51, p. 580; April, 1963.

where

σ^2 = mean-square value of phase error, radian^2

τ_c = time it takes the correlation function of phase to drop to $1/e$ of its zero argument value, sec.

Substituting (11) into (10) and (10) into (8), we finally obtain

$$P_0 = \exp(-\sigma^2) \int_{-\infty}^{\infty} \int_{-\infty}^{\infty} \cdot \left[\exp \left\{ -2k \left[x^2 + y^2 + \left(\frac{y - \Delta}{\eta} \right)^2 + \left(\frac{x - \Delta}{\eta} \right)^2 \right] + \sigma^2 \exp \left[- \frac{(x - y)^2}{\tau_c^2} \right] \right\} \right] \times \cos [2\Delta\beta(x - y)] dx dy. \tag{12}$$

Our task now is to evaluate the interesting integral of (12). Expanding the $\exp\{\sigma^2 \exp[-(x-y)^2/\tau_c^2]\}$ term in an infinite series, which is valid for all values of the exponent, (12) becomes

$$P_0 = \exp(-\sigma^2) \sum_{j=0}^{\infty} \frac{\sigma^{2j}}{j!} \int_{-\infty}^{\infty} \int_{-\infty}^{\infty} \cdot \left[\exp \left\{ -2k \left[x^2 + y^2 + \left(\frac{x - \Delta}{\eta} \right)^2 + \left(\frac{y - \Delta}{\eta} \right)^2 \right] - \frac{j(x - y)^2}{\tau_c^2} \right\} \right] \cdot \cos [2\Delta\beta(x - y)] dx dy. \tag{13}$$

Expanding the exponent in (13) and performing the excruciating double integration with the help of integral tables[6] give the following *exact* result for P_0

$$P_0' = \exp(-\sigma^2) \sum_{j=0}^{\infty} \frac{\sigma^{2j} \exp \left[-4 \left(\frac{x}{x_0} \right)^2 \left\{ \left[1 + \frac{j}{4 \left(\frac{x_c}{L_{\text{eff}}} \right)^2} \right]^{-1} + 4 \left(\frac{x_0}{L_{\text{eff}}} \right)^2 \frac{\eta^2}{(\eta^2 + 1)^2} \right\} \right]}{j! \left[1 + \frac{j}{4 \left(\frac{x_c}{L_{\text{eff}}} \right)^2} \right]^{1/2}} \tag{14}$$

[6] W. Grobner and N. Hofreiter, "Integraltafel, Zweiter Teil, Bestimmte Integral," Springer-Verlag, Vienna and Innsbruck, Austria, p. 143, 1958; Eqs. (1a), (1b).

where

P_0' = normalized system power response

x = distance along vehicle track, feet

$x_0 = (2/\pi)(\lambda R)/L_{\text{eff}}$, -4.34 db $(1/e)$ synthetic aperture response width, under no phase-error conditions, feet

$L_{\text{eff}} = L(2/(1+1/\eta^2))^{1/2}$, by definition, the effective synthetic aperture length, feet

$x_c = v\tau_c$, the *distance* to $1/e$ point of the phase-error correlation function, feet

η = (by definition) the fraction of available data processed. See (5) and (6).

Two key quantities defined above bear repeating as (15) and (16).[7]

$$x_0 = \frac{2}{\pi} \left(\frac{\lambda R}{L_{\text{eff}}} \right), \quad \text{feet} \tag{15}$$

$$L_{\text{eff}} = L \left(\frac{2}{1 + 1/\eta^2} \right)^{1/2}, \quad \text{feet.} \tag{16}$$

Note that, for small η, (14) reduces identically to the two-way power pattern of the real beam. This is expected since, for small η, little processing of the received data is performed.

For the case of considerable beam sharpening, $x/L \to 0$, the exact equation, (14), reduces to

$$P_0' = \exp(-\sigma^2) \sum_{j=0}^{\infty} \frac{\sigma^{2j} \exp \left\{ -4 \left(\frac{x}{x_0} \right)^2 \left[1 + \frac{j}{4 \left(\frac{x_c}{L_{\text{eff}}} \right)^2} \right]^{-1} \right\}}{j! \left[1 + \frac{j}{4 \left(\frac{x_c}{L_{\text{eff}}} \right)^2} \right]^{1/2}}. \tag{17}$$

[7] Note, in (14)–(19), that the time variable along the track has been replaced by the more conventional distance variable.

Note that in (17) the power response of the system depends only on 1) σ, the rms magnitude of the random phase error, 2) x_c/L_{eff}, the ratio of the correlation distance of phase errors to effective array length, and 3) x/x_0, the normalized distance from maximum output. Note also, as $x_c/L_{eff} \to \infty$ (complete correlation across the array), (17) becomes

$$P_0' = \exp\left[-4\left(\frac{x}{x_0}\right)^2\right] \tag{18}$$

and is completely uninfluenced by the rms magnitude of the phase error. This is expected. Eq. (18) demonstrates the significance of x_0 as the width to the $1/e$ (-4.34 db) power response with no phase error.

Figs. 3(a)–(n) are the results of IBM 7090 computations utilizing (17) for various values of x_c/L_{eff} and σ.[8] To generalize these results to the case of arbitrary beam sharpening (14), simply add the following decibel correction to the ordinates:

$$C = -\frac{80 \log_{10}(e)}{\eta^2 + 1}\left(\frac{x}{L}\right)^2, \quad \text{db.} \tag{19}$$

[8] Note that with Gaussian weighting no sidelobes exist. Therefore, if sidelobe suppression is important, Gaussian amplitude weighting is indicated.

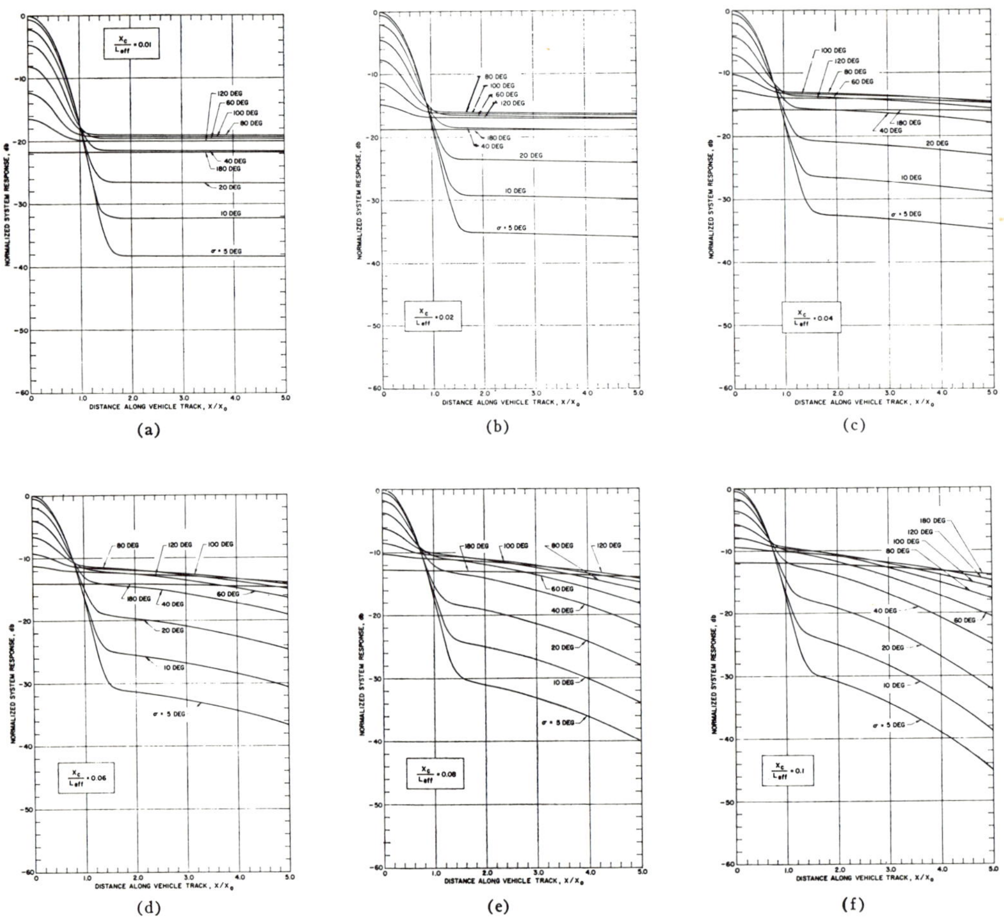

Fig. 3(a–n)—System power response vs distance along vehicle track.

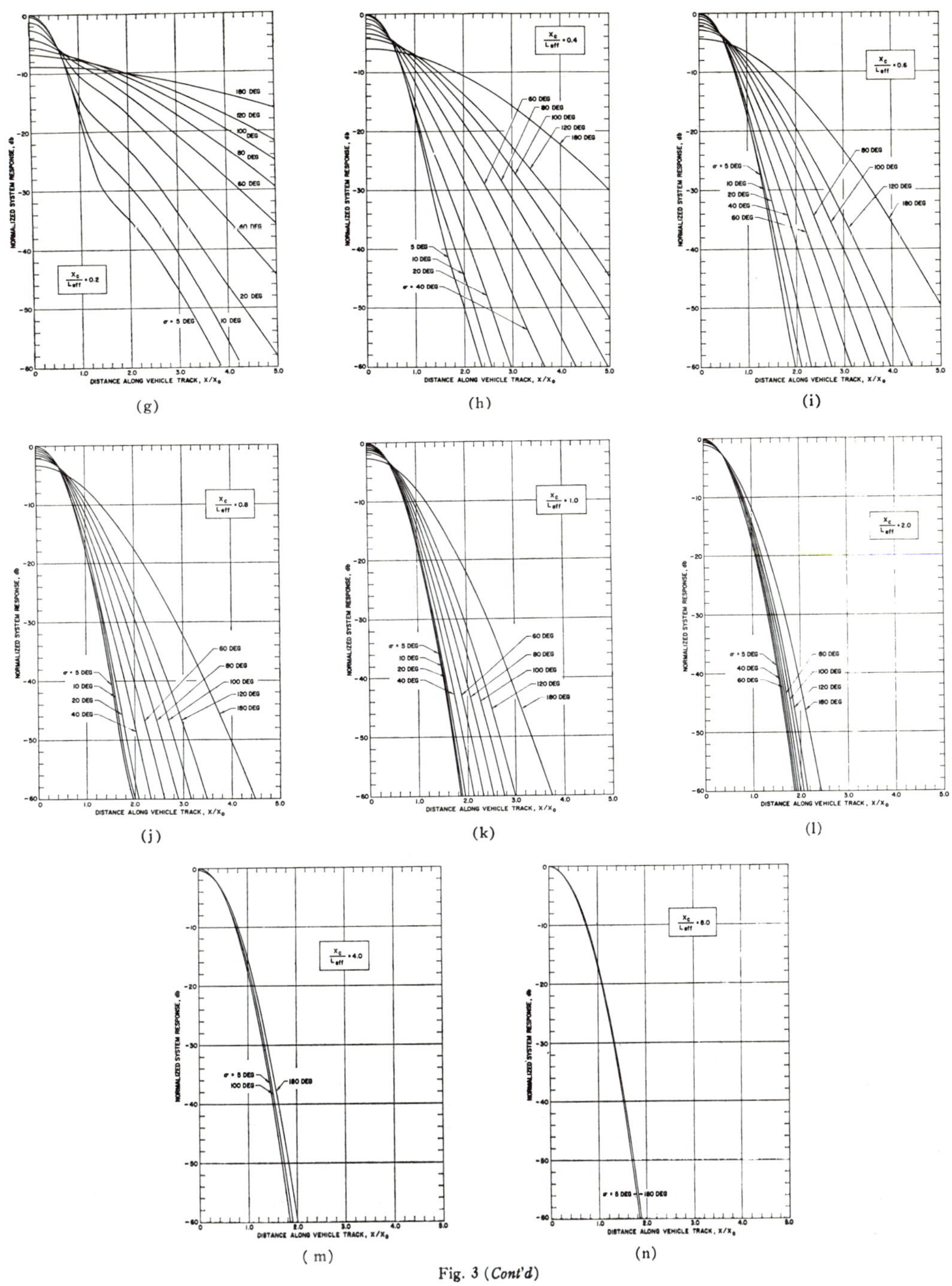
NORMALIZED SYSTEM RESPONSE, db
DISTANCE ALONG VEHICLE TRACK, X/X₀
(g)
(h)
(i)
(j)
(k)
(l)
(m)
(n)

Fig. 3 (*Cont'd*)

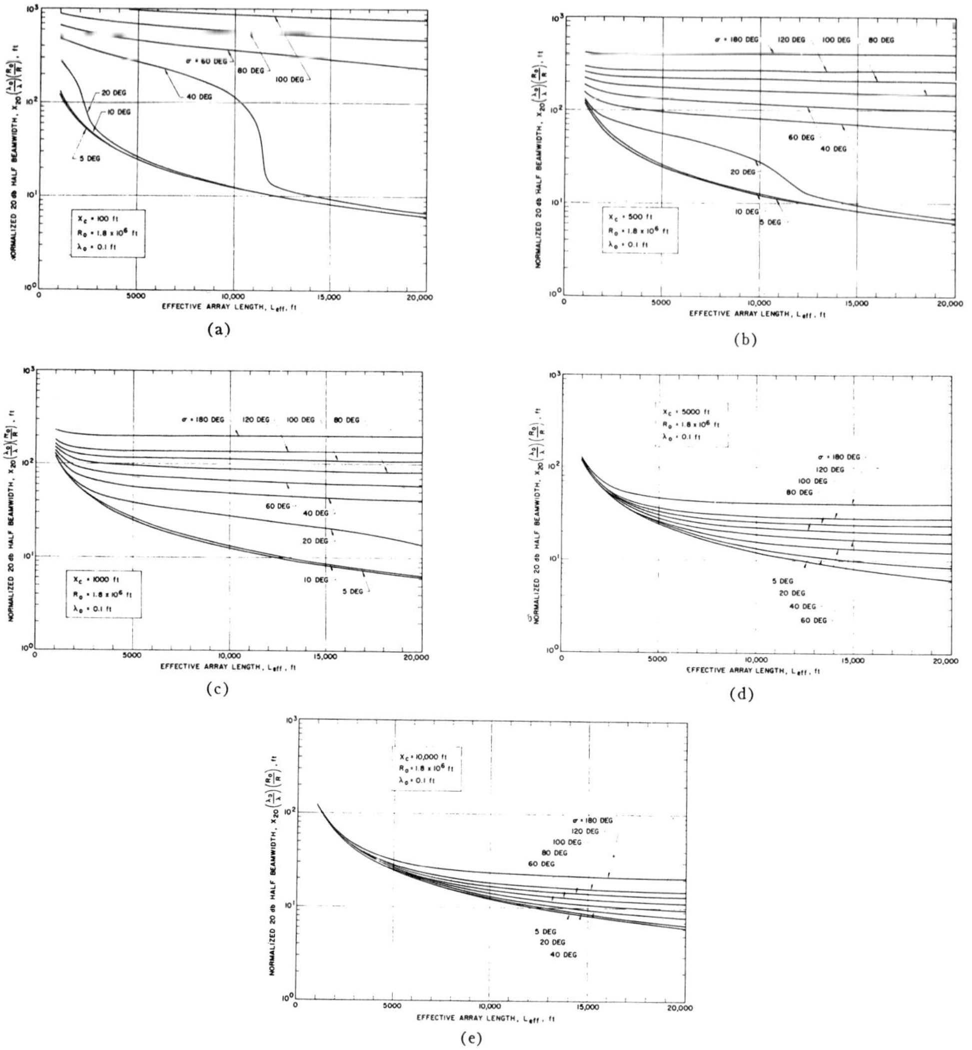

Fig. 4(a–e)—Twenty-decibel half beamwidth vs array length.

Figs. 4(a)–(e) are interesting and can assist in assessing the effective array length that one should use in the presence of phase errors to achieve a specified 20-db synthetic beamwidth. The figures were derived from IBM 7090 computations utilizing (17) and, as such, assume considerable beam sharpening, *i.e.*, neglect of the last term in the exponent of (14). These curves are normalized and can be used for any range R or system wavelength λ for which the assumption of considerable beam sharpening holds. For the special case of $R=R_0=1.8\times10^6$, feet, and $\lambda=\lambda_0=0.1$, feet, the 20-db half beamwidth appears directly as the ordinate.

Note the rapid expansion with phase error of the 20-db beamwidth as contrasted to the 3-db beamwidth. A reasonable definition of resolution is that it is the half beamwidth to the point equal in decibels to the dynamic range of the system. It is felt that this is a good engineering criterion considering 1) the sensitivity of the dynamic-range beamwidth to σ and x_c, and 2) the lack of precise knowledge of σ and x_c. For N strong targets in the vicinity and for short phase-error correlation distances, it is suggested that a safety factor of 10 $\log_{10}(N)$ be added to the dynamic range. [See Fig. 3(a).]

III. Conclusions

Certain assumptions regarding the physical system and random phase errors have yielded an exact result that predicts the expected power response for a synthetic aperture radar system.

Complete freedom to choose the fraction of the received data to be processed has been left to the system designer by his choice of η.

The assumed Gaussian weighting not only yielded a mathematically tractable system but, in addition, gave an expected response with *no sidelobes*.

The definition of resolution suggested by the author is the half beamwidth to the dynamic range point on the expected power response curve. A figure of 20 db was chosen for illustration in Fig. 4(a)–(e). The definition of resolution as the dynamic-range half beamwidth is consistent with 1) the high sensitivity of resolution to σ and x_c and 2) the inaccurate knowledge of these parameters.

It is felt that an exact decision theory attack on this problem would not be fruitful because of its mathematical formidability and the inaccurate available data on σ and x_c.

Eqs. (11) and (12) indicate that *multiple* sources of error with different correlation intervals make explicit solution of the integral equation for system power response a formidable undertaking. In this situation, a reasonable approach is to evaluate the system power response separately for each error. If one of the errors is clearly dominant, it may be regarded as bounding achievable performance.

Turbulence Induced Phase Errors In Synthetic-Aperture Radars

L.J. Porcello
(Radar and Optics Laboratory, Institute of Science and Technology, University of Michigan, Ann Arbor)

IEEE Transactions on Aerospace and Electronic Systems, Vol. AES-6, No. 5, (September 1970)
Pages 636-644

Abstract

Radar waves propagating through the turbulent troposphere acquire phase fluctuations which appear as aperture phase errors in synthetic-aperture radar systems. The phase error description follows directly from Tatarski's theory of wave propagation through turbulence; for most situations of interest, the phase error is Gaussian with stationary first increments. The structure function of the phase error is a function of the radar parameters and of the vertical profile $C_{nm}(h)$ of the microwave refractive index structure constant of the turbulence. A closed form expression for best attainable resolution can then be given as a function of these same parameters, via the results of a phase error analysis by Brown and Riordan. Published data on $C_{nm}(h)$ and its optical-region counterpart are reviewed. The data from a recent microwave propagation experiment by Thompson and Janes are then analyzed and are consistent with the order-of-magnitude estimates of C_{nm}. Some unsolved problems associated with applying the results to synthetic-aperture systems are discussed.

Refractive-index fluctuations associated with turbulence in the lower atmosphere impose performance limits on large antenna apertures and arrays. Synthetic-aperture radars are subject to these limitations, although the effects are different in certain details from those observed in large ground-based physical aperture systems. In this paper, the synthetic-aperture radar (henceforth SAR) is discussed in the context of Tatarski's theory of wave propagation through a turbulent medium; some theoretical results by Brown and Riordan [1] are combined with results of Tatarski to obtain a general parametric expression for attainable resolution. Some experimental microwave propagation data reported by Thompson and Janes [2] are then used to provide some insight into the magnitude of the phase error which is to be expected.

This work was supported in large part by the Avionics Laboratory, Air Force Systems Command, U. S. Air Force, under Contracts AF33(615)-1452 and F33615-67-C-1814.

I. Form of the Phase Error

The synthetic-aperture terrain-imaging radar concept has been described in a number of papers, for example, [3]–[5]. In the simplest sidelooking mode, a terrain strip parallel to the ground track of an aircraft is illuminated by a coherent radar as illustrated in Fig. 1, and reflected signals are stored and subsequently processed to generate a fine-resolution image of the microwave reflectivity properties of the terrain. A reflecting object at slant range R is viewed by the radar as the aircraft moves through a distance $L \approx \beta R$, where β is the along-track angular width of the antenna pattern. In the region between the radar's trajectory and the target, the refractive index $n(x, y, z, t)$ varies with respect to both time and position. The vertical profiles of temperature, pressure, and water vapor content cause the refractive index to decrease more or less systematically with increasing altitude, while weather movements, local heating effects, wind currents over the local topography, and other similar phenomena induce random temporal and spatial fluctuations in n. We restrict our attention to inhomogeneities which are on the order of L or smaller, and which are induced by tropospheric turbulence. The most important consequence of these inhomogeneities is that they induce random phase errors across the synthetic aperture; these, in turn, limit the attainable resolution of the radar system, perturb sidelobe levels, and introduce other undesirable distortions into the system's impulse response. The radar's response to a particular target at range R is determined by the propagation conditions within and close to a triangular region whose vertex is at the target and whose base, of length L, coincides with the aircraft trajectory as shown in Fig. 2. We neglect gross refractive effects resulting from the vertical gradient of refractive index. Furthermore, we assume for the present that the aircraft moves through the distance L so rapidly that time variations in refractive index do not occur during the time any specific point target is being observed.

The basic theory of wave propagation through a turbulent medium was set forth by Tatarski [6] and subse-

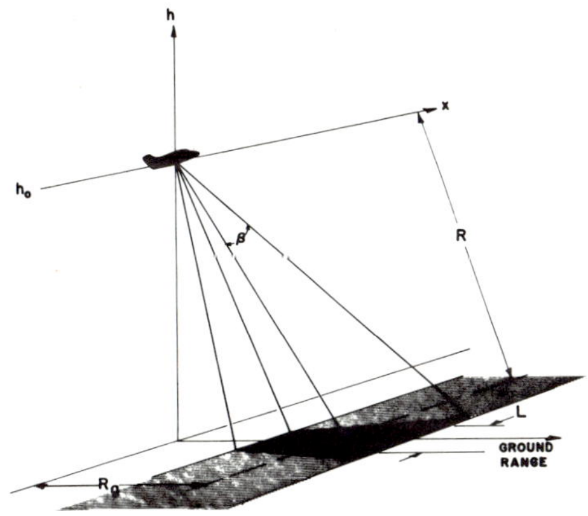

Fig. 1. Terrain-imaging synthetic-aperture radar.

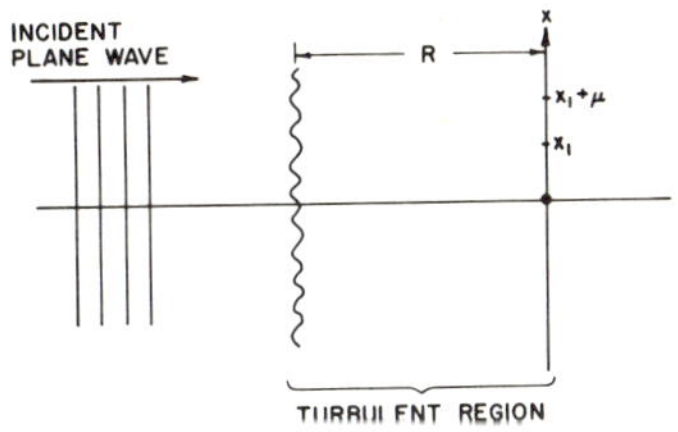

Fig. 3. Plane wave incident on turbulent region of thickness R.

Fig. 2. Triangular sector affecting synthetic-aperture's impulse response.

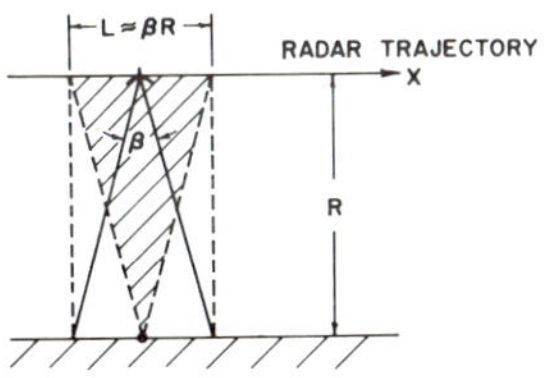

quently has been refined and extended by many workers in the field. A recent tutorial paper by Strobehn [7] summarizes some of these extensions. Tatarski's work, with a few refinements, provides a framework for the SAR propagation phase error problem; we summarize some of the required results in the next few paragraphs.

Tatarski studied the problem of wave propagation through turbulence which is homogeneous and isotropic within some region whose size L_0 is termed the *outer scale* of the turbulence. In such an atmospheric region, the instantaneous spatial distribution $n(\vec{r}, t)$ of refractive index is a nonstationary random function. However, the difference $n(\vec{r}_1, t_1)-n(\vec{r}_2, t_1)$ is stationary as long as $|\vec{r}_2-\vec{r}_1|<L_0$; i.e., $n(\vec{r}, t_1)$ is *random with stationary first increments.* Such a random process is conveniently described by a refractive-index *structure function* D_n, defined by

$$D_n(\vec{r}_1, \vec{r}_2) \triangleq E\{[n(\vec{r}_1) - n(\vec{r}_2)]^2\}. \qquad (1)$$

In the troposphere, mechanical energy is introduced into the turbulence by large anisotropic turbulent eddies, with scales larger than L_0. The energy is transferred to eddies with successively smaller scale lengths, and is finally dissipated by viscous friction in the smallest eddies, whose dimension l_0 is termed the *inner scale* of the turbulence. The degree of anisotropy diminishes as scale l becomes small compared with L_0; therefore, for $r \ll L_0$, the function $D_n(\vec{r}_1-\vec{r}_2)$ becomes a function of $r=|\vec{r}_1-\vec{r}_2|$. It can then be shown that

$$D_n(r) = C_n^2 r^{2/3}, \qquad l_0 \ll r \ll L_0, \qquad (2)$$

where the quantity C_n is called the *refractive-index structure constant.* Within the troposphere, l_0 is thought to be on the order of one to ten mm near the surface. L_0 appears to be a function of orientation and of altitude; near the surface the horizontal extent of the large eddies is thought to be of the same order as the height above the surface, while at higher altitudes it ranges from 100 to 1000 meters. There is some indication that the vertical extent of these eddies seldom exceeds a few meters [7].

Let a plane wave of wavelength $\lambda \ll l_0$ be incident on a turbulent region with structure constant C_n, as shown in Fig. 3. Then, at a distance R into the medium, the *transverse structure function of the instantaneous phase distribution* (henceforth referred to simply as the *phase structure function*)

$$D_\phi(\mu) \triangleq E\{[\phi(x+\mu) - \phi(x)]^2\} \qquad (3)$$

is given by[1]

$$D_\phi(\mu) = \alpha k^2 R C_n^2 \mu^{5/3} \qquad (4)$$

where

$$k = 2\pi/\lambda$$

$$\alpha = \begin{cases} 2.91, & \mu \gtrsim \sqrt{\lambda R} \\ 1.46, & l_0 \ll \mu \ll \sqrt{\lambda R}. \end{cases} \qquad (5)$$

Two available generalizations of this result are important

[1] The condition $\lambda \ll l_0$ is not strictly satisfied for microwave propagation in the lower atmosphere, where l_0 is on the order of 1 to 10 mm. The assumption $\lambda \ll l_0$ facilitates theoretical analysis, in the sense that it permits one to employ the scalar wave equation in formulating $D_\phi(\mu)$.

The maximum value of R for which (4) holds is also a matter of some controversy; see Strobehn [7] and references therein for detail. However, this theoretical difficulty applies primarily to the optical region and is of little practical consequence at microwavelengths.

(*Note added in proof:* A recent paper by Clifford and Strohbehn [17] shows that the theoretical expressions for phase fluctuation due to turbulence along a LOS path, derived under the condition $\lambda \ll l_0$, are equally valid when $\lambda \gtrsim l_0$.)

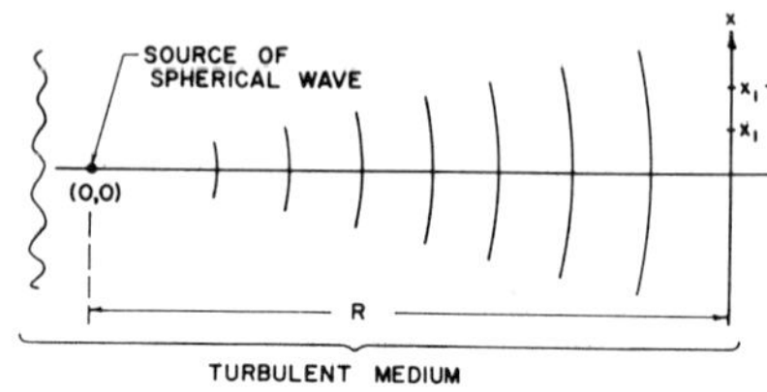

Fig. 4. Spherical wave propagating thrc region.

Fig. 5. Spherical wave propagating to elevated baseline.

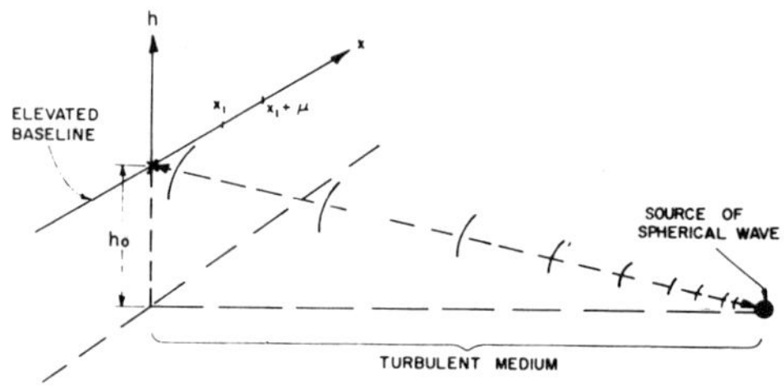

to the synthetic-aperture radar situation. First, if C_n changes slowly along the propagation path, (4) may be modified to handle this situation simply by replacing the quantity $C_n^2 R$ by the line integral

$$\int_0^R C_n^2(l)dl,$$

where the integral is taken along the nominal propagation path. Second, if the incident wave is spherical rather than plane as in Fig. 4, and if one excludes the systematic phase associated with geometric path length differences (due to curvature of the wavefront) then the corresponding phase structure function has the form

$$D_\phi(\mu) = \alpha k^2(\mu/R)^{5/3}\int_0^R C_n^2(l)l^{5/3}dl. \qquad (6)$$

In practice, one expects C_n to vary as a function of altitude, as well as with local climatic conditions. Let $C_n(h)$ be the vertical profile of C_n; let a spherical wave emanate from a point on the earth's surface and propagate to an elevated baseline normal to the line-of-sight (LOS) path as shown in Fig. 5. The phase structure function then is given by

$$D_\phi(\mu) = \alpha k^2(\mu/R)^{5/3}(\csc\psi)^{8/3}I \qquad (7)$$

where

$\psi = \sin^{-1}(h_0/R)$, the *elevation angle* of the baseline
h_0 = the altitude of the baseline

$$I = \int_0^{h_0} C_n^2(h)h^{5/3}dh$$

and μ is measured along the baseline. Therefore the mean-square behavior of the phase differences along the baseline can be related to system and atmospheric parameters. In addition, if $|\mu| \ll R$, the phase difference $[\phi(x+\mu)-\phi(x)]$ is Gaussian-distributed, a property which is analytically convenient and which has been confirmed experimentally.

In line with this approach, knowledge of $C_n(h)$ for the microwave region then permits one to calculate $D_\phi(\mu)$ for any extended-baseline system. We shall return to the discussion of $C_n(h)$ shortly; first, we divert briefly to examine the effects of phase errors whose form corresponds to (7).

II. Effects of the Phase Error

In a companion paper, Brown and Riordan [1] have calculated the limiting resolution of an aperture which is subject to phase error with a power-law structure function, as well as the length of the optimum aperture. We shall pursue the case in which the phase error is Gaussian and in which one makes a least-square approximation to the linear component of the phase error to remove the "tilt" associated with this linear component before calculating resolution. The "equivalent rectangle" measure of resolution is employed. Let $F(\omega)$ be the Fourier transform of the complex function $f(t)$ which is perturbed by phase error $\alpha(t)$:

$$F(\omega) = \int e^{-j\omega t}e^{j\alpha(t)}f(t)dt. \qquad (8)$$

Then, for the class of phase error cited, with structure function

$$D_\alpha(\mu) = b^n\mu^n, \qquad n = 5/3$$

the equivalent-rectangle width of $|F(\omega)|^2$ after least-squares tilt correction is

$$\rho_0 = 0.985b. \qquad (9)$$

The optimum aperture which yields this resolution is

$$(2a)_0 = \frac{13.4}{b} \qquad (10)$$

where the integration in $F(\omega)$ is taken over $(-a, +a)$.

These results are directly applicable to SAR systems. The along-track resolution of a SAR is

$$\rho_x = \frac{\lambda R}{4\pi}\rho_0. \qquad (11)$$

Taking the phase structure function for round-trip propagation to be four times the one-way structure function, and substituting from (7) and (9), we obtain

$$\rho_x = 0.985\left(\frac{\lambda R}{4\pi}\right)\frac{(11.6\,I)^{3/5}k^{6/5}(\csc\psi)^{8/5}}{R}, \qquad k = \frac{2\pi}{\lambda},$$

which reduces to

$$\rho_x \approx 3.3\,I^{3/5}(\csc\psi)^{8/5}\lambda^{-1/5}\text{ cm}. \qquad (12)$$

In this expression,

$$I = \int_0^{h_0} C_n^2(h) h^{5/3} dh$$

is given in cm² and λ is in cm. Neglecting a possible λ-dependence in $C_n(h)$, we observe that ρ_x is weakly dependent on λ, and improves slightly as λ increases. Furthermore, attainable resolution ρ_x varies as $R^{1.6}$, since $\csc\psi = R/h_0$. The optimum aperture is then given by

$$L_{opt} \approx 0.34 \frac{\lambda^{6/5} R}{I^{3/5} (\csc\psi)^{8/5}}. \quad (13)$$

As shown by Fig. 2 of Brown and Riordan, the optimum aperture—that which yields limiting resolution ρ_x for the case $n = 5/3$—is approximately twice the aperture which yields the same value of ρ_x in the absence of phase errors. The rms phase error between the center and either endpoint of the optimum aperture is $2[D_\phi(L_{opt}/2)]^{1/2}$ where D_ϕ and L_{opt} are given by (7) and (13), respectively:

$$2[D_\phi(L_{opt}/2)]^{1/2} \approx \frac{3}{2}\pi \text{ radians.}$$

III. The Profile $C_n(h)$

The refractive-index structure constant C_n appeared earlier as a coefficient in an equation relation $D_n(\mu)$ to μ:

$$D_n(\mu) = C_n^2 \mu^{2/3} \qquad l_0 \ll \mu \ll L_0. \quad (2)$$

This atmospheric parameter describes, in a sense, the degree of inhomogeneity of the refractive-index spatial structure. At microwavelengths, a static parcel of air has a refractive modulus N given by the Smith–Weintraub equation

$$N = \frac{77.6\,P}{T} + 3.73 \times 10^5 \frac{e}{T^2} \qquad (\text{"}N\text{-units"}) \quad (14)$$

where

P = total atmospheric pressure in m bars
T = temperature in °K
e = partial pressure of water vapor in m bars

and N is related to refractive index n by

$$N \triangleq (n - 1) \times 10^6. \quad (15)$$

In a turbulent atmosphere, the spatial distributions of P, T, and e exhibit random fluctuations; the random fluctuation of N is simply a manifestation of the fluctuations in these more fundamental quantities. In the microwave region and at longer wavelengths, N is strongly a function of water vapor content, but has no significant dependence on λ. At optical wavelengths, however, N is quite insensitive to water vapor content, but does vary with λ:

$$N = \frac{77.6\,P}{T} + \frac{0.584\,P}{T\lambda^2} \quad (16)$$

where P is total pressure in millibars and T is temperature in °K, and λ is now in microns. For $\lambda \gg 20\ \mu$, we may

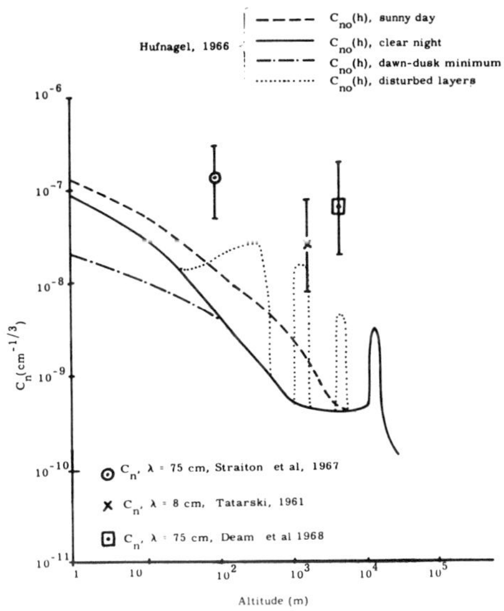

Fig. 6. C_n as a function of altitude.

neglect the second term in (16). Just as N in the microwave region behaves differently from N in the visible, the quantity C_n will behave differently in these two regions. We shall use the notation $C_{nm}(h)$ and $C_{n0}(h)$ to denote the refractive-index structure constant profiles in the microwave and optical regions, respectively.

Although the profile $C_{nm}(h)$ has not been determined to date, a few selected values C_{nm} have been deduced for specific values of h. However, a model for $C_{n0}(h)$ has been proposed by Hufnagel [8]; this model was constructed by combining optical scintillation data and an earlier model for $C_{n0}(h)$ (Hufnagel and Stanley [9]) which had been based on estimated profiles of the macroscopic meteorological variables which affect C_{n0}. Fig. 6 shows the Hufnagel model $C_{n0}(h)$ as well as some values of C_{nm} which have been reported in the literature. Using some experimental data on forward-scatter propagation taken from Chisholm *et al.* [10], Tatarski calculated a value $C_{nm} \approx 4.5 \times 10^{-9}$ to 8×10^{-8} cm$^{-1/3}$ for $\lambda = 8$ cm and a scattering volume at an approximate altitude of 1.5 km. This value was confirmed by Tatarski through use of some independent data by Bullington [11]. More recently, Straiton *et al.* [12] used spaced refractometers on an 85 meter high tower, at spacings of 0.25, 1, and 4 meters, at a wavelength of 75 cm, and inferred values of C_{nm} ranging from 5×10^{-8} to 3×10^{-7} cm$^{-1/3}$. Since that time, Deam *et al.* [13] have reported similar measurements atop Mt. Evans in Colorado at an altitude of 4.35 km. In this case the values of C_{nm} ranged from 1.4×10^{-8} to 1.4×10^{-7} cm$^{-1/3}$, values almost as large as those observed at 85 meters. One would expect, of course, that the presence of the air–earth boundary at the mountaintop would render C_{nm} larger than the corresponding value at that altitude in the absence of a mountaintop. A prominent feature of Fig. 6 is that C_{nm} exceeds C_{n0} by one order of magnitude or more for those values of h for which a comparison can be made. One could attribute the excess of C_{nm} over C_{n0} to the presence of inhomogeneously distributed water vapor in the troposphere. However, even at the earth's surface, water vapor seldom accounts for more than 20 percent of the measured value of N. Furthermore, the contribution of water vapor to N, relative to the dry air contribution, decreases with altitude; as a coarse approximation, the altitude profile $N(h)$ has the form

$$N(h) \approx D_0 e^{-h/H_d} + W_0 e^{-h/H_w} \tag{17}$$

where

D_0 = contribution of dry air constituents to N at surface
W_0 = contribution of water vapor to N at surface
H_d = "scale height" of the dry term
H_w = "scale height" of the wet term

and where H_d and H_w have typical values on the order of 10 and 2.5 km, respectively. If water vapor does account in fact for the striking difference between C_{nm} and C_{n0}, then:

1) the water vapor distribution must have a much greater degree of relative inhomogeneity than does the distribution of dry-air constituents;
2) the value of C_{nm} should approach the value of C_{n0}, at least to within 10 or 20 percent [this being associated with the term $0.584\,P/T\lambda^2$ in (16)], for altitudes well in excess of H_w.

This set of remarks is clearly only a conjecture at this point. However, it is consistent with an independent but somewhat related observation by Thompson and Waters [14], who compared the temporal fluctuations in one-way radio range (at 9.5 GHz) along a LOS path with the simultaneous fluctuations in one-way optical range (at 5650 Å) along the same nominal path, and found that the radio-range fluctuations exceeded the optical-range fluctuations by at least one order of magnitude. Some caution is required in extrapolating this result to C_n, since the one-way radio path length is affected by gross meteorological parameters whose effect on C_n may be negligible.

In the next section, we discuss some data which yields an estimate of $D_\phi(\mu)$ for a specific geometric configuration, and which is consistent with the estimate of relative magnitudes of C_{nm} and C_{n0} as depicted in Fig. 6.

IV. Empirical Determination of $D_\phi(\mu)$

We recall from Section I that $D_\phi(\mu)$ is, by definition, the statistical mean of $[\phi(x+\mu)-\phi(x)]^2$, where the spatial displacement μ is taken in a direction normal to the nominal LOS ray path. The LOS microwave propagation experiment reported by Thompson and Janes elsewhere in this issue [2] provides us with extensive data on the time-fluctuation of $[\phi(x+\mu)-\phi(x)]$ for several discrete values of μ ranging from 1.25 to 790 meters. In this section, we construct an estimate of $D_\phi(\mu)$ from the Thompson–Janes data, hence an estimate of the integral

$$\int_0^{h_0} C_{nm}^2(h) h^{5/3} dh$$

for their particular experimental site at Maui. We then compare this result with the data on $C_{n0}(h)$ and C_{nm} displayed in Fig. 6. Frequent reference is made to the figs. in [2].

The Thompson–Janes data were collected over a one-month period at Maui, Hawaii. A transmitter at sea level radiated, at $\lambda = 3.2$ cm, to a set of four spaced receivers which were deployed along an elevated baseline at an altitude of about 3 km on Mt. Haleakala as shown in Figs. 1–3 of [2]. The phase difference $\Delta\phi_{ij}(t)$ between receivers i and j was monitored as a function of time, for different values of receiver separation μ_{ij}. Each particular configuration of the four receivers was operated continuously for 24 hours, in order to provide several 15 minute long sample periods distributed throughout the diurnal cycle, and the overall experiment using nine different configurations required about one month. The problem then is one of using the collection $\{\Delta\phi_{ij}(t)\}$ to form an estimate of $D_\phi(\mu)$, taking into account diurnal effects and changing weather over the one-month period.

The primary data collected by Thompson and Janes were:

1) one-way radio range[2] $R(t)$ from the sea level site to some one receiver on the mountaintop baseline;
2) the range-differences $\Delta R_{ij}(t)$ observed on pairs of paths;
3) some second differences $\{\Delta R_{ij}(t) - \Delta R_{kl}(t)\}$.

The secondary data included records of atmospheric refractivity; wind speed and direction at the upper and lower terminals; wet and dry bulb temperatures; barometric pressures; and periodic photographs from both terminals. The primary data were subjected to spectral and correlation analyses. The more important results of the analysis are presented in Figs. 4–18 of [2].

The power spectra of the range samples $R(t)$ are shown in Fig. 4 of [2]. The temporal spectrum $S_R(f)$ displays the dependence

$$S_R(f) \sim f^{-5/2}$$

over the range $7\times 10^{-3} < f < 1.4$ Hz, where the 1.4 Hz limit is established by the data recording system and the 7×10^{-3} Hz limit is established by the data sample length. Although the amplitude of the spectrum varies from sample to sample, the f-dependence is reasonably well preserved.

Next, we examine the one-sided spectra $S_{\Delta R}(f)$ of the range-difference $\Delta R(t)$, Fig. 7–9 of [2]. In the case of the widely separated[3] antennas, $\mu = 790$ m, we observe

[2] Range R and range-difference ΔR are related to phase and phase-difference by $\phi = (2\pi/\lambda)R$, $\Delta\phi = (2\pi/\lambda)\Delta R$.

[3] Receiver separation is denoted by b in [2], rather than μ as in this paper.

$$S_{\Delta R}(f) \approx 2S_R(f).$$

In the minimum separation case, $\mu \approx 1.25$ m, we observe

$$S_{\Delta R}(f) \sim f^{-1/2}.$$

For intermediate values of antenna separations, $S_{\Delta R}(f) \approx 2S_R(f)$ for large values of f, but falls below $S_R(f)$ as f decreases and becomes asymptotic to $f^{-1/2}$ for very small f. The observed spectra are reasonably consistent with the model

$$S_{\Delta R}(f) \approx \begin{cases} 2S_R(f) & f \gg f_1 \\ (f/f_1)^2 S_R(f) & f \ll f_1 \end{cases} \tag{18}$$

where the transition frequency f_1 takes on successively smaller values as receiver separation μ is increased. The spectra in Figs. 7–9 of [2] are plots of log $[S_{\Delta R}(f)]$ versus log $[f]$; each of these was approximated by two straight lines representing the asymptotic behavior for $f \ll f_1$ and $f \gg f_1$, and the value of f at which these lines intersected was taken as an estimate of f_1. The plot of f_1 as a function of μ is shown in Fig. 7; the relation

$$f_1 \approx \frac{2.2}{\mu},$$

where f_1 is in Hz and μ is in meters, constitutes a good fit to these points, and is shown as the solid line in Fig. 7.[4]

The behavior of $S_{\Delta R}(f)$ as given by (18) averts certain potential integrability problems in $S_{\Delta R}$ which are present in the range-spectrum $S_R(f)$. We recall that $S_R(f) \sim f^{-5/2}$, a dependence which has been observed over a wide range of f. (See additional references in [2].) If we attempt to compute a true range-variance, under the assumption that $S_R(f) \sim f^{-5/2}$ for all f, we find that

$$\sigma_R^2 = \int_0^\infty S_R(f)df$$

is not bounded; a physically meaningful integration requires that we either limit the range of f or make some further assumptions regarding the small-f behavior of S_R. In the case of the range-difference variance,

$$\sigma_{\Delta R}^2 = \int_0^\infty S_{\Delta R}(f)df$$

is finite for $S_{\Delta R}$ as given in (18) with a smooth transition between the two regions and for $S_R(f) \sim f^{-5/2}$. The form

$$S_{\Delta R}(f) = 2S_R(f)[1 - \cos(f/f_1)] \tag{19}$$

has the asymptotic behavior of (18), provided we neglect the highly oscillatory character of the bracketed term and assume we observe, in any experimental evaluation, the envelope of $S_{\Delta R}(f)$ for large f. For $S_R(f) = A \cdot f^{-5/2}$, the integral

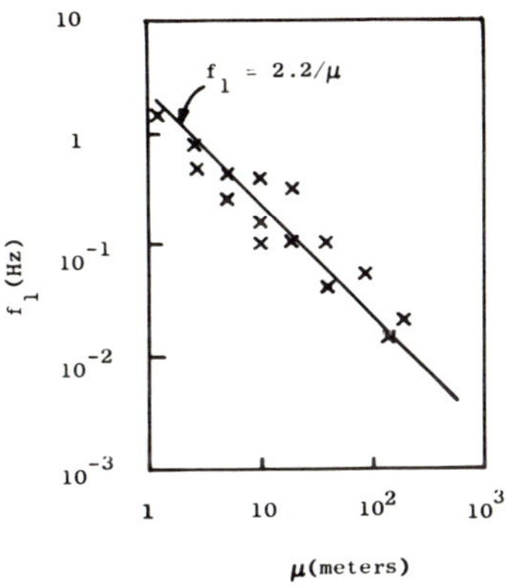

Fig. 7. Transition frequency f_1 as a function of μ.

$$\sigma_{\Delta R}^2 = 2A \int_0^\infty f^{-5/2}[1 - \cos(f/f_1)]df \tag{20}$$

accumulates most of its value in the vicinity of f_1; we need not know the precise behavior of $S_R(f)$ for very large or very small f, as long as the slope of S_R does not become too large as $f \to 0$ or too small as $f \to \infty$. As one practical aspect of this situation, a phase measuring system whose bandpass characteristic extends at least one decade to either side of f_1 can be used to measure the true phase difference variance directly. If this criterion is not met, the variance estimate will be biased; the measured variance will be smaller than the true variance since some of the contributing spectral content will have been ignored. This is clearly the case for the receivers separated by 790 meters, and is true to a lesser extent for some of the smaller separations. The variances discussed by Thompson and Janes are computed from the power spectra, as discussed in Section III of [2]. For our purposes, their "zero to 1.4 Hz" phase-difference variances are adequate estimates of the true variances for all separations but the 790 meter value.

Fig. 10 of [2] shows a plot of $\sigma_{\Delta R}^2$ as a function of μ; a distinction is drawn between data taken between 6 a.m. and 6 p.m. local time and those collected between 6 p.m. and 6 a.m. The general increase of $\sigma_{\Delta R}^2$ with μ is evident but, for any given value of μ, $\sigma_{\Delta R}^2$ tends to range over one to two orders of magnitude; this is not surprising in view of the variability of local meteorological conditions over the month-long observation time. Furthermore, there appears to be no consistent distinction between the nighttime and daytime values of $\sigma_{\Delta R}^2$. However, Fig. 18 of [2] indicates a strong correlation between $\sigma_{\Delta R}^2$ and the wind speed v_l at the lower terminal of the path; as a crude estimate, $\sigma_{\Delta R}^2 \sim v_l^2$. Furthermore, for small values of μ, the ratio $\sigma_{\Delta R}^2/v_l^2$ exhibits a diurnal effect (not shown); specifically, $\sigma_{\Delta R}^2/v_l^2$ tends to be consistently higher during the hours between 11 a.m. and 11 p.m. local time than it is during the complementary 12-hour interval. Fig. 8

[4] The relative behavior of $S_R(f)$ and $S_{\Delta R}(f)$ is consistent with (but does not necessarily imply) the concept of a "frozen" inhomogeneity structure which drifts across the LOS propagation paths. The effective drift velocity which follows from this model has a magnitude on the order of a few meters per second, which is consistent with observed wind speed.

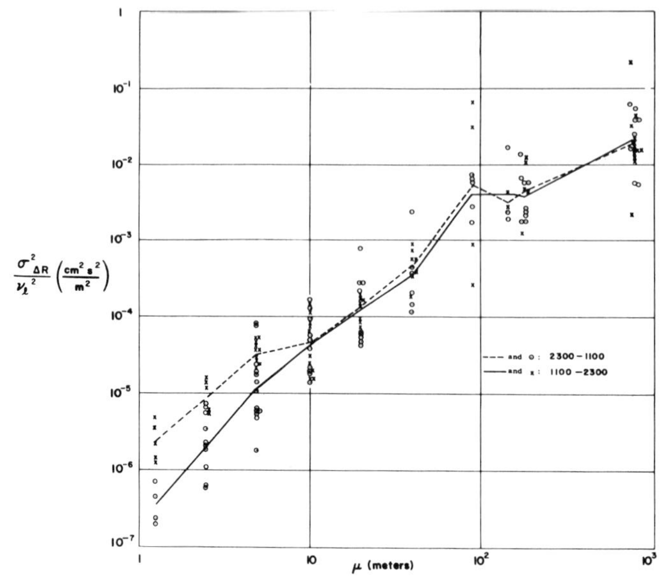

Fig. 8. $\sigma^2_{\Delta R}/v_l^2$ as a function of μ for the periods 11 a.m.–11 p.m. and 11 p.m.–11 a.m.

shows a plot of $\sigma^2_{\Delta R}/v_l^2$ versus μ, with a distinction made between "daytime" (now 11 a.m. to 11 p.m.) and "nighttime" values. The solid and dotted lines are drawn through the geometric means of the several values of $(\sigma^2_{\Delta R}/v_l^2)$ observed for each value of μ, for daytime and nighttime, respectively. For each value of μ, and within either the daytime or the nighttime interval, most values of the ratio $\sigma^2_{\Delta R}/v_l^2$ tend to be within a factor of 3 or 4 of the geometric mean. At very small separations, $\sigma^2_{\Delta R}/v_l^2$ is distinctly smaller by night than by day, but this phenomenon is not observed for the larger values of μ.

The data of Fig. 8 can now be used to plot the one-way phase structure function for the Maui path employed by Thompson and Janes, for any value of λ in the microwave region, and for an assumed value of wind speed. The measured values of $\sigma^2_{\Delta R}$ for $\mu = 790$ m are low estimates for the reasons discussed earlier in this section, but can be corrected via estimation of f_1 from Fig. 7, construction of a spectrum $S_{\Delta R}(f)$ for $\mu = 790$ m, and numerical integration. The corrected plot of $\sigma^2_{\Delta R}/v_l^2$ is then used to calculate the function $D_\phi(\mu)$ for $\lambda = 3.2$ cm and for $v_l = 8$ m/s (a typical wind speed observed during the Maui experiment). The results are shown by the solid and broken lines of Fig. 9. The boundaries of the shaded region in Fig. 9, which vary as $\mu^{5/3}$, correspond to plots of (7) with the values

$$R = 24.5 \text{ km}$$

$$h_0 = 3 \text{ km}$$

$$\int_0^{h_0} h^{5/3} C_n^2(h) dh = \begin{cases} 0.3 \text{ cm}^2, \text{ lower boundary} \\ 3 \text{ cm}^2, \text{ upper boundary.} \end{cases}$$

Fig. 9. The structure function $D_\phi(\mu)$ for Maui, $\lambda = 3.2$ cm, $v_l = 8$ m/s.

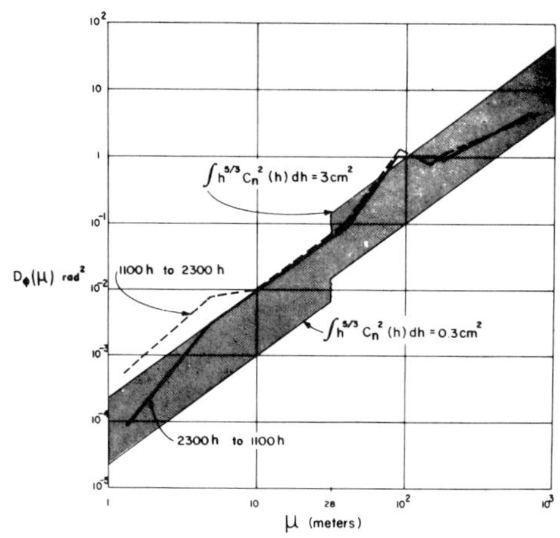

The discontinuities in the boundaries correspond to the factor-of-two change in α as we make the transition from $\rho \ll \sqrt{\lambda R}$ to $\rho \gtrsim \sqrt{\lambda R}$, where $\sqrt{\lambda R} \approx 28$ m for the Maui path.

The experimental structure function $D_\phi(\mu)$ displays the predicted $\mu^{5/3}$ dependence quite well, although there is some evidence of saturation for $\mu > 100$ m. Tatarski [6, Section 2.2] predicts a saturation effect in $D_\phi(\mu)$ which stems from saturation of the refractive-index structure function $D_n(\mu)$ when the condition $\mu \ll L_0$ is violated.

In Section III we discussed some properties of the profile $C_n(h)$. It is interesting to compare the values of the integrals corresponding to the boundaries in Fig. 9 with the data of Fig. 6. If we take Hufnagel's $C_{n0}(h)$ profiles in Fig. 6, and neglect the shape peaks associated with disturbed layers, then for the "sunny day" profile we obtain

$$\int_0^{3\text{ km}} h^{5/3} C_{n0}^2(h) dh \approx 7 \times 10^{-4} \text{cm}^2,$$

while for the "clear night" profile we obtain

$$\int_0^{3\text{ km}} h^{5/3} C_{n0}^2(h) dh \approx 10^{-4} \text{ cm}^2.$$

Although the disturbed layers might locally increase C_{n0} by as much as two orders of magnitude, this would occur over a relatively thin h-increment, and probably would not dominate the integral. Acknowledging all the dangers involved in comparing Hufnagel's model C_{n0} profile with the $C_{nm}(h)$ profile at Maui, we nevertheless observe that the weighted integral of C_n^2 appropriate to the optical case is some three orders of magnitude smaller than the integral deduced from the Maui microwave results. This is consistent with Fig. 6, which indicated that, for those few values of h for which estimates of C_{nm} exist, C_{nm} is larger than C_{n0} by one order of magnitude or more. A word of caution is in order at this point: just as the results of Deam *et al.* [13] may have been affected to an unknown extent by the presence of a mountaintop, it is likely that the distribution $C_{nm}(h)$ along the Maui path was affected by the presence of Mt. Haleakaka.

V. Additional Remarks

Equation (12) of Section II permits an estimate of attainable resolution to be made, given some system parameters and knowledge of the profile $C_{nm}(h)$. The Maui experiment discussed in Section IV is not capable of providing us with the $C_{nm}(h)$ profile per se. It does provide an estimate of the weighted integral

$$\int_0^{h_0} h^{5/3} C_{nm}^2(h) dh$$

for $h_0 \approx 3$ km, for the path in the vicinity of Mt. Haleakala. The presence of the mountaintop probably causes this estimate to be higher than the true value of the integral over relatively flat terrain. Subject to this bias, the Maui data also provide a direct estimate of the one-way phase structure function $D_\phi(\mu)$ along an elevated baseline at the 3 km altitude. Extrapolation to other inclination angles (hence, ranges) is straightforward. Extrapolation of these 3.2 cm results to other wavelengths requires some caution. For longer λ, $\lambda \gg l_0$, the theoretical form for $D_\phi(\mu)$ may not be valid (see [17]). At significantly shorter wavelengths, a λ-dependence in C_n would be expected. Our ability to extrapolate to other baseline altitudes is severely limited, however; in the absence of a reliable $C_{nm}(h)$ profile, the best we can do is to make an educated

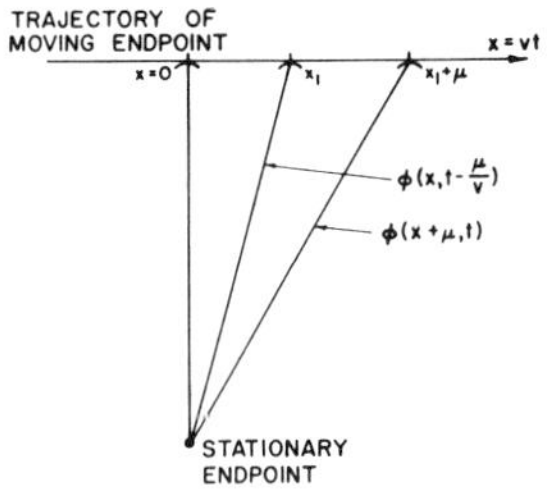

Fig. 10. Phase fluctuation for a LOS path with one moving endpoint.

guess based on the data of Fig. 6 coupled with the exponential decline of water vapor content as a function of altitude. Likewise, the extent to which the integral

$$\int_0^{3\text{ km}} h^{5/3} C_{nm}^2(h) dh$$

for the Maui climate resembles the corresponding integral at other geographic locations is not known.

The discussion of Section I, and all that followed from it, was restricted to the description of phase fluctuations induced by turbulence. However, the amplitude of the wave reaching the extended baseline also exhibits random fluctuations in its amplitude. Fried [15], [16] has treated situations in which the structure functions of both amplitude and phase are incorporated into the calculation of resolution. The variance of the log-amplitude fluctuation goes as $(R^{11/6}/\lambda^{7/6})$; it is negligible for the microwave propagation experiment at Maui and we will ignore it for the SAR problem.

Finally, the structure function $D_\phi(\mu)$ pertains to the statistics of the spatial distribution of phase along the baseline, under the assumption that the turbulent medium is "frozen" over the observation time. If a synthetic-aperture radar moves through distance $L \approx \beta R$ in a time sufficiently short that no significant rearrangement of $n(x, y, z)$ occurs along the propagation path, then the results of Section II apply. If this condition is not met, as might be the case with a low-speed aircraft and/or long-range operation, then additional phase error accrues as a consequence of time variation of refractive index. Fig. 10 illustrates the problem for an LOS path with one endpoint moving along the x coordinate with speed v. The turbulence-induced phase error is now

$$\Omega(x, \mu, t) = \phi(x + \mu, t) - \phi(x, t - \mu/v) \tag{21}$$

which can be written in the form

$$\Omega = \phi_i + \phi_R \tag{22}$$

where

$$\phi_i = \phi(x + \mu, t) - \phi(x, t)$$

and

$$\phi_R = \phi(x, t) - \phi(x, t - \mu/v).$$

The term ϕ_i is the instantaneous phase fluctuation treated earlier, while ϕ_R describes the time-fluctuation of phase along an LOS path from the stationary endpoint to some fixed point in the extended baseline. The structure function of Ω is then given by

$$D_{\Omega}(\mu) = D_{\phi_i}(\mu) + D_{\phi R}(\mu/v) + 2E\{\phi_i \phi_R\}. \quad (23)$$

The term $D_{\phi_i}(\mu)$ is that which we treated earlier. The term D_{ϕ_R} can be approached via the range-spectrum $S_R(f)$ discussed in Section IV. It can be shown that

$$D_{\phi_R}(\mu/v) = 2\left(\frac{4\pi}{\lambda}\right)^2 \int_0^{\infty} S_R(f)\left[1 - \cos\frac{2\pi f \mu}{v}\right] df \quad (24)$$

and considerable data are available on $S_R(f)$. To the best knowledge of the author, data which yield adequate insight into the form and parameter dependences of $E\{\phi_i \phi_R\}$ are not available at this time. However, we observe that

$$D_{\Omega}(\mu) \leq 2D_{\phi_i}(\mu) + 2D_{\phi_R}(\mu/v). \quad (25)$$

Therefore, an upper bound on resolution can be estimated even in the absence of data on the third term of (23).

In summary, Tatarski's formulation of the phase error structure function can be combined with Brown and Riordan's analysis of the effects of phase errors having a power-law structure function, to yield an expression for best attainable SAR resolution in the presence of tropospheric phase errors. Data collected by Thompson and Janes at Maui provide some insight into the magnitude of C_{nm} at altitudes below 3 km, although extrapolations from this experiment to SAR systems at other altitudes and locations are tenuous at best.

Acknowledgment

The author wishes to acknowledge many helpful suggestions made by M. C. Thompson and H. B. Janes of ESSA, and by W. M. Brown and other staff members of the Radar and Optics Laboratory at the University of Michigan.

References

[1] W. M. Brown and J. F. Riordan, "Resolution limits with propagation phase errors," *IEEE Trans. Aerosp. Electron. Syst.*, this issue, pp. 657–662.
[2] M. C. Thompson, Jr. and H. B. Janes, "Measurements of phase-front distortion on an elevated line-of-sight path," *IEEE Trans. Aerosp. Electron. Syst.*, this issue, pp. 645–656.
[3] W. M. Brown, "Synthetic aperture radar," *IEEE Trans. Aerosp. Electron. Syst.*, vol. AES-3, pp. 217–229, March 1967.
[4] W. M. Brown and L. J. Porcello, "An introduction to synthetic-aperture radar," *IEEE Spectrum*, vol. 6, pp. 52–62, September 1969.
[5] R. O. Harper, *Synthetic-Aperture Radar Systems: Theory and Design*. New York: Academic Press, 1970.
[6] V. I. Tatarski, *Wave Propagation in a Turbulent Medium*, R. A. Silverman, Trans. New York: McGraw-Hill, 1961.
[7] J. W. Strobehn, "Line-of-sight wave propagation through the turbulent atmosphere," *Proc. IEEE*, vol. 56, pp. 1301–1318, August 1968.
[8] R. E. Hufnagel, "An improved model turbulent atmosphere," in *Restoration of Atmospherically Degraded Images*, vol. 2, App. III, Woods Hole Summer Study. National Acad. Sci./National Res. Council, July 1966.
[9] R. E. Hufnagel and J. R. Stanley, "Modulation transfer function associated with image transmission through turbulent media," *J. Opt. Soc. Am.*, vol. 54, p. 52, 1964.
[10] J. H. Chisholm, P. A. Portmann, J. J. deBettencourt, and J. F. Roche, "Investigation of angular scattering and multipath properties of tropospheric propagation of short radio waves beyond the horizon," *Proc. IRE*, vol. 43, pp. 1317–1335, October 1955.
[11] K. Bullington, "Radio transmission beyond the horizon in the 40- to 4,000 mc band," *Proc. IRE*, vol. 41, pp. 132–135, January 1953.
[12] A. W. Straiton, A. P. Deam, and J. L. Dodd, "Amplitude distribution of refractive index differences," *J. Geophys. Res.*, vol. 72, p. 4051, 1967.
[13] A. P. Deam, A. W. Straiton, and J. L. Dodd, "Surface refractive index differences at 14,250 feet elevation," *IEEE Trans. Antennas Propagat.* (Communications), vol. AP-16, pp. 365–368, May 1968.
[14] M. C. Thompson and D. M. Waters, "Radio-optical dispersion observed over a 25-km tropospheric path," *IEEE Trans. Antennas Propagat.* (Communications), vol. AP-15, pp. 487–488, May 1967.
[15] D. L. Fried, "Optical resolution through a randomly inhomogeneous medium for very long and very short exposures," *J. Opt. Soc. Am.*, vol. 56, p. 1372, 1966.
[16] ——, "Limiting resolution looking down through the atmosphere," *J. Opt. Soc. Am.*, vol. 56, p. 1380, 1966.
[17] S. F. Clifford and J. W. Strohbehn, "The theory of microwave line-of-sight propagation through a turbulent atmosphere," *IEEE Trans. Antennas Propagat.*, vol. AP-18, pp. 264–274, March 1970.

Leonard J. Porcello (S'57–M'59–SM'66) was born in New York, N. Y., on March 1, 1934. He received the B.A. degree from Cornell University, Ithaca, N. Y., and the M.S., M.S.E., and Ph.D. degrees from the University of Michigan, Ann Arbor, in 1957, 1959, and 1963, respectively.

Since 1955 he has been affiliated with the Institute of Science and Technology, University of Michigan. He has headed the Radar and Optics Laboratory since 1968 and is currently an Associate Professor of Electrical Engineering.

Dr. Porcello is a member of the Optical Society of America, the American Association for the Advancement of Science, Sigma Xi, and Eta Kappa Nu.

CHAPTER V
Motion Compensation in Synthetic Aperture Radar

The extended usefulness of the synthetic aperture radar is dependent on its ability to isolate the production of radar imagery from the effects of aircraft motion. This motion is caused by both weather disturbances and intentional maneuvering of the aircraft. If the motion is left uncompensated, the radar imagery becomes geometrically distorted — straight roads become curved and the resultant imagery is smeared. The early demonstration of the synthetic aperture radar imagery (see Figure 2 in Sherwin, Ruina, and Rawcliffe in Chapter I) is an example of uncompensated radar imagery. This first synthetic aperture map was generated as the pilot of the aircraft attempted to fly a straight line, but, due to buffeting of the aircraft by atmospheric turbulences, the aircraft veered from a straight path. Since that time, more conventional airborne synthetic aperture radar imaging systems have employed tight autopilots which restricted the requirements for motion compensation provisions to small deviations from a straight path or a constant radius turn. As such, the resultant SAR designs could employ analog-optical processing implementations restricted to mapping at a fixed squint angle or at broadside. With advances in digital processing of motion-sensed parameters and the need for real-time digital processing of the coherent signals, newer SAR radar designs permit motion compensation for varying aircraft maneuvers. The papers in this section treat these advanced techniques for motion compensation with aircraft maneuverability while producing variable squint, high resolution mapping.

J.H. Mims and J.L. Farrell, in the first paper, describe a digital approach for performing motion compensation in SAR in the presence of speed change, turn, and climb or dive type maneuvers of the aircraft; they consider both side-looking and squint mapping. Included are the results of simulation runs to verify their unified scheme for accommodating motion compensation.

In the second paper, Farrell and Mims, along with A. Sorrell, extended the development of their first paper to include the effects of navigation error. The error sources investigated included platform servo transients, noise and quantization in accelerometers, interaction of angle pick-off uncertainty with the platform displacement to radar antenna, and uncertainty in this displacement. Their treatment of these error sources, corrected by their motion compensation scheme, is general enough to be applied to a wide variety of possible systems.

In the third paper, J.C. Kirk, Jr. presents a generalized motion compensation approach which is applicable to all SAR mapping modes. He introduces the concept of a unit vector and the slaving of the real antenna and the process synthetic antenna to this vector. He shows how the amount of motion compensation can be developed in terms of transfer curves for the main motion reduction paths.

These three papers represent the limited amount of published works on this essential and complex topic. Each concludes with recommendations for extended study tasks.

Synthetic Aperture Imaging With Maneuvers

J.H. Mims and J.L. Farrell
(Westinghouse Electric Corporation, Baltimore, Maryland)

IEEE Transactions on Aerospace and Electronic Systems, Vol. AES-8, No. 4 (July 1972)
Pages 410-418

Abstract

A digital processing approach has been devised for performing motion compensation in a high-resolution airborne synthetic aperture radar in the presence of simultaneous longitudinal (speed change), lateral (turn), and vertical (climb or dive) maneuvers. Both side-look and squint are accommodated in a unified scheme, which is validated by various simulation runs reported herein. Present attention is focused on theoretical verification, irrespective of mechanization or specific parameter values.

Introduction

Synthetic aperture imaging constitutes a method for achieving much finer resolution than is normally obtainable in radar mapping [1], [2]. Conventional airborne synthetic aperture radar (SAR) imaging systems typically employ a tight autopilot, with motion compensation provisions restricted to small deviations from a reference path such as a straight line or a constant radius turn. It has long been known that much greater generality is theoretically attainable but quite difficult to achieve in an analog/optical mechanization. The difficulty has been largely associated with changing geometric phenomena; for any chosen reflector in the mapped area, the locus of scatterers lying in the same range cell will vary dynamically with aircraft motion. Analog RF phase compensation, dependent on autopilot steering and an analog film drive mechanism, would be a taxing objective in the presence of maneuvers. Changes in range cell boundaries vary between image cells, in general nonlinearly; moreover, even the functional form of this variation will depend upon aircraft motion. It is crucial to note, however, that conventional imagining performance is limited only by mechanization constraints; variations in range boundaries are not inherently detrimental to the image, so long as these variations can be characterized properly while processing the data.

With recent advances in digital processing a far more flexible procedure for motion compensation in a general flight path is feasible. The functional form of range boundary variation [easily determined with the aid of the inertial navigation system (INS)] can be taken into account for each separate image cell, and as a natural benefit of conversion to a digital approach, ramifications such as nonlinearity and motion-dependent changeable functional forms no longer present any hardship. Significantly, these advances in imaging capability are not at all accompanied by a corresponding growth in complexity. Mapping no longer imposes any autopilot requirements whatever, and there is no blind "settling time" following sharp changes in flight path direction. High resolution is achieved with equal facility in either forward squint or side look; the latter is merely a special case of the unified overall approach.

Fundamental to the ability to map during maneuvers is the necessity for applying motion compensation as a function of azimuth angle within the physical beamwidth and during the coherent integration time. This subdivision is accomplished by the first of two stages of azimuth correlation. The first stage correlation reference function is based on the actual flight path rather than some finite order of phasing approximation (e.g., linear plus quadratic phase shift, generally employed by conventional motion compensation systems) and thus constitutes complete compensation for the image center, automatically tailored to the flight path. The first correlation stage produces separate image cells (subbeams) of intermediate resolution within the physical beamwidth at each range interval. At this point, a second stage of motion compensation (i.e., correction as a function of azimuth angle and range) is readily applied to the data of each subbeam. The fully compensated time histories of each subbeam are then submitted to the second stage of correlation to produce the final azimuth resolution. The second stage of motion compensation performs an interpolation which effectively redistributes the signal energy while providing the corresponding phase adjustment for each image cell, according to its location relative to the map center. Changing geometries are readily accommodated by the two-stage correlation, since interpolation is carried out separately for small subsegments of the aircraft excursion during each signal data block collection.

All of these advances would of course be purely academic if they led to unrealistic computational requirements. Fortunately, all motion compensation quantities can be computed from truncated Taylor series, in which the range from each ground reflector to each aircraft position is expanded about reference locations (aircraft and map reference points denoted $\mathbf{R}_A$ and $\mathbf{b}_0$, respectively). Performance of the imaging system is not sensitive to the selection of these two vectors, provided that they reasonably correspond to some point near the current flight path and the mapped area, respectively.

Many other considerations which are important for design but not for theoretical validation are not scrutinized here. Although each frame in a mechanized display will generally contain multiple superimposed processed signal data blocks, the present objective is satisfactorily attained with attention restricted to single blocks. The antenna is restricted to spot mode, with no attempt at demonstrating strip map construction. The only motion compensation errors included within the present scope are fixed velocity and acceleration uncertainties. Roundoff errors in motion compensation computations are ignored. Extensions involving these factors, however, are logical outgrowths of the scope under immediate consideration.

This paper describes a simulation which verifies high-image-quality performance in the presence of simultaneous maneuvers in three dimensions for a target removed in altitude (elevation 50 feet off the reference ground map plane) and in lateral position (diagonally displaced, 1000 feet in both horizontal directions) from the center of a mapped area nominally 45 degrees off the aircraft velocity vector, with fixed uncertainties in velocity and acceleration. Following is a complete analytical background and presentation of sample simulation data, with subsequent interpretation and conclusions.

Background Analysis for SAR Performance Evaluation

A digital simulation program has been chosen as an accurate and flexible means of obtaining a wealth of detailed performance data under a variety of conditions. Standard relationships are first used to determine the response of an airborne radar to any specified array of ground reflectors in the presence of errors (e.g., imperfections in motion compensation) under realistic aerodynamic flight conditions for a radar antenna at any specified location relative to the aircraft mass center. This receiver response is then subjected to a computational sequence of the same form intended for airborne processing. A conversion routine then reconstructs the image, and the result is compared against the input target array.

In its present form, the program simulates a spot mode (in contrast to a strip map mode with frame continuity). The simulation includes variable pulsewidth, adapted to altitude and sight-line depression angle for fixed nominal range gate ground span. No consideration is presently given to any implications of differential refraction and Earth curvature, chirp mechanization (the development is in terms of *effective* pulsewidth throughout), aircraft flexure, or additive low-level noise.

Coordinate Conventions and Kinematics

The first step in describing the time history of radar signals is to define the relative geometry of the target array and the time-varying antenna position. Fig. 1 shows a right-hand coordinate frame with the z axis defined positive downward and the x axis defined such that the x-z plane contains the initial[1] velocity (corresponding to standard geographic axes with initial North bearing). The vector $\mathbf{b}_0$ defines the ground plane ($z = 0$) intersection of the initial[1] antenna boresight. The SAR selects signals from an even integer K of range gates, ideally centered (i.e., to within the antenna pointing accuracy, continuously centered for spot mode) at this boresight intersection. Any combination of simulated reflectors, each with its own magnitude (denoted S_4) and scintillation phase[2] (denoted S_5, expressed in radians) as well as its own coordinates (denoted S_1, S_2, S_3), can be defined within the range and Doppler boundaries shown (which lie inside the one-way half-power beamwidth θ of the radar).

At any time t the position vector of the antenna, expressed in the map (Fig. 1) coordinate frame, is

$$\mathbf{R} = \begin{bmatrix} x_1 \\ x_2 \\ x_3 \end{bmatrix} = \mathbf{R}_{(t_0)} + \begin{bmatrix} \int_{t_0}^{t} V_1\, dt \\ \int_{t_0}^{t} V_2\, dt \\ \int_{t_0}^{t} V_3\, dt \end{bmatrix} + [B]\mathbf{d}_A \qquad (1)$$

[1] In general, "initial" refers to the time t_0 when the SAR is first switched to the spot mode. For single frame simulation, however, it is permissible to set $t_0 = t_1$, the time at the leading edge of the first pulse-repetition frequency (PRF) pulse in the frame.

[2] Individual reflector scintillation phase angles are held fixed during any one SAR frame. Note, however, that scintillation occurring within a frame can easily be simulated, merely by placing multiple reflectors in a cluster.

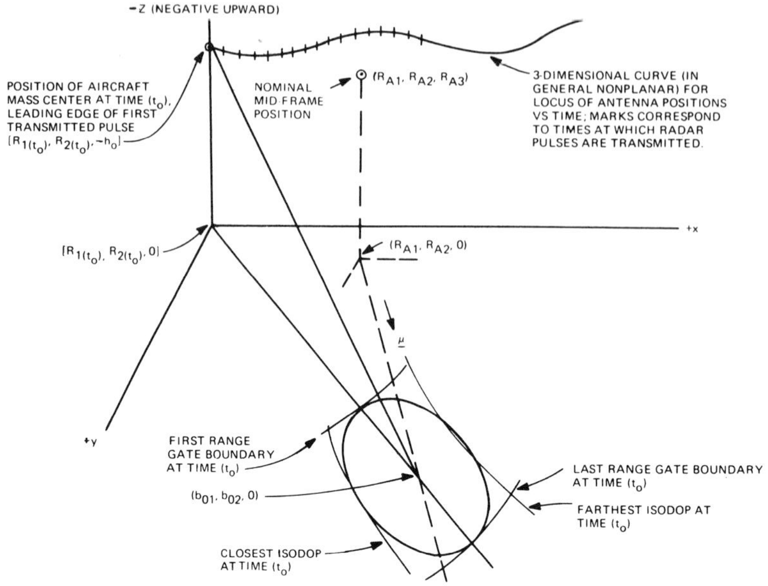

Fig. 1. Geometry for single SAR frame.

where

$\mathbf{R}_{(t_0)} \triangleq$ initial position vector of aircraft mass center

$V_j \triangleq$ jth component of aircraft velocity expressed in map coordinates ($j = 1, 2, 3$, for x, y, z, respectively)

$\mathbf{d}_A \triangleq$ constant displacement from aircraft mass center to antenna, expressed in aircraft coordinates

$[B] \triangleq$ orthogonal transformation from aircraft to map coordinates.

Ignoring the slow Earth rotation, the map coordinate frame is inertial; therefore, $[B]$ is the solution of the matrix differential equation

$$[\dot{B}] = [B]\,[\omega] \tag{2}$$

subject to the initial conditions

$$\begin{bmatrix} \cos\eta_{0d} & \sin\eta_{0d} & 0 \\ -\sin\eta_{0d} & \cos\eta_{0d} & 0 \\ 0 & 0 & 1 \end{bmatrix} \begin{bmatrix} \cos\eta_{02} & 0 & \sin\eta_{02} \\ 0 & 1 & 0 \\ -\sin\eta_{02} & 0 & \cos\eta_{02} \end{bmatrix} \begin{bmatrix} 1 & 0 & 0 \\ 0 & \cos\eta_{01} & -\sin\eta_{01} \\ 0 & \sin\eta_{01} & \cos\eta_{01} \end{bmatrix} = [B_0] \tag{3}$$

where η_{01}, η_{02}, and η_{0d} are initial roll, pitch, and drift angles, respectively, and $[\omega]$ is the skew-symmetric matrix of roll (ω_1), pitch (ω_2), and yaw (ω_3) rates,

$$[\omega] = \begin{bmatrix} 0 & -\omega_3 & \omega_2 \\ \omega_3 & 0 & -\omega_1 \\ -\omega_2 & \omega_1 & 0 \end{bmatrix} \tag{4}$$

It follows that the antenna velocity, expressed in map coordinates, is

$$\frac{d\mathbf{R}}{dt} = \begin{bmatrix} x_4 \\ x_5 \\ x_6 \end{bmatrix} = \begin{bmatrix} V_1 \\ V_2 \\ V_3 \end{bmatrix} + [B]\,[\omega]\,\mathbf{d}_A . \tag{5}$$

For the simulation to provide SAR performance under actual dynamic conditions, values of V_j and ω_j can be computed from time histories of translational and rotational motion. The resulting values of x_1 through x_6, combined with simulated errors $\tilde{x}_j$ in apparent position and velocity components, fix the quantities

$$\hat{x}_j = x_j - \tilde{x}_j, \quad 1 \leqslant j \leqslant 6 \tag{6}$$

and it is this apparent state that is used for all motion compensation operations in the airborne data processor (and thus, of course, throughout the simulation).

First Stage Motion Compensation

The first task to be performed in the name of motion compensation is proper timing of the range gates. From Fig. 1, it is immediately apparent that the distance from the aircraft to an arbitrary ground reflector

$$r_n = \sqrt{(S_{1n}-x_1)^2 + (S_{2n}-x_2)^2 + (S_{3n}-x_3)^2} \quad (7)$$

will change every time a new radar pulse is transmitted. Clearly, in an effort to keep the antenna boresight vector at the center of the SAR map, proper closing[3] time for the $(K/2)$th range gate in a contiguous set of length K will be (ignoring the aircraft excursion during pulse transit)

$$t_{(c)} = t_m + 2\hat{r}_0/c \quad (8)$$

where t_m represents the time at the leading edge of the mth transmitted pulse, c is the speed of light, and $\hat{r}_0$ is computed from indicated antenna boresight coordinates $(\hat{b}_{01}, \hat{b}_{02}, 0)$ and instantaneous aircraft position as

$$\hat{r}_0 = \sqrt{(\hat{b}_{01} - \hat{x}_1)^2 + (\hat{b}_{02} - \hat{x}_2)^2 + (\hat{x}_3)^2}\ . \quad (9)$$

More generally, the kth range gate $(1 \leqslant k \leqslant K)$ will be opened and closed[3] according to its position relative to this center. Prior to complete explanation of range gate timing, however, certain items pertinent to the above development are first noted.

1) The analysis pertains to unambiguous radar ranges throughout.

2) From (8), the $(K/2)$th range gate (K is an even integer) closes when the leading edge of the transmitted pulse reaches the point $(b_{01}, b_{02}, 0)$ as desired, only in the ideal case.

3) Departures from ideal timing are taken into account by using $\hat{r}_0$ instead of the actual distance,

$$r_0 = \sqrt{(b_{01} - x_1)^2 + (b_{02} - x_2)^2 + x_3^2}\ . \quad (10)$$

While the timing error can be regarded as traceable to the inertial platform (producing imperfections in $\hat{x}_1, \hat{x}_2, \hat{x}_3$), the deviation of actual from apparent boresight is worth describing. The airborne processor uses precise knowledge of initial range gate timing and less precise knowledge of initial antenna boresight direction to compute fixed values of $\hat{b}_{01}$ and $\hat{b}_{02}$ used to control range gate timing for every transmitted pulse.[4] Rather than degrading the SAR image, the primary effect of boresight uncertainty will be to shift the entire map translationally; this is not serious, since estimates for map coordinates can be refined by navigation update.

The method of general range gate timing can now be defined. A reference position vector $\mathbf{R}_A$ (again, expressed in map coordinates) is chosen near a predicted antenna location at SAR frame center time (among other reasons, for easy and accurate computation of associated motion compensation quantities, display recognition, etc.). (Fig. 1 illustrates that the choice of $\mathbf{R}_A$ is not critical; it need not lie in the x-z plane and the antenna need not actually pass through this position.) Directly below this point, a vector is drawn parallel to the dotted line from its ground projection to the boresight position; in the actual aircraft processor, a unit vector $\underline{\mu}$ is computed from the x, y coordinates of $\mathbf{R}_A$ relative to the apparent boresight,

$$\underline{\mu} = \begin{bmatrix} \mu_1 \\ \\ \mu_2 \end{bmatrix} = \frac{1}{\sqrt{(\hat{b}_{01} - R_{A1})^2 + (\hat{b}_{02} - R_{A2})^2}} \begin{bmatrix} \hat{b}_{01} - R_{A1} \\ \\ \hat{b}_{02} - R_{A2} \end{bmatrix}. \quad (11)$$

At this point, the basis for the first motion compensating operation can be made clear. Each range gate is timed on the basis of distance from $(\hat{b}_{01}, \hat{b}_{02}, 0)$, expressed in integral multiples of a chosen length increment s and measured along the direction of $\underline{\mu}$. Thus, the kth range gate closes at

$$t_{(k)} = t_m + \frac{2}{c}\left|\hat{\mathbf{R}} - \left\{\hat{\mathbf{b}}_0 + \left(k - \frac{K}{2}\right)s\underline{\mu}\right\}\right| \quad (12)$$

[in which $\hat{\mathbf{b}}_0$ and $\underline{\mu}$ are treated as 3×1 vectors having zero z-axis component, and $\hat{\mathbf{R}}$ consists of the first three elements of the array $\hat{\mathbf{x}}$ of (6); note that (8) is a special case of this relation]. Thus, to the best of the motion compensation system's ability, the center of the mapped area in Fig. 1 will lie in a consistent range gated position, regardless of flight path excursions accompanying the transmitted pulses.[5]

After controlling the timing of all range gates, the first stage motion compensator provides phase angle correction separately for each gate. Radar response to the nth reflector will contain a multiplicative term

$$\exp\left\{-i4\pi r_n/\lambda\right\}, \quad (13)$$

which obviously varies with every transmitted pulse [in this expression λ is the radar wavelength expressed in the same units as the range r_n; see (7)]. This is compensated through multiplication of the kth range gate response by the phase term

$$\exp\left\{+ i4\pi r_{T(k)}/\lambda\right\} \quad (14)$$

where $r_{T(k)}$ is associated with the fixed computed coordinates at the range gate center [i.e., a half-interval ahead of the value used in (12)]

$$\mathbf{R}_{c(k)} = \hat{\mathbf{b}}_0 + \left(k - \frac{K+1}{2}\right)s\underline{\mu} \quad (15)$$

by the simple relation

$$r_{T(k)} = |\hat{\mathbf{R}} - \mathbf{R}_{c(k)}|\ . \quad (16)$$

It is the pulse-to-pulse variation in the resultant phase

[3]The concept of opening and closing of range gates is introduced here only for illustrative purposes; actual mechanization, of course, calls for instantaneous sampling at the center of each gate interval.

[4]As a likely alternate, the apparent boresight ground intersection can be used as the origin of the coordinate frame, and the coordinates of all other points can be referenced accordingly. In any case, the equations contained herein remain applicable, regardless of origin translation.

[5]A similar statement cannot yet be made for points remote from this center. Both the reason and cure for this occurrence (generally referred to as "range curvature" and "second stage motion compensation," respectively) are described subsequently.

argument

$$\exp\{i4\pi[r_{T(k)} - r_n]/\lambda\} \qquad (17)$$

that defines instantaneous Doppler shift of the nth reflector relative to the frequency associated with its range gate center. Note that the PRF required for unambiguous determination of this Doppler frequency shift is dictated, not by the largest range rate observed within the mapped area, but by the spread across the antenna beam. From the classical interpretation this implies, for any value above this lower PRF boundary, a theoretical capability of ideal Shannon waveform reconstruction from the sampled response of point reflectors provided that: 1) no spectral energy was present outside the aforementioned band (antenna main beam spread) of Doppler frequencies; 2) these sampled responses were defined over an infinite duration. Of course, neither of these conditions can be met in practice. Furthermore, the point reflector idealization, however closely it may be approximated in a physical array of scatterers, will be undone by finite bandwidth; it suffices here to say the following:

1) The responses of the kth range gate to a reflector at range r_n is a bandwidth-limited function (denoted F_k) of the distance $r_{T(k)} - r_n$.
2) This response was chosen for favorable (low) sidelobe properties.
3) The nominal width of this response function is a multiple (h_1, not necessarily integral) of the slant range increment [see (11)],

$$W_G = s[\underline{\mu} \cdot (\hat{b}_0 - \mathbf{R}_A)/|\hat{b}_0 - \mathbf{R}_A|] \, . \qquad (18)$$

It is of interest to express, in terms of the foregoing relations, the processed (first stage motion compensated) range gated response to the complete array of ground scatterers. Several radar parameters that influence this response (e.g., peak transmitted power, wavelength, etc.) are common to all range gates and all scatterers; thus, they do not affect mapping, which relies only on contrast. The normalized, first stage compensated response of the kth range gate, corresponding to the mth transmitted pulse, is

$$v_{mk} = \sum_n \sigma_n g_n [\exp\{i4\pi(r_{T(k)} - r_n)/\lambda\}]^* F_{(r_{T(k)}, r_n)} \qquad (19)$$

where all quantities (r_n, $r_{T(k)}$, etc.) are computed using position at the time of the mth transmitted pulse [see (7), (16), and (17)]; g_n represents normalized antenna gain; σ_n is the complex amplitude (corresponding to magnitude S_{4n} and phase S_{5n}) of the nth reflector [refer to the discussion preceding (1)]; and * denotes discrete convolution.

Actually, to be precisely correct, (19) should theoretically contain one correction factor directly proportional to effective pulsewidth (which slightly varies for different range gates) and another inversely proportional to the fourth power of range (which varies for different reflectors). These effects are not particularly important, however. Moreover, they could be almost completely eliminated by amplitude compensation. Consequently, (19) is used in the SAR simulation.

Range Curvature and Second Stage Motion Compensation

A large array of values for v_{mk} might be thought of as a matrix, with each column representing a single range gate time history, the Fourier transform of which could provide Doppler (therefore azimuth) separation of scatterers within the beam. Actually, this is true for brief intervals within the SAR frame, but in general is not true for the complete set (typically several hundred for a whole frame) of transmitted pulses. It is recalled that thus far the motion compensator has fixed only the image center at a consistent range gated position; the locus of scatterers lying at the same range as this central point is obviously the ground intersection of a sphere centered at the instantaneous antenna location. The center range gate boundary therefore evolves in time essentially (or, if the first stage motion compensation were perfect and the ground profile were flat, precisely) as a sequence of incrementally rotated nonconcentric circular arcs, each arc having a radius equal to the current range from antenna to map center, and all arcs passing through a common pivot point at the map center. All other range cell boundaries undergo similar changes, accompanied by additional translation. A somewhat simplified description of compensation required in the second stage would dictate a redistribution of signal energy and a corresponding phase adjustment for each image cell in accordance with its position relative to the image center, using the chosen radar position $\mathbf{R}_A$ as a convenient (but uncritical) reference. More specifically, the processor performs the following computations.

1) Block segmentation of limited time histories (e.g., groups of successive values from 32 or 64 transmitted pulses) of v_{mk}, for calculation of the discrete Fourier transform (DFT)[6] over intervals of sufficient length for coarse azimuth separation but sufficiently brief to ignore aircraft excursion traversed therein, is performed.

2) Selection of Cartesian position and velocity vector components x_{sj} of the antenna under some reference conditions, is performed, e.g.,

$$x_{sj} = R_{Aj}, \qquad 1 \leq j \leq 3 \qquad (20)$$

$$x_{sj} = \hat{x}_{j(t_0)}, \qquad 4 \leq j \leq 6. \qquad (21)$$

3) For these reference conditions the closing Doppler of antenna relative to each range gate center is calculated (note: $r_{v(k)} = |\mathbf{R}_{c(k)} - \mathbf{R}_s|$),

$$f_{v(k)} = \frac{2}{\lambda} \begin{bmatrix} x_{s4} \\ x_{s5} \\ x_{s6} \end{bmatrix}^T \frac{\mathbf{R}_{c(k)} - \mathbf{R}_s}{r_{v(k)}} . \qquad (22)$$

4) Calculation of total Doppler $F_{v(k,\,l)}$ for each reflector (DFT index l) within the kth gate is made,

[6] In practice, this calculation can utilize the fast Fourier transform (FFT) algorithm.

$$F_{v(k,\,l)} = f_{v(k)} + f_l \tag{23}$$

[since, it should be noted, the operation identified by (14) ensures that the DFT of v_{mk} will produce only Doppler shifts f_l relative to range gate centers].

5) Solution of the two simultaneous equations, using the observed range and closing Doppler (see the Appendix),

$$(X_P - x_{s1})^2 + (Y_P - x_{s2})^2 + x_{s3}{}^2 = r_{v(k)}{}^2 \tag{24}$$

$$\frac{1}{r_{v(k)}}\left[(X_P - x_{s1})x_{s4} + (Y_P - x_{s2})x_{s5} - x_{s3}x_{s6}\right] = \frac{\lambda F_{v(k,\,l)}}{2} \tag{25}$$

to provide coordinates for the computed position vector $\mathbf{R}_P = (X_P,\ Y_P,\ 0)$ of each reflector cell center to be used in obtaining the differential range

$$\delta_{kl} = |\hat{\mathbf{R}} - \mathbf{R}_P| - r_{T(k)} \tag{26}$$

which is the basis of second stage motion compensation.

6) Use of this differential range to shift each spectral component separately to its own repositioned range gate and to its corrected phase as developed below.

The DFT computations provide sets of complex spectral coefficients A_{klj} which represent the discrete time history from $l = 1, 2, \cdots, L/I_S$ separate but overlapping, synthetically generated subbeams within the actual antenna beamwidth for the jth DFT block at each range gate $k = 1, 2, \cdots, K$ and are defined by

$$A_{klj} = \sum_{m=1+(j-1)\Delta m}^{L+(j-1)\Delta m} g_1\{m - (j-1)\Delta m\} \cdot V_{mk} \exp(-i2\pi f_l t_m), \quad \Delta m = L/I_S \tag{27}$$

where Δm is the sample spacing between DFT data block centers, g_1 is the pretransform weighting for shape control of the subbeam response, and I_S is the DFT redundancy factor (presently restricted to integral values). From the preceding discussion, however, it is clear that this spectral subdivision is associated with estimated antenna location at the time of each separate DFT block center.

These time intervals are designated by a time index j, which defines the reference position for the relocation distance calculations δ_{klj}. In general, of course, δ_{klj} is not an integral multiple of the range sampling interval W_G and thus requires a phase and envelope interpolation of complex data samples on the range index k of the form

$$w_{lkj} = \exp(-i4\pi\delta_{klj}/\lambda)(A_{klj} * H\{\delta_{klj}/W_G\}),$$
$$l = 1, 2, \cdots L/I_S;\ j = 1, 2, \cdots \tag{28}$$

where * denotes discrete convolution with respect to index k and $H\{x\}$ is a suitable interpolation function[7] having the approximate form

$$H\{x\} = \begin{cases} \dfrac{\sin \pi x}{\pi x}, & |x| < L_x \text{ (constant)} \\ 0, & |x| \geq L_x\,. \end{cases} \tag{29}$$

From standard, widely accepted techniques, 1) pretransform weighting is employed for filter response shaping, 2) continuity is enhanced by allowing successive DFT blocks to overlap while removing the resulting redundancy through selective discarding of spectral components, and 3) the amplitudes A_{klj} are complex, mathematically defined to correspond to its associated frequency f_l for all l within the limits

$$1 \leq l \leq L/I_S \tag{30}$$

where the above set of index numbers l are assigned to those frequencies and complex amplitudes that are retained (although, within this set extending to L/I_S, any arbitrary spectral ordering is permissible).

Upon completion of the operation defined by (28), many time samples (each responding to the full antenna beamwidth illumination) have been transformed into sets of fewer time samples from each of several fully compensated subbeams. The spectrum of each fully compensated subbeam history, over the complete frame, provides a fine grid in terms of range and Doppler for the center of each finally resolved grid element. It should be noted that much of this spectral computation is already performed in preparation for the second stage and, therefore, instead of a full (M-point) DFT computation for w, only the spectral subdivisions need be obtained at the end of a data block as follows:

$$u_{m'lk} = \sum_{j=1}^{J} g_{2j} w_{lkj} \exp(-i2\pi(f_l t_j + m'j/J)),$$
$$J = M/\Delta m;\ m' = 1, 2, \cdots, L/I_S \tag{31}$$

where g_{2j} is an appropriate pretransform weighting function (e.g., Taylor) for the desired resolution shape. The second term of the complex exponential in (31) is the Fourier kernel; the first term frequency translates each lth data subset of w_{lkj} so that all subbeam centers will be at the same frequency and thus have consistent azimuthal relationships to the fully resolved spectral coefficients.

Judicious choice of parameters simplifies many of these operations; for example, the resolution indices m' and l can be combined to form the same data set length as the original input data set M.

All that remains is to clarify the relation between the time argument of the sequence u and the corresponding radar data that influenced it, via the short-term DFT. To cite an example, with the values $L = 64, I_S = 4$ substituted into (27), each regularly overlapping 64-point block of A_{klj} for v_{mk} would contribute to its first 16 values of w. This removes the redundancy artificially introduced into the short-term DFT for the prevention of spectral ambiguities. (It should be noted that a more general version of this

[7] When argument x is large, it is necessary to redefine the limits of $H\{x\}$ to $L_1 \leq x \leq L_2$, $L_1 = -L_x + [x]$, $L_2 = L_x + [x]$, where $[x]$ is the largest interger value of x, in order to maintain proper centering of the interpolation function.

approach could be postulated, with overlap defined by an arbitrary integral number of points separating successive blocks; the present formulation restricts this number to an integral submultiple of L.) At first glance the procedure might appear to introduce a time bias into w. This is not the case, however, since the array w is treated simply as the response to M pulses (typically hundreds), in chronological sequence.

Sample Simulation Results

Several program runs were made with $M = 1024$, $L = 64$, $I_S = 4$, a PRF of 1 kHz, $\lambda = 1.77$ cm, an initial speed of 500 ft/s, and a nominal ground increment to slant range ratio of $s/r_0 = 5 \times 10^{-5}$ (nominal values chosen only for illustrative purposes, not in any way related to ultimate limits of capability). Specific run trials to be discussed here are categorized as initial (covered by verbal description only) or substantiated (described by actual computed image data). The initial simulation runs first verified expected performance under familiar conditions, such as straight and level flight with side-look geometry to isolated point targets at various known positions relative to map center. For these initial runs, velocity and acceleration uncertainties were set to zero and $\mathbf{R}_A$ was chosen at the central flight path position.

Various departures from simple conditions were then simulated, such as 45-degree squint, separate 1-g maneuvers in longitudinal, lateral, or vertical directions, and multiple reflector arrays. By providing successful image reconstruction under these conditions, these runs verified the basic validity of the generalized procedures derived in the preceeding section. At this point, for economy of computation, further attention was restricted to the single reflector case; it is noted that this procedure still allows more general conclusions to be drawn provided that the single reflector is located along a crosswise position (i.e., remote from Doppler center loci) in the mapped area.

Program runs to be substantiated by actual computed image data are defined in terms of departures from a standard set of conditions. This standard run (Case 1) simulates a single reflector at zero altitude relative to the reference ground plane, displaced nominally 1000 feet in both horizontal directions from a map center at a nominal squint angle of 45 degrees, with the aircraft at an initial altitude of 30 000 feet undergoing simultaneous 1-g longitudinal (speed change), 2-g lateral (turn), and 1-g vertical (climb) maneuvers with $\mathbf{R}_A$ at flight path center and no motion uncertainties. Results of this case are shown in Fig. 2(A), to be interpreted as follows. Only the responsive portion of the reconstructed image is shown; correct location of this response, relative to the image center, has been verified. Each grid position contains either a numeral or a dash. Numerals represent energy levels in each range-Doppler cell, expressed in multiples of 5 dB below the input reflector strength, and rounded to the nearest integer. Dashes signify responses at nominal levels of 40 dB or lower. All grid positions not shown are occupied by dashes.

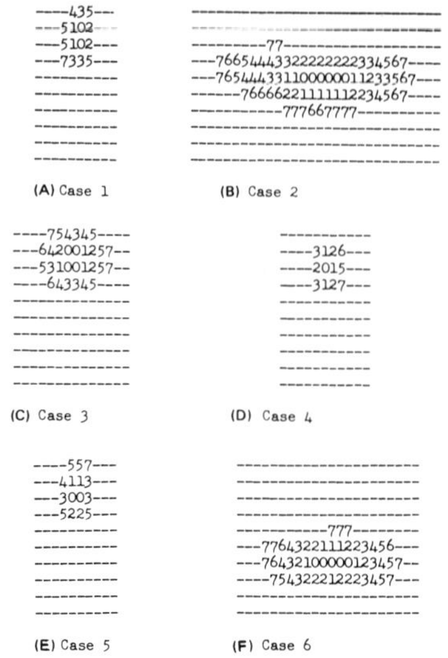

Fig. 2. Simulation results. (A) Case 1, standard. (B) Case 2, 5-ft/s velocity uncertainty in all axes. (C) Case 3, 5×10^{-3} g defocus. (D) Case 4, 50-foot altitude displacement. (E) Case 5, remote $\mathbf{R}_A$ location. (F) Case 6, all departures from standard case combined.

Cases 2 through 5 each simulate a single departure from Case 1, defined in order as follows: Fig. 2(B)—simultaneous velocity uncertainties of 5 ft/s in all three axes; (C)—defocusing due to 5×10^{-3} g acceleration error; (D)—reflector altitude of 50 feet off the reference ground plane (it can be shown that this value does not exceed depth of focus limitations); and (E)—remote $\mathbf{R}_A$ location (250, −100, and 50 feet from flight path center in x, y, and z directions, respectively). Case 6 simulates all of these conditions in combination. (Note that the degradations from velocity uncertainties in this case oppose, rather than reinforce, the acceleration defocusing.) Results are shown in Fig. 2(B) through (F), respectively.

These results actually provide only the beginning of a full system verification. As mentioned in the next section, several additional phenomena need to be studied; there are also several parameter combinations to be investigated. Quite conceivably, some of these parameter sets may require interpolation in Doppler cells as well as second stage range interpolation for successful imaging. At any rate, however, the results presented here provide adequate verification of the basic overall approach.

Conclusions and Recommendations

A general means of obtaining high-resolution SAR images over a broad class of aircraft-map geometries in the presence of simultaneous three-dimensional maneuvers has been derived and verified by simulation. Possible future extensions include parametric tradeoffs for various geometric and dynamic (and realistic aerodynamic) conditions, simulation of actual typical short-term motion compensation errors, modulation from antenna pointing error dynamics, scintillation, and mechanization of superimposed data blocks for display.

Appendix

Reduction of Second Stage Algorithm

Repeated solution of (24) and (25) for separate image cells would impose excessive computing requirements; furthermore, (26) is extremely sensitive to roundoff as it stands. The following analytical reduction circumvents these problems; the resulting expression, as evident from the discussion accompanying the derivation, is applicable to a broad class of conditions. It is noted that this reduced expression was used to obtain all simulation results presented in this paper.

The first step in the derivation is to impose the definitions.

$$\mathbf{A} \triangleq \mathbf{R}_A - \mathbf{R}_{c(k)} \tag{32}$$

$$\mathbf{B} \triangleq \mathbf{R}_{c(k)} - \mathbf{R}_{P(k,\, l)} \tag{33}$$

$$\mathbf{C} \triangleq \hat{\mathbf{R}} - \mathbf{R}_A \tag{34}$$

so that (26) can be rewritten as

$$\delta = |\mathbf{A} + \mathbf{B} + \mathbf{C}| - |\mathbf{A} + \mathbf{C}| \tag{35}$$

which can be approximated by the truncated Taylor series,

$$\delta \doteq \frac{1}{|\mathbf{A}|}\left\{\mathbf{A} \cdot \mathbf{B} + \mathbf{B} \cdot \mathbf{C} + \frac{1}{2}|\mathbf{B}|^2 - \frac{(\mathbf{A} \cdot \mathbf{B})(\mathbf{A} \cdot \mathbf{C}) + \frac{1}{2}(\mathbf{A} \cdot \mathbf{B})^2}{|\mathbf{A}|^2}\right\}. \tag{36}$$

The task remains to solve for $\mathbf{R}_P$. Tentatively omitting the subscript from $\mathbf{X}_S$ and its components and writing the definitions,

$$\mathbf{L} \triangleq \mathbf{R}_P - \mathbf{R}$$

$$Y_1 \triangleq r_{v(k)}$$

$$Y_2 = (\lambda/2)F_{V(k,\, l)} \ . \tag{37}$$

Equations (24) and (25) can be written as

$$|\mathbf{L}|^2 = Y_1{}^2 \tag{38}$$

$$(x_4, x_5, x_6) \cdot \mathbf{L} = Y_1 Y_2 , \tag{39}$$

In combination with the relation

$$(0, 0, 1) \cdot \mathbf{L} = -x_3 \tag{40}$$

the solution can be written as a classical expression for a finite length element common to two intersecting cones,

$$\mathbf{L} - \frac{Y_1}{x_4{}^2 + x_5{}^2}\left\{\left(\frac{-x_3(x_4{}^2 + x_5{}^2 + x_6{}^2)}{Y_1} \quad Y_2 x_6\right)\begin{bmatrix}0\\0\\1\end{bmatrix} + \left(Y_2 + \frac{x_3 x_6}{Y_1}\right)\begin{bmatrix}x_4\\x_5\\x_6\end{bmatrix} + G\begin{bmatrix}x_5\\-x_4\\0\end{bmatrix}\right\} \tag{41}$$

where

$$G \triangleq \pm \sqrt{(x_4{}^2 + x_5{}^2)(1 - x_3{}^2/Y_1{}^2) - (Y_2 + x_3 x_6/Y_1)^2} \ . \tag{42}$$

Therefore,

$$\mathbf{p} \triangleq \begin{bmatrix}R_{P1}\\R_{P2}\end{bmatrix} = \frac{1}{V_H{}^2}\begin{bmatrix}V_H{}^2 x_1 + x_4 Y_1 Y_2 + x_3 x_4 x_6 + G Y_1 x_5\\V_H{}^2 x_2 + x_5 Y_1 Y_2 + x_3 x_5 x_6 - G Y_1 x_4\end{bmatrix} \tag{43}$$

where $V_H{}^2 \triangleq x_4{}^2 + x_5{}^2$, and R_{P3} of course vanishes. At this point it is convenient to adopt additional two-dimensional vector notations $\mathbf{r}$, $\mathbf{c}$, and $\mathbf{v}$ for horizontal components of $\mathbf{R}$, $\mathbf{R}_{c(k)}$, and velocity, respectively. Equation (42) is then written as

$$G Y_1 = \left\{|\mathbf{v}|^2|\mathbf{r} - \mathbf{c}|^2 - [\mathbf{v} \cdot (\mathbf{r} - \mathbf{c} + \mathbf{c} - \mathbf{p})]^2\right\}^{1/2} \tag{44}$$

Except for squint angles approaching forward direction this can be approximated by

$$G Y_1 \doteq |\mathbf{v} \times (\mathbf{r} - \mathbf{c})| - \frac{1}{|\mathbf{v} \times (\mathbf{r} - \mathbf{c})|} \cdot [\mathbf{v} \cdot (\mathbf{r} - \mathbf{c})]\,[\mathbf{v} \cdot (\mathbf{c} - \mathbf{p})] \tag{45}$$

which, in combination with (43) using the relations

$$|\mathbf{v} \times (\mathbf{r} - \mathbf{c})| = x_2 x_4 - x_1 x_5 + c_1 x_5 - c_2 x_4 \tag{46}$$

and [note (22), (23)]

$$\mathbf{R} \cdot \mathbf{V} + Y_1 Y_2 = \mathbf{v} \cdot \mathbf{c} + \frac{\lambda}{2} f_l Y_1 , \tag{47}$$

leads to the expression

$$\mathbf{p} = \frac{1}{V_H{}^2}\left\{\begin{bmatrix}x_4\\x_5\end{bmatrix}\left(\mathbf{v} \cdot \mathbf{c} + \frac{\lambda}{2} f_l Y_1\right) + \xi\begin{bmatrix}x_5\\-x_4\end{bmatrix}\right\} \tag{48}$$

where

$$\xi = \frac{1}{x_2x_4 - x_1x_5 + c_1x_5 - c_2x_4}\Big\{c_1(c_1 - x_1)x_5^2$$

$$+ c_2(c_2 - x_2)x_4^2 + (c_1x_2 + c_2x_1 - 2c_1c_2)x_4x_5$$

$$+ \frac{\lambda}{2} f_l Y_1 [(x_1 - c_1)x_4 + (x_2 - c_2)x_5]\Big\}. \qquad (49)$$

This computation is readily made and substituted into (33) in preparation for (36).

Acknowledgment

The authors are indebted to A. Sorrell of Westinghouse Electric Corporation for programming the digital simulation.

References

[1] W.M. Brown, "Synthetic aperture radar," *IEEE Trans. Aerospace and Electronic Systems,* vol. AES-3, pp. 217-229, March 1967.

[2] *IRE Trans. Military Electronics,* Special Issue on "Signal Processing Radar," vol. MIL-6, April 1962.

James H. Mims (S'52–A'54–M'61) was born in Cleburne, Tex., on July 24, 1933. He received the B.S.E.E. degree from the Texas A&M College, College Station, in 1954, and attended evening courses in mathematics at the Ohio State Graduate Extension School during 1956 and 1957.

After graduation he joined the Aerophysics Group of Convair, Fort Worth, Tex., and participated in the development of the B-58 Defensive Subsystem until being called to active duty in the Air Force in 1955. His Air Force assignments at Wright-Patterson AFB, Ohio, included a project Officer for electronic countermeasures on the WS125A and B-70 programs. In 1959 he joined the Weapon System Synthesis Group, Westinghouse Defense and Space Centèr, Baltimore, Md., where he has performed electronic countermeasures effectiveness studies on advanced ground and airborne systems. Since 1965 he has concentrated on the application digital signal processing techniques to airborne tactical and reconnaissance radars and communication and sonar systems.

James L. Farrell (S'60–M'61–SM'67) was born in Chicago, Ill., on October 12, 1934. He received the B.E.E. degree from Marquette University, Milwaukee, Wis., in 1957, the M.S. degree from the University of California, Los Angeles, in 1961, and the Ph.D. degree in electrical engineering from the University of Maryland, College Park, in 1967.

From 1957 to 1959 he was employed by Minneapolis-Honeywell, where he worked on the development of electronic autopilot subsystems and in quality control. He then joined the Underwater Ordnance Circuit Design Group of Bendix-Pacific Division. In 1960 while at the University of California he was a Teaching Assistant. In 1961 he joined the Analytical Section of Westinghouse Defense and Space Center, Baltimore, Md., where he has been engaged in the analysis of advanced radar techniques, data processing, navigation, communications, and satellite dynamics.

Dr. Farrell is a member of Tau Beta Pi, Eta Kappa Nu, and Pi Mu Epsilon.

Effects of Navigation Errors In Maneuvering SAR

J.L. Farrell, J.H. Mims, and A. Sorrell
(Westinghouse Electric Company, Baltimore, Maryland)

IEEE Transactions on Aerospace and Electronic Systems, Vol. AES-9, No. 5 (September 1973)
Pages 758-776

Abstract

A maneuvering synthetic aperture radar in squint mode, during a loosely piloted maneuver, is simulated with presence of various navigation system errors. The error sources investigated place emphasis on short-term effects, involving platform servo transients, noise and quantization in accelerometers, interaction of angle pickoff uncertainty with the displacement from platform to radar antenna, and uncertainty in this displacement itself. Simulation results are accompanied by interpretive discussion, and followed by suggested areas for further study.

I. Introduction

This paper describes a detailed simulation for a high resolution, variable squint, maneuverable, synthetic-aperture radar (SAR), whereby the development in [1] is extended considerably to include effects of navigation error. The presentation is general, without regard to specific systems and parameters, but the motion patterns (aerodynamically coupled translation and rotation), as well as error time histories (generated in accordance with realistic navigation system behavior), used herein have a widely applicable form. In fact, it is within the present contex that full benefit is realized from the formulation approach in [1]. The kinematical details provided therein, while not strictly required for a limited scope investigation, establish a firm foundation on which the present general development is based.

This work was supported by the Air Force Avionics Laboratory, WPAFB, under Contract No. F33615-71-C-1842.

Manuscript received October 5, 1972; revised June 12, 1973.

In regard to the two-stage configuration for motion compensation, a rather detailed background for the full adaptive range sampling and range curvature correction approach appears in [1]. Fundamentally, the method provides high resolution SAR motion compensation in the presence of maneuvers, with variable squint, as well as side look, by automatically tailoring the compensating commands to the flight path. Different image cells receive different compensation, while, at the same time, total instrumenting and computing requirements are maintained at reasonable levels by taking advantage of modern algorithms and mechanization techniques.

Basically, each image point (denoted here as **P** for illustrative purposes) should have its radar return processed in accordance with the time history of its range $|\mathbf{R} - \mathbf{P}|$ from the radar antenna (R denotes the time-varying position vector of the antenna itself, and $||$ denotes absolute magnitude; thus $|\mathbf{R} - \mathbf{P}| =$ range). Since this is, of course, an idealization, **P** will first be replaced for each region within the image area by $\mathbf{R}_{p(k,l)}$, a chosen location for the kth range cell and the lth Doppler cell. Also, since the true value of **R** is unknown to the airborne data processor, $\hat{\mathbf{R}}$ will be used to denote the *apparent* position vector of the antenna. The time history of range, used as the basis of motion compensation in the (k, l) range-Doppler cell, can then be written as

$$|\hat{\mathbf{R}} - \mathbf{R}_{p(k,l)}| \equiv |\hat{\mathbf{R}} - \mathbf{R}_{\text{ref}}| + \left\{ |\hat{\mathbf{R}} - \mathbf{R}_{p(k,l)}| - |\hat{\mathbf{R}} - \mathbf{R}_{\text{ref}}| \right\}.$$

The purpose in writing this function in the form on the right side of this equation is to reflect the manner in which actual computation will be realized. With appropriate choice of a reference position $\mathbf{R}_{\text{ref}}$, the first quantity on the right can be used to govern range sampling and the first stage of RF phase compensation, while the bracketed quantity can be implemented in a second stage which redistributes the received radar energy in accordance with proper image and flight path geometry. In both stages, of course, there are interpolations to be performed in order to maintain a smooth distribution of signal energy in discrete cells. For the appropriate convolution operations, the reader is again referred to [1].

Actual implementation of the motion compensation scheme under consideration raises a question as to inertial navigation system (INS) output data rates, as well as computational methods. For the former, it is immediately recognized that accelerometer data must be made available *between* Navigation computer iterations due to the high PRF's typically encountered. A mechanization which serves the purpose quite well is a cyclic accumulation, described as follows.

Standard INS output includes geographic velocity components, typically current to within 1/50 second. Between the 0.02-second iterations, the effects of gravity plus specific force, taken directly from the accelerometers, are accumulated to keep the velocity data current from one PRF pulse to the next. These accumulations are, of course, then reset at every Navigation computer iteration. The instantaneous velocity is also transformed into the desired ground track (map) reference frame. This transformed output, instantaneous velocity in map coordinates, is accurate to within the precision dictated by the integrating accelerometer *at each PRF pulse.* To this INS velocity, integrated for INS position, is added the INS-to-radar antenna displacement, denoted $\mathbf{d}_{N/R}$. This vector must, of course, be transformed through the aircraft-to-map coordinate rotation matrix, denoted $[B]$. The vector sum is then used for the components of $\hat{\mathbf{R}}$.

Given the necessary time history of $\hat{\mathbf{R}}$, there remain certain computational considerations. While much of the computational approach (e.g., FFT, convolution, and interpolation, etc.) is implicit in [1], different applications (range, swath size, etc.) will call for slight modifications of ground rules. First, some (and, in fact, several) applications will allow $\mathbf{R}_{ref}$ to be fixed at map center for all image cells, although it is somewhat more accurate (and required somewhat greater complexity of implementation) to vary $\mathbf{R}_{ref}$ as a function of range index K. It should be noted that both methods have been successfully simulated here, even for noncentrally located reflectors. Secondly, there are various ways to compute the chosen points $\mathbf{R}_{P(k,l)}$. The method defined in [1] suffices for the cases considered therein. Alternatively, a direct expansion approach produces a rather intuitively pleasing "incremental x versus incremental y" locus, used herein. In addition to these items, the choice of a second-stage algorithm will bear some scrutiny for various sets of system requirements. Again, the formulation in [1] is a candidate, but there are alternative series expansion references, requiring addition of certain terms or allowing cancellation of others. In any application, the best choice will be the simplest algorithm which provides the required accuracy throughout the entire range of parameters anticipated for actual operation.

The next section indentifies principal error sources in a practical system implementation, and also describes the aerodynamic flight path used for all simulation results; these results are accompanied by appropriate interpretive discussion. This is followed by consideration of implications regarding system mechanization and recommended future study. A mathematical background is provided in the Appendix.

II. Description of Simulation and Results

The computer program used to evaluate SAR performance is based on the scheme described [1]. The overall approach is quite flexible, since it allows insertion of any time histories for actual aircraft motion and for uncertainties in the instantaneous position vector. No matter what source generated these time histories, the actual radar response to each target is governed by true antenna position (e.g., note actual instantaneous range given by (7) of [1]), while all motion compensation commands are determined from the apparent position (e.g., note $\hat{\mathbf{x}}$ in accordance with (6) of Ref. [1], used as the basis for both range sample timing and the range curvature correction). For the runs to be described here, the actual time histories are in agreement with the first equation of [1] as fed by realistic aerodynamic flight paths in the presence of gusts as well as maneuvers. The time histories for errors that produce degradations in motion compensation are computed in accordance with an INS model, as in the Appendix. It is of fundamental importance to note that this model, as appropriate for the SAR application, is concerned with short-term variations in antenna location uncertainty, and, thus, it places emphasis on parameters other than those often used as INS figures of merit. For example, questions regarding relative performance of a 3 mi/h platform versus a 0.5 mi/h platform are not the predominant issue here; drift rates reflect upon long-term errors, which are counteracted by updating of the Navigation system. Even under the best updating conditions, of course, image quality is still heavily dependent on error transients within intervals on the order of a few seconds or less. The situation is depicted in Fig. 1, which illustrates the instantaneous position vector $\mathbf{R}$ between an airborne antenna and a point in the mapped area. When there is a nonzero displacement ($\mathbf{d}_{N/R}$, expressed in aircraft coordinates) between the platform and the radar antenna, then, physically, the total position vector $\mathbf{R}$ should be thought of as the vector sum of 1) the platform position vector $\mathbf{R}_N$ and 2) $\mathbf{d}_{N/R}$, as transformed into the same coordinate frame (via $[\mathbf{B}]$, the transformation from aircraft to map coordinates), as illustrated in the figure. By taking variations, the total error $\tilde{\mathbf{R}}$ is then expressed in a form that allows interpretation of error sources as they are manifested in the actual mechanization.

For the first constituent of error in Fig. 1, it is easy to see how errors in $\mathbf{R}_N$ (or, more specifically, its derivatives, since a constant component produces an essentially fixed RF phase shift, of no consequence for Doppler beam splitting) are produced by accelerometer noise. Also, whenever the aircraft rotates, the platform undergoes a delayed miniature distorted replica of the aircraft angular maneuver (due to servo imperfections). This momentarily couples unwanted acceleration into the motion compensating commands. In addition, quantization of velocity

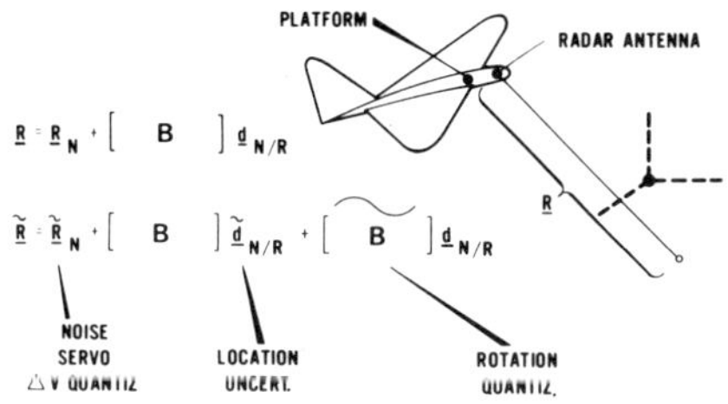

Fig. 1. Motion compensation error sources. Symbols with an underbar correspond to boldface symbols in the text.

increments introduces sudden changes in motion compensation errors.

Displacement vector error in Fig. 1 arises from 1) uncertainty $\tilde{\mathbf{d}}_{N/R}$ in the location of the antenna reference point itself, relative to the INS, and 2) attitude uncertainties.

To illustrate typical aircraft motion patterns in the presence of lateral and vertical gusts alone, a 28-second flight at 894 ft/s and a 30 000-foot altitude was simulated, with rms gust velocity components set at 3 ft/s. Various translational and rotational parameters were accumulated, and standard deviations were computed at the end of the 28-second simulated duration; results are summarized in Table I.

TABLE I

Standard Deviation Of Aircraft Responses to 3 ft/s Lateral And Vertical Gust

		Standard Deviation
Angular acceleration	roll pitch yaw	0.00831 rad/s² 0.00334 rad/s² 0.00237 rad/s²
Angular rate	roll pitch yaw	0.00653 rad/s 0.00168 rad/s 0.00121 rad/s
Angular displacement	roll pitch heading	0.00721 rad 0.00168 rad 0.00102 rad
Translational acceleration	longitudinal lateral vertical	0.0936 ft/s² 1.1513 ft/s² 1.7591 ft/s²
Translational velocity	longitudinal lateral vertical	1.2642 ft/s 1.3661 ft/s 1.6486 ft/s

Note: Longitudinal velocity and (due to nonzero angle of attack) vertical velocity components, as well as pitch angle, are rms deviations from nonzero mean values.

A 2-*g* turn command was then activated with the same rms gust levels (i.e., 3 ft/s in the lateral and vertical directions). All components of position and velocity vectors, along with the aircraft angular orientation time history, are shown in Fig. 2. The origin for the position components is placed at the center of an SAR map area, at nominal 45-degree squint and 90 000-foot total range. The coordinate frame is defined with $+Z$ downward and the initial velocity vector in the X-Z plane. Initial ground track angle was set at 1 radian. Standard units for these curves are feet, seconds, and radians, unless otherwise specified. The anticipatory autopilot, in this case, produced an initial upward altitude transient. The pitch angle undergoes a long transient, seeking a slightly changed quiescent value in the steady-state turn. Heading increases almost linerly with time, in accordance with the 2-*g* command at 894 ft/s, after a slight delay. The roll angle rises to a level of roughly 60 degrees.

The corresponding angle of attack, sideslip, and specific force components (i.e., as would be measured by body-mounted accelerometers) are shown in Fig. 3. As expected, the sideslip undergoes an initial transient before returning to a zero-mean oscillatory response, and the angle of attack seeks a new level for the turn. The lateral specific force experiences a transient along with the sideslip. Vertical specific force begins at 1 *g* upward for straight and level flight, but changes to approximately 2 *g* in the turn. A forward component is also present, essentially equal to the product of the angle of attack and the specific force upward along the yaw axis.

The above time histories are applicable to all SAR simulation cases to be described in this section. Motion compensation errors were derived on the basis of these conditions; standard parameter values for SAR simulation are given in Table II.

Fig. 4 shows the angular rate time history of the aircraft and of the platform (the designation "HOMOG" for the latter denotes exclusion of any wideband excitation), plus the platform angular orientation error. Note that the roll rate time history is dominated by a large transient that (due to the initial ground track angle) drives the east axis farther than the north (very roughly, peak north and east platform rates are the cosine and sine of the ground track angle, respectively, multiplied by the product of peak roll rate and platform servo gain for this run). Pitch and yaw rates progress toward values approximately equal to the appropriate components (proportional to the sine and cosine of the roll angle, respectively) of the heading rate. Peak azimuth misalignment is relatively small (since it is not driven by the larger components of angular rate transients), but, due to the lingering turn command, it approaches a nonzero steady-state level.

The platform servo error phenomenon is described in Figs. 5, 6, and 7. On the left of Fig. 5 is shown a simplified torque loop. The Appendix shows that the transfer function (labeled "TF" in the figure), in combination with the gyro, can produce an open-loop transfer function, shown at the upper right of Fig. 5, yielding platform-to-aircraft angular rate transfer function asymptotes, as shown at the lower right of the figure. This effect combines with the specific force, as measured by the accelerometers (sketched in Fig. 6, and derived in greater detail in the Appendix), to produce the interaction depicted in Fig. 7. The products of

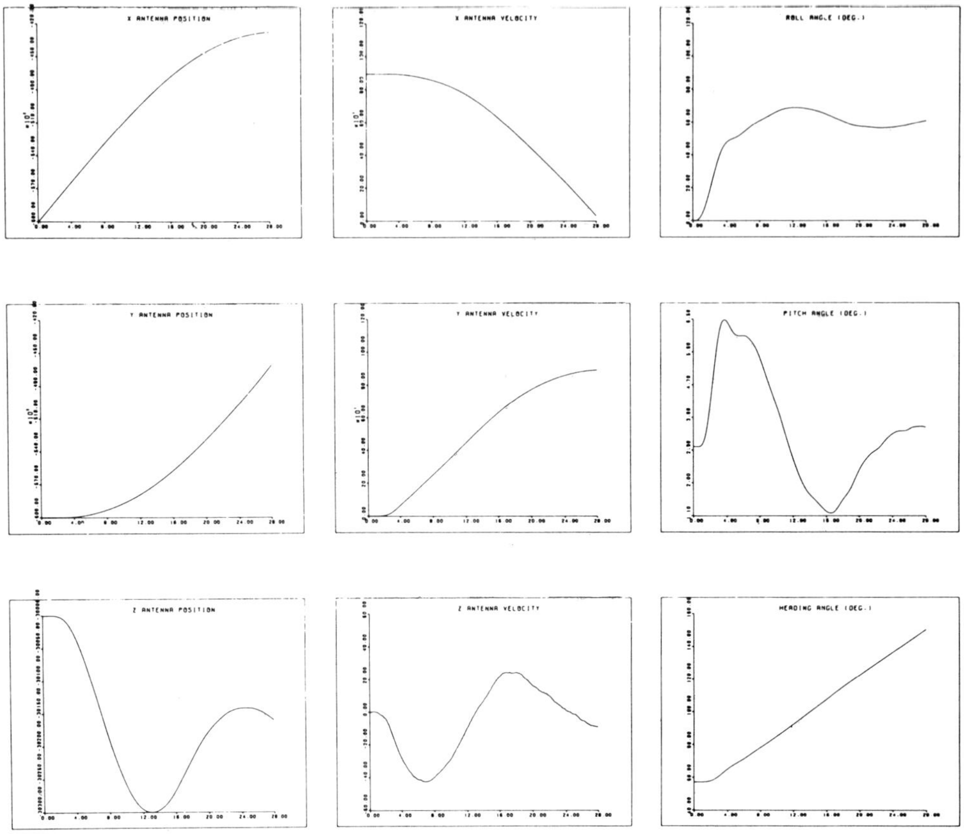

Fig. 2. Trajectory variables for piloted 2-*g* turn.

platform misorientation angles, with the appropriate acceleration components, reflect the momentary coupling of unwanted apparent derivatives into the motion compensation commands.

The time history of errors for the standard case is shown in Fig. 8. Velocity and position errors at the left of the figure refer to INS location; to this must be added the vector of uncertainty in the platform-to-antenna displacement (shown at the right of the figure) for total motion compensation' input error. For the standard conditions, however, this lever arm error is relatively small; most of the motion compensation error for this case is due to the platform servo effect. Velocity error increases most rapidly during the rotational transients. Steady-state slope of horizontal velocity error is commensurate with steady-state azimuth platform misorientation in the turn. This, of course, does not affect the vertical velocity error trend.

These time histories for true antenna position vector and accompanying uncertainties were fed into the simulation scheme described in [1]. A reflector was displaced at (+1000, −800) feet from the origin (map center) in the x and y directions, respectively, 25 feet off the reference ground plane. The coherent integration time was roughly 0.5 second (1024 PRF pulses at 0.57 ms spacing), starting near the peak of rotational activity (at approximately 3 seconds after turn command initiation). Data processing parameters were set for an overlap factor of four, operating with 64-point FFT blocks to derive the second-stage correction. An expanded view of the pertinent time segment from Fig. 8 appears in Fig. 9, and the results of the single coherent integration interval simulation appear in Fig. 10(A). The results are interpreted as follows.

Numerals appear in the responsive portion of the reconstructed image; these represent energy levels in each range-Doppler cell, expressed in multiples of 5 dB below input reflector strength, and rounded to the nearest integer. Dashes signify responses at nominal levels of 50 dB or lower. In contrast to the results described in [1], range

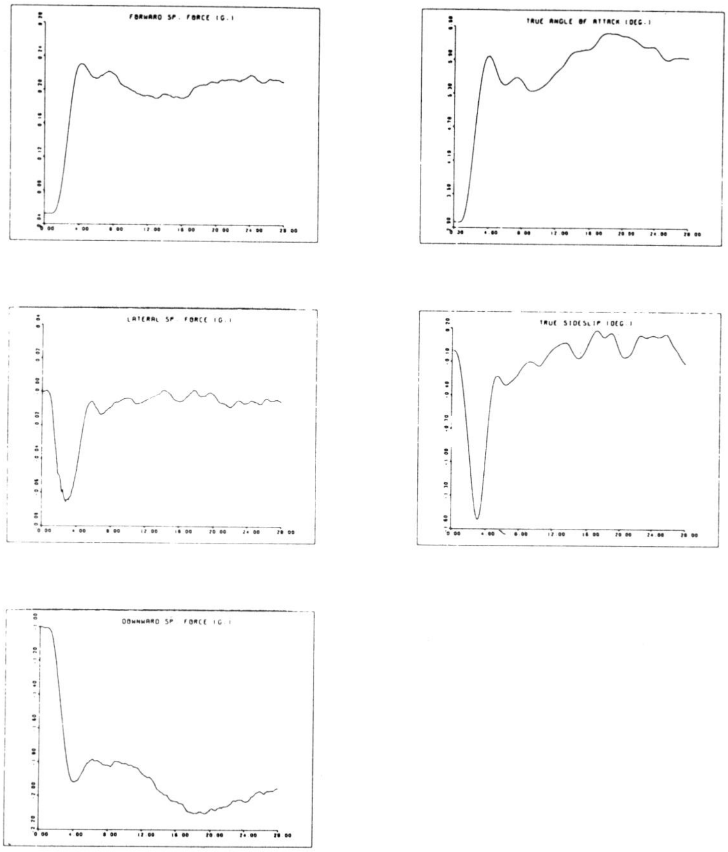

Fig. 3. 2-*g* turn aerodynamics.

centering computations have been applied to outputs here. A typical straight-line approximation to a three-dimensional response sketch appears in Fig. 11.

It is appropriate to note that a rerun of the standard case, with special provisions added to zero all motion compensation errors throughout, yielded the same overall image quality; thus, Table II represents negligible motion compensation error.

To demonstrate quantitatively the effect of reduced tightness in the platform servo, a second case was run with only one modified condition (K_ω changed to 24 dB). This naturally produced a factor of 200 increase in the platform response; the corresponding increase in motion compensation errors is shown in Fig. 12, and Fig. 10(B) exhibits the resulting image degradation. Most of this degradation is, of course, due to location error of the INS itself, but the increase in scale of the lever arm error is worth noting. The small angle quantization effect, clearly visible at left and right of Fig. 9, is too small to be discernible on the right of Fig. 12. However, the overall trends of these errors are naturally similar, but with the scale appropriately changed. All of this is, of course, consistent with the lever arm error analysis in the Appendix.

This analysis also provides for an additional platform servo error of relatively wide noise bandwidth (25Hz). A third case was run under conditions that conform to the standard, except for the presence of this wideband servo rate error at 10^{-4} $(\text{rad/s})^2/\text{Hz}$ spectral density. The resulting motion compensation errors and image degradation appear in Figs. 13 and 10(C), respectively. As expected, the image degradation affects more remote Doppler cells than the preceding case. An analytical interpretation for this case would proceed as follows. For a tilt rate noise bandwidth of 25 Hz, the autocorrelation time of the first-order integrator input error is 0.01 second. Since the Doppler cell bandwidth is essentially 1 Hz, rms tilt angle associated with one cell, accumulated in one autocorrelation interval, is 10^{-4} radian. In one coherent

TABLE II

Standard Run Conditions

Symbols (see Appendix)	Parameter	Units	Value
K_ω	Platform servo constant (0.7 damping ratio and 20 rad/s break)	Dimensionless	3.16×10^{-4} (i.e., 70 dB)
η_{P2}	Spectral density for wideband platform error (25-Hz noise bandwidth)	$(rad/s)^2/Hz$	0
η_{ea}	Spectral density for accelerometer noise (25-Hz noise bandwidth)	$(ft/s^2)^2/Hz$	0
Δ_V	Velocity quantization	ft/s	10^{-4}
Δ_ψ	Angle pickoff quantization	radians	10^{-6}
$\mathbf{d}_{N/R}$	Displacement vector from platform to radar antenna	feet	1 foot in each (x, y, z) direction
$\tilde{\mathbf{d}}_{N/R}$	Uncertainty in $\mathbf{d}_{N/R}$	feet	0 (all 3 components)

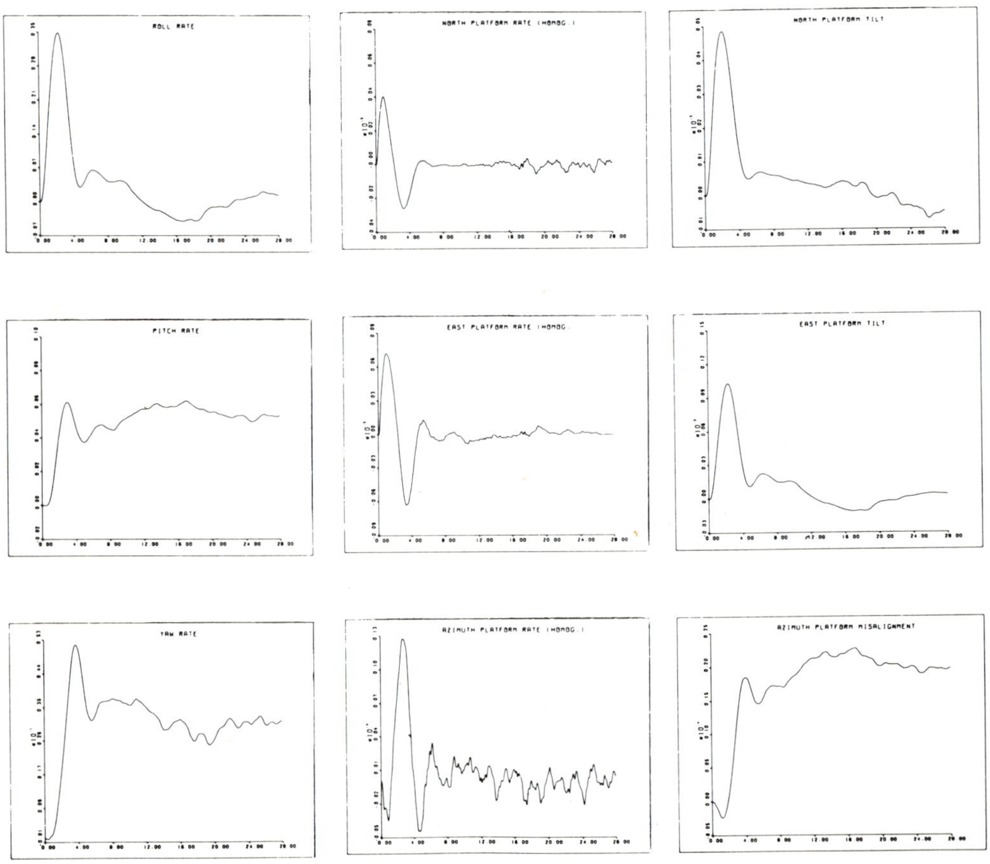

Fig. 4. Rotational dynamics for 2-g turn.

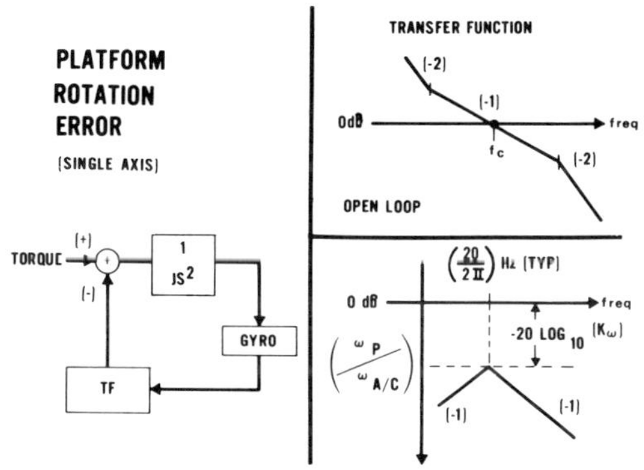

Fig. 5. Platform servo error.

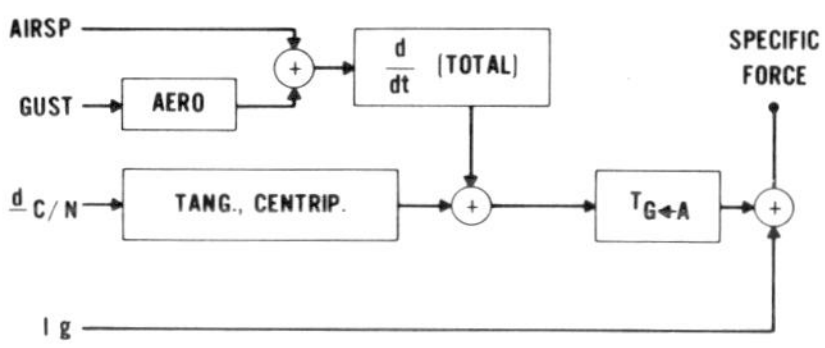

Fig. 6. Accelerometer triad forcing function.

integration interval (~ 0.6 second), then, the random walk produces an rms angle of less than 1 mrad. At the concurrent mean acceleration level, this yields roughly 0.05 ft/s^2 error in a cell. To look ahead for a moment, this is an order of magnitude lower than the value obtained in the next case; therefore, it is not sufficient to explain the degradation in Fig. 10(C). Note, however, that the rms tilt angle at midframe time is approximately $\sqrt{\eta_{P2}} \times 0.3s \doteq 5.5$ mrad. This yields appreciable spread from low-frequency blur, and the accompanying variations interact with $\mathbf{d}_{N/R}$ to produce sidelobes (again, looking ahead, note sidelobes in case 6 at only 1 mrad attitude uncertainty interacting with the same lever arm distance).

Another source of random error at 25 Hz noise bandwidth was simulated for a fourth case, differing from standard conditions only by presence of accelerometer noise at approximately 0.3 (ft/s^2)^2/Hz. The effect on image degradation, Fig. 10(D), is comparable to the case preceding, but the interpretation differs. At the nominal Doppler cell bandwidth (1 Hz), rms acceleration error in a cell is approximately 0.55 ft/s^2. The manner in which this produces sidelobes at any modulating frequency (denoted ω_m) is demonstrated with the aid of Fig. 14. The correlating function shown therein obviously gives a second time derivative with amplitude of order $\alpha\omega_m^{\ 2}$ for sidelobe generation, where α denotes peak phase modulation in radians; thus, $(\lambda\alpha\omega_m^{\ 2}/2\pi)$ defines the acceleration amplitude. The Nth Doppler filter away from any chosen cell [e.g., the first cell containing a zero, when reading from left

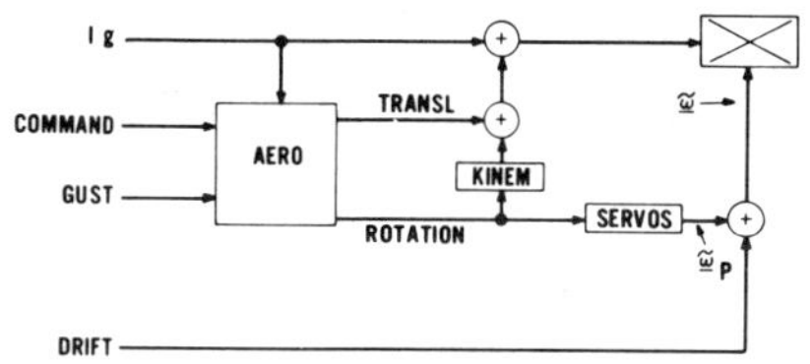

Fig. 7. Platform servo effect. The symbol with an underbar corresponds to the boldface symbol in the text.

to right in Fig. 10(D)] is associated with a nominal modulating frequency of $2\pi N$ rad/s for sidelobes pertaining to the response in that chosen cell. With $N = 3$ and $\lambda = 0.058$ feet substituted into the acceleration amplitude above, set equal to $0.55\sqrt{2}$ (from the previously established rms value of 0.55 ft/s^2), it is found that α is approximately 0.24. The phase modulation energy is split into two sidelobe frequencies; thus, the decibel level in the third filter ($N = 3$) from the Doppler cell chosen should not deviate too far from a statistical level of $20 \log_{10}(\alpha/2) \doteq 19$ dB. The value in the third cell to the left of the chosen filter in Fig. 10(D) shows a nominal level of 25 dB.

Corresponding velocity and position errors arising from accelerometer noise, shown in Fig. 15, are in statistical conformance to theoretical rms levels at time t of $\sqrt{\eta_{ea}t}$ and $\sqrt{1/3(\eta_{ea}t^3)}$, respectively. Lever arm errors are unaffected.

The fifth simulation case shows the effect of replacing the standard velocity resolution by 0.03 ft/s. This is the size of the sudden changes in velocity error components on the left of Fig. 16, and their integrated effect constitutes a random walk that outweighs the other error sources previously present (i.e., present in the standard case). The lever arm errors are unchanged. Despite the increase in INS position error wrought by velocity quantization, the image quality in Fig. 10(E) is substantially the same as the standard. This attributed to the wide distribution of error throughout the spectral region covered by the image. Especially in the presence of gusts, the effective noise energy cannot find sufficient concentration in any one cell to comprise a large false response.

For case 6, the standard gimbal pickoff angle resolution was set at 1 mrad. Fig. 17 demonstrates the result on motion compensation errors; INS position and velocity errors are naturally unaffected, while the lever arm position errors exhibit the combined effects of rotation and indicated angle quantization. Even with only 1 foot of displacement in each direction (standard $\mathbf{d}_{N/R}$), appreciable image degradation appears in Fig. 10(F). To emphasize the importance of this item, the 1-mrad resolution was retained for a seventh case, with the longitudinal displacement component changed to 30 feet. The effect is a large increase in the pertinent motion compensation error components (right of Fig. 18) and, as shown in Fig. 10(G), a totally unacceptable image degradation. The eighth case, Figs. 19 and 10(H), demonstrates the relative unimportance of fixed uncertainties in antenna phase center location.

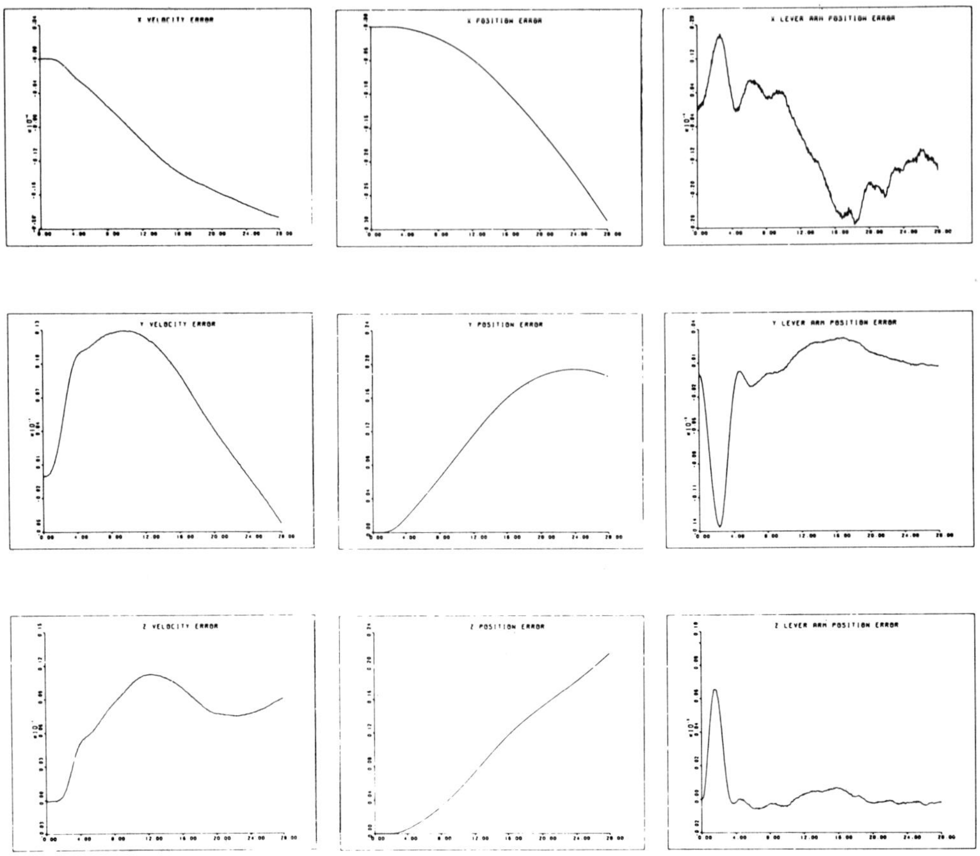

Fig. 8. Motion compensation error time histories.

III. Conclusions Regarding Practical Mechanization and Suggested Further Study

With the plurality of possible aircraft/map geometries, system parameters, and performance specifications, these representative simulation trials do not, of course, speak the last work regarding exact parameter values. Before parameters, can be pinned down for any set of performance specifications, there are several additional cases to be considered. Not only geometry and parameters must be varied, but additional phenomena must be investigated (as discussed in greater detail near the end of this section); these include scintillation from multiple reflectors in a cluster, antenna gain variations, small azimuth angle squint, high-frequency characteristics of gusts, etc. In additional to further simulation, supporting analysis for imaging with maneuvers must be extended concurrently; the foregoing examples verify only that the simulation results have the correct orders of magnitude. The input parameters adopted for the simulation cases just described are quite pessimistic an many cases; this was often necessary to exhibit appreciable error accumulation with the time allotted for a single SAR frame. A more typical set of inputs might have included a platform servo constant of 40 dB, platform servo and accelerometer noise spectral densities several orders of magnitude below the extremely pessimistic values used for demonstration (cases 3 and 4), velocity quantization at 0.01 ft/s, and pickoff angle quantization at a few tenths of a milliradian. It can safely be concluded that these values (particularly the wideband noise sources) would be acceptable in many mapping situations. Before any broader generalization could be made, however, verification would be required for the range of conditions included in a system's specifications.

Despite the dependence described above, some further general insights can be gained even from a limited number of simulation trials. One fundamentally important aspect of this investigation has been the identification of primary

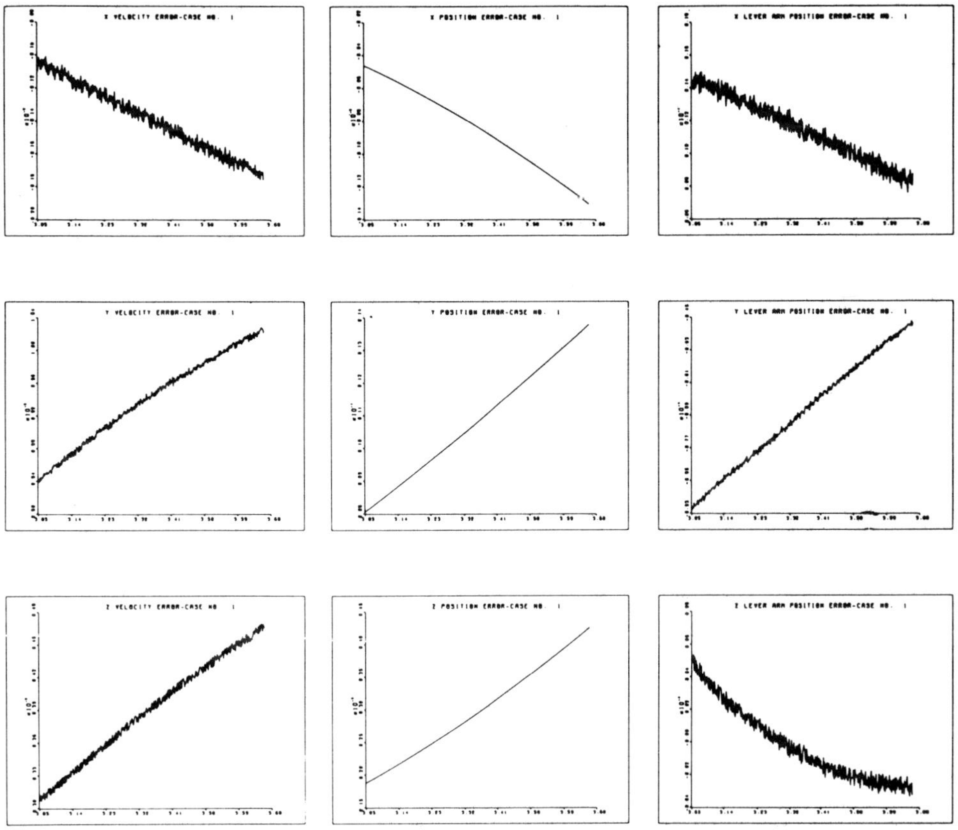

Fig. 9. Motion compensation errors for standard case.

considerations involved in SAR configurations, and how they differ from conventional systems (i.e., either conventional Navigation systems without SAR, or conventional analog/optical SAR restricted to sidelook or, at most, fixed squint, with a tight autopilot that precludes maneuvers in mapping). Perhaps the most obvious conclusion regarding advanced (i.e., digital, variable squint, high resolution, maneuverable, with optimal updating) SAR mechanization at this point is INS proximity to the radar antenna. Attitude uncertainties alone would forbid large displacements; furthermore, these degradations would be compounded in practice by unknown relative motion (due to nonrigidity) that would arise with long lever arm distances in practice. At a pickoff resolution of a few tenths of a milliradian, displacements on the order of 1 foot are permissible from the standpoint of attitude uncertainties alone. The question of nonrigidity remains, however, and, in fact, it immediately opens a broader discussion of interacting considerations, outlined below.

The inertial platform is shock mounted to prevent limit cycling at frequencies that would cause inertial instrument errors to be appreciably amplified. For the SAR system under scrutiny here, either 1) shock mounting isolation must be restricted to frequencies for which excursion amplitudes are sufficiently low (e.g., appreciably less than 1 percent of the wavelength) for acceptable image degradation, or 2) the antenna and the inertial measurement unit must be mounted in common. The first alternative necessitates a compromise between overall INS performance and image quality. The second approach necessitates a compromise between performance of the imaging system and the antenna pointing accuracy, because, clearly, the azimuth and elevation gimbals cannot maintain as tight a pointing control when the antenna is less rigidly attached to the airframe. Furthermore, when the second approach is adopted, it must be verified that the accompanying amplitude modulation due to directivity variations within a SAR frame can be tolerated. Theoretically, a concident

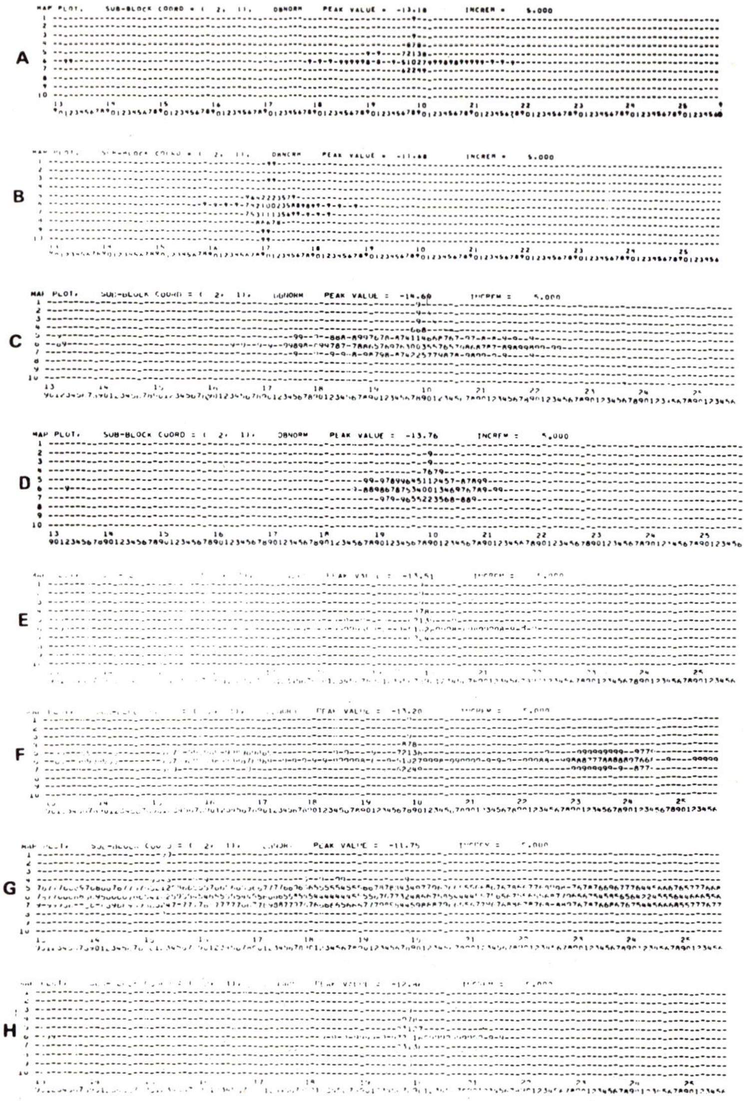

Fig. 10. Single coherent integration time simulation results.

location of INS and antenna phase center could simplify the above problems, but with finite size instruments and physical mounting constraints, some verification is prudent and even necessary. In addition, the location of the antenna phase center in an actual physical situation is a task in itself. Although it has been shown (in the last simulation case) that some fixed uncertainties are permissible, the question of phase center wandering needs to be answered, and there is little data currently available to pin this down.

Another phenomenon that needs further investigation for SAR applications is the dynamic response to gusts. Although the representation used here is the widely accepted model (see [2]), its original purpose was aimed primarily at the study of aerodynamic behavior. Details of high-frequency spectral response were not important for that purpose. In SAR studies, however, the influence of possible further attenuation (beyond the break at approximately 1 Hz in the commonly accepted gust response

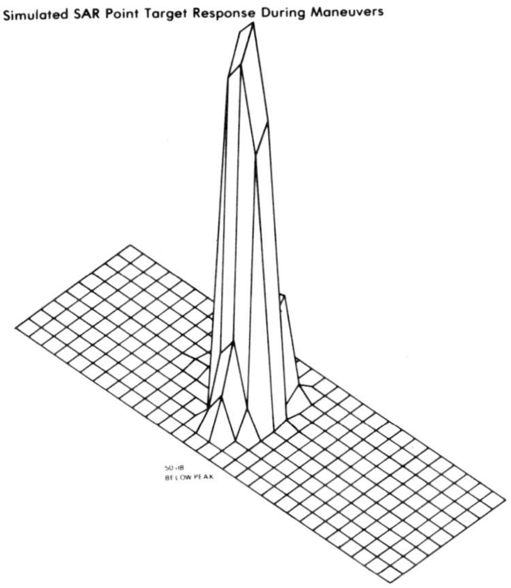

Fig. 11. Three-dimensional response approximation.

model), aeroelastic effects, and any other phenomena affecting high-frequency response should be further studied for implications upon imaging system performance.

One example of an aerodynamic phenomenon affecting high-frequency performance involves the equation for pitch angular acceleration in the standard aerodynamic equation. One small term, arising from wing tip vortices,[1] produces a slight flat spectral component in the tangential acceleration [denoted $\dot{\omega}_A \times \mathbf{d}_{C/N}$ in the Appendix]. While small motions of this type do not heavily influence the aerodynamics, they can exert influence on SAR imaging. Another consideration associated with motion spectra is the possibility of resonance in the nonrigid structure. Realistic motion patterns, in combination with finite navigation data rates (e.g., in the presence of the quantization effects studied here), determine the amount of image degradation in any given system.

In summary, the present study has identified, interpreted, and illustrated various sources of SAR degradation due to imperfect motion compensation, modeled in a way that reflects the physical mechanization. There are additional study tasks to be recommended, some of which involve appreciable extension of present work, while others can be accomplished without major revisions. The suggested tasks include:

1) Performance of production simulation runs for many combinations of aerodynamic conditions (altitude and speed, which involve different stability derivatives; gust levels; turn rates), aircraft/map geometries (range, squint angle, etc.), and system parameters.

2) Additon of drag equation to aerodynamics.

3) Addition of detailed models for acceleration errors (e.g., scale factor, misalignment, vibropendulosity, etc.; see [4]).

4) Inclusion of antenna pointing zero effects on amplitude modulation (due to directivity variations within a frame).

5) Investigation of scintillation from multiple reflectors within a cluster.

6) Assessment of high-frequency effects of aircraft response in the presence of turbulence.

7) Additional supporting analysis, such as direct spectral analysis (via FFT) of motion compensation error time histories. In combination with the corresponding geometric data, this could provide a rationale for predicting image quality as a function of system parameters. The rationale would then provide added insight beyond the results of single frame simulation.

[1] See, for example [3, pp. II-17 and II-23]. Although not discussed therein, from the standpoint of gust response, it is clear from the physics that the pertinent component of pitching moment is proportional to the time derivative of an oscillatory attack angle.

Appendix

Analytical Basis for Simulation of Motion Compensation Errors

A realistic assessment of imperfections in airborne SAR motion compensation must take into account the process whereby the compensation commands are generated; both the dynamic environment of the system and the sensitivity of instrumentation errors to this environment must be realistically modeled. Inevitably, this reflects on the overall system design (e.g., without constraining the minute details of INS specifications, the present investigation will impose definite requirements on overall platform performance and will limit allowable INS/antenna displacement).

To characterize the time history of actual motion, the antenna position vector is simply expressed as a vector sum of aircraft location and mass center-to-antenna displacement [1]. A somewhat different subdivision is needed to characterize motion compensation errors; to this end, the antenna position vector, in map coordinates, is written as

$$\mathbf{R} = \mathbf{R}_N + [B](\mathbf{d}_{N/G} + \mathbf{d}_{G/R}) \tag{1}$$

where $\mathbf{d}_{N/G}$ and $\mathbf{d}_{G/R}$ represent, respectively, the platform-to-antenna gimbal axis intersection and gimbal axis intersection-to-radar antenna displacement vectors; $[B]$ denotes the orthogonal transformation from aircraft to map coordinates; and, $\mathbf{R}_N$ denotes the platform position vector, expressible as

$$\mathbf{R}_N = \mathbf{R}_{C(t_0)} + \int_{t_0}^{t} \mathbf{V}_C \, dt + [B]\mathbf{d}_{C/N}, \tag{2}$$

in which $\mathbf{R}_C$ and $\mathbf{V}_C$ represent aircraft mass center position and velocity vectors, respectively, expressed in map coordinates; t_0 is the SAR frame initialization time; and $\mathbf{d}_{C/N}$ denotes the aircraft mass center-to-platform displacement vector, expressed in aircraft coordinates.

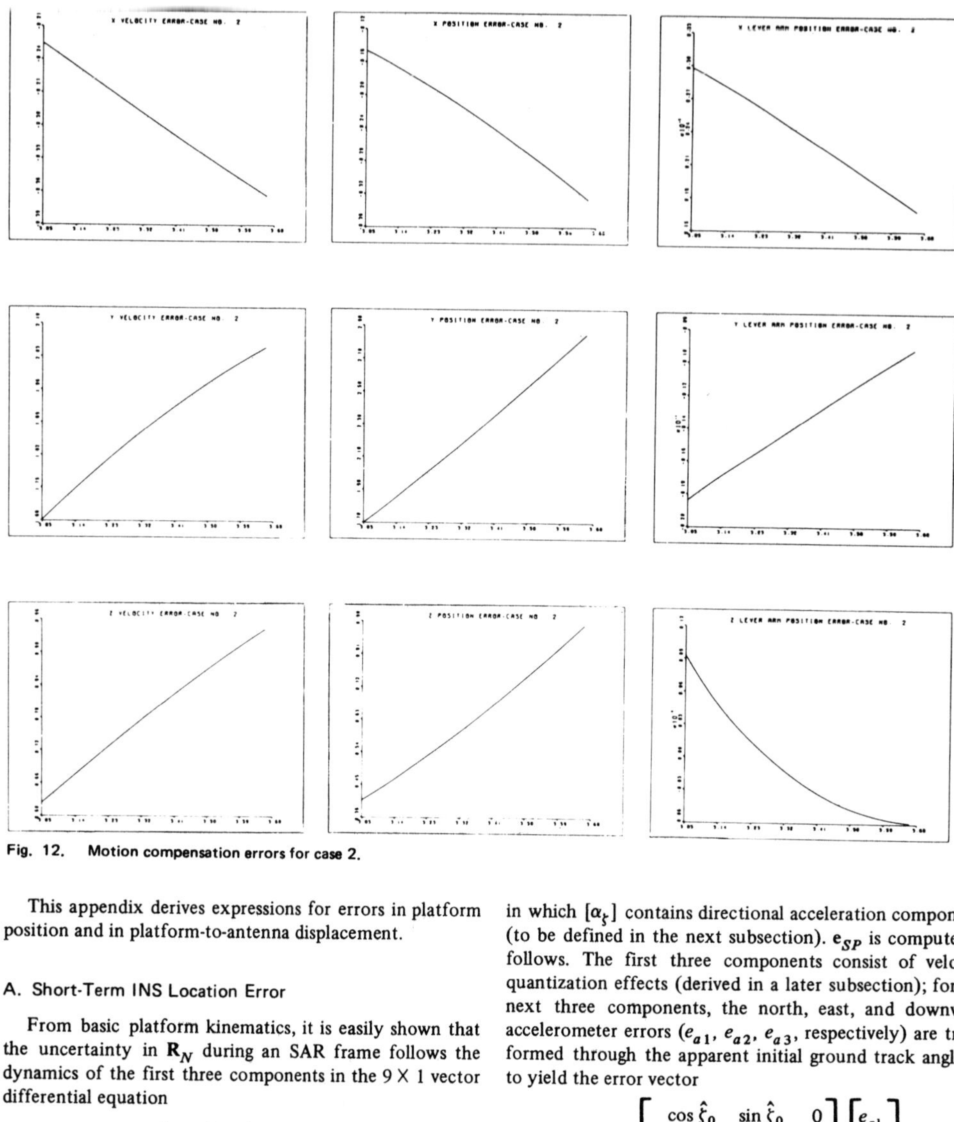

Fig. 12. Motion compensation errors for case 2.

This appendix derives expressions for errors in platform position and in platform-to-antenna displacement.

A. Short-Term INS Location Error

From basic platform kinematics, it is easily shown that the uncertainty in $\mathbf{R}_N$ during an SAR frame follows the dynamics of the first three components in the 9 X 1 vector differential equation

$$\dot{\mathbf{s}}_P = [A_{SP}]\,\mathbf{s}_P + \mathbf{e}_{SP} \tag{3}$$

where $\mathbf{s}_P$ consists of position, velocity, and orientation error vectors, respectively, expressed in map coordinates; and

$$[A_{SP}] = \begin{bmatrix} [0_{3\times 3}] & [I_{3\times 3}] & [0_{3\times 3}] \\ & [0_{3\times 6}] & [\alpha_{\zeta}] \\ & [0_{3\times 9}] & \end{bmatrix}, \tag{4}$$

in which $[\alpha_{\zeta}]$ contains directional acceleration components (to be defined in the next subsection). $\mathbf{e}_{SP}$ is computed as follows. The first three components consist of velocity quantization effects (derived in a later subsection); for the next three components, the north, east, and downward accelerometer errors (e_{a1}, e_{a2}, e_{a3}, respectively) are transformed through the apparent initial ground track angle ζ_0 to yield the error vector

$$\mathbf{e}_{Sa} = \begin{bmatrix} \cos\hat{\zeta}_0 & \sin\hat{\zeta}_0 & 0 \\ -\sin\hat{\zeta}_0 & \cos\hat{\zeta}_0 & 0 \\ 0 & 0 & 1 \end{bmatrix} \begin{bmatrix} e_{al} \\ e_{a2} \\ e_{a3} \end{bmatrix}, \tag{5}$$

while the gyro drift vector ($\mathbf{e}_{\omega}$, components defined in order about north, east, and azimuth axes) is added to the platform rotational response error vector ω_P to yield the total forcing function for angular misorientation. More detailed descriptions for the acceleration components and error forcing function are given in the next three Subsections. These are followed by an analysis of coupled

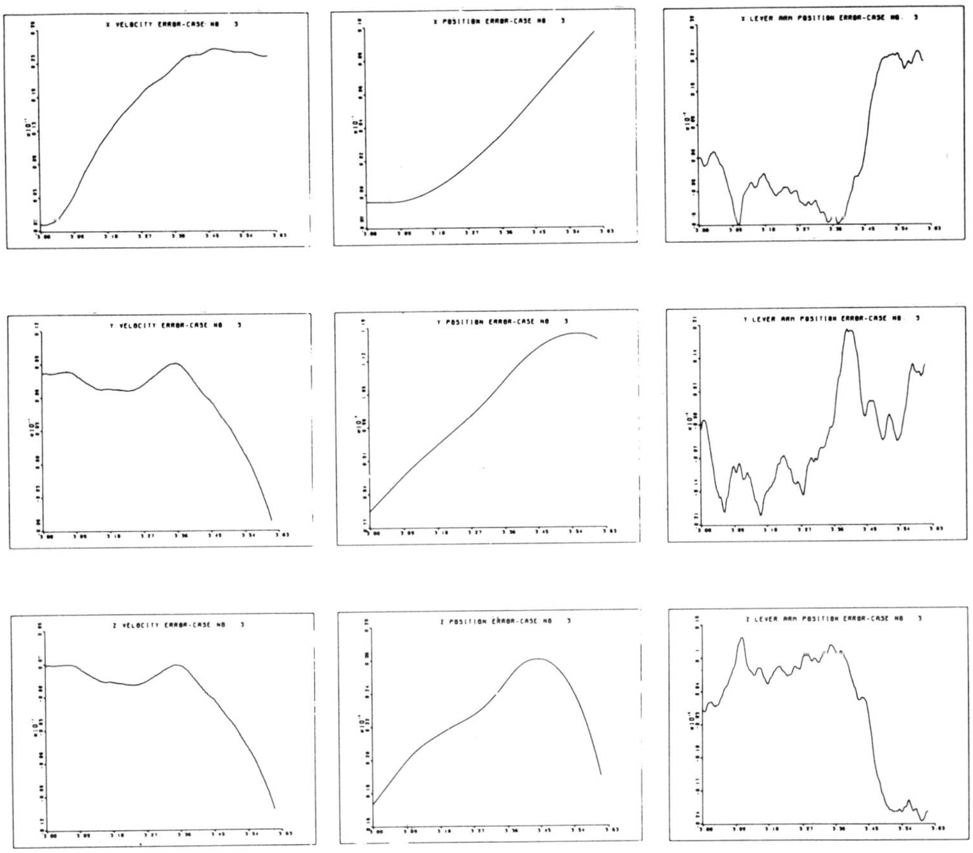

Fig. 13. Motion compensation errors for case 3.

dynamics for $\mathbf{s}_P$ and ω_P expressing the total INS position error in canonical form for multidimensional dynamical systems.

Specific Force Components in Gust Environments

The accelerometers measure specific force (i.e., total nongravitational acceleration with respect to inertial space). In a north-east-downward coordinate frame, the specific force vector over the short-term (SAR frame duration) is, essentially,

$$\mathbf{A}_G = \begin{bmatrix} A_N \\ A_E \\ A_V \end{bmatrix} = \begin{bmatrix} 0 \\ 0 \\ -g \end{bmatrix} + \frac{d}{dt}\{[\hat{\zeta}_0]_Z{}^T [B] (\mathbf{V}_A + \omega_A \times \mathbf{d}_{C/N})\} \tag{6}$$

where $[\hat{\zeta}_0]_Z{}^T$ is the transpose of the matrix shown in (5). g represents total gravity (i.e., the resultant of gravitational attraction and Earth rate centripetal acceleration); ω_A denotes the aircraft absolute angular rate (consisting of aircraft roll, pitch, and yaw rates as conventionally defined); and $\mathbf{V}_A$ is defined as the aircraft velocity vector, expressed in aircraft coordinates. By differentiation,

$$\mathbf{A}_G = \begin{bmatrix} 0 \\ 0 \\ -g \end{bmatrix} + [\hat{\zeta}_0]_Z{}^T [B]\{\dot{\mathbf{V}}_A + \omega_A \times \mathbf{V}_A + \dot{\omega}_A \times \mathbf{d}_{C/N} + \omega_A \times (\omega_A \times \mathbf{d}_{C/N})\}. \tag{7}$$

The specific force components needed for (4) are expressed in map coordinates, so that

$$[\alpha_\zeta] = [\hat{\zeta}_0]_Z \begin{bmatrix} 0 & A_V & -A_E \\ -A_V & 0 & A_N \\ A_E & -A_N & 0 \end{bmatrix}. \tag{8}$$

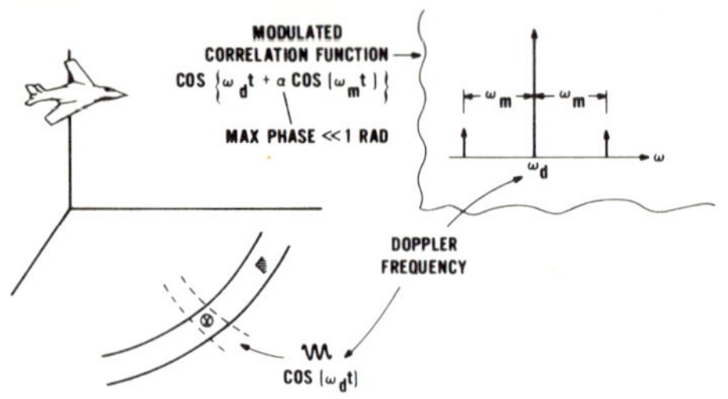

Fig. 14. Sidelobe degradation.

Errors Traceable to Accelerometer Imperfections

A triad of integrating accelerometers with digital outputs will produce a quantization error vector $\mathbf{e}_{sv}$ derived as follows. The platform velocity $\mathbf{V}_P$, expressed in map coordinates, is first transformed (via the ground track azimuth reference) into the frame in which its measurement is obtained. It is then related to the time integral of specific force [by accounting for gravity effect; see (6)], and this time integral vector is replaced by integral multiples of the velocity resolution increment Δ_V:

$$\mathbf{N}_V \triangleq \text{Trun}\left\{\frac{1}{\Delta_V}\int_{t_0}^{t} \mathbf{A}_G\, dt\right\} = \text{Trun}\left\{\frac{1}{\Delta_V}[\hat{\xi}_0]_Z^{\,T}\left(\mathbf{V}_P - \begin{bmatrix}0\\0\\g\end{bmatrix}(t-t_0)\right)\right\} \tag{9}$$

where the operator "Trun" denotes integer truncation, and

$$\mathbf{e}_{sv} = \mathbf{V}_P - \begin{bmatrix}0\\0\\g\end{bmatrix}(t-t_0) - (\Delta_V)[\hat{\xi}_0]_Z \mathbf{N}_V, \tag{10}$$

which, of course, constitutes the upper 3 × 1 partition of the Navigation system error forcing function in (3). In these expressions, $\mathbf{V}_P$ is equivalent to

$$\mathbf{V}_P = [B](\mathbf{V}_A + \omega_A \times \mathbf{d}_{C/N}), \tag{11}$$

as easily verified by differentiation of (2), noting the definitions of $\mathbf{V}_C$ and $\mathbf{V}_A$.

For the acceleration error $\mathbf{e}_a$, expressed in geographic coordinates, appearing on the right of (5), the results of case 4 of Section II were obtained with values exceeding typical misalignment, bias, vibropendulosity, and scale factor errors, in combination with the effects of gravitational anomalies and deflections of the vertical.

Platform Rotational Error Dynamics

The driving function for the last three components of the 9 × 1 vector in (3) results from gyro drift in combination with platform command response imperfections. In regard to the former, it is noted that even drifts that are high by conventional standards (e.g., 0.1 deg/h, which would yield poor long-term navigation performance) exert negligible effects over the duration of an SAR frame; gyro drifts can, therefore, be omitted from further investigation. For the latter, characterization begins with consideration of conventional INS gimbal servo loop mechanization. Fundamentally, the platform may be regarded as a rotating body subjected to a net torque vector $\mathbf{T}_P$ and responding with an absolute angular rate vector ω_P, governed by Euler's classical relation

$$[I_P]\,\dot{\omega}_P + \omega_P \times [I_P]\,\omega_P = T_P \tag{12}$$

where $[I_P]$ represents the principal inertia matrix (therefore, ω_P is expressed in platform principal axes; for purposes of this analysis, these may be considered coincident with the inertial instrument input axis directions).[2] For a net torque vector $\mathbf{T}_P$, consisting of independent components (i.e., generated with no cross-axis coupling) applied to a rigid platform in perfect balance (both statically and dynamically), the preceding relation could be written as three separate scalar equations.[2] In terms of a platform offset angle component, denoted ϕ_P, each equation would take the simple form

$$J\ddot{\phi}_P = T_E \tag{13}$$

where J and T_E represent the elements of $[I_P]$ and $\mathbf{T}_P$ corresponding to the same axis as ϕ_P, the latter now being expressed as an error torque resulting from incomplete concellation of a disturbance torque T_D by a feedback torque T_F. Thus,

$$T_E = T_D - T_F. \tag{14}$$

Furthermore, the transfer function for feedback torque/ϕ_P is the product of a feedback network transfer function F_F and a gyro transfer function F_G. From this, in combination with the last two equations, it is readily deduced that the angular rate/disturbance torque transfer function is

$$\frac{\dot{\phi}(S)}{T_D(S)} = \frac{S}{JS^2 + F_F F_G}. \tag{15}$$

For the next step in advancing this investigation, it is convenient to regard F_G simply as a constant of proportionality (ratio of angular momentum to damping coefficient) for a rate integrating gyro; dynamics of the gyro loop

[2] In practice, there are departures from these conditions caused by mass unbalance, anisoinertia, anisoelastic platform deformation, various misalignments (including that of the azimuth resolver), nonorthogonal gimbals, etc. To avoid undue preoccupation with small effects, all motion due to these departures, in combination with any incompletely suppressed resonant responses of the platform structure, can be collectively modeled as a wideband noise vector ω_{PI} to be superimposed upon the angular rate vector ω_{P2} arising from the servo response error derived above. Further details of this analysis can be obtained from inertial navigation texts, such as [5].

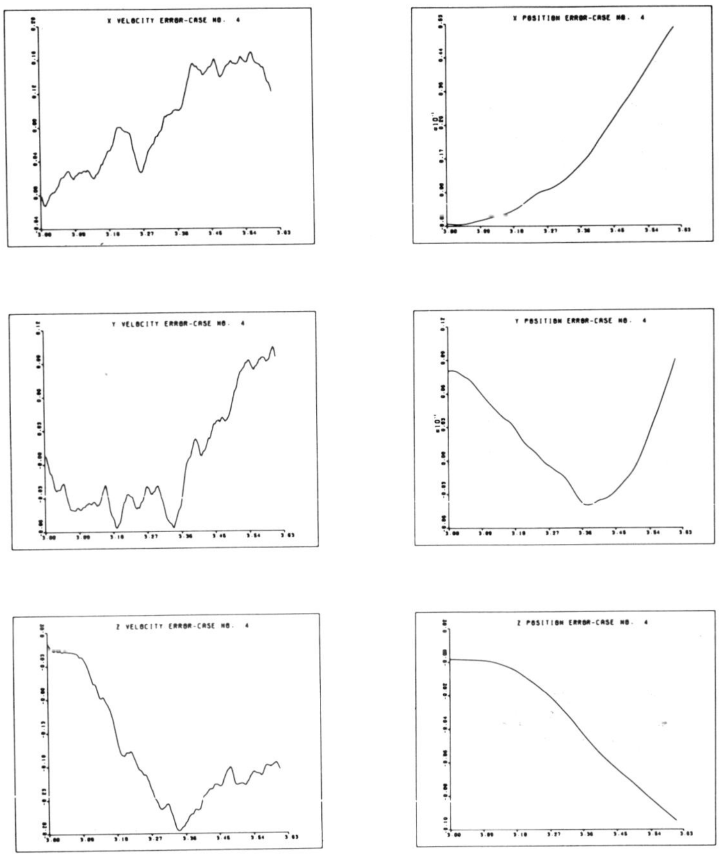

Fig. 15. Motion compensation errors for case 4.

itself can be considered included in the overall drift, which, with proper overall system design and not too severe a dynamic environment, exerts negligible influence here, as previously noted.[3] The transfer function F_F can then take a simple lead-lag form, and the open-loop transfer function of the platform is easily tailored to a common type (e.g., −1 slope in the vicinity of 0-dB crossover, and −2 slope at higher and lower frequencies of interest). Another compensating function can then be inserted to improve the low-frequency stiffness. Again, however, the aforementioned open-loop transfer function can readily be obtained from the design. For disturbance torques generated from gimbal friction, (15) is replaced by a more convenient form, in terms of an effective proportionality factor K_ω, damping ratio ζ_d, and natural frequency ω_0:

[3]For severe dynamic environments, gyro loop dynamics can exert an amount of influence comparable to the short-term gimbal servo outer loop effect analyzed here. The present analysis, however, will still provide an indication of overall behavior for environments up to the allowable limits of the INS itself.

$$\omega_{P1}(S) = \left[K_\omega \frac{S}{1 + \zeta_d(S/\omega_0) + (S/\omega_0)^2} \right] \omega_a(S) \quad (16)$$

where ω_a denotes the aircraft angular rate, expressed in geographic coordinates. The vector notation used in this equation indicates that all three axes are governed by the corresponding differential equation; noting that the derivative of ω_a is essentially

$$\frac{d}{dt}\left\{[\hat{\zeta}_0]_Z{}^T [B]\, \omega_A\right\} \doteq [\hat{\zeta}_0]_Z{}^T [B]\, [\Omega_A]\, \omega_A + [\hat{\zeta}_0]_Z{}^T [B]\, \dot{\omega}_A = [\hat{\zeta}_0]_Z{}^T [B]\, \dot{\omega}_A, \quad (17)$$

the response conforms to

$$\ddot{\omega}_{P1} + (\zeta_d \omega_0)\, \dot{\omega}_{P1} + (\omega_0{}^2)\, \omega_{P1} = (\omega_0{}^2 K_\omega) [\hat{\zeta}_0]_Z{}^T [B]\, \dot{\omega}_A \quad (18)$$

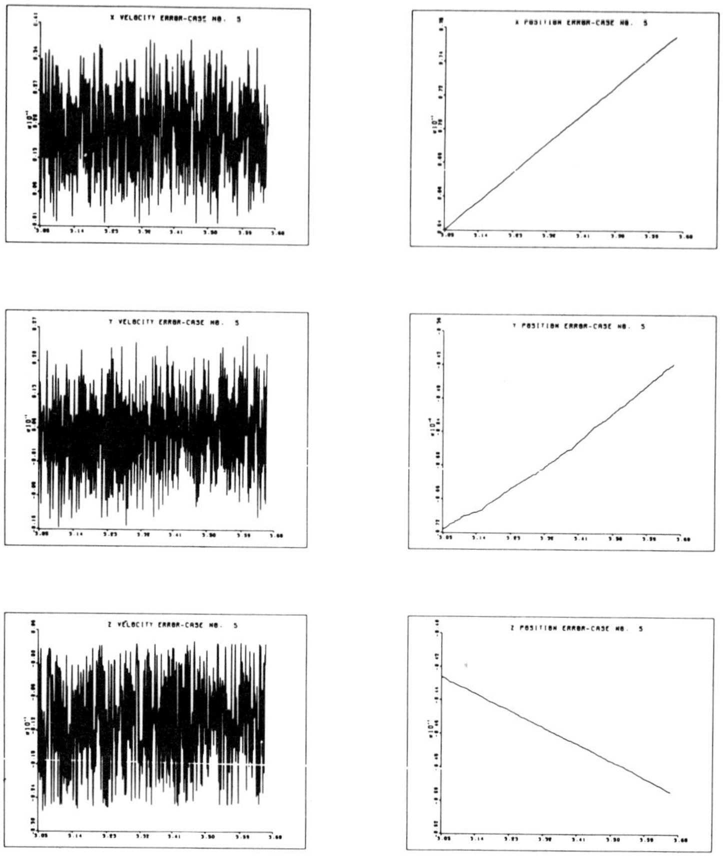

Fig. 16. Motion compensation errors for case 5.

as described in [6, p. 306], for direct drive.

Short-Term Dynamic Coupling

The contribution of $\boldsymbol{\omega}_{P1}$ to e_{SP} of (3) is appropriately transferred to the left side of the equation, producing the higher order relation

$$\dot{\mathbf{s}}_P^* = [A_{SP}^*]\mathbf{s}_P^* + \mathbf{e}_{SP}^* \tag{19}$$

where the augumented state and forcing function take the partitioned form

$$\mathbf{s}_P^* = \begin{bmatrix} \mathbf{s}_P \\ \omega_{P1} \\ \dot{\omega}_{P1} \end{bmatrix}, \quad \mathbf{e}_{SP}^* = \begin{bmatrix} \mathbf{e}_{Sv} \\ \mathbf{e}_{Sa} \\ \omega_{P2} \\ \mathbf{0}_{3 \times 1} \\ \omega_0{}^2 \mathbf{K}_\omega \dot{\omega}_a \end{bmatrix} \tag{20}$$

and the augmented matrix is

$$[A_{SP}^*] = \begin{bmatrix} [A_{SP}] & \begin{matrix} [0_{6 \times 6}] \\ [I_{3 \times 3}] \; [0_{3 \times 3}] \end{matrix} \\ [0_{3 \times 12}] & [I_{3 \times 3}] \\ [0_{3 \times 9}] & [-\omega_0{}^2 I_{3 \times 3}] \; [-\zeta_d \omega_0 I_{3 \times 3}] \end{bmatrix} \tag{21}$$

Thus, the coupled platform servo response and INS short-term error dynamics are expressible in canonical form. It should be noted that, by reason of the initial apparent ground track azimuth reference, initial conditions should include a zero value for the ninth state component.

B. Antenna Displacement Vector Error

Equation (1) indicates that total motion compensation error is derived from the uncertainty in $\mathbf{R}_P$, which

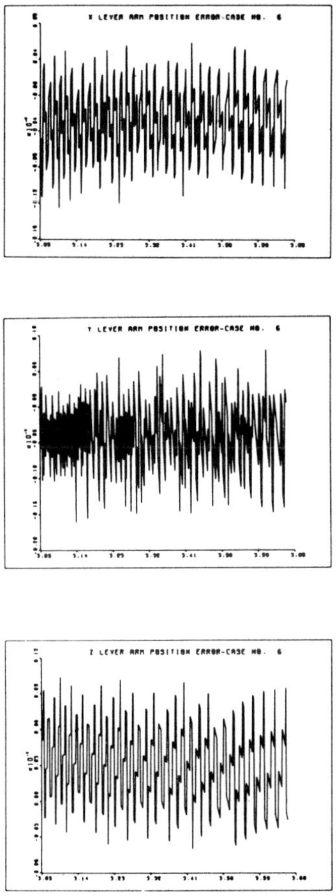

Fig. 17. Motion compensation errors for case 6.

conforms to the dynamics of the upper 3 X 1 partition in (20) combined with the error

$$\{\widetilde{[B]\mathbf{d}_{N/R}}\} \doteq [B]\tilde{\mathbf{d}}_{N/R} + [\tilde{B}]\mathbf{d}_{N/R} \qquad (22)$$

where the tilde denotes uncertainty and $\mathbf{d}_{N/R}$ denotes total displacement from platform to radar antenna. This total displacement vector and its uncertainty are both regarded as constants for the present analysis, implying adequate structural rigidity between antenna and platform (note implications upon common shock mounting isolation or restriction to low-amplitude components of vibration) and ignoring the small change in the vector resultant arising from rotation of $\mathbf{d}_{G/R}$ within a frame (since this rotation remains small within the brief frame interval, and only the dynamic variations in position errors can degrade the SAR image). The two terms on the right of (22) will now be expressed in terms of familiar parameters.

Interaction of Orientation Dynamics with Dispacement Uncertainty

The first term on the right of (22) is a constant vector premultiplied by the angular orientation matrix

$$[B] \equiv [\hat{\zeta}_0]_Z [\eta_3]_Z^T [\eta_2]_y^T [\eta_1]_x^T \qquad (23)$$

where η_1, η_2, η_3 represent aircraft roll, pitch, and heading angles, respectively. With 0 subscripts denoting initial values for these angles, and the notation η_{0d} used for initial drift angle $\hat{\zeta}_0 - \eta_{03}$, the matrix can also be defined as the solution to the differential equation

$$[\dot{B}] = -[B][\Omega_A]$$
$$[B_0] = [\eta_{0d}]_Z [\eta_{02}]_y^T [\eta_{01}]_x^T \qquad (24)$$

where $[\Omega_A]$ is the matrix operator corresponding to $-\omega_A \times$.

Interaction of Orientation Uncertainty Dynamics with Displacement

For the second term on the right of (22), it is noted that the apparent transformation $[\hat{B}]$ from aircraft to map coordinates contains an additonal intermediate multiplicative matrix (i.e., the transformation from apparent platform to geographic coordinates)

$$[\psi_T] \doteq \begin{bmatrix} 1 & \psi_{P3} + s_{P9} & -\psi_{P2} - s_{P8} \\ -\psi_{P3} - s_{P9} & 1 & \psi_{P1} + s_{P7} \\ \psi_{P2} + s_{P8} & -\psi_{P1} - s_{P7} & 1 \end{bmatrix} \qquad (25)$$

in which ψ_{Pj} denotes the jth attitude readout error and $s_{P(j+6)}$ denotes the corresponding platform misorientation error [(3) and (19)]. Thus, the second term of (22) becomes

$$[\tilde{B}]\mathbf{d}_{N/R} = [\hat{\zeta}_0]\{[I_{3\times 3}] - [\psi_T]\}[\hat{\zeta}_0]^T [B]\mathbf{d}_{N/R}. \qquad (26)$$

For a gimballed platform, the elements ψ_{Pj} can be regarded as components of a 3 X 1 vector $\boldsymbol{\psi}_p$, containing time varying pickoff errors, with magnitude dependent on the least significant bit of the readout ($\Delta_\psi \triangleq$ angular quantization increment). The formulation can also be modified for a strapdown (pulse torqued, rate integrating gyro triad) mechanization, with Δ_ψ redefined as the attitude computation increment. In any case, the transformation from initial to current aircraft coordinates can, over the short term (single frame duration), be approximately represented as a matrix with skew-symmetric, off-diagonal elements,

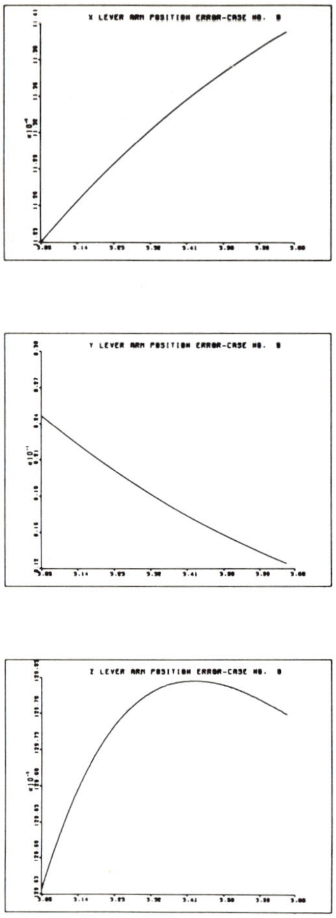

Fig. 18. Motion compensation errors for case 8.

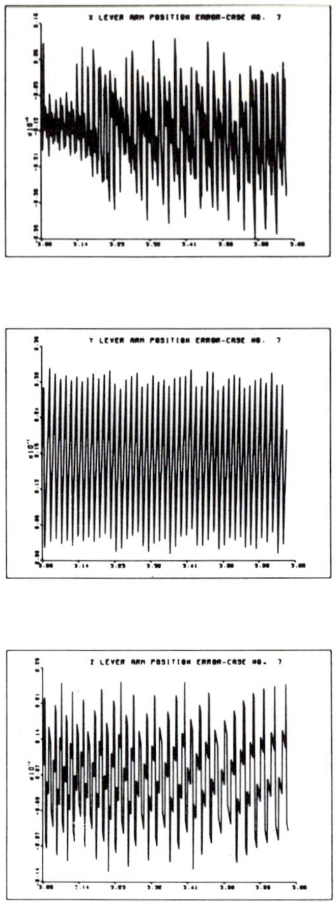

Fig. 19. Motion compensation errors for case 7..

$$[B]^T[B] = [I_{3\times 3}] - \begin{bmatrix} 0 & \beta_{12} & -\beta_{31} \\ -\beta_{12} & 0 & \beta_{23} \\ \beta_{31} & -\beta_{23} & 0 \end{bmatrix}, \quad (27)$$

and the skew-symmetric matrix on the right corresponds to a vector product operator denoted here as $\boldsymbol{\beta}x$. The elements of this vector $\boldsymbol{\beta}$, transformed into geographic coordinates, form the basis for defining angular quantization error in the present analysis:

$$\psi_P = [\hat{\zeta}_0]^T[B]\boldsymbol{\beta} - (\Delta_\psi)\mathbf{N}_\psi \quad (28)$$

where $\mathbf{N}_\psi$ is computed by an expression analogous to (9),

$$\mathbf{N}_\psi \triangleq \text{Trun}\left\{\frac{1}{\Delta_\psi}[\hat{\zeta}_0]_Z{}^T[B][\beta]\right\}. \quad (29)$$

Acknowledgment

The authors are indebted to R. Wancowicz, R. Tucker, and G. Bendor for providing the aerodynamic model and pertinent computational routine.

References

[1] J.H. Mims, and J.L. Farrell, "Synthetic aperture imaging with maneuvers," *IEEE Trans. Aerospace and Electronic Systems*, vol. AES-8, pp. 410-418, July 1972.
[2] B. Etkin, *Dynamics of Flight*. New York: Wiley, 1959.
[3] "Dynamics of the Airframe," U.S. Navy, BU AER Rep. AE-61-4-II, September, 1952.
[4] J.L. Farrell, "Analytic platforms in cruising aircraft," *AIAA J. Aircraft*, vol. 4, pp. 52-58, January-February 1967.
[5] G.R. Pitman, Ed., *Inertial Guidance*. New York: Wiley, 1962.
[6] M. Kayton and W.R. Fried, Eds., *Avionics Navigation Systems*. New York: Wiley, 1969.

James L. Farrell (S'60–M'61–SM'67) was born in Chicago, Ill., on October 12, 1934. He received the B.E.E. degree from Marquette University, Milwaukee, Wis., in 1957, the M.S. degree from the University of California, Los Angeles, in 1961, and the Ph.D. degree in electrical engineering from the University of Maryland, College Park, in 1967.

From 1957 to 1959 he was employed by Minneapolis-Honeywell, where he worked on the development of electronic autopilot subsystems and in quality control. He then joined the Underwater Ordnance Circuit Design Group of Bendix-Pacific Division. In 1960, while at the University of California, he was a Teaching Assistant. In 1961 he joined the Analytical Section of the Westinghouse Defense and Space Center, Baltimore, Md., where he has been engaged in the analysis of advanced radar techniques, data processing, navigation, communications, and satellite dynamics.

Dr. Farrell is a member of Tau Beta Pi, Eta Kappa Nu, and Pi Mu Epsilon.

James H. Mims (S'52–A'54–M'61) was born in Cleburne, Tex., on July 24, 1933. He received the B.S.E.E. degree from the Texas A&M College, College Station, in 1954, and attended evening courses in mathematics at the Ohio State Graduate Extension School during 1956 and 1957.

After graduation he joined the Aerophysics Group of Convair, Fort Worth, Tex., and participated in the development of the B-58 Defensive Subsystem until being called to active duty in the Air Force in 1955. His Air Force assignments at Wright-Patterson AFB, Ohio, included Project Officer for electronic countermeasures on the WS125A and B-70 programs. In 1959 he joined the Weapon System Synthesis Group of the Westinghouse Defense and Space Center, Baltimore, Md., where he has performed electronic countermeasures effectiveness studies on advanced ground and airborne systems. Since 1965 he has concentrated on the application of digital signal processing techniques to airborne tactical and reconnaissance radars and communication and sonar systems.

Almon T. Sorrell was born in Baltimore, Md., on April 6, 1946. He received the B.S. degree in physics-engineering from Loyola College in 1969.

After graduation he joined the Advanced Radar Systems Group of the Westinghouse Defense and Space Center, Baltimore, where he has been involved in digital equipment design and computer programming. Since 1971 he has concentrated on digital computer programming of advanced radar systems simulations and radar digital signal processing techniques.

Motion Compensation For Synthetic Aperture Radar

J.C. Kirk, Jr.
(Raytheon Company, Goleta, California)

IEEE Transactions on Aerospace and Electronic Systems, Vol. AES-11, No. 3 (May 1975)
Pages 338-348

Abstract

A generalized motion compensation approach applicable to all SAR modes, i.e., strip mapping (side-looking or squint), spotlight (or telescope) mapping, and Doppler beam sharpened mapping (DBS), is described. The basic concept is the formation for unit vector $\bar{u}$ and the slaving of the real illuminating antenna and the processed synthetic antenna to this unit vector. The amount of motion compensation which is required is developed in terms of transfer curves for the main motion reduction paths, i.e., translational, rotational (lever arm), and real antenna stabilization. The transfer curves are obtained by dividing the expected motion spectrum by the required sensitivity spectrum. The most critical motion reduction path for typical parameters is shown to be the translational path. The lever arm and real antenna stabilization paths are less critical, but must also be implemented.

Symbols

$\varphi(t)$ = motion compensation reference phase
$R(t)$ = range from aircraft to map point
λ = RF wavelength
V_L = line-of-sight velocity to map point
$\bar{V}$ = aircraft velocity vector
$\bar{u}$ = unit pointing vector
$\bar{P}$ = position vector of aircraft relative to map point
H_M = heading angle to map relative to north
E_S = elevation angle to map
V_A = velocity component along projected synthetic aperture
V_K = velocity normal to V_L and V_A
Δ_a = incremental azimuth angle about central map point
Δ_e = incremental elevation angle about central map point
ψ = heading gimbal angle
φ = roll gimbal angle
θ = pitch gimbal angle
A/C = aircraft (coordinates)
I = inertial (coordinates)
ANT = antenna (coordinates)
λ_A = antenna azimuth gimbal angle
λ_E = antenna elevation gimbal angle
φ_L = lever arm phase correction
$\bar{l}$ = lever arm vector
φ_{ANT} = reference phase measured at antenna
φ_I = reference phase measured at inertial navigation system (INS)
$\bar{a}$ = acceleration vector
D = Doppler navigator
CT = clutter tracker
V_F = velocity component in direction of map ground track
A_{S_0} = initial azimuth angle between map ground track and LOS
LOS = line of sight
ΔR_S = amount of range gate shift in strip map
ΔR_F = map ground track displacement
L = synthetic aperture length
K_S = synthetic array sidelobe weighting constant
r_a = azimuth resolution
θ_S = synthetic beamwidth
θ_R = real antenna beamwidth
B = beam sharpening ratio θ_R/θ_S
$\dot{\theta}$ = antenna scan rate
ΔT = coherent integration time
N_F = number of parallel Doppler channels in correlator
V_g = aircraft ground speed
A_S = azimuth angle between $\bar{V}$ and $\bar{u}$
ω = angular rate of A/C motion
r = aircraft displacement due to disturbance
v = aircraft velocity change due to disturbance
a = acceleration due to disturbance
R = peak displacement due to disturbance
V = peak velocity due to disturbance

A = peak acceleration due to disturbance
Δr = allowable uncompensated displacement
Δv = allowable uncompensated velocity
$\Delta\phi$ = allowable uncompensated phase error
ϕ_0 = peak value of allowable phase error
Φ = quadratic phase error at end of synthetic aperture
f = frequency of sinusoidal disturbance
SL = sidelobe gain
ML = main lobe gain
ΔV = amplitude of Δv for sinusoidal motion
g = force of gravity
Ψ = amplitude of angular A/C rotation due to disturbance
$\dot{\psi}, \dot{\Psi}$ = instantaneous value and amplitude of angular rate
$\dot{\psi}, \ddot{\Psi}$ = instantaneous value and amplitude of angular acceleration
l = length of lever arm
α = angle between $\bar{l}$ and $\bar{u}$
ΔR = required range sensitivity
CT BW = clutter track bandwidth
AM = amplitude modulation.

I. Introduction

Extensive literature exists describing the synthetic aperture concept and the processing necessary to form the synthetic aperture. Recently, digital processing has enabled real-time synthetic aperture imagery to be obtained for a variety of applications. A discussion of digital processing was included in the companion paper which precedes this one. To be fully useful, the processed synthetic aperture image must be isolated from any effects of aircraft motion. This paper will describe the motion compensation which provides the isolation of the processed image from effects of aircraft motion.

The formation of the synthetic aperture requires a coherent phase history be maintained during the time it takes to fly the synthetic aperture. As the aircraft is flown, the radar antenna successfully becomes the elements in the synthetic array. Variation in the location of the synthetic array elements, if uncompensated, can cause degradation of the processed synthetic antenna pattern. Without motion compensation the aircraft is required to fly at a constant velocity (magnitude and direction). Motion compensation involves the electronic phase adjusting of the synthetic array element to achieve the coherent phase history.

To look at it another way, SAR mapping involves the line-by-line buildup of the ground reflectivity as measured in range gates and Doppler filters. This is illustrated in Fig. 1. The Doppler filters are the synthetic antennas which provide the high angular resolution. If the range gates and Doppler filters move on the ground during the integration time, the resultant image will be smeared. If the line-by-line buildup is not performed in the proper fashion, the resultant image will be geometrically distorted, i.e., straight runways will be curved, etc.

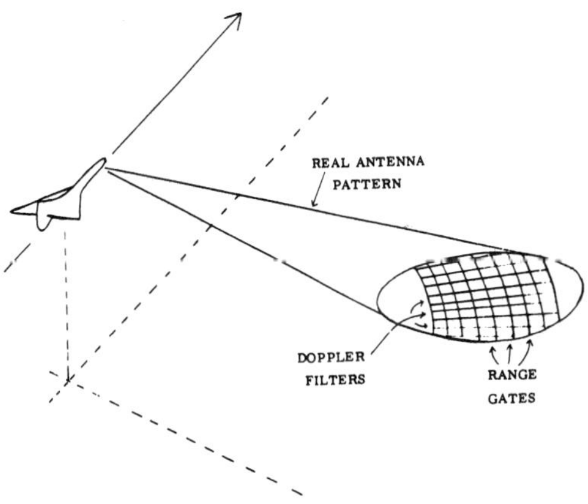

Fig. 1. Range gating and Doppler filtering.

Nonmotion compensated processing assumes no nonlinear motion; i.e., the aircraft flies a straight line at a constant speed. The range gates and Doppler filters are rigidly registered with respect to the aircraft and a good image will be obtained if there are no perturbations in the constant-velocity straight-line flight. There is a degree of motion compensation even in this steady-state case, in that the range gates must be drifted and the Doppler filters focused.

If the velocity vector changes, either in orientation or magnitude, and the reference function does not track this change, then the range-Doppler cells which are rigidly locked to the aircraft will also move on the ground. Thus, the key to motion compensation is to accurately predict the aircraft's motion and dynamically adjust the time gates and Doppler reference so that the processor cells are stationary on the ground.

Aircraft motion is caused by both weather disturbances and intentional maneuver. Higher frequency motion causes a loss in dynamic range; i.e., degraded sidelobes, and resolution loss. The effect of very low frequency motion is to cause geometric distortion and poor map accuracy. Motion compensation refers to corrections for any change in the velocity vector caused by either atmosphere disturbances or intentional maneuvering.

This paper will describe a generalized approach to motion compensation using the concept of a unit pointing vector. This approach is applicable to all SAR modes, namely strip mapping, spotlight (or telescope) mapping, and Doppler beam sharpened mapping. A quantitative analysis will be performed using transfer functions to show how much motion compensation is required.

II. Motion Compensation Implementation

To provide motion compensation the location of the aircraft with respect to the ground being imaged must be accurately known and differential motion changes must be

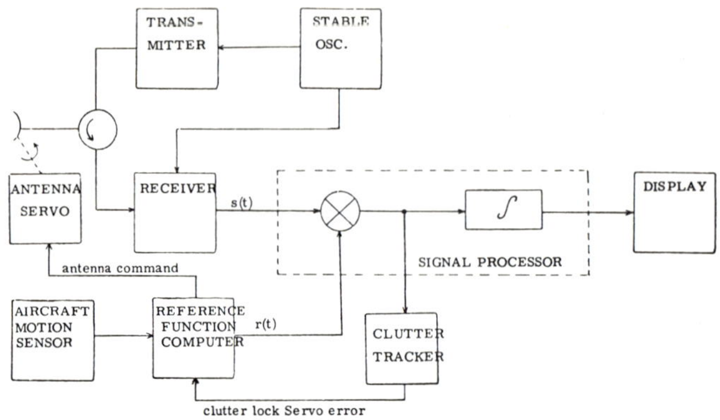

Fig. 2. SAR block diagram.

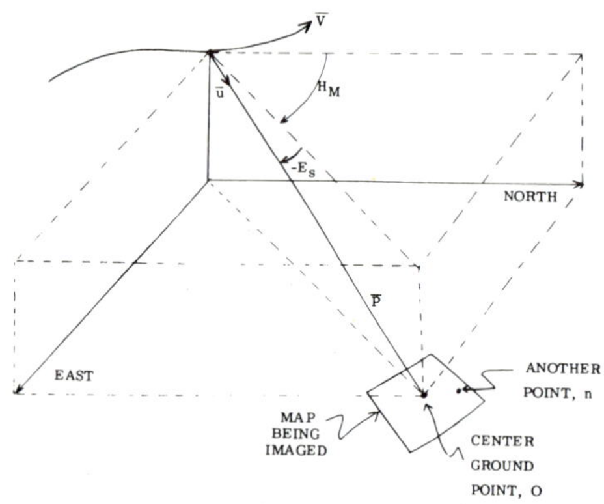

Fig. 3. Mapping geometry in geographic coordinates.

measured very accurately. Motion compensation is applied to the received signal by way of the reference function $r(t)$, as illustrated in Fig. 2. This is the central motion compensation path. Other functions, such as integration time, range gating, and the display must also be compensated.

In the SAR processing the reference function must accurately track the input signal so that there is no time-varying phase on the signal out of the mixer. The effect of time-varying phase is to displace energy from the main beam of the filter (synthetic antenna) into the sidelobes causing a loss in amplitude, resolution, and higher sidelobes. An offset error in predicting the location of a point on the ground or a small constant ΔV error will cause an azimuth shift in the location of the target.

The phase return of the ground point which is to be imaged is given by

$$\varphi(t) = (4\pi/\lambda)\, R(t) = 4\pi/\lambda\ [R_0 - \int V_L\, dt]$$

where V_L is the line-of-sight velocity between the antenna and the ground point and R_0 is the initial range. The line-of-sight velocity is found from

$$V_L = \bar{u} \cdot \bar{V}$$

when $\bar{u}$ is the unit vector to the ground point and $\bar{V}$ is the aircraft velocity vector. Thus, both velocity vector and pointing vector must be known. Notice that the range is computed to find the phase reference and, thus, the range gate position is adjusted by slaving the range gates to $R(t)$.

The velocity vector is generally provided by an inertial reference, i.e., three gyro-stabilized orthogonal accelerometers, and is usually in the geographic coordinates of north, east, and vertical. The unit vector is determined by tracking the ground point, as illustrated in Fig. 3. That is, the position vector between antenna and ground point is dynamically computed and the unit vector is thus defined. The position vector is given by

$$\bar{P} = \bar{P}_0 - \int \bar{V}\, dt$$

where $\bar{P}_0$ is the initial position, and the unit vector is given by

$$\bar{u} = \bar{P}/|\bar{P}|\,.$$

The unit vector is defined by two angles, one azimuth and the other elevation. The azimuth angle is called the map heading angle H_M, and is the angle from the inertial reference north to the ground point being imaged, measured in a plane parallel to the Earth's surface. The elevation angle F_S is measured from the horizontal plane of the aircraft to the ground point in a vertical plane. These angles are illustrated in Fig. 3.

The reasoning so far has been to develop the phase reference for a single ground point. But a ground map is made up of many such points or cells and, in general, each cell must be motion-compensated differentially. If the velocity vector is rotated through H_M and E_S, rather than using the dot product, the velocity vector

$$V = \begin{bmatrix} V_L \\ V_A \\ V_K \end{bmatrix}$$

in map coordinates is obtained. That is, V_L is the line-of-sight velocity to a map point, V_A is the velocity at which the projected synthetic aperture is flown (normal to V_L), and V_K is the velocity normal to the slant range plane determined by V_L and V_A. For another point n, located at an azimuth angle Δa and an elevation angle Δe with respect to the main point 0, the line-of-sight velocity is

$$V_{L_n} = \begin{bmatrix} \cos\Delta a \ \cos\Delta e \\ \sin\Delta a \ \cos\Delta e \\ -\sin\Delta e \end{bmatrix} \begin{bmatrix} V_{L_0} \\ V_{A_0} \\ V_{K_0} \end{bmatrix}$$

For small angular displacement, this becomes

$$V_{L_n} = V_{L_0} + \Delta a\, V_{A_0} - \Delta e\, V_{K_0}.$$

Thus, all the motion compensation points are easily and linearly related, and only one point need be tracked for typical mapping situations.

The unit vector, besides determining the velocity vector rotation, also stabilizes the real antenna and determines the lever arm correction for the displacement between antenna phase center and inertial navigation system (INS) axis.

The antenna is servoed to the unit vector. Its stabilization involves a series of coordinate rotations. The inertial reference is related to the aircraft axis through the heading, pitch, and roll gimbal angles, ψ, θ, and φ. Thus, the unit vector in aircraft coordinates is related to the unit vector in inertial (geographic) coordinates by

$$[u]_{A/C} = [\psi, \theta, \varphi]\,[u]_I$$

where

$$[u]_I = \begin{bmatrix} \cos H_M \ \cos E_S \\ \sin H_M \ \cos E_S \\ -\sin E_S \end{bmatrix}$$

and

$$[\psi, \theta, \varphi] = [\psi]\,[\theta]\,[\varphi]$$

are the 3 X 3 coordinate rotation matrixes

$$[\psi] = \begin{bmatrix} \cos\psi & \sin\psi & 0 \\ -\sin\psi & \cos\psi & 0 \\ 0 & 0 & 1 \end{bmatrix}$$

$$[\theta] = \begin{bmatrix} \cos\theta & 0 & -\sin\theta \\ 0 & 1 & 0 \\ \sin\theta & 0 & \cos\theta \end{bmatrix}$$

$$[\varphi] = \begin{bmatrix} 1 & 0 & 0 \\ 0 & \cos\varphi & \sin\varphi \\ 0 & -\sin\varphi & \cos\varphi \end{bmatrix}.$$

For an electronically scanned antenna, $[u]_{A/C}$ can be used to calculate the beam-steering phase shifts which electronically point the antenna beam. For a mechanically scanned antenna, $[u]_{A/C}$ must further be rotated into antenna coordinates:

$$[u]_{ANT} = [\lambda_A, \lambda_E]\,[u]_{A/C} = \begin{bmatrix} 1 \\ \Delta AZ \\ \Delta EL \end{bmatrix}$$

where λ_A and λ_E are the antenna gimbal angles, with the coordinate transform matrixes being

$$[\lambda_A, \lambda_E] = [\lambda_A]\,[\lambda_E]$$

$$[\lambda_A] = \begin{bmatrix} \cos\lambda_A & \sin\lambda_A & 0 \\ -\sin\lambda_A & \cos\lambda_A & 0 \\ 0 & 0 & 1 \end{bmatrix}.$$

$$[\lambda_E] = \begin{bmatrix} \cos\lambda_E & 0 & -\sin\lambda_E \\ 0 & 1 & 0 \\ \sin\lambda_E & 0 & \cos\lambda_E \end{bmatrix}$$

Also, ΔAZ is the azimuth servo error and ΔEL is the elevation servo error. These servo errors should be zero when the antenna is aligned to the unit vector.

The antenna and inertial sensor are physically displaced with respect to each other. The inertial platform measures displacement between its axis and the image point. It is necessary that the motion compensation be applied with respect to the antenna phase center, since this is how the signal phase shift occurs. Therefore, the motion as measured at the center of the inertial sensor must be translated to the center of the antenna. Let the physical displacement between the inertial sensor and antenna as measured in the aircraft axis be $\bar{l}_{A/C}$. Then the lever arm phase correction is

$$\varphi_L = (4\pi/\lambda)\bar{u}_{A/C}\,\bar{l}_{A/C} = (4\pi/\lambda)\, l \cos\alpha$$

and the phase reference as measured at the antenna is

$$\varphi(t) = \varphi_{ANT} = \varphi_I - \varphi_L$$

where φ_I is the phase reference computed from the inertially

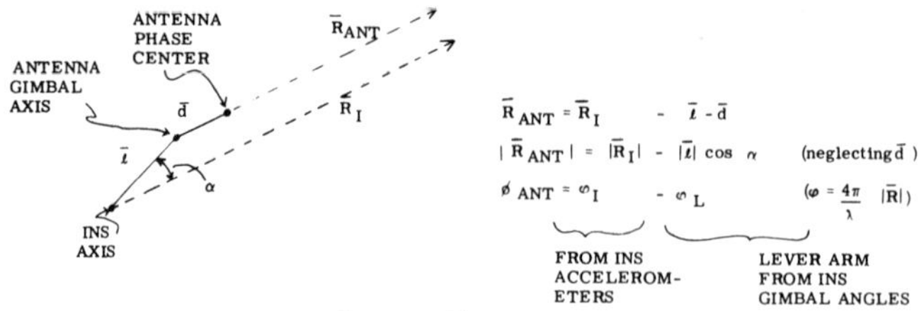

Fig. 4. Lever arm correction. Here $\bar{R}_{ANT}$ and $\bar{R}_I$ are the ranges to the map from the antenna phase center and inertial axis, respectively. The phase is proportioned to the magnitude of the displacement.

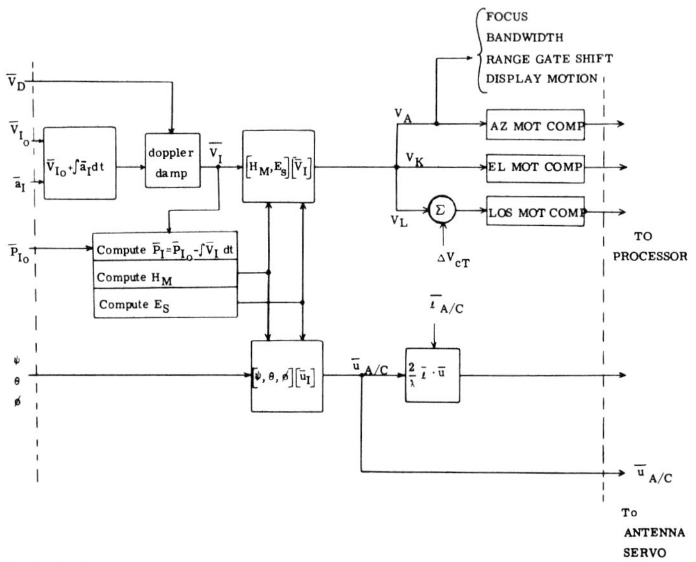

Fig. 5. Motion compensation computation block diagram.

measured velocity vector. This is illustrated in Fig. 4, for a mechanically scanned antenna. The effect of the displacement between antenna gimbal axis and antenna phase center ($\bar{d}$ in Fig. 4) can generally be neglected, since this vector is parallel to the unit vector. The lever arm is constant in A/C coordinates and easily measured.

The block diagram for the motion compensation is shown in Fig. 5. It involves motion compensation to the center of the map being processed, correction for delta points about the center, antenna stabilization, and lever arm computation.

The velocity reference is provided by the INS ($\bar{V}_I$). If the velocity reference is too noisy, then the raw accelerometer data $\bar{a}_I$ must be used to compute changes in velocity after the mapping has begun. A highly filtered velocity signal from a Doppler navigation system $\bar{V}_D$ can be used to dampen out long-term velocity errors as computed from the INS inputs. The map point is tracked in inertial coordinates by subtracting the integral of the velocity vector from the original designated map point. The map angles H_M and E_S are then computed from this dynamic position vector $\bar{P}_I$, as illustrated in Fig. 3.

The equations discussed above stabilize the cells on the ground during the integration time. Focussing is automatically accomplished by tracking the image point. But how is the map built up? How are the successive map lines tied together?

Spot Mapping

For a spotlight map, the above equations apply directly, since the spotlight map involves a small area about some central point. If the map is built up by stepping the processor azimuth filters across the ground area in azimuth, then this stepping is accomplished by stepping the incremental angle Δa about the central point defined by H_M, E_S.

Strip Mapping

For a strip map, the central point must travel along the ground in a straight line at the speed of the aircraft. For a fully stabilized map, the central point moves along a straight

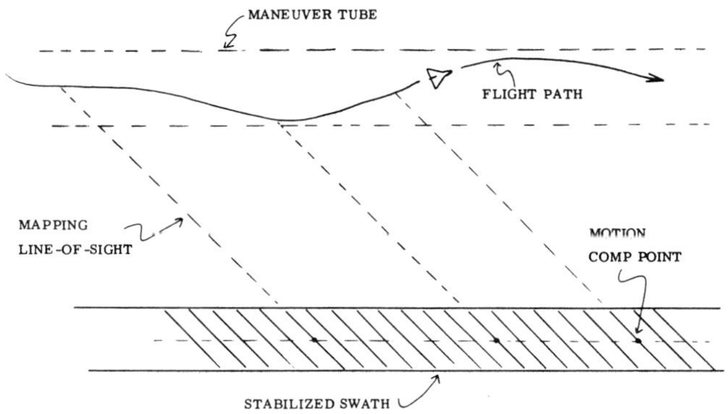

Fig. 6. Stabilized strip map.

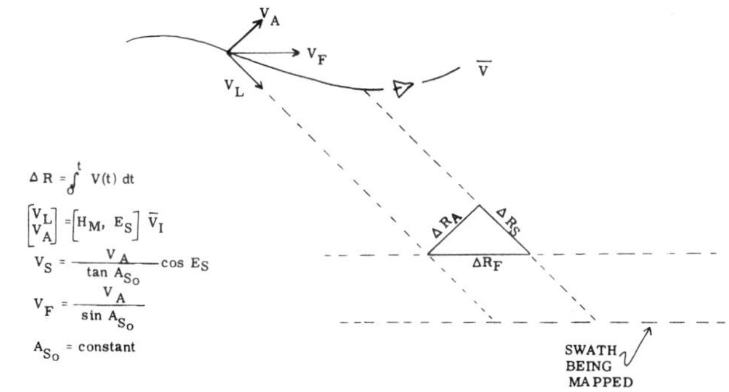

Fig. 7. Range gate resetting and display movement.

line in inertial space regardless of the motion of the aircraft, as illustrated in Fig. 6. A tube of some reasonable diameter is established to allow reasonable maneuvering and yet not unduly complicate the motion compensation problem.

The motion compensation for the strip map case is accomplished by freezing the map heading angle H_M. Focus is applied by realizing that the phase shift reference can be expressed as

$$\varphi = (4\pi/\lambda) \int [V_L + \Delta a V_A]\, dt$$

where Δa is the incremental azimuth angle which varies as a function of time from $+L/2R$ to $-L/2R$ during the integration time according to the equation

$$\Delta a(t) = 1/R \int_{-\Delta T/2}^{t} V_A(t)\, dt$$

where ΔT is the coherent integration time. Normally, V_A is nearly constant during the integration time, and the above equation corresponds to a linear angle change and, hence, a linear velocity change; i.e.,

$$\varphi \approx (4\pi/\lambda) \int [V_L + (V_A^2/R)t]\, dt.$$

This is the more familiar linear FM form of traditional SAR mapping. Notice here that V_L and V_A are orthogonal velocity components. For side-look mapping, V_L is zero and V_A is the aircraft ground speed. Non-motion-compensated SAR side-look mapping therefore reduces to the equation

$$\varphi = (2\pi/\lambda)(V_A^2/R)t^2$$

which is simply the linear FM or quadratic phase-shift compensation.

The range gates must be drifted during the integration time and then reset. The drifting is accomplished by integrating V_L and then shifting by an amount

$$\Delta R_S = \int_0^t V_A(\cos E_S/\tan A_{S_0})\, dt$$

where A_{S_0} is the initial azimuth angle between the map ground track and the line-of-sight velocity. The movement of the map along its ground track is given by

$$\Delta R_F = \int_0^t (V_A/\sin A_{S_0})\, dt.$$

This is used to adjust the display or map recording device to properly register the processed image. The computation for ΔR_S and ΔR_F is illustrated in Fig. 7. Notice the key role that the aperture velocity plays.

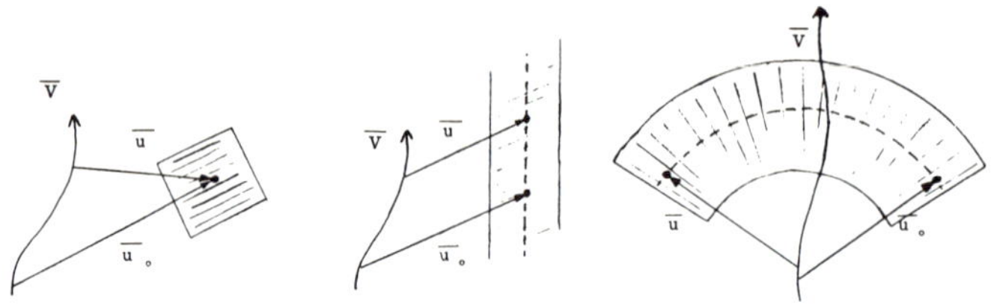

(A) spot (B) strip (C) DBS

Fig. 8. Illustration of unit vector. (Actually, the position vector is shown here, and the unit vector is the normalized position vector.)

The coherent integration time is determined by integrating the aperture velocity. When the integral of V_A is equal to L, the synthetic aperture length, then the integration is complete and the required azimuth resolution is achieved; i.e.,

$$\int_0^{\Delta T} V_A \, dt = L = K_S \lambda R / 2r_a .$$

Doppler Beam Sharpening (DBS) Mapping

As long as the map is inertially registered, full motion compensation can be applied. In general, DBS mapping is registered with respect to the velocity vector and, thus, only limited motion compensation can be applied as a rotating $\bar{V}$ will smear the map. If a large angular sector with respect to an inertial reference, i.e., north, is selected, then motion compensation can be applied. In this case the unit vector is rotated through the angular sector to be imaged. This is accomplished by changing the map heading angle H_M.

DBS works on the basis of synthetic beamwidth θ_S as opposed to displacement r_a. This is strictly a terminology preference as

$$r_a = R\theta_S .$$

A beam sharpening ratio is defined as $B = \theta_R / \theta_S$, where B is the resolution improvement achieved and θ_R is the real antenna beamwidth. Thus, a DBS mode of $B = 10/1$ or $20/1$, etc., is mechanized.

The real antenna must dwell on an area to be processed for at least the integration time. Thus, the maximum scan rate is

$$\dot{\theta} = \theta_R / \Delta T .$$

If the processor has N_F filters, which is less than B, then this rate decreases to

$$\dot{\theta} = (N_F/B)(\theta_R/\Delta T) .$$

The coherent integration time is given by

$$\Delta T = L/V_A = (K_S \lambda / 2\theta_S)/V_A .$$

Thus, the angular scan increment ΔH_M as a function of time is

$$\Delta H_M(t) = \int_0^t \dot{\theta} \, dt = (2N_F / K_S \lambda B^2) \int_0^t V_A \, dt .$$

The aperture velocity can be written as $V_A = V_g \sin A_S$, where V_g is the aircraft ground speed and A_S is the azimuth angle between $\bar{V}$ and $\bar{u}$. The scan rate is fast at wide angles, but decreases as the map vector approaches the velocity vector. Thus, no SAR mapping can be done on the velocity vector and the unit vector must be slewed across $\bar{V}$. There will be degraded resolution imagery along the velocity vector.

In summary, the key to motion compensation is the movement of the unit vector $\bar{u}$. It is fixed on the ground for spot, moves in a straight line for strip, and is rotated for DBS. This is illustrated in Fig. 8.

III. Motion Compensation Analysis

The question of how much motion compensation is required is answered by analyzing the expected aircraft motion and the motion compensation sensitivity requirements. The ratio of the aircraft motion divided by the allowed uncompensated motion tells how much motion reduction the motion compensation must apply. The transfer function of the mechanized motion compensation path must therefore encompass this required motion reduction ratio, and with a noise contribution that does not degrade the required sensitivity. There are three major motion compensation paths: the basic translational motion line-of-sight path, the real antenna stabilization path, and the lever arm correction path.

Translational Path

The aircraft's motion model can be obtained by measuring the aircraft's motion under typical weather and maneuver conditions. Another approach is to take the transfer curves for the aircraft and multiply them by a weather gust model or a maneuver command. A simpler approach will be taken here.

If the aircraft motion is assumed sinusoidal, then the displacement, velocity, and acceleration can be expressed as

$$r = R \cos \omega t$$

$$v = -V \sin \omega t$$

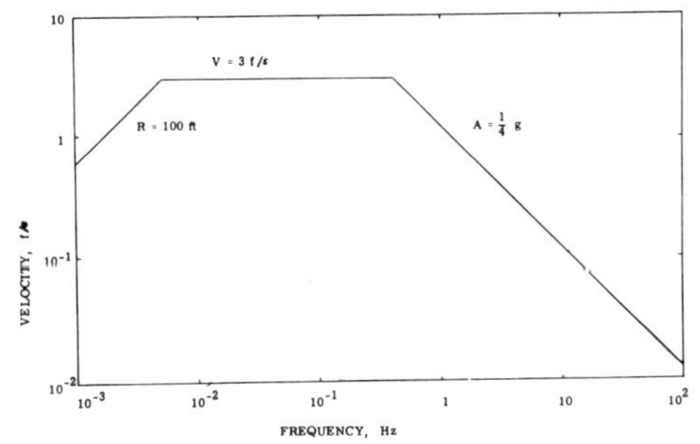

Fig. 9. Translational motion model.

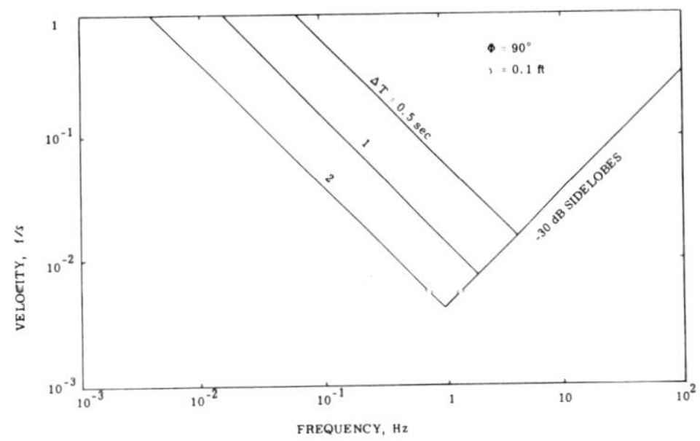

Fig. 10. Required motion compensation sensitivity.

Fig. 11. Minimum motion compensation transfer curve.

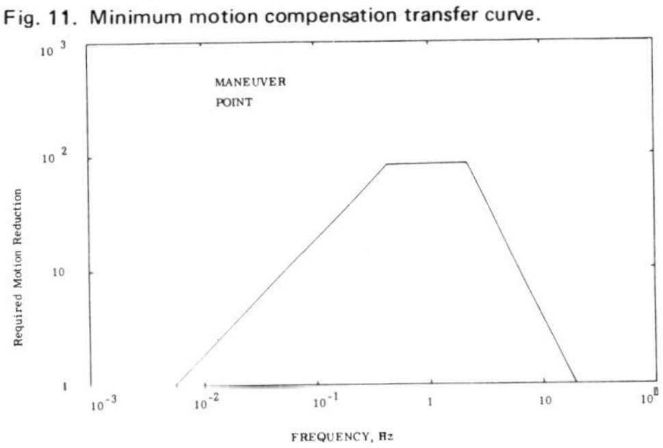

$$a = -A \cos \omega t.$$

The magnitudes can be written as

$$R = V/\omega = A/\omega^2$$

$$V = \omega R = A/\omega$$

$$A = \omega V = \omega^2 R\ .$$

Thus, if reasonable estimates on R, V, and A can be made, a plot encompassing the expected motion can be obtained. This is illustrated in Fig. 9.

For the intentional sinusoidal maneuver, the frequency and velocity are given by

$$\omega = \sqrt{A/R}$$

$$V = \sqrt{AR}\ .$$

If limits of $A = 1\ g$ and $R = 2$ nmi are chosen, then the aircraft line-of-sight velocity magnitude is 620 ft/s at a frequency of 0.0082 Hz.

The synthetic antenna pattern is the Fourier transform of the weighted output of the correlation mixer over the coherent integration time; i.e., it is the transform of

$$y(t) = W(t)e^{-j\Delta\phi(t)}$$

where $\Delta\phi(t)$ is the difference in phase between the return signal and the reference signal and $W(t)$ is the synthetic aperture weighting function. If $\Delta\phi(t)$ is assumed sinusoidal, then

$$\Delta r = (\lambda/4\pi)\ \Delta\phi = (\lambda/4\pi)\ \phi_0 \cos \omega t$$

and

$$\Delta v = -(\lambda\phi_0/4\pi)\ \omega \sin \omega t\ .$$

Thus, the amplitude is

$$\Delta V = (\lambda\phi_0/4\pi)\ \omega\ .$$

To evaluate the required sensitivity in ΔV, ϕ_0 must be computed. Alternately, the amplitudes ΔR or ΔA could also be used.

Two regions of frequency can be defined, high and low. The transform of $y(t)$ for high-frequency sinusoidal phase errors yields a pair of echoes each $\phi_0/2$ down from the main lobe. Thus, for -30-dB sidelobes, for example, ϕ_0 is 4 degrees.

For low-frequency motion where the frequency is less than the integration time, either one of two effects are noticed depending upon which part of the cosine term is encountered; either a quadratic term

$$\Delta\phi = \phi_0(1 - \omega^2 t^2/2)$$

or a linear term

$$\Delta\phi = \phi_0(\pi/2 - \omega t)\ .$$

The quadratic term defocusses the synthetic antenna degrading resolution, and the linear term mispoints the synthetic antenna.

The peak quadratic phase error at the ends of the synthetic aperture, i.e., at $t = \Delta T/2$, is

$$\Phi = \phi_0 \omega^2 \Delta T^2/4$$

or

$$\phi_0 = 4\Phi/\omega^2 \Delta T^2\ .$$

Evaluation of the Fourier transform of $y(t)$ for various values of Φ indicates a resolution which is approximately

$$\theta_{S\,\text{effective}}/\theta_{S\,\text{ideal}} = [1 + (\Phi/180°)^2]^{1/2}$$

where Φ is in degrees. Thus, taking a peak quadratic phase error at the ends of the aperture of 90 degrees corresponds to a resolution loss of approximately 10 percent.

The sensitivity in ΔV or the allowable uncompensated motion becomes

$\Delta V = \lambda(\text{SL/ML})f$ (high frequency)

$\Delta V = (\lambda/\pi)(\Phi/180°)(1/\Delta T^2)(1/f)$ (low frequency)

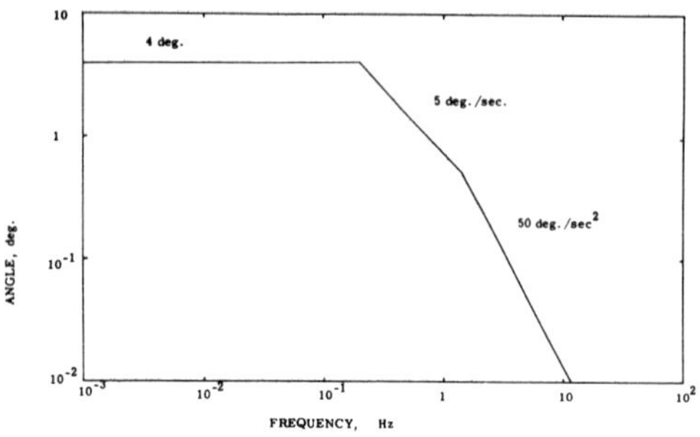

Fig. 12. Angular motion model.

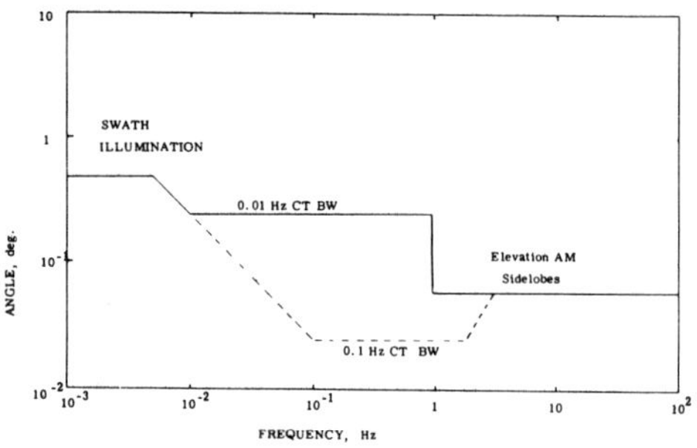

Fig. 13. Angular stabilization requirements for two values of the clutter track (CT) loop bandwidth (BW).

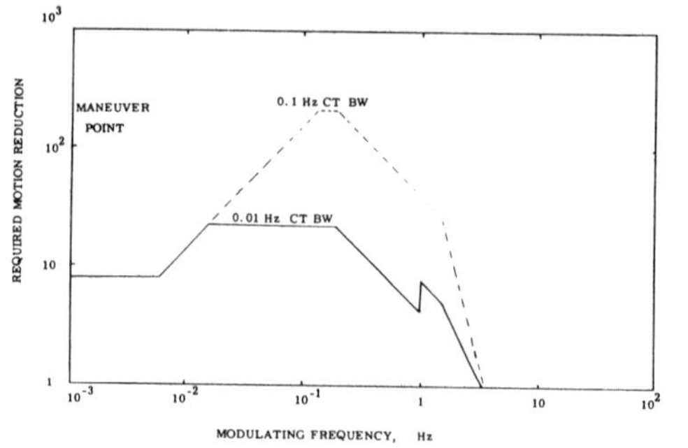

Fig. 14. Minimum antenna stabilization transfer curve for two values of clutter track (CT) loop bandwidth (BW).

where $f = (1/2\pi)\,\omega$ is the modulation frequency and SL/ML is the peak sidelobe to main lobe ratio. Plots of the required sensitivity are shown in Fig. 10 using typical parameters. The ratio of the aircraft motion (Fig. 9) to the sensitivity (Fig. 10) yields the amount of motion compensation which is required. This is plotted in Fig. 11.

Real Antenna Stabilization

If it is assumed that the rotational motion of the aircraft due to a weather disturbance is sinusoidal, then, for a particular axis, i.e., roll, ptich, or yaw, the motion can be expressed as

$$\psi = \Psi \cos \omega t$$

$$\dot{\psi} = -\dot{\Psi} \sin \omega t$$

$$\ddot{\psi} = -\ddot{\Psi} \cos \omega t$$

where

$$\psi_{max} = \Psi = \dot{\Psi}/\omega = \ddot{\Psi}/\omega^2 .$$

For typical values of Ψ, $\dot{\Psi}$, and $\ddot{\Psi}$, the expected rotational motion is shown plotted in Fig. 12.

During a typical maneuver the aircraft will be banked or rolled to achieve a resultant change in yaw. The roll angle during a coordinated turn is approximately

$$\phi = \tan^{-1} A/g$$

where A is the crosstrack acceleration and g is the force of gravity. For a 1-g crosstrack maneuver the roll angle is 45 degrees. For the sinusoidal 1-g 2-nmi maneuver previously mentioned the aircraft will roll and yaw approximately ±45 degrees in a sinusoidal fashion at a period of $\omega = (A/R)^{1/2} =$ 0.0082 Hz. This analysis allows a maneuver point to be obtained. Further analysis needs to be performed to obtain a maneuver spectrum.

The effect of antenna motion with respect to the ground image is to put amplitude modulation on the return signal. This AM will produce echo pairs if the frequency is high enough. The antenna must also be pointed well enough to keep the area to be imaged properly illuminated. In addition, antenna motion (mainly azimuth) will produce FM modulation on the output of the clutter loop discriminator. The clutter loop filter will attenuate high-frequency motion, but will pass low-frequency motion. If this error signal is added to the phase reference, the AM will translate to FM. The particular antenna shape must also be taken into account. Thus, the requirements are highly dependent on both antenna shape and clutter loop design. An example of typical angular stabilization sensitivity requirements, as derived in the Appendix, is shown in Fig. 13, and the resultant stabilization gain or motion reduction factor is shown in Fig. 14.

Lever Arm Path

The projection of the lever arm down the mapping line of sight is given by

$$l_{LOS} = \bar{l} \cdot \bar{u} = l \cos \alpha$$

where l is the length of the lever arm and α is the angle between the lever arm and the line of sight, as shown in Fig. 4. The change in l_{LOS} due to an angular rotation is

$$\Delta l_{LOS} = -l \sin \alpha \Delta\alpha .$$

When α is small, the lever arm effect is negligible. The worst

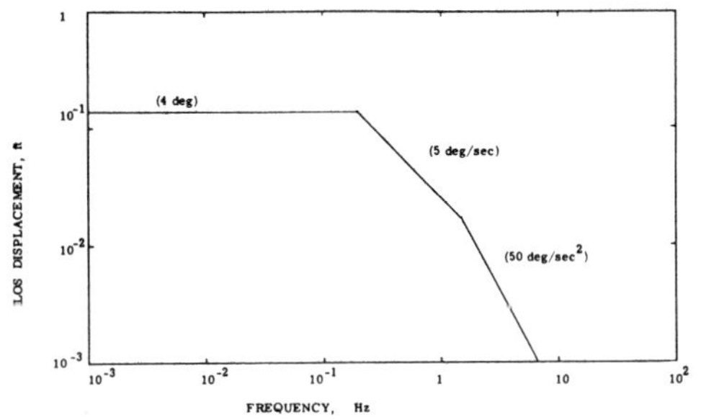

Fig. 15. Lever arm motion model.

Fig. 16. Lever arm compensation sensitivity.

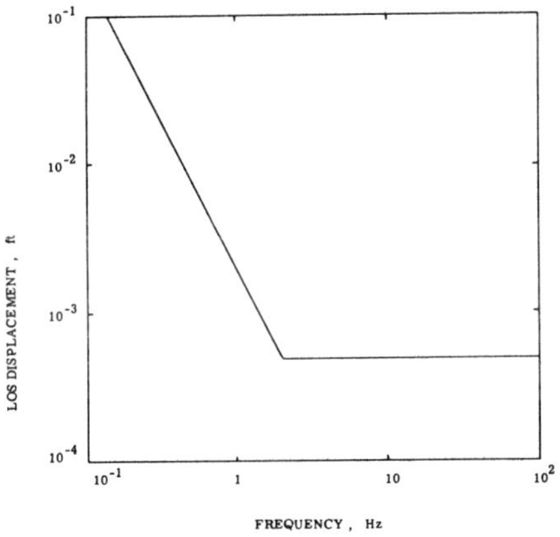

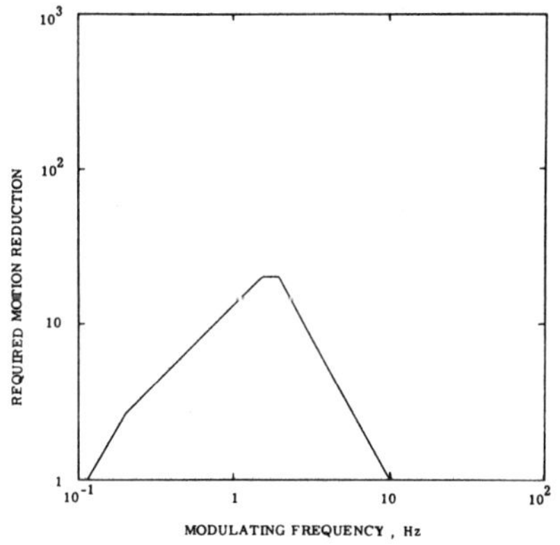

Fig. 17. Minimum lever arm compensation transfer curve.

Fig. 18. Required angular sensitivity.

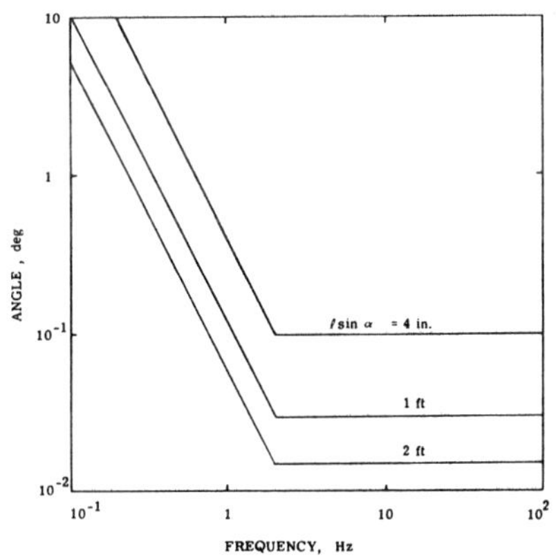

case occurs when the lever arm is perpendicular to the line of sight. The expected lever arm motion is obtained by letting α be the angular motion of Fig. 12. For a perpendicular displacement of two fect, the expected lever arm motion is shown in Fig. 15.

The required sensitivity is identical to that required for translational motion, and is

$$\Delta R = \Delta V/\omega = (\lambda/2\pi)(\mathrm{SL/ML}) \qquad \text{(high frequency)}$$

$$\Delta R = (\lambda/2\pi^2)\,(\Phi/180°)\,(1/\Delta T^2)\,(1/f^2) \qquad \text{(low frequency)}$$

This is plotted in Fig. 16, and the required minimum lever arm compensation transfer curve is shown in Fig. 17.

One additional factor should be mentioned in conjunction with the lever arm, and that is the required measurement accuracy on the rotation angles. The required angular sensitivity is given by

$$|\Delta\alpha| = \Delta l_{LOS}/l \sin\alpha \, .$$

This is plotted in Fig. 18 for various values of the perpendicular (to the LOS) displacement. Notice that, for a two-foot displacement, the required measurement sensitivity (not absolute accuracy) is 0.01 degree. This is why the inertial sensor and antenna must be located in close proximity.

The conclusion drawn from asking the question, "How much motion compensation is required?", is that the translational path is the most critical. The real antenna stabilization and lever arm path are about an order of magnitude less critical, though still very important. And, finally, the inertial sensor must be located near the antenna.

IV. Conclusions

Motion compensation is the key to isolating the SAR image from effects of aircraft motion. A uniform motion compensation approach can be implemented for all SAR modes using a unit pointing vector which inertially stabilizes the SAR processed image.

The amount of motion compensation which must be applied can be analyzed from the point of view of transfer

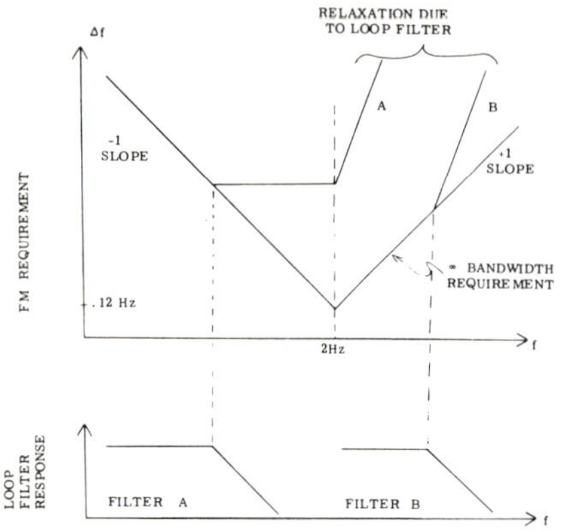

Fig. 19. Effect of loop filter on allowable FM modulation (VCO clutter tracking).

curves which are obtained by taking the ratios of expected aircraft motion to required sensitivity. A simplified approach for obtaining these curves has been presented and typical transfer curves for each of the basic motion reduction paths, i.e., translational, lever arm, and antenna stabilization, have been computed. For typical SAR parameters and typical values of aircraft motion, the required motion reduction is on the order of 100 to 1 for the translation path and 20 to 1 for the lever arm and antenna stabilization paths.

The motion reduction transfer curves were based mainly on sinusoidal motion caused by weather disturbances. Further analysis should be performed on adding the effect of aircraft maneuvers to the transfer curve. One example maneuver point was calculated in this paper.

The motion compensation equations are easily programed in a small standard minicomputer. A typical program will use less than 4 K of memory.

Appendix

Angular Stabilization Sensitivity Requirements

The effect of antenna motion is to amplitude-modulate the target signal. For a pencil beam with a shape which is approximately Gaussian, the two-way voltage gain is

$$G = e^{-a(\theta^2/\theta^2_{3\,\mathrm{dB}})} \qquad (a = 2.77).$$

An error in angle produces an error in gain, given by

$$G = (2a\theta/\theta^2_{3\,\mathrm{dB}})G\Delta\theta$$

and

$$\Delta G/G = (2a\theta/\theta^2_{3\,\mathrm{dB}})\Delta\theta.$$

If $\Delta\theta$ is sinusoidal, $\Delta G/G$ is sinusoidal and the result is a pair of echo sidebands of amplitude $\Delta G/2G$. Thereforc,

$$\Delta\theta = (\theta^2_{3\,\mathrm{dB}}/a\theta)(\mathrm{SL/ML})$$

where

$$\mathrm{SL/ML} = (1/2)\,(\Delta G/G)$$

is the sidelobe to main lobe requirement. Notice that for a target near the center of the antenna ($\theta = 0$), the required antenna stabilization is very relaxed. However, for a target at the edge of the antanne ($\theta = \theta_{3\,\mathrm{dB}}/2$), the requirement becomes more severe.

For azimuth motion, the effect is not severe, since the Doppler filters are at the center of the antenna and a target must be at the edge of the pattern to produce serious AM. In order to fall in a Doppler filter, the modulating frequency would have to be very high (~ 100 Hz), and this is not likely due to the antenna size. In elevation, however, due to the extent of the range swath, a target at the edge of the swath could be illuminated by the edge of the pattern and AM sidelobes could result. Taking $\theta = \theta_{3\,\mathrm{dB}}/2$, the elevation requirement is

$$\Delta E_{\mathrm{PK}} = (\theta_{3\,\mathrm{dB}}/a)2(\mathrm{SL/ML}).$$

For a 30-dB sidelobe requirement, this becomes

$$\Delta E_{\mathrm{PK}} = 0.06°.$$

To illuminate the swath properly the antenna should be pointed accurately to within a fraction of the beam width.

In clutter tracking, amplitude modulation can be coupled into Doppler modulation if the loop is closed through a VCO. The clutter tracker measures the center of the main beam clutter spectrum and any antenna motion is coupled to FM by the Doppler equation

$$\Delta f = (2V/\lambda)\sin\theta\Delta\theta.$$

This error can be compensated for by the loop filter, depending on the filter width. The required stabilization is therefore given by

$$\Delta\theta = \Delta f/[(2V/\lambda)\sin\theta].$$

For $V = 500$ ft/s, $\lambda = 0.1$ foot, and $\theta = 45°$, this becomes

$$\Delta\theta(\mathrm{deg}) = 0.008\Delta f.$$

The allowed FM, Δf, is illustrated in Fig. 19, showing the relaxation caused by narrowing the filter width. Using this value for Δf, the required antenna stabilization can be calculated.

Acknowledgment

Some of the work described in this paper was done with the help of discussion with R.C. Weber of the Westinghouse Electric Corporation.

John C. Kirk, Jr. (M'69) received the B.S.E.E. degree from North Carolina State University in 1964 and the M.S.E.E. degree from Marquette University in 1966.

He has been with Westinghouse Corporation, where he was involved in the design of synthetic aperture and MTI radars, and with the U.S. Air Force, where he performed flight-line maintenance of radar, navigation, and electronic countermeasures systems. He is now with the Advanced Systems Department of the Electromagnetic Systems Division of the Raytheon Company, Santa Barbara, Calif., where he is involved in the design of advanced systems and the application of digital processing techniques.

Mr. Kirk is a member of Tau Beta Pi, Phi Kappa Phi, Pi Mu Epsilon, and Eta Kappa Nu. He is a registered Professional Engineer in the state of Maryland.

CHAPTER VI
Processing for Synthetic Aperture Radar

The data processor is the "heart" of the synthetic aperture radar. True, the coherent RF radar is an essential instrument which is required to transmit and receive the signal; but, because of the long integration time over which data must be stored, it is only after the data is processed that the synthetic aperture is manifest.

The first synthetic aperture radar stored the coherent data on magnetic tape the loops of which were continuously and mechanically analyzed by a commercial frequency analyzer to extract the resolved Doppler signals and thereby yield radar imagery. Subsequently, processors such as a range-gated filter bank, recirculating delay line, and electronic storage tube integrators (see Kelly et al. (1964) and Sherwin, Ruina, and Rawcliffe in Chapter I) were only stepping stones towards the more fruitful processors. It was the optical processor system which made the synthetic aperture radar and its photo-like quality in radar mapping imagery a reality. And, it will be the digital processor which will extend SAR systems' versatile mapping types to near-real time applications in the future.

L.J. Cutrona, E.N. Leith, L.J. Porcello, and W.E. Vivian applied the techniques of coherent optical processing to synthetic aperture radar. The optical processor is an optical analog to the total SAR system; with its inherent multi-channel capability, it has provided an elegant and simple way to yield radar imagery. This paper describes the basic implementation of the optical processor. More importantly, the Fresnel-lens interpretations of the recorded radar signals are introduced as one-dimensional microwave holograms which have self-focusing properties. For a greater understanding of optical data processing and filtering systems, see Cutrona, Leith, Palermo, and Porcello (1960). The dynamic range characteristics of films which are the data storage media of optical correlators can be found in Nuttall (1966).

Initially, SAR systems transmitted uncoded radar pulses; the unprocessed data were azimuthal single line holograms or one-dimensional Fresnel zones. With the introduction of chirped transmitted pulses, the SAR data exhibited two-dimensional Fresnel zone plates. The techniques for optical processing – simultaneous pulse compression and beam-sharpening – are described by E.N. Leith in the second paper. He describes the optical processor configured as both a two-dimensional matched filter and a two-dimensional cross-correlator. The optical processor is a very versatile instrument and its use in optical simulations of radar ambiguities can be found in Leith, Friesem, and Funkhouser (1970). The ability of the optical processor to accomplish radar-Doppler imaging with motion through resolution cells is described in Brown and Fredricks (1969). A short discussion of the images of a flat plate by a SAR as related to the processing of parts of the one-dimensional Fresnel zone is found in Johansen (1970). A Fresnel zone-plate processing system which uses a flying spot scanner to read the radar signal film is found in Moore and Rouse (1967). An optical analyzer is described in Revillon (1965).

In the third paper, W.M. Brown, C.G. Hauser, and R.E. Jenkins consider the method of presumming a number of received pulses before processing. Some SAR systems over-sample the Doppler spectrum, thereby taxing the storage capabilities of core or magnetic drums, for example. Pre-summing to match the limited storage is a well-established method to meet the desired design constraints. Another approach to reducing the storage requirement in a SAR is suggested in Clarke (1971) – which is the spatial domain version of the chirp pulse compression technique known as *STRETCH.*

The modern approach of employing digital signal processors in SAR systems is discussed by J.C. Kirk, Jr. in the fourth paper. He develops the digital processor design using digital correlation algorithms, such as multiple parallel pre-filters, two-stage correlation, and Fast Fourier Transform (FFT) processing. Trade-offs are made between functional complexity and reduced requirements for arithmetic and memory. Considerations of digital SAR technology are found in Kirk (1975); for more discussions of parallel processing using an associative memory, see Ellis (1973).

In addition to work on the processing of SAR data directed to applications where near real-time use of the radar imagery is a requirement, processing facilities for in-depth interpretation are being established. The interested reader can learn about a signal processing and image exploitation facility from Ausherman et al. (1975). This facility utilizes coherent optical, hybrid optical-digital, and wholly digital processing technologies to generate and effectively display SAR imagery. Also, an interactive SAR interpretation facility is described in Hermann, Kerr, and Tammy (1975). Raw SAR data is collected by an airborne wideband digital tape recorder; a dual-mode digitizing photometer/densitometer digitizes film recorded data; raw radar data may be processed in either an optical correlator or a digital computer; image manipulation is under interactive computer control; and final image output is on a pseudo-color TV display.

On The Application Of Coherent Optical Processing Techniques To Synthetic-Aperture Radar

L.J. Cutrona, E.N. Leith, L.J. Porcello, and W.E. Vivian
(Cutrona – Conductron Corporation and the University of Michigan, Ann Arbor; Leith and Porcello – University of Michigan, Ann Arbor; Vivian – US Representative from the 2nd Congressional District in Michigan [formerly with the University of Michigan and Conductron Corporation])

Proceedings of the IEEE, Vol. 54, No. 8 (August 1966)
Pages 1026-1032

Abstract—Coherent optical systems are useful for the processing of data collected by sidelooking synthetic-aperture radars. This type of radar, which is useful for generating fine-resolution terrain imagery, has the capability of providing along-track, or azimuth, resolution which is much finer than the azimuth dimension of the radiated beam itself; however, the attainment of this finer resolution requires extensive processing of the received radar signal. In this paper the form of the returning radar signals is discussed, as are the general reasons why optical processing of these signals might be attractive. The required processing can be viewed as a multichannel weighted vector summation, matched filtering operation, or cross-correlation operation; the cross-correlation viewpoint is developed, as is a corresponding optical processor implementation. The role of a conical lens in simplifying reference-function realization is discussed. Finally, a Fresnel-lens interpretation of the recorded radar signal leads to a discussion of the strong physical analogy between a coherent radar system and a coherent optical system.

Introduction

One of the most successful applications of coherent-optics technology to the processing of data generated by an electronic system is to be found in the synthetic-aperture radar system. Within the synthetic-aperture class of radar, a data processor is utilized to convert stored video data into a fine-resolution radar image; it is our purpose in this paper to discuss the manner in which a coherent optical system may be used to carry out the particular linear operations required in such a radar.

Consider a sidelooking, terrain-imaging radar system which is borne by an aircraft in level flight (Fig. 1). Let us term the cross-track coordinate of the radar image obtained by such a radar the "ground-range" coordinate, and the along-track linear coordinate the "azimuth" coordinate. Similarly, it will be convenient to define a "slant-range" coordinate, in the plane joining the radar trajectory and any target under consideration, again in a direction normal to the aircraft track. If the radar is of the conventional variety, then its azimuth resolution at any range is determined by the azimuth dimension of the physical radar beam, measured at that range; i.e., if λ is the wavelength and if a physical aperture of along-track dimension D is employed, then an azimuth resolution of the order of $r\lambda/D$ at range r is available. Because radar wavelengths are several orders of magnitude larger than optical wavelengths, very large values of D must be used if conventional radars are to realize angular resolutions $\beta \approx \lambda/D$ comparable with photo-reconnaissance systems. If one wishes to have fine (linear) azimuth resolution at long ranges, the required antenna length can be of the order of hundreds or even thousands of feet, obviously incapable of being carried by aircraft.

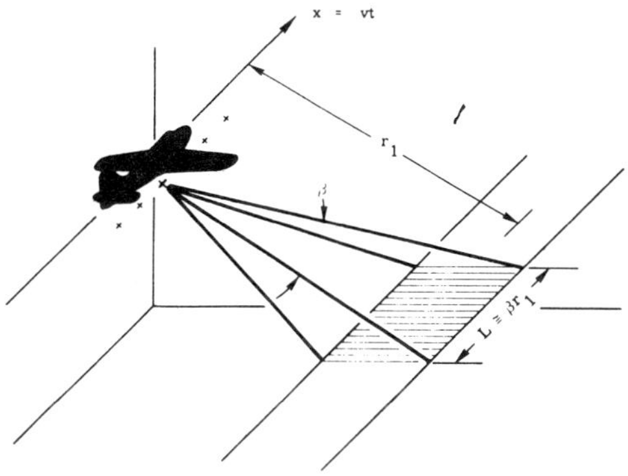

Fig. 1. Sidelooking radar configuration.

The synthetic-aperture technique offers a way around this apparent impasse: The aircraft carries a small, sidelooking antenna, producing a beam that is relatively wide in the azimuth direction, which scans the terrain by virtue of the aircraft motion. The antenna is carried by the aircraft to a sequence of positions which can be treated as if they were the positions occupied by the individual elements of a linear antenna array. At each such position, the antenna radiates a pulse, then receives and stores the reflected signal. These stored data are then processed in a manner analogous to the coherent weighted summation carried out in a large linear array. The processed radar signals bear a quantitative similarity to those which would be obtained if a large antenna were used; in particular, the resolution and the signal-to-noise ratio are greatly improved by the signal processing. Synthetic-aperture radars may well become the most

Manuscript received December 20, 1965; revised May 9, 1966. This paper was prepared under the joint sponsorship of the U. S. Army Electronics Command through Project MICHIGAN, Contract DA 28-043-AMC-00013(E), and the Air Force Avionics Laboratory, Wright-Patterson Air Force Base, Ohio, Contract AF33(615)-1452. The technical concepts and achievements discussed herein have evolved over a number of years, under the sponsorship of Army and Air Force contracts which have culminated in the two mentioned above.

important of the terrain-imaging types of radars, since they have an azimuth-resolution potential at long ranges which is superior to that of other types of imaging radars.

An introductory discussion of synthetic-aperture radar systems was given by Cutrona et al.[1] in 1961. Reference was made in that paper to the use of analog processing of the video return to generate a fine-resolution image, but no further details regarding the processing were provided. The early development of synthetic-aperture radars was recounted in a paper by Sherwin, Ruina, and Rawcliffe[2] published in 1962, again without discussion of optical processing techniques applicable to this task.

The rather straightforward theory underlying the use of coherent optical systems to carry out linear integral transforms was discussed in detail in 1960 in an independent paper by Cutrona et al.[3] It was shown in that paper that, through use of suitable combinations of spherical and cylindrical optical elements, one could synthesize a multichannel optical analog computer capable of realizing an independent one-dimensional integral transform operation in each of its many channels; the one-dimensional imaging process itself, necessary to preserve channel separation, was shown to consist of a sequence of two one-dimensional Fourier transformations. Similarly, one could eliminate all cylindrical elements and obtain a system which would carry out all its transform operations in both spatial variables rather than in one alone.

The mating of these two techniques, that of the synthetic-aperture and that of coherent-optical analog processing, is the primary topic to be discussed in this paper. Experience has shown that coherent optical processing is eminently well suited to the processing tasks which arise in synthetic-aperture radar systems.

Finally, a clarifying remark must be included in this introduction. The synthetic-aperture falls into the array class of antenna; array-type antennas add, or integrate, the returns realized on each of the array elements. One can readily generate synthetic apertures which are so long that they will only realize their full gain and azimuth-resolution capabilities if they are focused at the range at which the radar is operating. The focusing operation represents a phase adjustment of the signal received on each element of the array so that, in the summation process, the contributions from all array elements are combined in phase. We will restrict ourselves to the discussion of focused synthetic-aperture radars for the remainder of this paper.

[1] L. J. Cutrona et al., "A high-resolution radar combat-surveillance system," *IRE Trans. on Military Electronics*, vol. MIL-5, pp. 127–131, April 1961.

[2] C. W. Sherwin, J. P. Ruina, and R. D. Rawcliffe, "Some early developments in synthetic aperture radar systems," *IRE Trans. on Military Electronics*, vol. MIL-6, pp. 111–115, April 1962.

[3] L. J. Cutrona et al, "Optical data processing and filtering systems," *IRE Trans. on Information Theory*, vol. IT-6, pp. 386–400, June 1960.

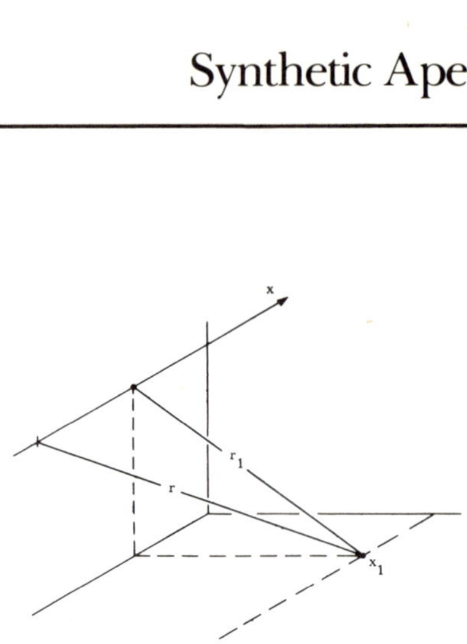

Fig. 2. System geometry.

The Radar Signal

As shown in Fig. 1, an aircraft carries an antenna which illuminates a ground-swath parallel to the flight path; the radar beam is oriented in azimuth roughly normal to the direction of flight; at range r, the azimuth lineal beamwidth βr is much larger than the desired azimuth resolution at that range. Finally, the radar is "coherent;" that is, the receiver has available a reference signal from which the transmitted signal was derived.

Synthetic-aperture radars derive range information through pulsing, and derive fine azimuth resolution by processing Doppler-shifted radar returns which lie in a spectral band which is adequately sampled by the pulse rate of the radar. For our purposes, we may neglect this intermediate sampling process and consider only the reconstructed azimuth histories which are easily derived from the samples. We will assume that the entire radar receiver and processor behave as a linear system; we can then investigate the response of the radar to a single "point" target,[4] and by superposition extend the results to apply to realistic reflective complexes.

To find the form of the signal returned to the radar antenna, consider a point target located at a position x_1 along the ground track, at a range r_1 from the aircraft flight path (Fig. 2). Let a sinusoid of radian frequency ω be transmitted. The signal returned to the aircraft then has the form

$$s_1(t) = \sigma_1 \exp\left\{j\left[\omega\left(t - \frac{2r}{c}\right)\right]\right\} \tag{1}$$

where r is the range from aircraft to target, c is the velocity of light, and σ_1 is the complex amplitude of the signal as received by the radar. Into σ_1 we have absorbed such factors as the radiated power level, the target reflectivity and phase shift, and the inverse fourth-power law. By use of the approximation

$$r = \sqrt{r_1^2 + (x - x_1)^2} \approx r_1 + \frac{(x - x_1)^2}{2r_1} \tag{2}$$

[4] "Point" targets are defined to be those targets which have physical dimensions that are small compared with the resolution capability of the radar system.

(1) can readily be written as

$$s_1(t) = \sigma_1(x_1, r_1) \exp \cdot \left\{ j \left[\omega t - \frac{4\pi r_1}{\lambda} - \frac{2\pi}{\lambda} \frac{(x - x_1)^2}{r_1} \right] \right\}. \quad (3)$$

The motion of the aircraft links the temporal behavior of the signal with the spatial coordinates of the target, through the relation

$$x = vt$$

where we have chosen the origins of both systems so that $x=0$ at $t=0$. If the terrain at range r_1 is regarded as a collection of numerous point scatterers, the return signal is of the form

$$s = \sum_n s_n(t) = \sum_n \sigma_n(x_n, r_1) \exp \cdot \left\{ j \left[\omega t - \frac{4\pi r_1}{\lambda} - \frac{2\pi}{\lambda r_1} (vt - x_n)^2 \right] \right\} \quad (4)$$

and after synchronous demodulation, the signal assumes the form

$$s(t) = \sum_n | \sigma_n(x_n, r_1) | \cos \cdot \left[\omega_c' t - \frac{4\pi r_1}{\lambda} - \frac{2\pi}{\lambda r_1} (vt - x_n)^2 + \phi_n \right] \quad (5)$$

where ω_c' is an arbitrary carrier frequency remaining after synchronous demodulation.[5]

Some means of storing this return signal is now needed. The reason is apparent: In a long physical-array system, all the elements coexist in time, and their outputs can appropriately be phased, weighted, and summed without intermediate storage. However, in a synthetic-aperture system, the single radar antenna must sequentially play the role of each element; therefore, sufficient storage capacity is necessary to permit the antenna to occupy an adequate sequence of positions before summing.

The returning radar signals will have a bandwidth determined by the range-resolution capability built into the radar system. A photographic process permits recording of wide-band signals with greater ease than is possible with certain other recording systems; consequently, it is well suited for use with synthetic-aperture radars, provided one can tolerate some of the film-processing inconveniences. Photographic recording also permits high-density storage, another quality which is desirable in fine-resolution imaging systems.

The radar receiver output is a sequence of reflected range pulses; these are used to intensity-modulate a cathode-ray tube, the electron beam of which is swept in synchronism with the returning pulses. Successive range traces are recorded side by side, producing a two-dimensional format in which the dimension across the film represents range, and the dimension along the film corresponds to the along-track dimension. This format is shown in Fig. 3. The signals in the along-track dimension are the sampled versions of $s(t)$ as given by (5); the sampling is carried out in such a way that, by the time the samples have been recorded on film, the sampled version is essentially indistinguishable from the unsampled version. In the recording process, the time variable is converted to a space variable defined in terms of distance along the film. The recording process itself introduces a pair of scale factors, one in azimuth and one in range, into the data record; however, for our purposes, all relationships are handled in "radar space," and the presence of these scale factors is neglected. The radar signal is also placed on a bias level s_b prior to recording; thus the recorded signal is now in a form which is a spatial analog to that given by (5)

$$s_b + s(x) = s_b + \sum_n | \sigma_n(x_n, r_1) | \cos \cdot \left[\omega_x x - \frac{4\pi r_1}{\lambda} - \frac{2\pi}{\lambda r_1} (x - x_n)^2 + \phi_n \right] \quad (6)$$

where

$$\omega_x = \omega_c'/v.$$

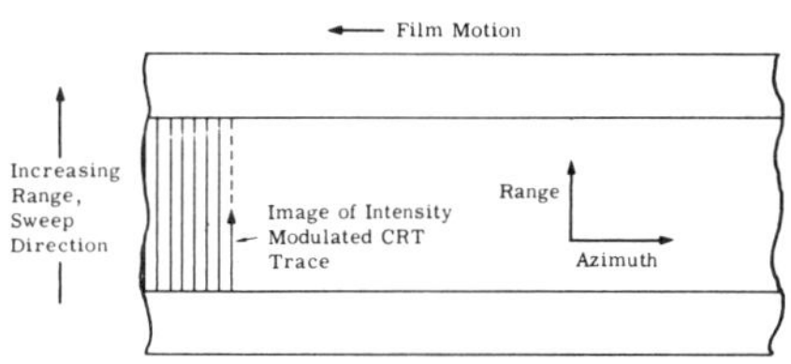

Fig. 3. Radar signal storage format.

Reasons for Processing Optically

There are several methods for carrying out the data-processing operation which is required to generate a fine-resolution radar image from the raw radar data. Coherent optical processing is one method which is particularly well suited to this task for the following reasons.

a) The video data are generally stored on photographic film because of its large storage capacity; therefore, the video data record is available for direct insertion into an optical channel.
b) Because the weighted vector-summation in azimuth is range dependent, each range interval must be processed differently; this creates a multichannel requirement. Optical systems, because of their two-dimensional nature, are readily adaptable to multichannel configurations through the use of anamorphic lens elements.
c) The required processing in azimuth can be viewed as a weighted vector-summation operation, a

[5] Equations (3), (4), and (5) above omit amplitude weighting which is inserted by the azimuth cross section of the radiated beam; this weighting is not essential to the present discussion.

matched-filter operation, or a cross-correlation operation. For which ever way it is viewed, there is a certain configuration of optical elements which, when coherently illuminated, automatically will carry out a linear integral-transform operation that is equivalent to it.

d) Finally, there is a striking similarity between the coherent radar system which acquires the data, and the coherently illuminated processor which converts the data to a radar image. The coherent optical system, therefore, fits its application in a very natural way, and it tends readily to perform precisely those operations which are required of it. Hence, the optical system is both simple and elegant. This analogy will be developed later in this paper.

The Required Processing

We will treat the required processing as a cross-correlation operation, although we will keep in mind the complete equivalence among the three viewpoints, as mentioned. The radar video from a particular range r_1 is cross correlated with an appropriate reference function, which is a replica of the expected return from a point target at the same range. The resulting operation is

$$\chi(x') = \int_{-L/2}^{L/2} \{[s_b + s(x + x')]f(x)\}dx \tag{7}$$

where

$$f = \cos\left(\omega_x x - \frac{2\pi}{\lambda}\frac{x^2}{r_1}\right)$$

is the two-dimensional reference function, and L is the aperture over which the integration takes place; L is numerically equal to the length of the synthetic aperture which is to be generated. When this operation is carried out for the case in which the σ's are constant, we obtain

$$\chi(x) = \frac{1}{2}\sum_n L\,|\,\sigma_n(x_n, r_1)\,|\,\text{sinc}\left[\frac{2\pi L(x' - x_n)}{\lambda r_1}\right]\cos \cdot\left[\omega_x x' - \frac{4\pi r_1}{\lambda} - \frac{2\pi(x' - x_n)^2}{\lambda r_1} + \phi_n\right] \tag{8}$$

where sinc $x = \sin x/x$ and where the bias term and terms in $2\omega_x$ have been ignored. Here, the signal from each point target has been converted into a sinc function with amplitude proportional to the strength of the signal reflected from that target, and centered about a value of x corresponding to the location of the target. The width of the sinc envelope is exactly that which corresponds to the resolution available from an aperture of $2L$. Thus, the resolution is better by a factor of 2 than that which would result from a real antenna of similar aperture. This factor is characteristic of synthetic apertures, but we will not describe its origin here. The cross-correlation process has separated the overlapping signals, compressing each to a diffraction-limited spot. The result of this process is a high-resolution image similar to that which one would obtain by using a large antenna.

Implementation of the Processor

The basic optical system used to carry out the multichannel data processing is shown in Fig. 4, and was described in a previous paper by the authors.[3] The general operation performed by this optical system is

$$\chi(y) = \int f_1(\alpha, y)f_2(\alpha, y)d\alpha \tag{9}$$

where the variable of integration is α, and y is a variable which takes on a range of discrete values, each corresponding to an independent data channel. The functions f_1 and f_2, written as transparencies, are placed in planes P_1 and P_3, respectively, and the imaging of one transparency onto the other effects the multiplicative operation given in (9).

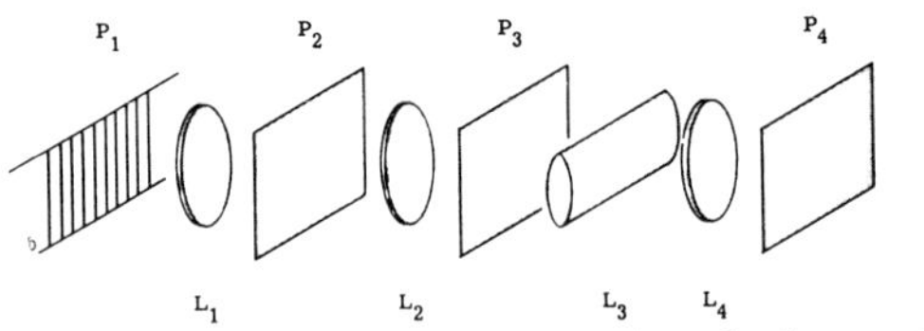

Fig. 4. Basic optical system for processing radar data.

The integration process is carried out between planes P_3 and P_4. The basis for performing this operation arises from the spatial Fourier transform relation that exists between the complex light-amplitude distributions at the front and back focal planes of a lens; it is assumed that P_3 and P_4 are, respectively, these planes of lens L_4. If f_1f_2 is the light-amplitude distribution at P_3, then the light-amplitude distribution at P_4 is given by

$$\chi(x, y) = \int f_1(\alpha, y)f_2(\alpha, y) \exp\left\{j\left(\frac{\pi}{\lambda F}\alpha x\right)\right\}d\alpha \tag{10}$$

where F is the focal length of the lens, and λ is the wavelength of the light. The cylindrical lens L_3, by imaging plane P_3 onto P_4 in the y-dimension, causes the Fourier transformation to occur only in the x-dimension. If the observation is made along the line $x = 0$, and if provision is made for moving one of the transparencies through the aperture, the above equation becomes

$$\chi(x', y) = \int f_1(\alpha - x', y)f_2(\alpha, y)d\alpha \tag{11}$$

which is a cross correlation of f_1 with f_2.

One important point must be made here. The functions f_1 and f_2 are light-amplitude transmission functions having two independent spatial variables. The magnitude-squared of each one would normally be termed its intensity transmission. Each f has a magnitude and phase associated with it; in practice, however, it is often preferable to generate a two-dimensional density pattern with uniform phase. In such a situation, f_1 and f_2, if bipolar, must be written about a bias term, since only positive values of transmittance can be obtained in the uniform-phase case. Thus, in practice, (11) assumes the form

$$\chi(x', y) = \int [f_{b_1} + f_1(\alpha - x', y)][f_{b_2} + f_2(\alpha, y)]d\alpha \quad (12)$$

where f_{b_1} and f_{b_2} are the bias levels needed to allow all negative values of the bipolar functions f_1 and f_2 to be recorded.

Expansion of the integrand of (12) yields four terms, three of which are extraneous. However, these three terms can be eliminated by utilizing the Fourier transform properties of the optical channel. First, the functions f_1 and f_2 will be made bandpass functions via their modulation onto a carrier which is sufficiently large to ensure that f_1 and f_2 have only negligible dc components. Second, the function $f_{b_1}+f_1$ is spatially filtered at plane P_2 to remove the dc bias term f_{b_1}. This removes two of the three extraneous terms, while the remaining one, the correlation of f_1 with f_{b_2}, can be made negligible, since f_1 and f_{b_2} can be made to occupy essentially nonoverlapping frequency bands.

Adaptation of the Optical System to Radar Data Processing

Equation (12) is of the same form as the required operation on the radar data, which was given by (7). The recorded signal $s_b+s(x, y)$ corresponds to $f_{b_1}+f(x, y)$; it is moved through the aperture at plane P_1, filtered at P_2 so as to remove the bias term s_b, and then imaged onto the reference function. A slit is placed along the line $x=0$ at the output plane, and a continuously moving photographic film records the illumination on the slit as the signal film moves through the input aperture. Thus a fine-resolution image is produced from the raw data film.

The optical system, by virtue of its two-dimensional nature, carries out a separate cross-correlation process for each range interval, thus producing a synthetic aperture that is effectively focused at all ranges simultaneously.

Simplification of the Reference Function

The two-dimensional reference function

$$\cos\left(\omega_x x - \frac{2\pi}{\lambda}\frac{x^2}{r}\right)$$

assumes, for each value of r, the proper form for cross correlation with the signal from that range. The reference function can be synthesized as a two-dimensional transparency mask; it is superimposed on a bias term f_b, and the range variable r is represented by the variable y used previously.

It is difficult to construct a shaded transparency which has a light-amplitude transmissivity that is given by the two-dimensional reference function. It is much easier to construct a transparency that is quantized to two levels: opaque and transparent. Such a substitution is acceptable, because such a mask will have, as the fundamental component of its Fourier series expansion, just the sinusoidal function which is needed. The higher harmonics can be removed by spatial filtering; but, even if they are not removed, they will not correlate with the signal.

The recorded signal $s(x, r)$ is a real function; therefore, its Fourier spectrum has an upper and a lower sideband (positive and negative frequencies) which contain identical information and are thus redundant. Hence, it is acceptable to pass only one of the two sidebands in the spatial-filtering operation which is carried out in plane P_2 (Fig. 4). After such filtering, the signal from each point target has the complex form

$$\exp \pm j\left[\omega_x x - \frac{2\pi}{\lambda r_1}(x - x_1)^2\right],$$

where the exponential has either the (+) or the (−) sign, depending on the sideband selected. This sideband will correlate with the conjugate component of the reference function and produce, as before, a fine-resolution radar image.

The significance of the single-sideband method is that it is possible, because of a fortuitous circumstance, to synthesize the reference function in the single-sideband form

$$\exp j\left(\omega_x x - \frac{2\pi}{\lambda r}x^2\right),$$

and it is highly advantageous to do so.

There are two advantages to a synthesis of the reference function in this form. First, the multiplication of signal and reference functions can be carried out by directly superimposing the two, rather than by imaging one onto the other. The imaging process was required in order to facilitate removal of the signal bias term by spatial filtering. However, if one synthesizes the reference function directly as a complex function, without the bias term, there is no need for removal of the signal bias. Two lenses and a spatial-filtering operation are thereby removed, and the optical system is simplified and shortened.

The other advantage is that the light-utilization efficiency of the system is greatly increased. The complex mask ideally transmits all of the light incident on it, while the other mask transmits only about $\frac{1}{2}$ of the incident light, the greater part of which corresponds to the bias term f_b. By the time all unwanted spatial components of the light are discarded, the amplitude-type mask has an effective insertion loss of 10 to 13 dB.

The complex mask is formed by figuring a piece of glass so that its thickness at any point is such as to produce the phase retardation indicated by the two-dimensional argument of the exponential. In general, such phase plates can seldom be fabricated; however, in this particular instance, the fabrication is possible.

For any particular value of the range variable r, the complex phase plate can be regarded as a cylindrical lens with curvature in the x-dimension and with focal length proportional to r. The plate, which can be obtained by slicing a pie-shaped section from an axicon-type lens,[6] is called a conical lens.

Fresnel-Lens Interpretation of Recorded Signals

That the reference function against which we correlate the radar signal is a lens carries a significant implication: a lens does not create images but only transports an image to a new position. The fine-resolution image which we seek to construct ought, therefore, to exist even if the conical lens and the other lenses are removed. This is indeed the case; it can be understood most readily by recognizing that the recorded signals are one-dimensional Fresnel zone-plate lenses. Illumination of the signal produces three diffracted waves, which emanate in different directions, as shown in Fig. 6. One wave, corresponding to the dc term, emerges from the signal transparency as an attenuated duplicate of the incident light. Another wave, corresponding to the primary real focus of the lens-like signal, produces a real image. A divergent wave is propagated in a third direction and produces the primary virtual focus of the signal. The real and the virtual image each independently represent the high-resolution image. These images are, in fact, exactly the fine-resolution images that are obtained by carrying out the cross-correlation process described earlier.

The Radar Optics Analogy

The production of a fine-resolution image of the radar field by mere illumination of the signal record with a coherent light beam does not imply that the associated optics and the reference function are unnecessary; they are indeed necessary, but their roles in the system take on a new interpretation.

The diffracted waves produced by illuminating the recorded radar signals are essentially optical replicas, on a reduced scale, of the radar waves that impinged on the radar receiver moving along the flight path; the image produced by the diffracted waves is an optical reconstruction of the reflectivity distribution of the original target field. Just as the target field in the radar system lies in a tilted position with respect to the aircraft flight path, so the reconstructed image is tilted with respect to the signal film record; i.e., signals originating from targets at near ranges produce signal records that have shorter focal lengths.

These focal properties apply only in the azimuth dimension. In the range dimension, resolution is obtained by pulse modulation of the radar, and resolution in

[6] J. McLeod, "The axicon: a new type of optical element," *J. Opt. Soc. Am.*, vol. 44, pp. 592–597, August 1954.

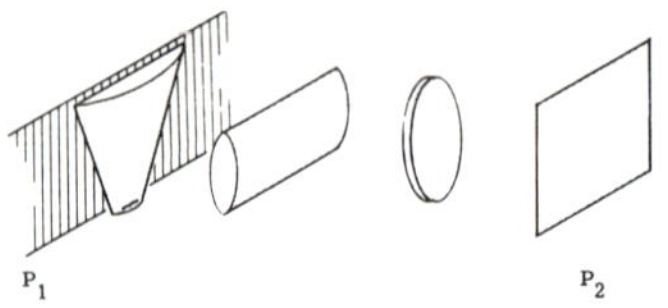

Fig. 5. Optical processor using a conical lens.

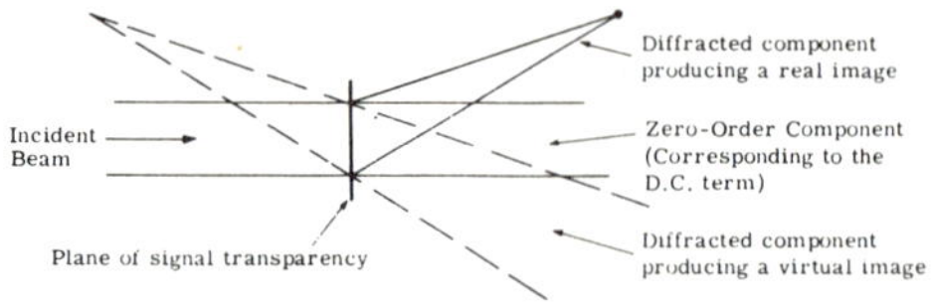

Fig. 6. The three diffracted components.

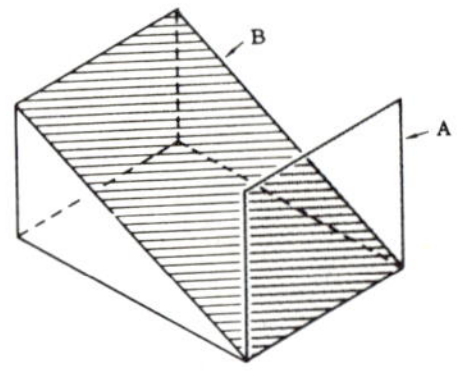

Fig. 7. Geometry of optical reconstruction of the radar field. Plane *A* is that of the signal record, and also the focal plane of the range data; Plane *B* is the azimuth-dimension focal plane.

range is therefore present on the film record itself. It is necessary only to preserve this range resolution throughout the optical processing. Thus, the locus of azimuth focus is a plane tilted with respect to the optical-system axis, while the plane of range focus is the signal record itself. An anamorphic lens system is required to bring these two planes, which are illustrated in Fig. 7, into coincidence. The three lenses of Fig. 5, one spherical, one cylindrical, and one conical, accomplish this. The conical lens, having a focal length equal and opposite to that of the signal film, operates on the azimuth focal plane and moves it to infinity. The azimuth focal plane thus becomes erected. The cylindrical lens operates only in the range dimension and images the signal film plane at infinity. The spherical lens, operating in both dimensions, takes the image at infinity and reimages it at its focal plane, where the image is now sharply focused in each dimension.

This viewpoint represents a considerable departure from the cross-correlation viewpoint which opened our discussion and upon which most of the development work was based; nonetheless, both views are equivalent in their mathematical formulation.

Experimental Results

An example of radar imagery obtained from optically processed data derived from a synthetic-aperture radar is shown in Fig. 8. The radar image, which is of the Monroe area, located south of Detroit, Mich., shows a variety of targets, including city, wooded areas, and

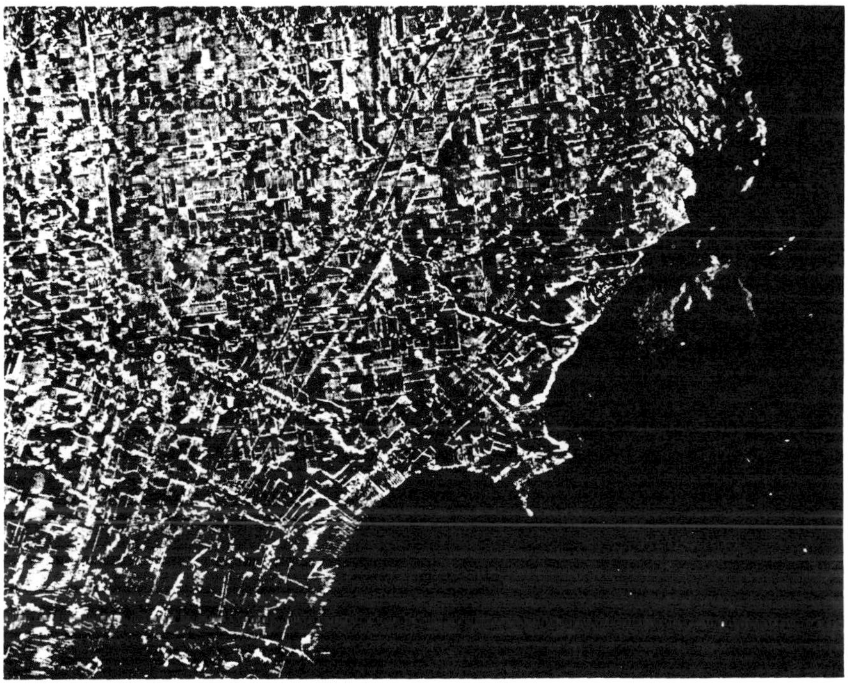

Fig. 8. Synthetic-aperture radar image of Monroe, Mich., area.

farmland. Highways and other landmarks are clearly visible. Lake Erie, with broken ice floes, is seen on the right side of the picture.

Concluding Remarks

It has been shown that a coherent optical-processing system quite naturally carries out the processing operations required in a coherent synthetic-aperture radar system. Initially we approached the application of optics to the problem from the viewpoint of a data-processing implementation which was required to provide a multichannel capability. The resulting processor, upon further analysis, was seen to behave as an analog to the total radar system and, doubtlessly, the elegant and simple manner in which the optical system serves its purpose must be attributed in no small way to this close parallelism. The reader can readily conceive of nonoptical implementations for the data-processing system; these will be found to be highly complex in comparison with the optical one. Certainly, the requirement of carrying out a separate cross correlation for each of a great multiplicity of channels constitutes a big problem, for which one has no right to expect a simple implementation; that the optical approach does indeed yield one is very gratifying and should spur the search for new vistas for coherent-optics technology.

Acknowledgment

The development of the synthetic-aperture radar concept was carried out by many persons, too numerous to list in entirety. A few of the most significant contributors were C. Wiley, W. Hausz, C. W. Sherwin, J. P. Ruina, D. Sunstein, and J. Koehler. Sherwin et al.[2] gives a more complete listing. The major contributions of the authors to synthetic-aperture radar technology are the use of optical processing and incorporation of focusing of the synthetic aperture at all ranges, which the former made practical. We also wish to acknowledge the contributions of the late R. Varian, who participated in the initial discussions that led to the use of optical processing, as well as those of A. Ingalls, who directed the design and construction of the optical processors that were built in accordance with the concepts presented here.

Optical Processing Techniques For Simultaneous Pulse Compression And Beam Sharpening

E.N. Leith
(Willow Run Laboratories, University of Michigan, Ann Arbor)

IEEE Transactions on Aerospace and Electronic Systems, Vol. AES-4, No. 6 (November 1968)
Pages 879-885

Abstract

A technique is described in which the separate techniques of beam sharpening (by synthetic antenna methods) and pulse compression are converted into a single two-dimensional operation, which is carried out with a coherent optical processor.

I. Introduction

Coherent optical data-processing methods have played a significant role in synthetic antenna radar technology. A coherent optical system for processing synthetic aperture radar data was described earlier [1]. The power of the optical system lies in its two-dimensional nature, which permits either the processing of two-dimensional signals or the simultaneous processing of a large number of one-dimensional signals. The latter is accomplished by using the additional dimension to stack a multiplicity of signals, which are then maintained in separation throughout the optical system by the use of cylindrical optics.

Use of an optical system for processing two-dimensional signals is hampered by the fact that two-dimensional signals, outside the optical domain, are uncommon; signals to be processed are generally of electrical origin, and being temporal, are thus one-dimensional. In situations where the electrical signal has been derived from the scanning of an inherently two-dimensional signal, as occurs, for example, in the case of an aerial reconnaissance radar system, the signal can be stored on a two-dimensional medium in such a manner that its original two-dimensional nature is restored. The data may then be processed as a two-dimensional structure if such processing is required.

An important example occurs when a pulse-compression technique is employed in a synthetic-antenna beam-sharpening radar system. The operations of pulse compression and beam sharpening are normally separate operations, carried out in widely separate portions of the radar system. By recording the uncompressed pulses photographically in a form for insertion into an optical processor, the two operations may be treated as orthogonal aspects of a single two-dimensional operation. An optical system, because of its inherent capability for treating two-dimensional signals, can readily perform the combined operation [2].

Manuscript received June 13, 1968.

The work reported in this paper was conducted by the Willow Run Laboratories of the Institute of Science and Technology for the U.S. Army Electronics Command under Project Michigan and subsequent development was funded with joint support from the U.S. Air Force.

II. The Stored Signal

A waveform cos $[\omega t-\phi(t)]$ is radiated from a coherent side-looking radar, where $\omega/2\pi$ is the radio-frequency carrier and $\phi(t)$ is the pulse-coding term. For convenience, we assume unity amplitude. A point reflector a distance r_1 from the aircraft returns the signal

$$\sigma \cos\left[\omega\left(t-\frac{2r_1}{c}\right)+\phi\left(t-\frac{2r_1}{c}-nT\right)\right], \quad (1)$$

where σ is a measure of the target cross section and contains other appropriate factors, such as the round-trip attenuation (we ignore, for simplicity, the x-dependence of σ), n designates the nth pulse, and T is the pulse repetition period. Successive pulses are recorded adjacently; thus, a two-dimensional signal record is constructed, with range as one coordinate (e.g., distance across the film)

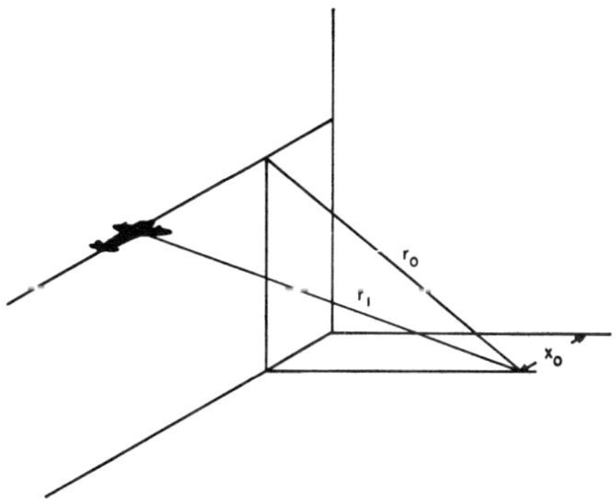

Fig. 1. Radar system geometry.

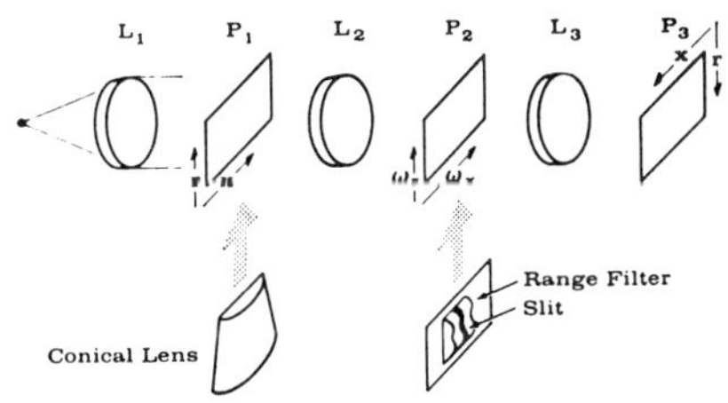

Fig. 2. Matched filter technique for optical beam-sharpening pulse compression.

and the along-track or azimuth dimension as the other coordinate. Such a record, for an isolated point target, has the form

$$s_b + \sigma \cos\left[\omega_c x + \frac{k(x-x_0)^2}{r_0} + \phi\left(r - r_0 - \frac{1}{2}\frac{(x-x_0)^2}{r_0}\right)\right] \quad (2)$$

where

s_b is a bias term, introduced to permit photographic recording of negative values of the signal,
ω_c is a spatial carrier term,
$k=2\pi/\lambda_R$, with λ_R the wavelength of the transmitted signal,
x_0, r_0 are the coordinates of the target,
x, 0 are the coordinates of the aircraft.

Note that $r_1^2=r_0^2+(x-x_0)^2$, as shown in Fig. 1. The usual approximation $r_1=r_0+\frac{1}{2}(x-x_0)^2/r_0$ has been made in (2) and constant phase terms dropped.

Equation (2) shows that the pulse-coding term ϕ contains the factor $(x-x_0)^2$; this represents a coupling between the beam-sharpening and pulse-coding terms, and imposes a limit on the combined azimuth and range resolution attainable, unless special compensation is provided. For the present, we ignore the coupling term.

The required operation in the range coordinate can be formulated either as a correlation of the data record with a stored replica of the radiated waveform or as a linear filtering operation with a matched filter, which is the Fourier transform of the stored replica. The viewpoints are, of course, fully equivalent; however, the implementations are different in the two cases. Similarly, the operation in the aximuth or x-dimension can be formulated as a cross correlation with a reference function having the form of an ideal point target or as a matched-filtering operation.

III. Implementation of an Optical Processor: The Two-Dimensional Matched Filter

The two-dimensional matched filter optical processor, using coherent illumination, is shown in Fig. 2. The signal is inserted at plane P_1 and, at P_2, a distribution of light representing the Fourier transform of the P_1 signal is displayed. At plane P_2, we place a mask containing a transmittance function representing the matched filter for the signal. Lens L_3 then takes a second transform and the filtered signal is displayed at the image plane P_3.

The Fourier transform of the signal given by (2) is

$$F\{s_b + s(x,r)\} = s_b[\delta(\omega_x,\omega_r) + s_1(\omega_x,\omega_r) + s_2(\omega_x,\omega_r)], \quad (3)$$

where δ is a delta function,

$$s_1 = \frac{\sigma}{2} F\left\{\exp j\left[\omega_c x + \frac{k}{r_0}(x-x_0)^2 + \phi(r-r_0)\right]\right\},$$

$$s_2 = \frac{\sigma}{2} F\left\{\exp j\left[\omega_c x + \frac{k}{r_0}(x-x_0)^2 + \phi(r-r_0)\right]\right\},$$

and $F\{\ \}$ designates a two-dimensional Fourier transformation on the variables x, r.

Each of the three terms is displayed in a separate position; this separation is a direct consequence of the offset frequency ω_c, introduced during the detection and recording process. The offset thus permits removal of the bias term and permits the two exponential components, s_1 and s_2, to be filtered separately.

Before proceeding further, we observe that the azimuth or x-varying part of the signal is a function of range; thus each range element requires a different filter. To meet this difficulty, an equalizer is introduced into the plane of the signal record; this supplies a phase factor exp $j(k/r)x^2$. Such an equalizer takes the form of a conical lens [1], [2]. The term s_2 becomes

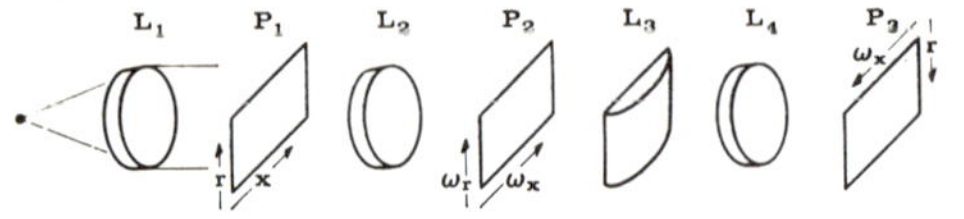

Fig. 3. Optical system for the combined processing, employing anamorphic elements.

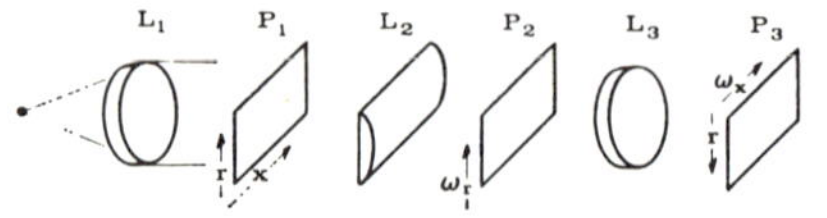

Fig. 4. A modification of the processor given in Fig. 3.

$$s_2(\omega_x, \omega_r) = \frac{\sigma}{2} F\left\{\exp - j\left[\omega_c x - \frac{2k}{r_0} x_0 x + \frac{k}{r_0} x_0^2 + \phi(r - r_0)\right]\right\}, \quad (4)$$

which is the component of interest. We have assumed $\exp j(kr_0)x^2 \approx \exp j(k/r)x^2$. Each target position x_0 produces a spatial frequency $\omega_x = \omega_c - (2k/r_0)x_0$. A narrow slit at position $\omega_x = \omega_c$ is a suitable approximation to the required matched filter. The slit passes only the spatial frequency corresponding to the target at $x_0 = 0$. Other spatial frequencies corresponding to other targets are rejected. As the signal record is moved through the aperture, different target positions x_0 in turn assume the proper frequency for passage by the filter.[1]

For the range dimension, a filter function

$$G(\omega_r) = F\{e^{j\phi(r)}\} \quad (5)$$

is introduced at the frequency plane. This filter is synthesized as a transparency with complex transmittance. The two-dimensional matched filter, therefore, consists of a narrow slit oriented along the ω_r axis and with transmittance $G(\omega_r)$ along its length.

At plane P_3, the transformation back to the spatial domain is made. A detector in the plane records the high-resolution output.

IV. An Anamorphic Configuration

The conical lens equalizer introduced at the signal plane can be regarded as a reference function, which is cross correlated with the recorded signal. The signal at the frequency plane can be written

$$\iint [s_b + s(x, r)] \exp\left(j \frac{k}{r_0} x^2\right) \cdot \exp [j(\omega_x x + \omega_r r)] dx dr. \quad (6)$$

If only that portion of the signal that is of interest is considered [s_2 of (3)], the expression becomes

$$\frac{\sigma}{2} \iint \exp - j\left[\omega_c x + \frac{k}{r_0}(x - x_0)^2 + \phi(r - r_0)\right] \cdot \exp\left(j \frac{k}{r_0} x^2\right) \exp [j(\omega_x x + \omega_r r) dx dr. \quad (7)$$

If only the line $\omega_x = \omega_c$ in the frequency plane is observed, the integral becomes

$$\frac{\sigma}{2} \int \left[\int \exp - j \frac{k}{r_0}(x - x_0)^2 \exp\left(j \frac{k}{r_0} x^2\right) dx\right] \cdot \exp [-j\phi(r - r_0)] \exp (j\omega_r r) dr. \quad (8)$$

The first bracketed term is a cross correlation of the signal with the conical lens reference function. This integral represents the beam-sharpening operation and is displayed in the frequency plane at position $\omega_x = \omega_c$. A slit is positioned along this line to pass this portion of the signal and reject the remainder. The range-dimension filter $G(\omega_r)$ is positioned, as before, along the slit, and the decoded or compressed signal is displayed at the spatial plane P_3.

Such an interpretation reveals that the beam-sharpening effect is displayed at plane P_2 and the pulse compression at plane P_3. To display both in the same plane, anamorphic optics are required. To this end, a cylindrical lens is inserted between planes P_2 and P_3 (Fig. 3). The presence of the cylindrical lens results in a one-dimensional Fourier transform occurring between planes P_2 and P_3; this transformation is with respect to the variable ω_r. The coordinates of plane P_3 are, therefore, ω_x and r. Plane P_3 now displays both the beam-sharpening and pulse-compression operations.

An alternative configuration is shown in Fig. 4. The cylindrical lens performs a one-dimensional Fourier transformation between planes P_3 and P_2. The range-dimension filter is placed at plane P_2. Lens L_3 produces a two-dimensional Fourier transformation between planes P_2 and P_3. The high-resolution output is displayed at plane P_3 (which has ω_x and r as coordinates). The output image is identical with that of Fig. 4.

V. A Two-Dimensional Cross Correlation

In yet another configuration, the entire operation is instrumented as a two-dimensional cross correlation of the signal with a reference function. Since cross correlation implies a scanning, or movement, of one signal across

[1] The azimuth term $(2k/r_0)x_0x$ is still a function of range r_0. The effect now, however, is simply to cause the scaling factor between x_0 and ω_x to be a function of range. Analysis shows that this causes no difficulty.

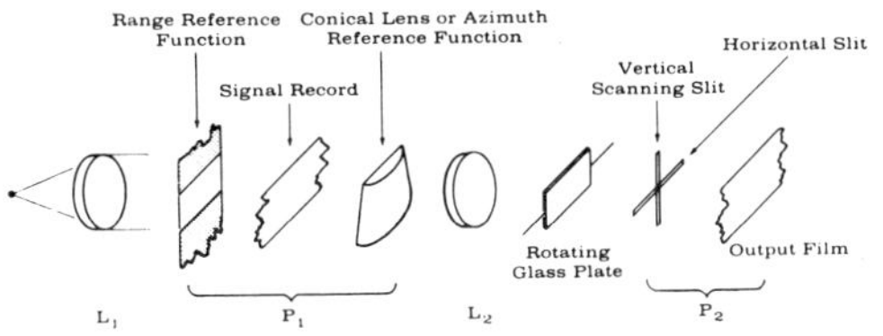

Fig. 5. A two-dimensional cross-correlation system.

another, provision must be made for a two-dimensional mechanical scan. The usual movement of the signal film through the aperture provides the azimuth-dimension scan; to this must be added the range-dimension scan. This could be done by imparting a vertical scanning motion to the signal film and a similar synchronized motion to the recording film. It appears more feasible, however, to produce a synchronized motion of the range-dimension reference function, the output image, and the output slit, as shown in Fig. 5.

The cross-correlation integral generated is

$$\chi(x', r') = \iint s(x - x', r)g(x, r - r')dxdr, \quad (9)$$

where the x-dimension scan is incorporated into the signal and the r into the reference function g. The reference function assumes the form

$$\exp j\left[\frac{k}{r}x^2 + \phi(r)\right]. \quad (10)$$

The scanning motion in the range dimension needed in this implementation makes it unwieldly and, to date, this configuration has never been constructed.

VI. The Linearly Frequency-Modulated Pulse

When the transmitted pulse is linearly frequency modulated, the system is greatly simplified, since a linearly frequency-modulated structure, when recorded (for example, on a medium such as photographic film), acquires lens-like properties; the structure is, in fact, akin to a Fresnel zone plate. Similarly, the matched filter for the pulse also becomes a lens, so we may compress the pulse using the configurations of Figs. 2, 3, or 4, using the appropriate cylindrical lens for the spatial matched filter.

Alternatively, we may discard the matched filter element entirely and utilize only the optical properties of the signal itself to carry out the compression, for the light distribution at the focal point is exactly the high-resolution image that we seek. And, of course, since the azimuth-dimension phase variation is also quadratic, the signal has focal properties in both of its dimensions.

The radiated pulse is of the form

$$\cos\left(\omega t + \frac{s}{2}t^2\right), \quad (11)$$

where s is the frequency-modulation rate. The recorded signal from a point scatterer becomes

$$s_b + \sigma\cos\left[\omega_c x + \frac{k}{r_0}\left(\frac{x}{p} - x_0\right)^2 + \frac{2s}{c^2}\left(\frac{r}{q} - r_0\right)^2\right], \quad (12)$$

where p and q are the azimuth and range-scaling factors, respectively, between radar space and the signal record. Heretofore, we have ignored these factors and have referred the recorded signal to the coordinates of the radar-terrain geometry, since this is generally convenient. Now, since we wish to examine some of the optical properties of the recorded signal, we should use the coordinates of the optical system geometry.

Analysis shows that for the azimuth dimension, the focal length is

$$f_x = \frac{p^2}{2}\frac{\lambda_R}{\lambda_L}r_0, \quad (13)$$

while for the range dimension, the focal length is

$$f_r = \frac{\pi c^2 q^2}{\lambda_L s}, \quad (14)$$

where λ_L is the wavelength of the light used in illuminating the record and c is the velocity of light.

The zone plate is generally anamorphic, with the degree of anamorphism being a function of the range to the target. Proper selection of the FM rate s can make the focal lengths equal, but for one range element only. For extended range intervals, anamorphic compensating elements are required.

For the azimuth dimension, the signals have focal lengths proportional to the range to the targets. For the range dimension, the signals have the same focal length. The optical system is required to compensate for these two anomalies: the range variation of focal length and the unequal focal properties in the two dimensions.

The corrective operation is done in two steps. First, a conical lens is overlaid on the signal record. This lens has a focal length that varies along its length and compensates for the varying focal length of the signal. Now, the focal properties in the two dimensions are yet unequal. A cylindrical lens introduces the compensation required to equalize the focal properties. This process is analogous to that described previously [3] for the case of azimuth processing only, i.e., when a noncoded pulse is radiated. The zone-plate signals are then cylindrical, with focal power in the azimuth dimension only. The signal record itself is then the locus of range focus.

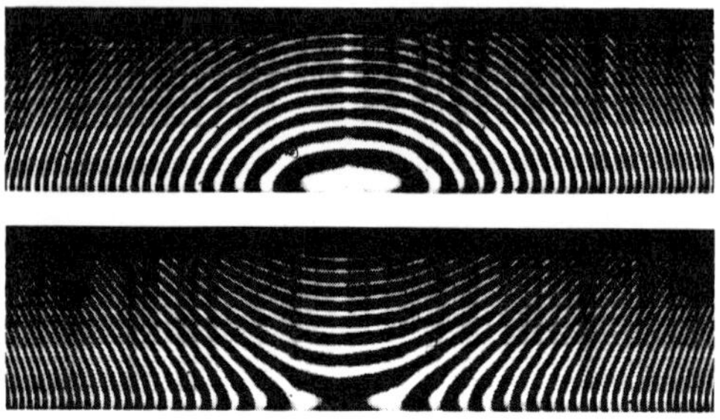

Fig. 6. Hyperbolic and elliptical zone plates produced from a pulse Doppler radar system employing a chirp pulse.

The optical system is thus similar to that used for azimuth-only processing, except that the cylindrical lens is in a different position [1], so that the range focal plane of the signal record, instead of the signal record itself, is in the front focal plane of the cylindrical lens.

The quadratic coefficients, k/r_0 and $2s/c^2$, may have like or unlike signs, depending on whether the frequency-sweep rate s is positive or negative. If the coefficients have like signs, the constant-phase contours are ellipses; if the signs are unlike, the contours are hyperbolas. Signals of each type are shown in Fig. 6. The cylindrical type is conceptually simpler. One order has a positive focal length in each meridian; that is, f_x, and f_r although not equal, are each positive for that order. The other order has a negative focal length in each meridian.

The hyperbolic type has f_x and f_r with unlike signs; one order has a focal length that is positive in one meridian and negative in the other, while the other order has the signs reversed.

Either type of zone plate can be handled, but the cylindrical type is preferable, since the anamorphic properties of the correcting optical system can be less severe.

VII. Realization of the Masks

Successful implementation of the various configurations implies an ability to construct the various masks g or G. These problems have indeed been overcome, although the development of techniques for so doing have been a major effort extending over a period of several years.

The reference function g may be readily implemented as the real function

$$T_g = \tfrac{1}{2} + \tfrac{1}{2}\cos\left[\omega_c x - \frac{kx^2}{r} + \phi(r)\right], \qquad (15)$$

where T_g is the transmittance of, for example, a photographic transparency, the coefficient $\frac{1}{2}$ is a bias term, and the remainder is the replica of a signal from range r.

Implementation of g as a phase function

$$\exp j\left[\frac{kx^2}{r} + \phi(r)\right]$$

is preferable, since such a transparency utilizes almost all the light incident on it, instead of the 5 percent that is in practice realizable from the mask described by (15). The portion $\exp j(kx^2/r)$ can be constructed; this is the conical lens [1]. The portion $\exp j\phi(r)$ is not realizable except in its real form, as in (15), because of limitations of present technology.

Similar arguments apply to the matched filter realization G. In general, neither the matched filter for the range channel nor that for the azimuth or Doppler channel can be realized in the complex form, but both can be readily realized in the real-function form. The range-channel matched filter $F\{\exp j\phi(r)\}$ for a shift-register pseudo-random-sequence waveform was produced in 1961 by Kozma and Kelly [4] in the real-function form

$$\tfrac{1}{2} + \mathrm{Re}\left[e^{ja\omega_r}F\{\exp j\phi(r)\}\right] \qquad (16)$$

with the aid of a digital computer. For convenience, the signal was quantized to two levels (i.e., hard clipped) to ease the problem of construction. This results in only about 2 dB loss of S/N in the system output. Similar results were subsequently obtained by the interferometric method of Vander Lugt [5].

The use of a linear FM pulse code enables the pulse compression to be achieved without the need for a range-code mask, since such signals are self-focusing. We may interpret free space as a distributed-parameter dispersive spatial filter with a phase retardation factor varying quadratically with frequency, i.e., a chirp filter. Obviously, a linear FM waveform is, from the signal processing standpoint, the most attractive pulse code to use when the signal is processed optically.

VIII. Examination of the Cross-Product Term

Let the radiated pulse have a duration τ_0; its length, in terms of round-trip radar range, is then $D = \frac{1}{2}c\tau_0$. Let the radar beam width at range r_0 be L. D and L define the extent of the recorded signal from a point scatterer.

The term $\frac{1}{2}(x-x_0)^2/r_0$ in the pulse-coding term represents a coupling between the pulse-compression and beam-sharpening operations. Thus far, we have assumed that the coupling is negligible. For large values of D or L, this assumption fails.

For convenience, consider only the case of the linearly frequency-modulated pulse. Then, for an FM rate s, the target return is

$$\sigma \cos\left\{\omega_c x + \frac{k(x-x_0)^2}{r_0} + \frac{2s}{c^2} \cdot \left[r - r_0 - \frac{1}{2}\,\frac{(x-x_0)^2}{r_0}\right]^2\right\}. \qquad (17)$$

Expanding the pulse-coding term and dropping the terms in x^4 and x_0^4 produces

$$\sigma \cos\left[\omega_c x + \frac{k(x-x_0)^2}{r_0} + \frac{2s}{c^2}(r-r_0)^2 - \frac{2s}{c^2}\frac{(r-r_0)(x-x_0)^2}{r_0}\right]. \quad (18)$$

For the cross-product term to be negligible, we require

$$\frac{2s}{c^2}\frac{(r-r_0)(x-x_0)^2}{r_0} \leqq \frac{\pi}{2}. \quad (19)$$

Since $(r-r_0)_{max}=D/2$ and $(x-x_0)_{max}=L/2$, the worst-case value of the expression is

$$\frac{2s}{c^2}\frac{DL^2}{8r_0} \leqq \frac{\pi}{2}. \quad (20)$$

The pulse-compression ratio is

$$C_r = \frac{2sD^2}{\pi c^2} \quad (21)$$

and the azimuth-resolution improvement over the beam width (the beam-sharpening ratio) is

$$C_x = \frac{2L^2}{\lambda r_0}. \quad (22)$$

Combining (20), (21), and (22) produces

$$C_r C_x \leqq \frac{8D}{\lambda}. \quad (23)$$

Thus, a limit is imposed on the compression factors achievable. Alternatively, these constraints may be expressed as the limit

$$x_{res} r_{res} \geqq \frac{L\lambda}{8} \quad (24)$$

on the resolution product, where x_{res} is the azimuth resolution and r_{res} is the range resolution.

As is well known, a system that radiates a short pulse instead of using pulse-compression techniques has a similar but more obvious limitation on the combined azimuth and range resolutions, arising from the requirement that the change in range as a target moves across the beam must be smaller than half the pulse length. Otherwise, the target return will not remain within the same range-resolution interval. The change in range is

$$\Delta r = \frac{x^2}{2r_0} \quad (25)$$

and, for a beam width of L, the change in range between center and edge of the beam is

$$\Delta r = \frac{L^2}{8r_0}. \quad (26)$$

This limitation will seriously manifest itself if we attempt to obtain range resolution finer than about twice this quantity or

$$r_{res} \geqq \frac{L^2}{4r_0}. \quad (27)$$

Now, combining (24) with the expression

$$x_{res} = \frac{\lambda r_0}{2L} \quad (28)$$

for azimuth resolution produces the range-resolution limitation

$$r_{res} \geqq \frac{L\lambda}{8x_{res}} = \frac{L^2}{4r_0}. \quad (29)$$

This is identical with (27). Hence the condition that the cross-product term of the pulse-coded signal be small (less than $\pi/2$) is exactly the condition that the change in range of the target as it passes through the beam be less than half the pulse length after compression; therefore, the process of pulse compression introduces no new limitations on the range resolution-azimuth resolution product.

Another interpretation of the cross-product term is possible. The Doppler frequency of the return signal is inversely proportional to the transmitted wavelength. However, the transmitter wavelength changes during the pulse. Therefore, different portions of the return pulse have different Doppler histories. It had, in the past, been argued that because of this effect the pulse-compression and beam-sharpening techniques are not compatible.

To investigate this effect, we rewrite (18) in the form

$$\sigma \cos\left[\omega_c x + \left(\frac{k}{r_0} - \frac{2s}{c^2}\frac{r-r_0}{r_0}\right)(x-x_0)^2 - + \frac{2s}{c^2}(r-r_0)^2\right]. \quad (30)$$

The middle term can be rewritten as

$$\frac{k'}{r_0}(x-x_0)^2, \quad (31)$$

where $k'=(2\pi/\lambda')$ and $(1/\lambda')=(1/\lambda)-(s/\pi c^2)\ (r-r_0)$. The cross-product term is, therefore, just the variation of the Doppler term caused by the change in transmitted frequency. Since the cross-coupling term is present regardless of the pulse form and, in particular, whether we radiate a chirp pulse or a simple noncoded pulse, the previously mentioned argument is invalid. This conclusion accords with the basic observation that a narrow noncoded pulse, having the same resolution capability, has the same power spectrum as the FM pulse, even though the instantaneous frequency of the former has but a single value throughout.

ACKNOWLEDGEMENT

The successful demonstration of the ideas presented here is due to a number of individuals, particularly H. Ristow and T. Van Sickle, for they did early work in incorporating the pulse-coding technique into the transmitter of The University of Michigan Experimental Sidelooking Radar System. I also wish to acknowledge the work of W. C. Curtis and A. Smith, of RCA, Inc., who also conceived the "orthogonal" optical-processing scheme of combined beam sharpening and pulse compression (private communication, 1959) and who subsequently carried out a development program paralleling ours.

REFERENCES

[1] L. J. Cutrona, E. N. Leith, L. J. Porcello, and W. E. Vivian, "On the application of coherent optical processing techniques to synthetic aperture radar," *Proc. IEEE*, vol. 54, pp. 1026–1032, August 1966.

[2] These ideas and their development were presented in the following reports: *5th Ann. Radar Symp. Rec.* (University of Michigan, Ann Arbor, January 1959), Rept. 2144-369-S; and *Proc. Symp. on Recent Advances in Matched Filter Theory and Pulse Compression Techniques*, Rept. RADC-TR-59-161, September 1959.

[3] E. N. Leith, *Appl. Optics*, vol. 7, p. 539, 1968.

[4] A. Kozma and D. Kelly, *Appl. Optics*, vol. 4, p. 385, 1965.

[5] A. Vander Lugt, "Signal detection by complex spatial filtering," *IEEE Trans. Information Theory*, vol. IT-10, pp. 139–145, April 1964.

Emmett N. Leith (SM'66) was born in Detroit, Mich., on March 12, 1927. He received the B.S. degree in physics in 1949 and the M.S. degree in physics in 1952, both from Wayne State University, Detroit, Mich.

He joined the staff of the University of Michigan, Ann Arbor, in 1952, where he is presently Professor of Electrical Engineering and Assistant Head of the Radar and Optics Laboratory. His work has been in the areas of synthetic aperture radar systems, communication theory, data processing, coherent optical systems, and holography.

Mr. Leith is a Fellow of the Optical Society of America. He has received various awards for his work in holography, including the Gordon Award from the Society of Photographic Industrial Engineers, the Liebmann Award from the IEEE, and *Industrial Research Magazine's* "Man of the Year" Award.

Synthetic Aperture Processing With Limited Storage And Presumming

W.M. Brown, G.G. Houser, and R.E. Jenkins
(Brown – Environmental Research Institute of Michigan, Ann Arbor; Houser and Jenkins – IBM Corporation, Owego, New York)

IEEE Transactions on Aerospace and Electronic Systems, Vol. AES-9, No. 2 (March 1973)
Pages 166-176

Abstract

The number of transmitted pulses associated with the Doppler histories of a side-looking radar may greatly exceed the desired azimuth compression ratio of the system. This discrepancy is taxing if the storage required for the azimuth processing is provided by cores, magnetic drums, and the like. Thus, as a practical matter, one considers presumming of the data prior to correlation in an attempt to achieve the desired performance with a minimum amount of digital storage. In this paper, the optimum (in terms of resolution) presummer is derived, along with the optimum apportionment of the available storage capacity between the presumming and correlation operations. Under the condition (or generally pessimistic approximation) that the illumination pattern of the antenna uniformly illuminates a Doppler bandwidth equal to the PRF of the radar, the optimum presumming coefficients are the first N_p Fourier coefficients of a function which is one of the Doppler bandwidth to be correlated and zero on the remainder of the PRF bandwidth, where N_p is the number of transmitted radar pulses over which presumming is provided. Increasing N_p reduces the degradation due to presumming, but may leave inadequate storage for correlation. Hence, we optimize the apportionment between the two operations and present the obtainable resolution as a function of total storage and the number of transmitted pulses in the received Doppler history.

I. Introduction

Let $A(\theta)$ be the two-way (electric field) pattern of the illuminating antenna of a side-looking radar. Then the Doppler history received[1] from a point target at range R is

$$f(x) = A\left(\frac{x}{R}\right) \exp\left(j\frac{2\pi}{\lambda R} x^2\right) \tag{1}$$

where x denotes along-track position, and $x = vt$ if v is the speed of the radar vehicle. The situation is depicted in Fig. 1. With a periodically pulsed radar, f is available only at the times of the (received) pulses. Let the distance traveled between transmitted pulses be X; then the observable data are $f(kX)$, where k is an integer index on the transmitted pulses.

The instantaneous frequency of the Doppler history is

$$\phi' = \frac{4\pi}{\lambda R} x \text{ rad/ft.}$$

If we process the data over the center part of the illumination pattern, specifically, data received over the pattern $|\theta| < \alpha/2$, then the half-bandwidth of this part of the Doppler history is approximately

$$B = \frac{2\pi}{\lambda R}(\alpha R).$$

The corresponding resolution potential is

$$\rho \cong \frac{\lambda}{2\alpha}.$$

Finally, the number of samples to be processed (at each range resolution cell) is

$$M = \frac{\alpha R}{X} = \frac{\lambda R}{2\rho X},$$

and the compression ratio is

$$C = \frac{\alpha R}{\rho} = \frac{\lambda R}{2\rho^2}.$$

With direct digital storage of the data, M is proportional to the amount of storage capacity required in a Doppler processor. On the other hand, if the pulses are "clumped" by a factor of ρ/X, then the required processor storage becomes proportional to C, which may be much less than M in that $\rho >> X$ is often the case in practical situations. Thus, we are led to consider presumming of the data; at first glance we might simply take $\gamma = \rho/X$ (nearest integer) as a presumming factor and correlate $q(\gamma kX)$ where

$$q(\gamma kX) = \sum_{m=-\gamma/2}^{\gamma/2} f(\gamma kX - mX). \tag{2}$$

[1] W.M. Brown "Synthetic aperture radar," *IEEE Trans. Aerospace and Electronic Systems,* Vol. AES-3, March 1967.

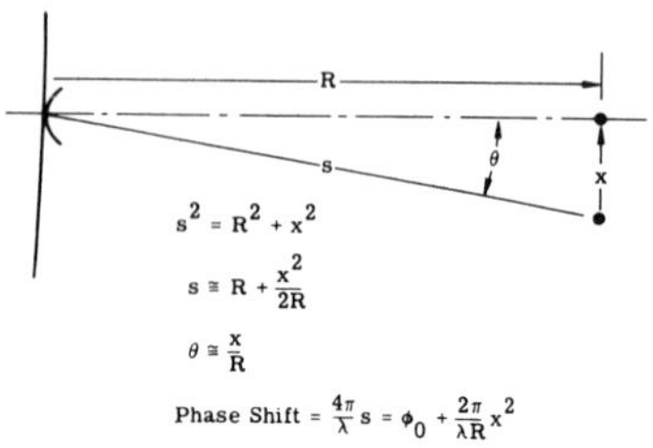

Fig. 1. Side-looking radar geometry.

In this fashion, only[2] C storage cells are required for correlation; however, additional storage is now required for presumming. That is, we now envision two stages of buffer storage; the first has γ storage elements for presumming, and the second has C elements for correlation. In turn, we can hope to get by with

$$\frac{M}{\gamma} + \gamma$$

storage cells which, for reasonable numbers, can save nearly a factor of γ in the required storage capacity.

Not so fast! This simple suggestion for presumming does not take into account Doppler ambiguity problems. The theory to be developed will consider a general presummer

$$q(\gamma kX) = \sum_{m=-\frac{1}{2}N_p}^{\frac{1}{2}N_p} b_m f(\gamma kX - mX). \qquad (3)$$

The number of pulses involved in the presummer, $N_p + 1$, and the presumming coefficients $\{b_m\}$ will be optimized as a function of the illumination pattern and total available storage capacity. If the illumination pattern is nearly constant on $|\theta| < \alpha/2$ and nearly zero on $|\theta| > \alpha/2$, then the simple presumming of (2) is appropriate,[3] and the great savings in storage requirements are realized. At another extreme, let the pattern illuminate the entire Doppler interval $\pm\pi/X$ (i.e., $A(\theta) = 1$ on $|\theta| < \lambda/4X$); then, much more elaborate presumming is in order. However, even in this case it will be seen that two-stage processing can greatly enhance the resolution obtainable with limited storage.

[2]The data are complex; hence, each range resolution cell provides $2C$ real numbers (binary quantized in the numerical calculations of Section III) in view of the need for in-phase and quadrature channels.

[3]In this simple case one could merely store every γth pulse of the radar and achieve the desired end, except for the associated waste of transmitter power. In the general case, the presummer synthesizes the equivalent of a properly tapered real antenna; then subsampling prior to focused processing will not give rise to serious Doppler ambiguity problems.

II. Formulation of Response with Presumming

The relations between continuous functions and continuous convolutions and the corresponding discrete (i.e., sampled) functions and discrete convolutions are central to a discussion of Doppler ambiguities in a periodically pulsed system. We are specifically concerned with these matters for linearly frequency-modulated signals and their corresponding matched filters. The underlying mathematics will be summarized briefly here.

First, two frequencies ω_1 and ω_2 are aliases of one another (or ambiguous) under sampling with period T, provided

$$\exp(j\omega_1 nT) = \exp(j\omega_2 nT)$$

for all integer n. It follows that ω_1 and ω_2 are ambiguous, provided

$$\omega_1 - \omega_2 = p\frac{2\pi}{T}$$

for some integer p.

For a given continuous signal $f(t)$, let $f(nT)$ denote the corresponding sample sequence.[4] The Fourier transform of $f(t)$ is

$$\mathscr{F}(f) = F(\omega) \equiv \int_{-\infty}^{\infty} f(t)\exp(-j\omega t)\,dt. \qquad (4)$$

The analogous transform for the sample sequence is the modified Z transform[5] and is defined by

$$Z_m(f) = F_m(\omega) \equiv \sum_{n=-\infty}^{\infty} f(nT)\exp(-j\omega nT)T. \qquad (5)$$

The factor T is placed in the definition of the Z transform to render the dimensions of F and F_m the same. This leads to nearly identical forms for the properties of Fourier and Z transforms; for example, the inversion equations are

$$f(t) = \frac{1}{2\pi}\int_{-\infty}^{\infty} F(\omega)\exp(j\omega t)\,d\omega \qquad (6)$$

$$f(nT) = \frac{1}{2\pi}\int_{-\pi/T}^{\pi/T} F_m(\omega)\exp(j\omega nT)\,d\omega. \qquad (7)$$

Simple manipulation of (6) with $t = nT$ yields

[4]W.M. Brown, *Analysis of Linear Time-Invariant Systems.* New York: McGraw-Hill, 1963, Ch. 7.

[5]For these basic observations, time will be used as the independent variable; when we return to the application, $x = vt$ will be used as the variable with $X = vT$, where T denotes the interpulse period of the radar.

$$f(nT) = \frac{1}{2\pi} \int_{-\pi/T}^{\pi/T} \sum_{k=-\infty}^{\infty} F\left(\omega + \frac{2\pi k}{T}\right) \exp(j\omega nT)\, d\omega \qquad (8)$$

which, with (7), gives the fundamental result.

Theorem 1

$$F_m(\omega) = \sum_{k=-\infty}^{\infty} F\left(\omega + \frac{2\pi k}{T}\right). \qquad (9)$$

Of course, this relation is as expected in that $F_m(\omega)$ is the sum of $F(\omega)$ over all the aliases of ω.

From theorem 1, the[6] sampling theorem follows, in that if $F(\omega) = 0$ for $|\omega| > \pi/T$, *then* $F_m(\omega) = F(\omega)$ for $|\omega| < \pi/T$; in turn, substitution of (5) into (6) yields

$$f(t) = \frac{1}{2\pi} \int_{-\pi/T}^{\pi/T} \left[T \sum_{m=-\infty}^{\infty} [f(nT) \exp(-j\omega nT) \right] \exp(-j\omega t)\, dt$$

$$= \sum_{n=-\infty}^{\infty} f(nT) \frac{\sin(\pi/T)(t - nT)}{(\pi/T)(t - nT)} \qquad (10)$$

Another theorem which follows from Theorem 1 is the isomorphism for continuous and discrete convolutions under the sampling operation. That is, let

$$g(t) = \int_{-\infty}^{\infty} h(t - \tau) f(\tau)\, d\tau$$

$$\hat{g}(nT) = \sum_{k=-\infty}^{\infty} h(nT - kT) f(kT) T.$$

Theorem 2

If $F(\omega)$ and $H(\omega)$ are zero except for $\omega \in \Omega$, where the measure of Ω is not greater than $2\pi/T$ and $\omega \in \Omega$ implies that $\omega' = \omega + (2\pi/T)p$ is not an element of Ω for any integer $p \neq 0$, then $\hat{g}(nT) = g(nT)$. That is, the sampled version of the continuous convolution equals the discrete convolution of the sampled functions.

As for the proof of Theorem 2, we have, by the convolution theorem of Fourier transforms,

$$G(\omega) = H(\omega)F(\omega)$$

$$g(t) = \frac{1}{2\pi} \int_{\Omega} H(\omega)F(\omega) \exp(j\omega t)\, d\omega$$

$$g(t) = \frac{1}{2\pi} \int_{\Omega} H_m(\omega)F_m(\omega) \exp(j\omega t)\, d\omega$$

$$g(t) = \frac{1}{2\pi} \int_{-\pi/T}^{\pi/T} H_m F_m \exp(j\omega t)\, d\omega$$

where the integral over Ω can be replaced by the integral from $-\pi/T$ to π/T by virtue of the fact that modified Z transforms are periodic with period $2\pi/T$ and Ω is alias free. Finally, the last expression is evaluated at $t = nT$,, and we have the inverse Z transform of $H_m F_m$. However, by the Z transform convolution theorem, this inverse transform is equal to the discrete convolution of the sample sequences $f(nT)$ and $h(nT)$;

$$g(nT) = \hat{g}(nT),$$

as was to be shown.

Of course, the point of Theorem 2 is that discrete convolutions can be calculated as samples of continuous convolutions in the bandwidth-limited case.

Example of Interest

Suppose

$$f(x) = A\left(\frac{x}{R}\right) \exp\left(j\frac{2\pi}{\lambda R} x^2\right);$$

then

$$F(\omega) = \int_{-\infty}^{\infty} A\left(\frac{x}{R}\right) \exp\left(j\frac{2\pi}{\lambda R} x^2\right) \exp(-j\omega x)\, dx. \qquad (11)$$

In a sense which can be made precise, the Fourier transform of the phase factor is

$$\mathscr{F}\left[\exp\left(-\frac{x^2}{2jc}\right)\right] = \sqrt{2\pi jc} \exp\left(-j\frac{c\omega^2}{2}\right)$$

where

$$c = \frac{\lambda R}{4\pi}.$$

Thus, let

$$P(\omega) = \int_{-\infty}^{\infty} A\left(\frac{x}{R}\right) \exp(-j\omega x)\, dx.$$

Then, by the convolution theorem,

[6] More properly a sampling theorem, in that similar theorems are obtained provided merely that $F(\omega) = 0$, except for $\omega \in \Omega$ where Ω is a set having measure not greater than $2\pi/T$ and Ω contains no two frequencies which are aliases of one another. In the general case, the interpolation function is the inverse Fourier transform of the indicator function of the set Ω.

$$F(\omega) = \sqrt{\frac{jc}{2\pi}} \int_{-\infty}^{\infty} \exp\left(-j\frac{c}{2}(\omega-\mu)^2\right) P(\mu)\,d\mu$$

$$= \sqrt{j\frac{c}{2\pi}} \exp\left(-j\frac{c}{2}\omega^2\right) \int_{-\infty}^{\infty} \exp(jc\omega\mu)$$

$$\exp\left(-j\frac{c}{2}\mu^2\right) P(\mu)\,d\mu, \tag{12}$$

which has a discouraging similarity to the problem we started with. However, the quadratic phase factor is now essentially one, since $A(x/R)$ is a smooth function that has a transform which is zero, except for values of μ very small compared to $c^{-1/2}$. Specifically, $A(\theta)$ is the Fourier transform of the illumination function of the physical aperture; in turn,

$$P(\omega) = 0 \quad \text{for } |\omega| > \frac{2\pi}{\lambda R}\,d$$

where d is the length of the physical aperture. Thus, where $P(\omega) \neq 0$,

$$\frac{c\omega^2}{2} \leqslant \frac{\lambda R}{8\pi}\left(\frac{2\pi}{\lambda R}\right)^2 d^2 = \frac{\pi}{2}\,\frac{d^2}{\lambda R}\,.$$

But, $d^2 << \lambda R$, so that, for all practical purposes, (12) reduces to

$$F(\omega) = \sqrt{\frac{jc}{2\pi}} \exp\left(-j\frac{c}{2}\omega^2\right) \int_{-\infty}^{\infty} \exp(jc\omega\mu) P(\mu)\,d\mu$$

$$= \sqrt{2\pi jc}\,\exp\left(-j\frac{c}{2}\omega^2\right) A\left(\frac{c\omega}{R}\right)$$

$$F(\omega) = \sqrt{j\frac{\lambda R}{2}} \exp\left(-j\frac{1}{2}\frac{\lambda R}{4\pi}\omega^2\right) A\left(\frac{\lambda}{4\pi}\omega\right). \tag{13}$$

Equation (13) depicts a well-known result. The Fourier transform of the Doppler history consists of a quadratic phase factor in ω weighted by $A[(\lambda/4\pi)\omega]$, and $\theta = (\lambda/4\pi)\omega$ is the angle from broadside at which a Doppler frequency of $2(v\theta/\lambda)$ Hz [or $(4\pi/\lambda)\theta$ rad/ft of travel] occurs.

If this Doppler history is processed with a (continuous) filter

$$H(\omega) = I_B(\omega)\exp\left(j\frac{c}{2}\omega^2\right),$$

where $I_B(\omega) = 1$ for $|\omega| < B$ and 0 for other ω, then the output is (the constant $\sqrt{j\lambda R/2}$ will be dropped)

$$g(x) = \frac{1}{2\pi}\int_{-B}^{B} A\left(\frac{\lambda}{4\pi}\omega\right)\exp(j\omega x)\,d\omega.$$

If A is nearly uniform on $|\omega| < B$, then the resolution obtained is $\rho \cong \pi/B$. By the same derivation discussed for the Doppler history, the impulse response of the filter is

$$h(x) = F^{-1}[H(\omega)] = 0 \quad \text{for } |x| > \frac{\lambda R}{4\pi}\,B$$

(to an excellent approximation). In turn, the sampled version of h has only $\lambda RB/2\pi X$ nonzero samples. If $F(\omega) \cong 0$ for $|\omega| > \pi/X$, and the observed sampled version of the Doppler history is discretely filtered with $h(nX)$, the discrete output is (by Theorem 2)

$$g(kX) = \frac{1}{2\pi}\int_{-B}^{B} A\left(\frac{\lambda}{4\pi}\omega\right)\exp(j\omega kX)\,d\omega.$$

If the band limit assumption on the Doppler history is removed, the response contains Doppler ambiguities which will now be described. Let the discrete filter bandwidth $B \leqslant \pi/X$; but now

$$F_m(\omega) = \sum_{k=-\infty}^{\infty} F\left(\omega + \frac{2\pi k}{X}\right)$$

is the Z transform of the received data. The sampled nature of the data now gives

$$g(nX) = \frac{1}{2\pi}\int_{-\pi/X}^{\pi/X} F_m(\omega)H(\omega)\exp(j\omega nX)\,d\omega.$$

It is convenient to express

$$g(nX) = \sum_{k=-\infty}^{\infty} g_k(nX)$$

with

$$g_k(nX) = \frac{1}{2\pi}\int_{-\pi/X}^{\pi/X} F\left(\omega + \frac{2\pi k}{X}\right)H(\omega)\,\exp(j\omega nX)\,d\omega$$

$$F\left(\omega + \frac{2\pi k}{X}\right) = A\left(\frac{\lambda}{4\pi}\omega + \frac{\lambda k}{2X}\right)\exp\left(-j\frac{\lambda R}{8\pi}\omega^2\right)\exp\left(-jk\frac{\lambda R}{2X}\omega\right)$$

except for a constant of unit magnitude. The linear phase factor $k(\lambda R/2X)\omega$ introduces a time shift on the inverse Z transform. Thus, let

$$\hat{g}_k(x) = \frac{1}{2\pi}\int_{-B}^{B} A\left(\frac{\lambda}{4\pi}\omega + \frac{\lambda k}{2X}\right)\exp(jx\omega)\,d\omega;$$

then

$$g_k(nX) = \hat{g}_k\left(nX - k\frac{\lambda R}{2X}\right).$$

The detailed shape of $\hat{g}_k(x)$ depends on the behavior of $A(\theta)$ in the neighborhood of $\theta = \lambda k/2X$. In any event, as k gets large, $|\hat{g}_k(x)|$ gets small, and the resolution associated with $\hat{g}_k(x)$ is approximately π/B. A typical situation is illustrated in Fig. 2.

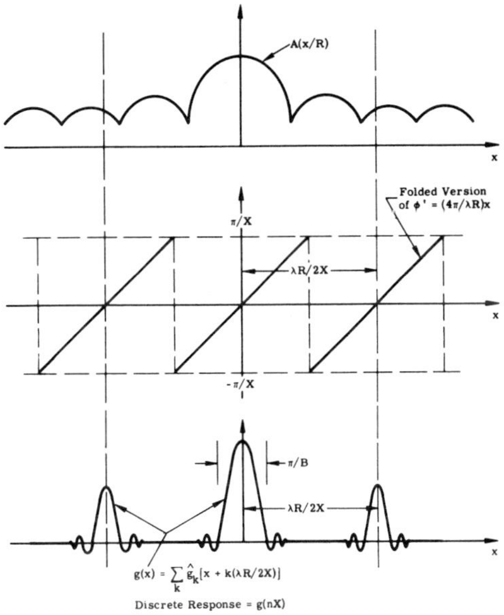

Fig. 2. Typical response in the presence of Doppler ambiguities.

Response With Presumming

Let

$$q(nX) = \sum_{k=-\frac{1}{2}N_p}^{\frac{1}{2}N_p} f(nX - kX)y(kX)X \qquad (14)$$

$$Y_m(\omega) \equiv \sum_k y(kX) \exp(-j\omega kX)X \qquad (15)$$

$$y(x) \equiv \frac{1}{2\pi} \int_{-\pi/X}^{\pi/X} Y_m(\omega) \exp(j\omega x)\, d\omega. \qquad (16)$$

Now, the finite sequence $\{y(kX)\}$ are the coefficients of the presummer. However, the objective of the presummer is to permit a reduced sampling rate; i.e., only every γth value of $q(nX)$ is utilized for correlation of the Doppler history. That is, we process $q(n\gamma X)$, where γ is an integer which depicts the factor by which the presummed data are subsampled. It is convenient to introduce

$$\hat{f}(x) = \frac{1}{2\pi} \int_{-\pi/X}^{\pi/X} F_m(\omega) \exp(j\omega x)\, d\omega. \qquad (17)$$

Then, by Theorem 2, $q(nX)$ is the sampled version of

$$q(X) = \int_{-\infty}^{\infty} \hat{f}(x-u)y(u)\, du,$$

since $\hat{f}$ and y were constructed so as to be appropriately bandwidth-limited. Finally, $q(n\gamma X)$ is the sampled version of $q(x)$. The Fourier transform of $\hat{f}(x)$ is $F_M(\omega)I_w(\omega)$, and the Fourier transform of $y(x)$ is $Y_m(\omega)I_w(\omega)$, where

$$I_w(\omega) = \begin{cases} 1 & \text{for } |\omega| < \dfrac{\pi}{X} \equiv W \\ 0 & \text{for } |\omega| > W. \end{cases}$$

Thus, the Fourier transform of $q(x)$ is

$$Q(\omega) = F_m(\omega)Y_m(\omega)I_w(\omega). \qquad (18)$$

The sampling interval of interest is now γX, and the unambiguous Doppler interval is $|\omega| < \pi/\gamma X \equiv B$. The discrete output now involves the folded version of $Q(\omega)$; however, it is easy to see that (since Y_m and F_M are periodic with period $(2\pi/X)$

$$\sum_{k=-\infty}^{\infty} Q\left(\omega + \frac{2\pi k}{\gamma X}\right) = \sum_{n=0}^{\gamma-1} \sum_{k=-\infty}^{\infty} F\left(\omega + \frac{2\pi k}{X} + \frac{2\pi n}{\gamma X}\right) Y_m\left(\omega + \frac{2\pi n}{\gamma X}\right). \qquad (19)$$

In turn, the Z transform of the overall response (with presumming and subsampling by a factor γ on the presummer output and on the correlator discrete impulse response) becomes

$$J(\omega) \sum_{n=0}^{\gamma-1} \sum_{k=-\infty}^{\infty} Y_m\left(\omega + \frac{2\pi n}{\gamma X}\right) A\left(\frac{\lambda}{4\pi}\omega + \frac{\lambda k}{2X} + \frac{\lambda n}{2\gamma X}\right) \exp\left[-j\left(\frac{k\lambda R}{2X} + \frac{n\lambda R}{2\gamma X}\right)\omega\right] \qquad (20)$$

where the correlator phase response again cancels the quadratic phase of factor on the folded version of F. Also, the impulse response of the correlator[7] has been taken as

[7] To an excellent approximation, ξ contains only C terms where $(\lambda/4\pi)\cdot(\pi/\gamma X) = CX/2R$, or $C = \lambda R/2\gamma X^2$, by the same derivation used to find $F(\omega)$. Hence, the storage required for correlation has been reduced by the factor γ.

$$\xi(\gamma nX) = \frac{1}{2\pi}\int_{-\pi/\gamma X}^{\pi/\gamma X} J(\omega)\exp\left(j\frac{\lambda R}{8\pi}\omega^2\right)$$

$$\exp(j\omega n\gamma X)\,d\omega$$

where $J(\omega)$ is the amplitude taper on the correlator and complex constants of unit modulus have again been dropped on the folded version of F.

Now, clearly, the response corresponding to (20) is similar in nature to that depicted in Fig. 2, except that:

1) the interval between Doppler ambiguities has been reduced to $\Delta x = \lambda R/2\gamma X$;
2) a simple folded version of the presummer transfer function can provide attenuation of the Doppler ambiguities; and
3) the resolution limit has been degraded to γX, which, in fact, would be achieved free of Doppler ambiguities provided $F(\omega) = 0$ for $|\omega| > \pi X$ and

$$J(\omega)Y_m(\omega)A\left(\frac{\lambda}{4\pi}\right)\omega = \begin{cases} 1 & \text{for } |\omega| < \dfrac{\pi}{\gamma X} \\ 0 & \text{for } |\omega| > \dfrac{\pi}{\gamma X}. \end{cases}$$

The above analysis provides an adequate basis for proceeding with the optimization of the presummer. The general optimization (for resolution and apportionment of available storage) is described in the Appendix; however, the useful results will now be derived.

III. Basic Optimization

If $A[(\lambda/4\pi)\omega] = 0$ for $|\omega| > B$, there is no problem with the presummer in that it is adequate to utilize only every γth pulse; or, in the interest of power utilization, we can take $y(kX) = 1$ for $|k| < (1/2)N_p = \gamma/2$. Also, if $B < W' < \pi/X = W$ and $A[(\lambda/4\pi)\omega] \cong 0$ for $|\omega| > W'$, comparable pre-presumming can be used (with very modest invasion of storage capacity) so that the subsamples from this pre-presummer have an along-track spacing of

$$X' = \frac{W}{W'}X.$$

This rather naive data reduction frames the problem[8] so that $A[(\lambda/4\pi)\omega]$ nearly fully occupies $|\omega| < \pi/X' = W'$. Simple sketches of folded versions of $A[(\lambda/4\pi)\omega]$ permit a judicious choice of W/W', so that (20) can be treated as though

$$A\left(\frac{\lambda}{4\pi}\omega\right) \cong \begin{cases} 1 & \text{for } |\omega| < W' \\ 0 & \text{for } |\omega| > W'. \end{cases} \qquad (21)$$

[8] If $A[(\lambda/4\pi)\omega]$ is significantly greater than zero for $|\omega| > \pi/X$, the system suffers from Doppler ambiguities even without presumming. Conceptually, this case is included in the Appendix.

The prime notation will now be dropped, and no taper will be used on the correlator. (The effects of $J(\omega)$ are nearly obvious in that J appears in the same way on each of the ambiguities.)

The Z transform of the response is now

$$R(\omega) \sim \sum_{n=0}^{\gamma-1} Y_m\left(\omega + \frac{2\pi n}{\gamma X}\right)\exp\left(-j\frac{n\lambda R}{2\gamma X}\omega\right) \qquad (22)$$

$$r(k\gamma X) = \frac{1}{2\pi}\int_{-B}^{B} R(\omega)\exp(j\omega k\gamma X)\,d\omega. \qquad (23)$$

The optimization will be with respect to the equivalent rectangle resolution criterion:

$$\rho \equiv \frac{\sum_{k=-\infty}^{\infty} |r(k\gamma X)|^2(\gamma X)}{|r(0)|^2}. \qquad (24)$$

Simple expressions are obtained for both the numerator and denominator of ρ. The simplicity can be viewed as a consequence of the separation of the terms corresponding to folded parts of Y_m, or, more directly,

$$r(0) = \frac{1}{2\pi}\int_{-B}^{B}\sum_{n=0}^{\gamma-1} Y_m\left(\omega + \frac{2\pi n}{\gamma X}\right)$$

$$\exp\left(-j\frac{n\lambda R}{2\gamma X}\omega\right)d\omega$$

$$\cong \frac{1}{2\pi}\int_{-B}^{B} Y_m(\omega)\,d\omega,$$

since, if we consider a term of $Y_m(\omega)$, say $y(kX)\exp(-j\omega kX)$, the corresponding contribution to $r(0)$ is

$$y(kX)\left[\frac{\sin kXB}{\pi kX} + \sum_{n=1}^{\gamma-1}\frac{\sin\left(kXB + \dfrac{n\lambda RB}{2\gamma X}\right)}{\pi\left(kX + \dfrac{n\lambda R}{2\gamma X}\right)}\exp\left(-j\frac{2\pi nk}{\gamma}\right)\right]$$

However,

$$\frac{\lambda R}{2\gamma X} \gg kX$$

even for the largest value of $k = (1/2)N_p$, in that if N_p is comparable to $\lambda R/\gamma X^2$, then enough storage has been allocated to presumming to accommodate direct correlation of the received data for the desired resolution γX. That is, the number of transmitted pulses to be processed for a resolution of γX is $\lambda R/2\gamma X^2$ (see Section I). Said in another way, we are not interested in presumming if it requires an amount of storage comparable to the storage required to directly correlate the received data for the desired resolution. As for the numerator, Parseval's theorem gives

$$\sum_{k=-\infty}^{\infty} |r(k\gamma X)|^2 \gamma X = \frac{1}{2\pi} \int_{-B}^{B} |R(\omega)|^2 \, d\omega$$

$$\cong \frac{1}{2\pi}$$

$$\int_{-B}^{B} \sum_{n=0}^{\gamma-1} \left| Y_m\left(\omega + \frac{2\pi n}{\gamma X}\right)\right|^2 d\omega$$

$$= \frac{1}{2\pi} \int_{-W}^{W} |Y_m(\omega)|^2 \, d\omega$$

where the cross terms in $\int |R|^2 \, d\omega$ are negligible by the same reasoning as that used on $r(0)$.

Thus, in the cases of practical interest, optimization for equivalent rectangle resolution consists of selecting the coefficients $y(kX)$ so as to minimize

$$\rho = 2\pi \frac{\int_{-W}^{W} |Y_m(\omega)|^2 \, d\omega}{\left|\int_{-B}^{B} Y_m(\omega) \, d\omega\right|^2} \tag{25}$$

where

$$Y_m(\omega) = \sum_{|k|<\frac{1}{2}N_p} y(kX) \exp(-jkX\omega) X$$

and

$$\frac{\pi}{X} = W = \gamma B. \tag{26}$$

Also, with apportionment of the available storage, we wish to optimize over the number of terms $N_p + 1$ as well. For the moment, let N_p be fixed; then substitution of (26) into (25) gives

$$\rho = X \frac{\sum_k |y(kX)|^2}{\left|\sum_k y(kX) \frac{\sin\left(\frac{k}{\gamma}\pi\right)}{\pi k}\right|^2} .$$

By the Schwarz inequality, the denominator is not greater than

$$\sum_k |y(kX)|^2 \sum_k \left[\frac{\sin\left(\frac{k}{\gamma}\pi\right)}{\pi k}\right]^2$$

with equality if (and only if)

$$y(kX) = \frac{\sin\left(\frac{k}{\gamma}\pi\right)}{\pi k} \text{(within a constant)}. \tag{27}$$

Thus, (27) depicts the optimum presumming weights, and the optimum resolution is

$$X\rho^{-1} = \sum_k \left[\frac{\sin\left(\frac{k}{\gamma}\pi\right)}{\pi k}\right]^2 \tag{28}$$

These results are eminently reasonable; i.e., for any fixed number of weights in the presummer, the optimum $Y_m(\omega)$ on $|\omega| < W$ consists of the first ($N_p + 1$) terms of the Fourier series for $I_B(\omega)$. As $N_p \to \infty$, the subsampling of the presummer output introduces no ambiguities.

It remains to optimize the apportionment of the storage between presumming and correlation. The storage capacity per range resolution cell, N_c, utilized for correlation is (proportional to)

$$\frac{4\pi}{\lambda R} \Delta x = 2B = \frac{2\pi}{\gamma X} \quad \text{(radian bandwidth)}$$

$$\Delta x = N_c(\gamma X)$$

$$N_c = \frac{\lambda R}{2\gamma^2 X^2} . \tag{29}$$

The storage required for presumming is in the same proportion to $(N_p + 1) \equiv \hat{N}_p$. If the same type of storage is used in both operations and the total available storage is N, then $\hat{N}_p = N - N_c$; thus, the optimum apportionment consists of selecting γ so as to maximize

$$X\rho^{-1} = \sum_{(N-C/\gamma^2) \text{ terms}} \left[\frac{\sin\left(\frac{k}{\gamma}\pi\right)}{\pi k}\right]^2 \tag{30}$$

where $C = \lambda R/2X^2$ is the compression ratio limit of the radar based on pulse repetition frequency (PRF).

For any fixed value of C it is a simple (numerical) problem to obtain the optimum γ (which fixes the apportionment) and the corresponding resolution limit as a function of available storage N. As a refinement, longer storage time is required in the correlation buffer store; in turn, storage for presumming might sometimes be done at less cost in some sense. In this case, equipment complexity might be taken proportional to $(N_p + \beta N_c)$, and the same procedure is followed with βC in place of C where the choice of β is geared to details of the implementation.

Figs. 3 through 5 show the optimum value of the presum integer γ as a function of total storage for three values of $C = \lambda R/2X^2$. The optimum attainable resolution as a function of total storage is also shown, along with the degraded resolution obtained in the absence of presumming. These data show attainable resolution divided by X, where a resolution of X would be obtained if a Doppler bandwidth equal to the PRF was unambiguously illuminated and correlated.

Note that the total storage N is for one quadrature channel only, and for complex signal processing the total processor storage required is $2N$. Fig. 6 presents obtainable resolution for a wide range of values for C.

Fig. 7 is a reproduction of a computer printout, after correlation of a point target response. The reproduction involves practical conditions, with long segments between ambiguities omitted. Notice that the ambiguity resulting from subsampling (presumming) is comparable to the first "true PRF" ambiguity.

The value of presumming is clear from Figs. 3 through 6. For example, at $N = 200$, Fig. 5 indicates an improvement in resolution of a factor of about four if the optimum amount of storage is devoted to presumming. For these figures, the resolution obtained with no presumming is given by C/N. The fact that this curve gets below the optimum resolution curve results from the simple form of processing used in the digital correlation. Specifically, binary phase factors[9] were used, which degrades performance slightly.

Appendix

General Optimization

If (20) is considered in place of (22), the expression corresponding to (25) is found to be

$$\rho = \frac{2\pi \int_{-B}^{B} |J(\omega)|^2 \sum_{n=0}^{\gamma-1} D\left(\omega + \frac{2\pi n}{\gamma X}\right) \left| Y_m\left(\omega + \frac{2\pi n}{\gamma X}\right)\right|^2 d\omega}{\left| \int_{-B}^{B} J(\omega) A(\omega) Y_m(\omega)\, d\omega \right|^2} \tag{31}$$

where

$$D(\omega) = \sum_{k=-\infty}^{\infty} \left| A\left(\frac{\lambda}{4\pi}\,\omega + \frac{\lambda k}{2X}\right)\right|^2 .$$

One can consider the problem of minimizing ρ over Y_m and J, as well as over the illumination of a finite aperture[10] which determines A. Optimization over any one of these factors is a not too formidable task; however, at this time it is not clear that a simple procedure can be formulated for the composite optimization. Thus, we will only treat the problem of optimizing Y_m (and the apportionment of storage) for fixed J and A.

Since $J(\omega)$ is periodic with period $2B$, the resolution can be written

$$\rho = \frac{2\pi \int_{-W}^{W} U(\omega) |Y_m(\omega)|^2 \, d\omega}{\left| \int_{-B}^{B} V(\omega) Y(\omega)\, d\omega \right|^2} \tag{32}$$

$$U(\omega) \equiv |J(\omega)|^2 D(\omega)$$

$$V(\omega) \equiv J(\omega) A(\omega).$$

The factors U and V are known. Let these functions be expanded in Fourier series on $|\omega| < W$:

$$U(\omega) = \sum_{k=-\infty}^{\infty} X u_k \exp(-j\omega k X)$$

$$V(\omega) = \sum_{k=-\infty}^{\infty} X v_k \exp(-j\omega k X)$$

$$Y(\omega) = \sum_{k=-1/2N_p}^{1/2N_p} X y_k \exp(-j\omega k X).$$

The expression for resolution becomes

$$X^{-1}\rho = \frac{\sum_n \sum_k u_{n-k} y_k \bar{y}_n}{\left| \sum_k y_k b_k \right|^2} \tag{33}$$

where

$$b_k = \sum_s v_s \frac{\sin \frac{\pi}{\gamma}(s+k)}{\pi(s+k)} .$$

[9] That is, the sine and cosine functions in the correlation were replaced with square waves.

[10] R.O. Harger, "An optimum design of ambituity function, antenna pattern, and signal for side-looking radars," *IEEE Trans. Military Electronics*, Vol. MIL-9, pp. 264-278, July-October 1965.

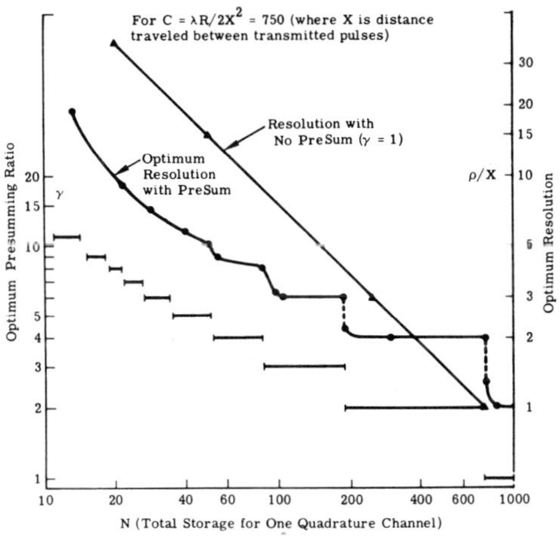

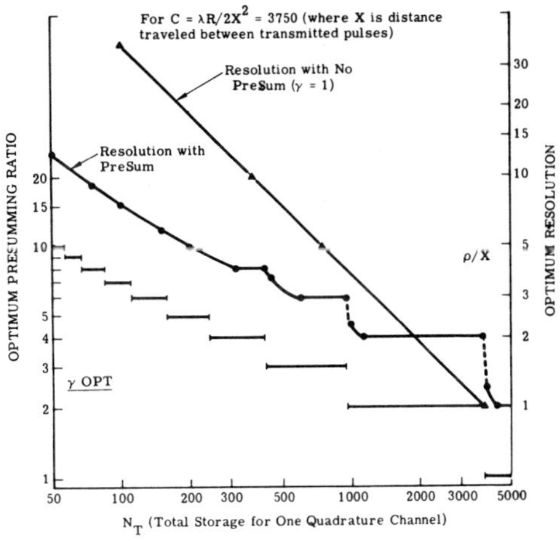

Fig. 3. Optimum resolution and presum ratio versus storage.

Fig. 5. Optimum resolution versus optimum presum ratio (γ optimum).

Fig. 4. Optimum resolution and presum ratio versus storage.

Fig. 6. Optimum resolution versus available storage.

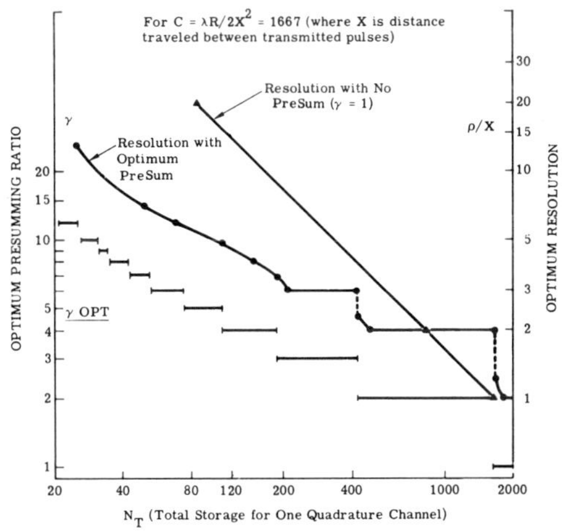

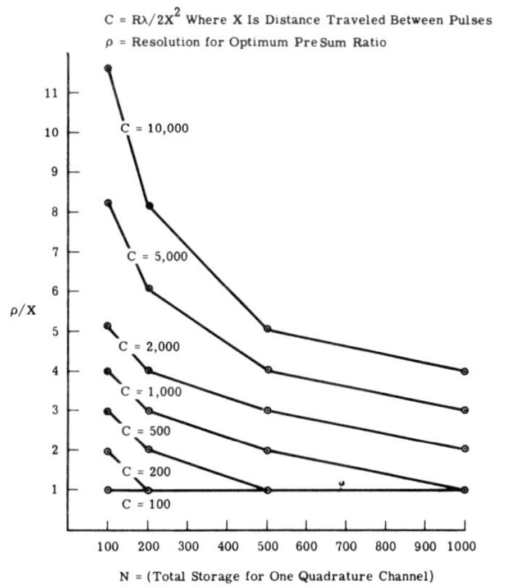

Fig. 7. Computer printout of point target response.

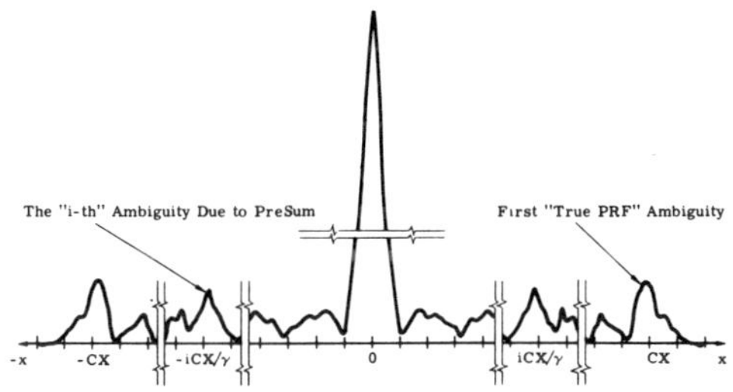

The u_k and b_k coefficients are known, so that we have the standard problem of minimizing the (finite) quadratic form in $\{y_k\}$ divided by the inner product (squared) of $\{y_k\}$ with $\{b_k\}$ By the way, the sequence u_n is easily shown to be non-negative definite, and, in fact, positive definite for the $U(\omega)$ function of interest.

Thus, let $\psi_1, \cdots, \psi_{\hat{N}_p}$ be the eigenvectors of the matrix $\{u_{n-k}\}$ where k and n run from $-(1/2)N_p$ to $(1/2)N_p$ and $\hat{N}_p = N_p + 1$. Let $\lambda_1, \cdots, \lambda_{\hat{N}_p}$ be the eigenvalues. Finally, let $\hat{N}_p$-dimensional vector $\{y_k\}$ be

$$Y = \sum \beta_k \psi_k$$

and let

$$\{b_k\} = \sum \alpha_k \psi_k$$

where the α_k are known and the minimization is now to be done on the β_k, which then determine the optimum $\{y_k\}$.

In terms of these eigenvectors and eigenvalues, (33) becomes

$$X^{-1}p = \frac{\sum_{k=1}^{\hat{N}_p} \lambda_k |\beta_k|^2}{\left|\sum_k \beta_k \bar{\alpha}_k\right|^2} . \tag{34}$$

By the Schwarz inequality,

$$\left|\sum_k \beta_k \frac{\alpha_k}{\lambda_k} \lambda_k\right|^2 \leqslant \sum |\beta_k|^2 \lambda_k \sum \frac{|\alpha_k|^2}{\lambda_k}$$

with equality provided by

$$\beta_k = \frac{\alpha_k}{\lambda_k} . \tag{35}$$

In turn, the optimum resolution with $\hat{N}_p$ terms in the presummer is

$$Xp^{-1} = \sum_{k=1}^{\hat{N}_p} \frac{|\alpha_k|^2}{\lambda_k} .$$

Thus, (35) depicts the optimum presummer for fixed $\hat{N}_p$, and (36) the corresponding minimum resolution. Optimization of the apportionment is now rather tedious in that the matrix $\{u_{n-k}\}$ must be diagonalized for various size restrictions, $|k| \leqslant (1/2)N_p$ and $|n| \leqslant (1/2)N_p$, and the minimum provided by (36) determined with $\hat{N}_p = N - C/\gamma^2$. That is, with C fixed (C is determined by the radar parameters), the resolution depicted by (36) is determined for various subsampling factors γ (which determines $\hat{N}_p$ if the storage capacity N is fixed). Conceptually, the general procedure called for here is much like the optimization given in Section III; however, it does not seem worthwhile to carry the analysis to numerical results unless C and the illumination A (and, perhaps, the processor taper J) are fixed for some radar of actual interest.

William M. Brown (M'57–SM'63–F'71) was born in Wheeling, W.Va., on February 14, 1932. He received the B.S.E.E. degree from the University of West Virginia, Morgantown, in 1952, and the M.S.E. and Dr. Eng. degrees from Johns Hopkins University, Baltimore, Md., in 1955 and 1957, respectively.

From 1952 to 1954 he worked in the Air Arm Division of Westinghouse Corporation. From 1954 to 1957 he was at the Johns Hopkins Radiation Laboratory. He then worked for a year at the Institute for Defense Analyses, Washington, D.C., and taught during the same period at George Washington University, Washington, D.C. Since 1958 he has been with the University of Michigan, Ann Arbor, where he is a Professor of Electrical Engineering. From 1960 to 1968 he was Head of the Radar and Optics Laboratory of the Willow Run Laboratories at the University of Michigan. He taught at the Imperial College in London, England, in the fall of 1968. During 1969 he was a Vice-President of the Chain Lakes Research Corporation, Detroit, Mich. From 1970 to 1972 he was Director of the Willow Run Research Laboratories, University of Michigan. He is currently President of the Environmental Research Institute of Michigan, Ann Arbor.

Dr. Brown is a member of sigma Xi, Tau Beta Pi, Eta Kappa Nu, and Sigma Pi Sigma. He is a registered Professional Engineer in Michigan. He is also Associate Editor for Radar/Sonar of this *Transactions.*

George G. Houser was born in Johnson City, N.Y., on March 31, 1937. He received the B.E.E. degree in 1959 from the Polytechnic Institute of Brooklyn, N.Y., and the M.S.E.E. in 1963 from Syracuse University, Syracuse, N.Y.

He has worked for the Federal Systems Division of IBM Corporation, Owego, N.Y. since 1959. From 1959 to 1966 he worked on airborne radar data processing activities. Since 1966 he has worked on electronic support measures (ESM) programs, both in systems engineering and equipment design. In 1971 he became Manager of ESM Programs.

Robert E. Jenkins was born in Waycross, Ga., on August 16, 1928. He received the B.S.E.E. and M.S.E.E. degrees from the Georgia Institute of Technology in 1957 and 1963, respectively.

From 1957 to 1959 he was with the Aero Equipment Division of Sperry Rand, Great Neck, N.Y. From 1960 to 1962 he was with IBM Corporation, working in the field of terrain avoidance computers. After an educational leave of absence, he returned to IBM in 1963, where he has since held a number of engineering and engineering management positions in matched filter design and radar processor and display development. His present position as 4 Pi Computer and Display Products Engineering Manager involves both present products and the advanced 4 Pi computers.

A Discussion Of Digital Processing In Synthetic Aperture Radar

J.C. Kirk, Jr.
(Raytheon Company, Goleta, California)

IEEE Transactions on Aerospace and Electronic Systems, Vol. AES-11, No. 3 May (1975)
Pages 326-327

Abstract

This is a summary paper describing the processing of synthetic aperture radar (SAR) data using digital correlation algorithms. Fundamental SAR theory as it applies to the various SAR modes, namely, strip mapping, spotlight mapping, and Doppler beam sharpened mapping, is described and a baseline design applicable to all SAR modes is presented. Digital processor design is developed, starting with a simple single filter mechanization and proceeding through more complex processing algorithms. Prefilter design is discussed, as is the more advanced processing algorithms, namely, multiple parallel prefilters, two-stage correlation, and FFT processing. The primary processor tradeoff is increased functional complexity versus reduced arithmetic and memory requirements. For high-resolution applications, the arithmetic requirements can be reduced by an order of magnitude or more by implementing the more advanced processing algorithms.

l = synthetic array length
L = projected synthetic aperture length
θ_S = synthetic beam width (3dB0
K_S = synthetic array weighting constant
r_a = azimuth resolution
r_r = range resolution
$\bar{V}$ = aircraft velocity
$\bar{u}$ = antenna pointing vector
ΔT = coherent integration time
f_r = pulse repetition rate
M_B = bulk memory size in bits
k_a = word length in bits
R_c = correlation arithmetic rate
N_R = number of range gates
K_R = range resolution oversample factor
B = beam sharpening ratio
N = number of radar interpulse periods during coherent integration time

Manuscript received May 31, 1974. Copyright 1975 by *IEEE Trans. Aerospace and Electronic Systems,* vol. AES-11, no. 3, May 1975.

C_R = range compression ratio
C_A = azimuth compression ratio
A/D = analog-to-digital convertor
I,Q = in phase and quadrature
λ = radar wavelength
R = range
N_F = number of Doppler filters in correlator
K_A = azimuth resolution oversample factor
T_F = map frame time in spot and DBS
N_A = number of high-resolution azimuth lines per map frame
Δf_D = Doppler band to be processed
Δf_S = Doppler filter width for synthetic antenna
f_p = sample rate out of prefilter
PRF = pulse repetition rate
K_{os} = prefilter oversample factor
N_p = number of parallel integrators per prefilter
ΔT_p = coherent integration time of prefilter
K_p = prefilter weighting constant
K_1 = frequency normalized to 3-dB width
R_p = prefilter arithmetic rate
m = number of parallel prefilter channels
TAR = total arithmetic rate
M = number of first-stage correlator filters
FFT = fast Fourier transform.

I. Introduction

Synthetic aperture radar (SAR) developed as a side-looking radar sensor using optical signal processing for all-weather ground mapping. This technology has developed over the years and good optically correlated SAR imagery has been obtained. Much literature over the last decade has been devoted to fundamental side-looking SAR theory and optical correlation principals [1]-[25]. SAR theory, however, is quite broad and has many applications other than just side-looking mapping using optical correlation, as is being discussed in current literature [26]-[31]. As with a real aperture, the synthetic aperture can be steered to forward

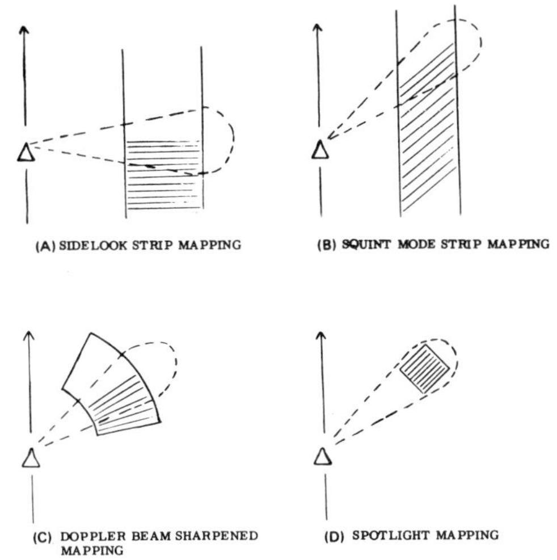

Fig. 1. SAR mapping modes.

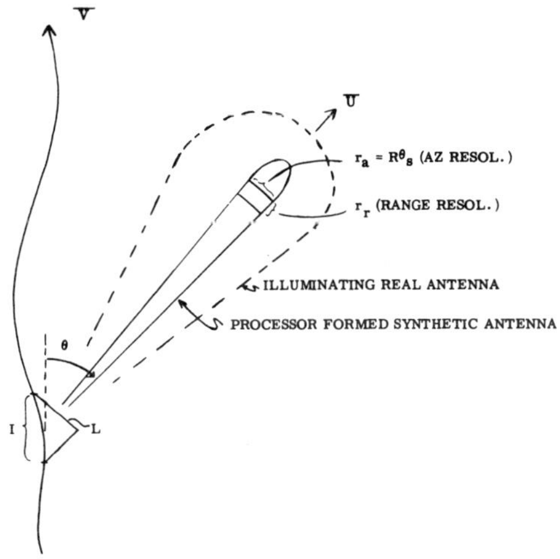

Fig. 2. Basic SAR geometry.

or aft directions and can even be scanned. Many new SAR modes are feasible and a wide variety of signal processing techniques are available to form and steer the synthetic antenna. It is the use of electronic signal processing and, in particular, digital techniques which makes these new modes feasible. It is the purpose of this paper to present a summary of the signal-processing algorithms which can be implemented with digital techniques.

The use of electronic signal processing in place of optical facilitates the steering of the synthetic array to positions other than broadside. In particular, the following "new" SAR modes, as illustrated in Fig. 1, can now be mechanized:

1) forward-looking (squint) strip mapping
2) Doppler beam sharpened (DBS) mapping
3) spotlight (telescope) mapping.

Squint mapping is similar to conventional side-look SAR, but at a variable squint angle. DBS is a beam sharpened version of PPI (planned position indicator) mapping. Spotlight mapping is the newest SAR mode, where a high resolution snapshot image is generated about designated map coordinates. Successive snapshot maps can be generated with each map centered about the same coordinates.

The design of the synthetic aperture radar involves range gating and implementing the equation

$$\int_I r(x) s(x)\, dx$$

in each range gate, where x is the displacement of the aircraft, $s(x)$ is the radar signal return, $r(x)$ is the radar reference function, and the interval I is the synthetic array length. The reference function, upon making the substitution $x = Vt$, is given by

$$r(t) = W(t) e^{-j\phi(t)}$$

where V is the aircraft velocity, t is time, $W(t)$ is the synthetic array weighting function which controls the sidelobes of synthetic antenna pattern, and $\phi(t)$ is the phase reference given by

$$\phi(t) = \int \omega_d(t)\, dt.$$

The phase reference is made up of a linear phase term which steers the synthetic beam, a quadratic phase term which focuses the synthetic aperture, and higher terms which are generally negligible. The Doppler reference frequency $\omega_d(t)$ is a dynamic variable, since the aircraft velocity $V(t)$ can change due to wind gusts or aircraft maneuvers; i.e., $\omega_d = 4\pi \overline{V(t)} \cdot \bar{u}/\lambda$, where λ is the RF wavelength and $\bar{u}$ is the synthetic antenna pointing vector.

The basic SAR geometry is illustrated in Fig. 2. The angular resolution is given by $\theta_S = K_S \lambda / 2L$, where K_S is the synthetic array weighting constant (approximately 1.2 for 30-dB sidelobes on the synthetic antenna pattern) and L is the projected synthetic aperture length given by $L = I \sin\theta$, where θ is the horizontal angle between $\bar{V}$ and $\bar{u}$. In distance, the azimuth resolution is $r = R\theta_S$, where R is the range to the resolution cell.

The synthetic aperture essentially sits inside the real aperture. When the synthetic aperture is steered, the real aperture must also be steered, but not as precisely. The two antennas are locked together by use of a clutter tracker or a clutter lock servo loop. The movement of the synthetic antenna is accomplished by rotating the unit vector $\bar{u}$. This changes the Doppler reference function ω_d, which can, for a constant aircraft velocity, be expressed as $\omega_d = (4\pi/\lambda) V$

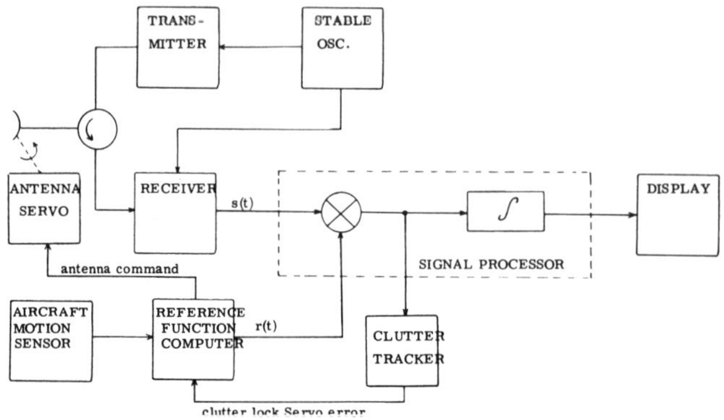

Fig. 3. SAR block diagram.

cos θ. The angle θ, as measured from the velocity vector, becomes the pointing angle for both synthetic and real antennas.

A block diagram for mechanizing the synthetic aperture radar is illustrated in Fig. 3. The radar is coherent so that the return signal phase history can be recorded. A stable oscillator which is common to both transmit and receive provides the required coherency. Formation of the synthetic aperture involves mixing the received signal on a range gate by range gate basis with a referenced function whose phase history hopefully is an accurate prediction of the return signal phase history and integrating the mixer output over many interpulse periods. The phase history prediction process is the essence of motion compensation. The mixing and integrating is the signal processing, which essentially forms Doppler cells in each range gate. The processing algorithm can be thought of in terms of Doppler filtering, Fourier transforming, cross correlation, or vector summation of synthetic array elements, depending upon which approach is most familiar and which approach is easiest to use. All of these approaches are equivalent.

Motion compensation is required to stabilize the processor range gates and Doppler filters on the ground. The basic motion compensation equation is the formation of the dot product $\overline{V} \cdot \bar{u}$. As the aircraft moves, the time range gating and the Doppler references are changed so that the processor cells are stationary on the ground. The motion compensation equations are generally solved in a general-purpose computer (GPC).

The digital signal processing will be discussed in greater detail in this paper. Particular attention will be paid to the various processing algorithms which can be employed to form the synthetic aperture and the memory and arithmetic requirements associated with each algorithm. A companion paper to this one, which follows, will discuss the motion compensation which is necessary to stabilize the synthetic aperture.

II. Basic SAR Processor Design

The two main components in the SAR signal processor are data storage for the elements of the synthetic array and arithmetic for the mixing operation. The optical correlator uses film for data storage and optics for arithmetic. The digital signal processor uses semiconductor devices for memory and arithmetic. The mixing operation is the key to the digital SAR signal processor, for it is here that the signal is phase shifted with the phase reference obtained from the motion compensation computer.

As illustrated in Fig. 3, the signal processing requires two basic operations, multiplication and integration. The multiplication is broken down into phase shifting and array weighting. The output of the integrator(s) on successive filter dumps is the synthetic antenna pattern. This very simplified signal processing is illustrated in Fig. 4.

To make a map, multiple range cells (and, usually, multiple filters per range cell) are required. The illustration in Fig. 4 shows the signal flow for a point target signal in a single range gate. The samples of the target occur at the pulse repetition frequency (PRF) of the radar, f_r. A number of samples N corresponding to the coherent integration time or desired array length are summed together after multiplication. The samples can be stored ahead of the mixer with the multiplication occurring after the array's length of data has been recorded, or after the mixer in the integrator. In the first case, the motion compensation is applied on a delayed basis and, in the latter case, it occurs in real time.

When multiple range gates are similarly processed, an azimuth line is formed. When successive azimuth lines are laid side by side, the map is formed. Each azimuth line is like a sample of the snythetic array pattern. The azimuth sample spacing must be at least equal to or less than the snythetic beamwidth to avoid loss in information.

The digital processor is mechanized at video with I and Q channels, as illustrated in Fig. 5. The video mixing operation

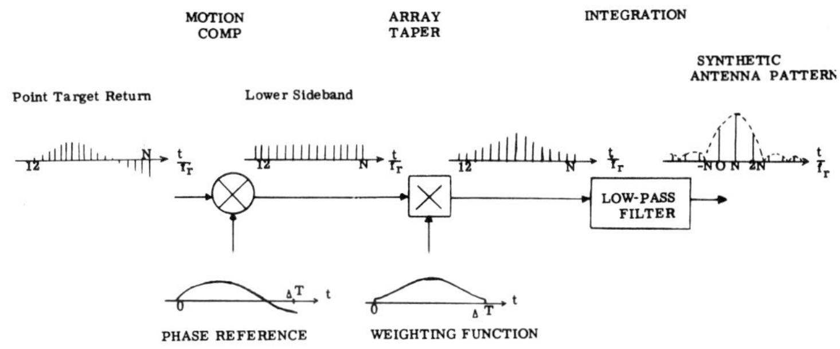

Fig. 4. Signal flow through a simplified SAR processor.

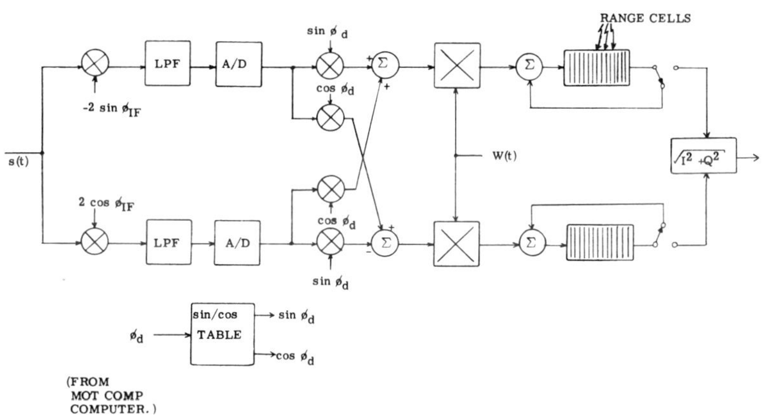

Fig. 5. Digital *I* and *Q* SAR processor—multiple range gate, single filter.

involves four multiplications and two additions. The adders select the lower sideband or difference signal and cancel the upper sidebands or sum signal. After integration the data is amplitude-detected and available for recording or display.

This example digital processor, as illustrated in Fig. 5, has one Doppler filter and, thus, could serve in an unfocussed SAR strip map mode. Successive dumps from this one filter would then be approximately a synthetic beam width apart.

A/D Convertor

The data comes into the processor by way of the analog-to-digital convertors (A/D). There are two convertors, one I and one Q. The amplitude or dynamic range of the A/D convertor is determined by the dynamic range inherent in the video signal and the compression gain of the signal processor. Each digital bit corresponds to 6 dB of dynamic range, and, thus, a four-bit A/D would have 24 dB of dynamic range. The conversion speed of the A/D's is determined by the bandwidth of the radar pulse. A range resolution of 100 feet corresponds to an RF bandwidth of approximately 5 MHz and, thus, two 5-MHz A/D convertors would be required. Approximate expressions for the A/D dynamic range (DR) and speed (S), based upon the above considerations, are

$$\mathrm{DR}_{\mathrm{A/D}} = \mathrm{DR}_{\mathrm{desired}}/C_R B$$

$$\mathrm{S}_{\mathrm{A/D}} = K_R c/2r_r$$

where $\mathrm{DR}_{\mathrm{desired}}$ is the desired linear dynamic range of the processed image, C_R is the range compression if digital pulse compression is used, B is the beam sharpening ratio (ratio of real beam width to synthetic beam width), K_R is a range resolution oversampling factor, usually near unity, c is the speed of light, and r_r is the range resolution.

The desired dynamic range is a function of the ground reflectivity variation. The synthetic aperture weighting function should be designed to provide sidelobes consistent with the desired dynamic range. In addition, this desired dynamic range must be compressed to fit the visual display or recording device, which usually has a dynamic range of approximately 20 dB. If the desired dynamic range is 50 dB and the processor compression gain $C_R B$ is 1000 or 30 dB, then the required A/D dynamic range is 20 dB. Thus, four-bit A/D's would be required for this case. The radar receiver must provide the required gain control, by way of AGC and IF limiting, to match the A/D dynamic range and prevent A/D saturation.

Memory and Arithmetic

The two main elements in designing a digital processor are the memory and processing speed requirements. For the simple single-filter processor illustrated in Fig. 5, the memory size and correlation rate are

$$M_B = 2k_a N_R \qquad R_C = N_R f_r$$

where k_a is the digital amplitude in bits per word, N_R is the number of range gates, and f_r is the PRF. The factor of two accounts for the I and Q channels. The bulk memory requirement M_B is the number of bits of digital storage required and the correlation rate R_C is the average speed that the data is pipelined through the arithmetic unit. Thus, for the single-filter case, there are N_R complex mixer operations each interpulse period $1/f_r$.

The required number of range gates is given by

$$N_R = K_R(R_{SW}/r_r)$$

where R_{SW} is the line-of-sight range swath to be mapped. For a range swath of 10 nmi, a range resolution of 100 feet, and a weighting constant of unity, 600 range cells are desired. The PRF f_r is determined by the Doppler spread associated with the real antenna and must be high enough to avoid Doppler ambiguities.

The correlator rate, assuming a typical PRF of 1 kHz, is 600 kHz for this example, and this is much less than the 5-MHz A/D rate. Thus, to realize this lower arithmetic rate, a buffer must be inserted after the A/D's to read the data in at a 5-MHz rate and out at a 600-kHz rate. This is possible since the unambiguous range for a 1-kHz PRF is approximately 80 nmi and, with a 10-nmi range swath, there is ample time between each range swath return read in to process the data slower. Thus, functional complexity, in the form of an additional operation, is added in order to achieve a slower processing rate. This is typical of the design trend in SAR digital processor design.

For the simple processor of Fig. 5, the dynamic range required in the integrator is the signal dynamic range out of the arithmetic unit times the gain of the integrator. For a point target, this gain is N, the number of samples coherently integrated. For distributed clutter which is noise-like, the gain is $N^{1/2}$ for a PRF approximately equal to the clutter width. For an N of 1000, the filter gain is between 30 and 1000. For a four-bit input, the integrator dynamic range is somewhere between 9 bits and 14 bits. Taking 12 bits as the design choice, the correlator memory is $2 \cdot 12 \cdot 600 = 14.4$ Kbits.

The alternate approach to store the data ahead of the correlator will require N lines of 600 range gates to be stored. The total memory would be $2 \cdot 4 \cdot N \cdot 600$ and, thus, the precorrelator storage approach will require more storage. For $N = 1000$, this storage is 4.8 Mbits and, thus, precorrelation storage is not a good approach for this simple single-filter processor case.

Parallel Correlator Channels

The single filter is generally not adequate for most cases and parallel channels must be formed. In strip mapping the map must keep up with the forward motion of the aircraft and, thus, the synthetic aperture length cannot exceed the azimuth resolution for the single-filter processor. The ratio of the aperture length to the azimuth resolution is called the azimuth compression ratio; i.e.,

$$C_A = L/r_a = K_S \lambda R/2r_a^2$$

where L and r_a are both measured perpendicular to the line of sight. The number of required parallel Doppler filters is

$$N_F = K_A C_A$$

where K_A is an azimuth oversample factor generally taken to be unity. Thus, if the azimuth resolution is specified at a particular range, then the number of parallel Doppler channels is also specified. A two-to-one improvement in azimuth resolution increases the number of parallel channels by a factor of four. Notice also the linear dependence of the azimuth compression ratio on frequency. Thus, ten times as many Doppler channels are required at L band (1 GHz) than at X band (10 GHz) for the same resolution at the same range. If the processor size is fixed, i.e., C_A is constant, then the angular resolution varies with the square root of frequency. Thus, in going from X to L, the resolution will degrade by approximately a factor of 3 for a fixed N_F or C_A.

For spot and DBS, the number of Doppler channels is determined by frame time requirements. Assume a spotlight map is to be processed with a total number of azimuth lines equal to N_A. If N_F filters are processed each integration time, the total frame time to produce the N_A line map is

$$T_F = (N_A/N_F)\Delta T$$

or

$$N_F = N_A(\Delta T/T_F).$$

As an example, for $N_A = 600$, $\Delta T = 1$ second, and $T_F = 10$ seconds, the required number of Doppler filters is 60.

Similar reasoning holds for DBS, except that the maximum frame time is limited by the real beam width. The antenna must dwell for the coherent integration time and,

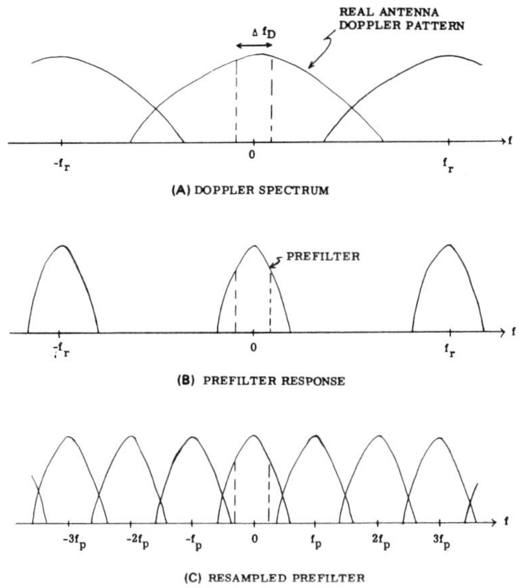

Fig. 6. Doppler considerations in SAR

thus, there is no need for more than B filters, where B is the beam sharpening ratio.

The multiple channels can be processed in either a line-by-line or batch process. In a line-by-line process, the filters are stagger tuned. The tuning of each filter and, hence, the time interval between the dump of adjacent filters is $\Delta T/N_F$. Thus, if filter one is turned on at time zero, filter two is turned on at $\Delta T/N_F$, filter three at $2\Delta T/N_F$, etc. In batch processing, a group of filters are tuned and, hence, dumped at the same time. It has been said that batch processing goes "bong, bong, bong," whereas line-by-line processing goes 'hmmmmmmmmmmmmm."

To improve the resolution or frame time performance of the single-filter processor, then, additional Doppler channels are mechanized. The bulk memory and correlation rate for the multiple Doppler channels are

$$M_B = 2k_a N_F N_R$$
$$R_C = N_F N_R f_r$$

where N_F is the number of Doppler channels in the correlator. For $k_a = 12, N_R = 600$, and $f_r = 1$ kHz, the memory and arithmetic rates are 14.4 Kbits and 0.6 MHz for one Doppler channel, 144 Kbits and 6 MHz for ten channels, and 1.44 Mbits and 60 MHz for 100 channels. The main thrust of the rest of this paper is discussing processing techniques which can be implemented to reduce the memory and arithmetic requirements.

Multilooks

Processing the SAR image with effectively one output sample per resolution cell, $r_a \times r_r$, yields a map which is somewhat noisy. This noisy effect has been given the name "coherent breakup," and the way to improve the image quality is to provide multilooks. That is, multiple independent coherent looks are made of each cell, and these coherent looks are noncoherently combined to provide a better quality image. These independent looks are obtained by processing different portions of the RF bandwidth, different look angles, or different polarizations. The effect on the processor is to increase its complexity by the number of looks. Thus, for four looks, the memory and arithmetic rate each increase by a factor of four. If 60 Doppler filters are required for one look, 240 are required for four looks. Alternately, resolution or frame time can be sacrificed for multiple looks, thus keeping the processor size nearly the same.

III. Prefiltering

The Doppler considerations involved in SAR processing are illustrated in Fig. 6. The Doppler bandwidth which is being processed is given by

$$\Delta f_D = N_F \Delta f_S$$

where Δf_S is the synthetic antenna filter width and N_F is the number of correlator Doppler channels. For many cases where the processed Doppler bandwidth is much less than the total clutter width as seen by the real antenna, the PRF excessively oversamples the Doppler band of interest and, hence, the correlator is running faster than necessary. The PRF can be lowered only to the point where adjacent spectral line clutter from the antenna main beam begins to fold into the processed Doppler band and degrade dynamic range. The problem is how to lower the effective PRF as seen by the correlator while maintaining a higher radar PRF.

The solution to lowering the effective PRF is to insert a prefilter ahead of the correlator. The width of the prefilter is selected to match the Doppler band to be correlated and output of the prefilter is then sampled at a rate lower than the radar PRF. If the prefilter were an ideal (rectangular) filter, the output sampling rate would be equal to Δf_D. However, due to practical filter design limitations, the prefilter sampling will be little greater than Δf_D, i.e., $f_p = K_{OS} \Delta f_D$, where K_{OS} is the oversample factor and f_p is the sample rate out of the prefilter. For example, assume that $\Delta f_S = 1$ Hz and $N_F = 10$, corresponding to a useful Doppler band of $\Delta f_D =$ 10 Hz. For $K_{OS} \approx 2$, the prefilter output sample rate is $f_p \approx 20$ Hz and, for a PRF of 1 kHz, the data have been slowed down in Doppler by about 50 to 1. Thus, the correlator rate need only run at a speed of 0.12 MHz instead of 6 MHz for the example case.

Thus, we have the second processor design alternative, a prefilter followed by a correlator. With a prefilter the choice of locating the bulk memory in the integrator or ahead of it can be reviewed. The number of samples which are integrated in the correlator is

$$N = f_p \Delta T = K_{OS} \Delta f_D \Delta T$$

$$= K_{OS} N_F \Delta f_S \Delta T$$

$$= K_{OS} K_S N_F$$

where K_S is the synthetic array weighting constant; i.e., $K_S \approx 1.2$ for 30-dB sidelobe weighting. If the factor $K_{OS} K_S$ is less than the dynamic range increase in the integrator, then storing the data ahead of the integrator is a feasible design choice. Typically, $K_{OS} K_S \approx 2$, and the integrator buildup in bits is about the same.

The effect of adding a prefilter is to reduce the correlator rate. Thus, the correlator rate has been reduced but functional complexity added. This is essentially the design trade involved.

Prefilter Design

The prefilter can be mechanized like the correlator, i.e., a multiplication followed by integration. A single integrator is not sufficient, as there will be excessive aliasing out of the prefilter. However, with two or more time-overlapped integrators, this problem is avoided. The processor organization with the prefilter is illustrated in Fig. 7. The basic motion compensation reference is applied in the prefilter mixer. Individual filter tuning, including additional azimuth motion compensation, if required, is then applied in the correlator.

The design of the prefilter is an interative process. In general, the hardware design variables are the number of prefilter integrators N_p, the weighting used in the prefilter for reducing sidelobes, and the integration time of the prefilter,

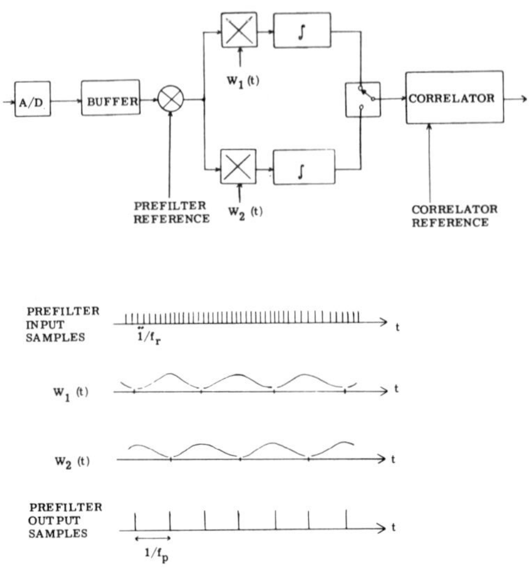

Fig. 7. Digital processor with prefilter and correlator.

ΔT_p. The design criterion is related to the dynamic range or base level in the prefilter, and this is determined by the prefilter main-beam and sidelobe folding or aliasing which occurs when resampling the prefilter output. An additional design criterion is the amplitude variation across the Doppler band caused by the prefilter shape. These prefilter design considerations will be illustrated in the following paragraphs.

In general, the sample rate out of the prefilter for N_p prefilter integrators is

$$f_p = N_p / \Delta T_p = K_{OS} \Delta f_D$$

where ΔT_p is the integration time of the prefilter. The prefilter 3-dB width is given by

$$\Delta f_p = K_p / \Delta T_p$$

where K_p is the prefilter weighting coefficient. To avoid prefilter foldover, the sample rate out of the prefilter (refer to Fig. 6) must be

$$f_p = (\Delta f_D / 2) + K_1 \Delta f_p$$

where K_1 is the normalized frequency, as measured from the prefilter center, related to the −3-dB width; i.e., −3-dB on the filter skirt corresponds to $K_1 = 0.5$, −20 dB on the filter skirt corresponds to $K_1 = 1.1$, etc. The value of the constants K_1 and K_p, as a function of the base level which can come from folding of either the main-beam skirt or the sidelobes, is illustrated in Fig. 8. The curve for K_1 in Fig. 8 was obtained by normalizing a synthetic antenna pattern

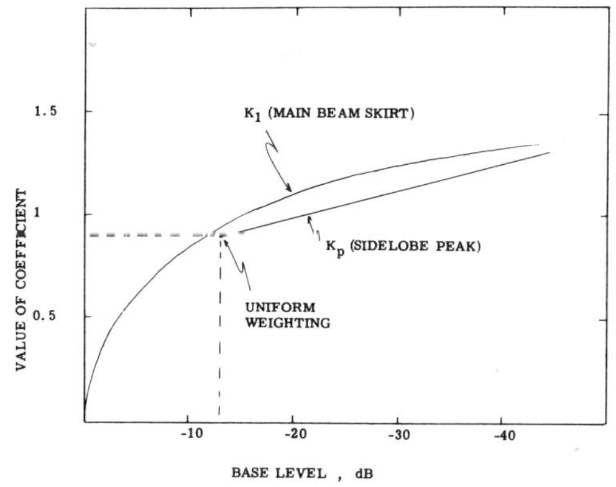

Fig. 8. Prefilter main-beam and sidelobe coefficients.

Fig. 9. Oversample ratio K_{OS} as a function of prefilter dynamic range.

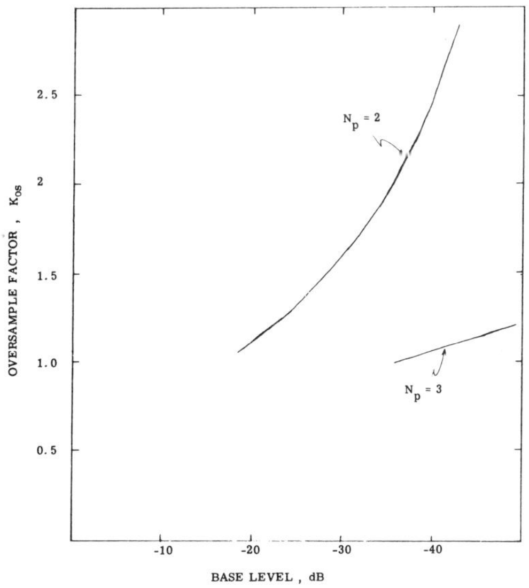

which had been Taylor-weighted for −30-dB sidelobes. The curve for K_p is obtained by computing the time-bandwidth product ($=K_p$) for various sidelobe level weighting functions. Essentially, K_1 determines the gain at which the main-beam skirt folds back into the useful Doppler band, and K_p determines the sidelobe level which folds back in.

From the above equations, the prefilter oversample ratio K_{OS} and the ratio of the Doppler band to the prefilter bandwidth (this determines the gain variation across the Doppler band which is caused by the prefilter) can be determined:

$$K_{OS} = f_p/\Delta f_D = N_p/2[N_p - K_1 K_p]$$

$$\Delta f_D/\Delta f_p = N_p/K_p K_{OS} \,.$$

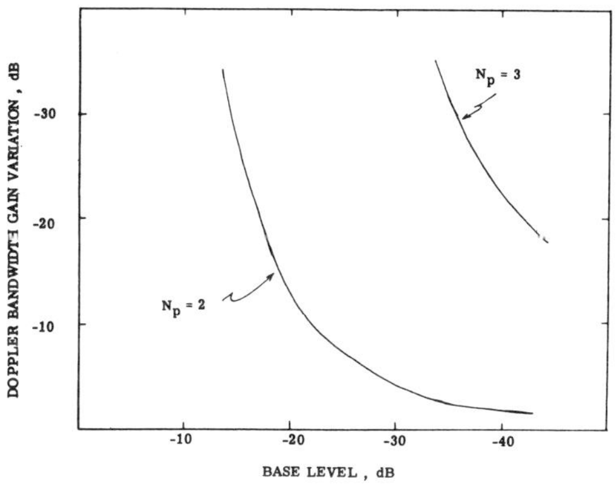

Fig. 10. Prefilter gain variation over Doppler band as a function of dynamic range.

The oversample ratio is shown plotted in Fig. 9 for values of K_1 and K_p from Fig. 8. It is desirable to keep K_{OS} as low as possible to minimize correlator memory and arithmetic requirements. The effect of the prefilter shape on the Doppler band to be processed can be determined by using Fig. 8 and K_{OS} and N_p from Fig. 9 to compute $\Delta f_D/\Delta f_p$. The gain variation is obtained by using K_1 of Fig. 8 and letting it be equal to $1/2\ (\Delta f_D/\Delta f_p)$. This weighting effect of the prefilter shape is shown plotted in Fig. 10, which shows the gain at the edge of the Doppler band relative to the center, assuming the Doppler band is centered in the prefilter.

This analysis of the prefilter design is based on the assumption that the dynamic range limitation from the prefilter skirt and sidelobes are roughly the same. Obviously, there are more design choices than this, but this case illustrates that two integrators $N_p = 2$ is a good design choice for 25- to 35-dB dynamic range. For more dynamic range at a reasonable oversampling, three or more integrators with heavier weighting must be employed. The gain variations across the Doppler band will also be large for three or more integrators. In addition, alternate prefilter design approaches can be taken, using multiple poles and zeros, etc.

Memory and Arithmetic

With a prefilter, the bulk memory and correlator rate requirements become

$$M_B = 2k_a N_F N_R$$

$$R_c = K_{OS} K_S N_F^2 N_R/\Delta T \,.$$

The memory is not really affected by adding the prefilter, although the bulk memory can now be placed ahead of the arithmetic unit if desired. The correlation rate varies with the square of the number of channels. For the example ten-channel case, the correlator rate is 6 MHz without the prefilter and 0.12 MHz with the prefilter. For 100 filters, the correlation rate is 60 MHz without the prefilter and 12 MHz with the prefilter. The prefilter rate is given by

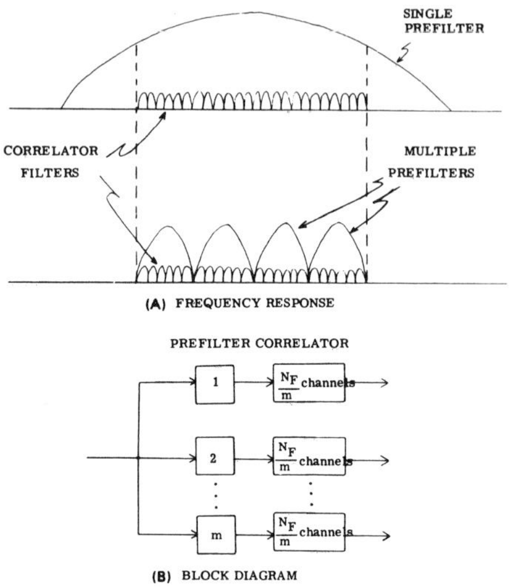

Fig. 11. Multiple-channel prefilter processor design.

$$R_p = N_R f_r$$

which, for the 600-range-gate, 1-kHz PRF case, is 0.6 MHz. A slight amount of additional arithmetic is required for multiple prefilter integrators and, also, a small amount of memory is required in the prefilter. In general, then, adding a prefilter will considerably reduce the required processor arithmetic rate, but will not significantly affect the required memory size.

IV. Advanced Processing Algorithms

Three advanced processing algorithms will be described showing how an increase in system functional complexity can result in greatly reduced arithmetic rate requirements and even a slight reduction in memory requirements. The concept to achieve the reduced arithmetic rate is to employ frequency division techniques.

Multiple-Channel Prefilter

For the 100-channel example case, the correlator rate when preceded by a prefilter is 12 MHz and the prefilter rate is 0.6 MHz. Without the prefilter, the 100-channel correlator rate would be 60 MHz. The question arises, can the arithmetic rate be further reduced? One approach to achieve a further improvement in arithmetic is to add multiple prefilter channels, as illustrated in Fig. 11. For an m-channel prefilter, the prefilter and correlator rates become

$$R_p = mN_R f_r$$

$$R_c = K_{OS} K_S N_F^2 N_R / m\Delta T$$

since there are now N_F/m correlation filters per prefilter channel. The total arithmetic rate (TAR) is

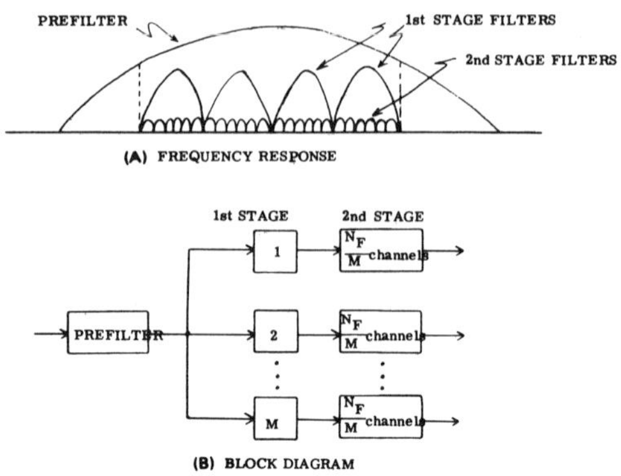

Fig. 12. Two-stage correlation.

$$\text{TAR} = R_p + R_c$$

$$= mN_R f_r + (K_{OS} K_S N_F^2 N_R / m\Delta T) .$$

The optimum m which minimizes the TAR is

$$m = (K_{OS} K_S N_F^2 / f_r \Delta T)^{1/2} .$$

For the 100-channel example, m is 4.45. For $m = 4$, the TAR is 5.4 and for $m = 5$, the TAR is also 5.4; thus, either a 4- or 5-channel prefilter would do. The TAR for $m = 1$ is 12.6 MHz and, thus, the TAR has been reduced by a factor of better than 2 to 1.

This additional improvement in the arithmetic rate has been achieved at the addition of extra prefilter memory and a little more functional complexity in the system design.

With a single prefilter, the processing can be performed either in a continuous line-by-line fashion or as a batch. With multiple prefilters, the correlator filters included in each prefilter must be batch-processed. The m prefilters, however, can be tuned sequentially, separated $\Delta T/m$ apart in time, or simultaneously. The advantage of sequential prefilter tuning is to allow a slight reduction in the correlator bulk memory if this memory is placed ahead of the correlator. The reason for this is that when the correlation is being done on the prefilter number one data, the data associated with prefilter number m is just beginning to be gathered. Thus, by appropriately sharing the memory among the m batches, a memory savings approaching 2 to 1 for large m can be realized. This memory savings is discussed in more detail for the two-stage correlator.

Two-Stage Correlator

The idea of frequency-domain splitting to reduce the arithmetic rate can be extended to the correlator. The two-stage correlator preceded by a prefilter is illustrated in Fig. 12. The correlator rate is the sum of the first- and second-stage correlator rates; i.e.,

$$R_c = R_{c_1} + R_{c_2} .$$

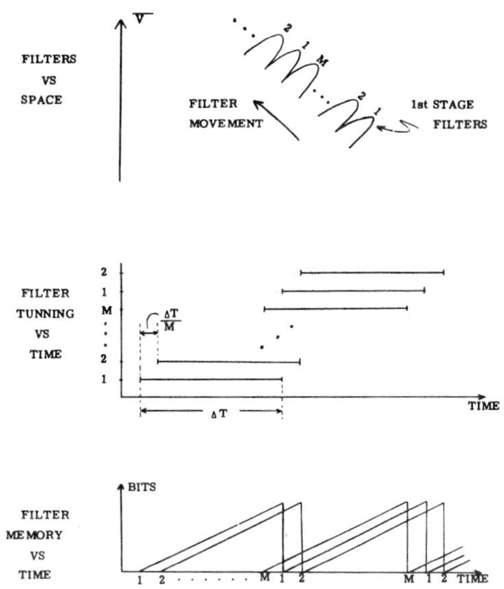

Fig. 13. Two-stage correlator memory savings.

The output of the first-stage correlator must be oversampled so as to avoid aliasing, as was the case for the prefilter. Thus, two or more integrators per first-stage correlator channel are required. If we assume the same mechanization for the prefilter as the correlator, i.e., two integrators, and the same weighting factor for both correlator stages, the correlator rate is

$$R_c = (K_{OS}K_SN_RMN_F/\Delta T) + (K_{OS}K_SN_RN_F^2/M\Delta T)$$

where M is the number of first-stage correlation channels. We are further requiring that the second correlator outputs N_F azimuth lines per coherent integration time, as would be necessary for a strip map mode or a scanning telescope mode. The minimum correlator rate occurs when $M = (N_F)^{1/2}$. For the 100-channel example case, M is 10 and the total correlator rate is 2.4 MHz. With a 0.6-MHz prefilter rate, the TAR is 3 MHz, which is a little better than a 4 to 1 improvement over the single-stage correlator TAR of 12.6 MHz, and not quite 2 to 1 better than the multichannel prefilter approach.

The first-stage correlator requires a little more arithmetic than the second-stage due to the necessity to form two or more individually weighted integrators per filter. Further, an additional bulk memory is required for the first-stage correlator. This additional bulk memory is

$$M_{B_1} = 2k_aMN_R.$$

The second-stage correlator memory stays the same, i.e., $2k_aN_FN_R$, although a slight savings may be realized in putting the bulk memory ahead of the second-stage correlator, instead of in the integrators.

Two-Stage Correlator Memory Savings

The two-stage memory savings is realized by sequentially tuning each of the M filters. Each batch of N_F/M final-stage filters is formed in $\Delta T/M$ seconds, as is illustrated in Fig. 13. The memory buildup in each filter follows a sawtooth pattern. Thus, at any one time only half the total memory is being used. By properly sharing the memory between the M batches, a memory reduction approaching 2 to 1 can be realized. The total bulk memory requirement for both stages, assuming precorrelator storage, is

$$M_B = M_{B_1} + M_{B_2}$$
$$= 2k_aK_{OS}K_SN_R\ [M + (M + 3)N_F/2M]$$

where we are assuming that the memory amplitude is the same for each correlator stage. For the 100-channel case, the memory improvement is

$$1 - \{M + [(M + 3)/2M]N_F\}/N_F = 1/4 \text{ or } 25\,\%.$$

Thus, there is a slight memory improvement in going to a two-stage correlator is a sawtooth precorrelator storage is used. The 1/2 memory size reduction is not reached in this example case, where $N_F = 100$, but for $N_F = 900$, the memory savings is 48 percent.

FFT Processing

The concept of multiple-stage frequency-domain splitting to minimize the arithmetic rate reaches its optimum in the fast Fourier transform (FFT). The FFT is composed of $\log_2 N_F$ stages to batch-process N_F filters. The correlator rate, if preceded by a prefilter, is

$$R_{c_{FFT}} = K_{OS}K_SN_F\log_2(K_{OS}K_SN_F)/2\Delta T.$$

The correlator rate improvement over a single correlator preceded by a prefilter is $2N_F/\log_2(K_{OS}K_SN_F)$. For a two-stage correlator preceded by a prefilter, the minimum arithmetic rate is

$$R_{c_{2SC}} = 2K_{OS}K_SN_RN_F^{3/2}/2\Delta T$$

and the FFT improvement is $4N_F^{1/2}/\log_2(K_{OS}K_SN_F)$. For $N_F = 64$ and $K_{OS}K_S = 2$, the FFT correlator rate is 18.3 times slower (improved) than the single-stage correlator rate and 4.6 times slower than the two-stage correlator rate.

The FFT has special considerations associated with its use. The number of points is generally binary; i.e., $K_{OS}K_SN_F = 32, 64, 128$, etc. The location of the filters is equally spaced throughout the PRF interval. No individual filter tuning can be accomplished. The memory must be stored ahead of the arithmetic.

TABLE I

SAR Processing Algorithm Summary

Algorithm	Bulk Memory	Total Arithmetic Rate
Correlator	$2k_a N_F N_R$	$N_F N_R f_r$
Prefilter plus correlator	$2k_a N_F N_R$	$N_F f_r + (K_{OS} K_S N_F^2 N_R / \Delta T)$
Multiple prefilters plus correlator	$2k_a N_F N_R$	$2(K_{OS} K_S f_r / \Delta T)^{1/2} N_F N_R$
Prefilter plus two-stage correlator	$2k_a K_{OS} K_S N_R [M + (M + 3) N_F / 2M]$	$N_R f_r + (2K_{OS} K_S N_R N_F^{3/2} / \Delta T)$
Prefilter plus FFT	$2k_a K_{OS} K_S N_F N_R$	$N_R f_r + (K_{OS} K_S N_F N_R \log_2 K_{OS} K_S N / 2\Delta T)$

Note: The number of amplitude bits k_a, will vary slightly among the algorithms and, also, depending upon whether the storage is precorrelator or in the integrator. The first three algorithms are for integrator storage nad the last two are for precorrelator storage.

As an example of these considerations, assume that the motion compensation point spacing in azimuth must be at least every eight filters. The 64-filter FFT process for this case can produce degraded imagery. This limitation can be overcome by dividing the process up so that smaller FFT batches are processed. For example, eight prefilters can be formed, with each prefilter followed by a 16-point FFT. Another approach would be to have a prefilter followed by two stages of FFT, with separate tuning applied prior to each FFT batch.

The FFT approach is not as flexible as the correlator approaches, but it does offer attractive arithmetic advantages. The memory requirements for the single- and two-stage FFT algorithms are similar to those for the single- and two-stage correlator approaches, including the 2 to 1 memory improvement possible with two-stage processing.

A summary of the bulk memory and correlator rate requirements for the various processing algorithms is presented in Table I.

V. Conclusions

The use of digital processing technology has expanded the flexibility of SAR. The synthetic antenna can be steered easily to forward directions, and even scanned. The new SAR modes of DBS, spotlight, and variable squint angle strip mapping can be accomplished using digital video processing technology.

Digital technology is continuing to move rapidly ahead as it has done over the last ten years. In the future, more and more capability will be provided in ever-shrinking hardware. Power consumption, size, weight, and cost will decrease, while performance and reliability will increase.

The advances in digital hardware are being matched by advances in processing techniques. More efficient processing algorithms enhance the digital approach even more and the variety of SAR modes are easily accommodated by essentially the same algorithms.

The use of this all-weather, long-range, high-resolution imaging technique is being made practical by the rapid advancement in digital technology. The future should see an ever increasing use of SAR in a variety of applications.

Acknowledgment

This paper was written using much of the general synthetic aperture knowledge as disclosed in the open literature and summarized in the references. However, much of the work reported here is drawn from the many fruitful working relationships which the author established while working at the Westinghouse Electric Corporation, Baltimore, Md., and special acknowledgment must go to the competent technical staff there.

References

[1] W.M. Brown and L.J. Procello, "An introduction to synthetic aperture theory," *IEEE Spectrum*, pp. 52-62, September 1969.

[2] C.W. Sherwin, J.P. Rauna, and R.D. Rawcliffe, "Some early developments in synthetic aperture radar systems," *IRE Trans. Military Electronics*, vol. MIL-6, pp. 111-115, April 1962.

[3] L.J. Cutrona, W.E. Vivain, E.N. Leith, and G.O. Hall, "A high-resolution radar combat-surveillance system," *IRE Trans. Military Electronics*, vol. MIL-5, pp. 127-131, January 1961.

[4] H.L. McCord, "The equivalence among three approaches to deriving synthetic array patterns and analyzing processing techniques," *IRE Trans. Military Electronics*, vol. MIL-6, pp. 116-119, April 1962.

[5] L.J. Cutrona and G.O. Hall, "A comparison of techniques for achieving fine azimuth resolution," *IRE Trans. Military Electronics*, vol. MIL-6, pp. 119-121, April 1962.

[6] R.C. Heimiller, "Theory and evaluation of gain patterns of synthetic arrays," *IRE Trans. Military Electronics*, vol. MIL-6, pp. 122-130, April 1962.

[7] W.M. Brown and C.J. Palermo, "Effects of phase errors on resolution," *IEEE Trans. Military Electronics*, vol. MIL-9, pp. 4-9, January 1965.

[8] J.A. Develet, Jr., "The influence of random phase errors on the angular resolution of synthetic aperture radar systems," *IEEE Trans. Aerospace and Navigational Electronics*, vol. pp. 58-65, March 1964.

[9] C.A. Greene and R.T. Miller, "The effect of normally distributed random phase errors on synthetic array gain patterns," *IRE Trans. Military Electronics*, vol. MIL-6, pp. 130-139, April 1962.

[10] J.A. Develet, Jr., "Performance of a synthetic-aperture mapping radar system." *IEEE Trans. Aerospace and Navigation of Electronics*, vol. xx, pp. 173-179, September 1964.

[11] W.M. Brown, "Synthetic aperture radar," *IEEE Trans. Aerospace and Electronic Systems*, vol. AES-3, pp. 217-229, March 1967.

[12] J.A. Develet, Jr., "The influence of random phase errors on the edge response of synthetic-aperture mapping radar systems," Aerospace Corp., El Segundo, Calif., Rept. TDR-269(9990)-4, March 4, 1964.

[13] W.M. Brown and R.J. Fredricks, "Range Doppler imaging with motion through resolution cells," *IEEE Trans. Aerospace and Electronic Systems,* vol. AES-5, pp. 98-162, January 1969.

[14] L.J. Cutrona *et al.*, "On the application of coherent optical processing techniques to synthetic-aperture radar," *Proc. IEEE,* vol. 54, pp. 1026-1032, August 1966.

[15] W.M. Brown and J.F. Riordan, "Resolution limits with propagation phase errors," *IEEE Trans. Aerospace and Electronic Systems,* vol. AES-6, pp. 657-662, September 1970.

[16] L.J. Procello, "Turbulance induced phase errors in synthetic-aperture radars," *IEEE Trans. Aerospace and Electronic Systems,* vol. AES-6, pp. 636-644, September 1970.

[17] M.C. Thompson, Jr., and H.B. Janes, "Measurements of phase front distortion on an elevated line-of-sight path," *IEEE Trans. Aerospace and Electronic Systems,* vol. AES-6, pp. 645-656, September 1970.

[18] L.J. Cutrona *et al.*, "Optical data processing and filtering systems," *IRE Trans. Information Theory,* vol. IT-6, pp. 386-400, June 1960.

[19] R.O. Harger, "An optimum design of ambiguity function, antenna pattern, and signal for side-looking radars," *IEEE Trans. Military Electronics,* vol. MIL-9, pp. 264-278, July-October 1965.

[20] J.F.A. Ormsby, "Analytic coherent radar techniques for target mapping," *IEEE Trans. Aerospace and Electronic Systems,* vol. AES-6, pp. 295-305, May 1970.

[21] R.K. Raney, "Synthetic aperture imaging radar and moving targets," *IEEE Trans. Aerospace and Electronic Systems,* vol. AES-7, pp. 499-505, May 1971.

[22] R.O. Harger, *Synthetic Aperture Radar Systems Theory and Design.* New York: Academic Press, 1970.

[23] R.W. Larsen *et al.*, "A microwave hologram system," *IEEE Trans. Aerospace and Electronic Systems,* vol. AES-8, pp. 208-217, March 1971.

[24] R.O. Pilon and C.G. Purver, "Radar imaging of oil slicks," *IEEE Trans. Aerospace and Electronic Systems,* vol. AES-9, pp. 630-636, September 1973.

[25] R.O. Harger, "Synthetic aperture radar systems design for random field classification," *IEEE Trans. Aerospace and Electronic Systems,* vol. AES-9, pp. 732-740, September 1973.

[26] J. Holahon, "Synthetic aperture radar," *Space/Aeronautics,* vol. 40, pp. 88-93, November 1963.

[27] R.M. Kelly *et al.*, "Data processing for synthetic aperture radar," 1964 *NAECON Rec.,* pp. 194-200.

[28] W.M. Brown *et al.*, "Synthetic aperture processing with limited storage and presumming," *IEEE Trans. Aerospace and Electronic Systems,* vol. AES-9, pp. 166-176, March 1973.

[29] J.H. Mims and J.L. Farrell, "Synthetic aperture imaging with maneuvers," *IEEE Trans. Aerospace and Electronic Systems,* vol. AES-8, pp. 410-418, July 1972.

[30] J.L. Farrell *et al.*, "Effects of navigation errors in maneuvering SAR," *IEEE Trans. Aerospace and Electronic Systems,* vol. AES-9, pp. 750-776, September 1973.

[31] W.G. Bruner *et al.*, "Radar and navigation systems," Ch. 27 in *Electronics Designers Handbook,* to be published by McGraw-Hill.

John C. Kirk, Jr. (M'69) received the B.S.E.E. degree from North Carolina State University in 1964 and the M.S.E.E. degree from Marquette University in 1966.

He has been with Westinghouse Corporation, where he was involved in the design of synthetic aperture and MTI radars, and with the U.S. Air Force, where he performed flight-line maintenance of radar, navigation, and electronic countermeasures systems. He is now with the Advanced Systems Department of the Electromagnetic Systems Division of the Raytheon Company, Santa Barbara, Calif., where he is involved in the design of advanced systems and the application of digital processing techniques.

Mr. Kirk is a member of Tau Beta Pi, Phi Kappa Phi, Pi Mu Epsilon, and Eta Kappa Nu. He is a registered Professional Engineer in the state of Maryland.

CHAPTER VII
Applications of Synthetic Aperture Radar

The early applications of synthetic aperture radar systems were directed towards military reconnaissance (see Cutrona, Vivian, Leith, and Hall in Chapter I) and targeting for air attack (Dulberger (1965)). However, the high resolution radar imagery which can be obtained any time – day or night, during adverse weather conditions, or through cloud cover – has permitted a broader and more useful application of this powerful remote sensor. This section includes a number of papers which demonstrate how the SAR systems are being used to develop and preserve earth resources. In addition to this direction of SAR usefulness, these papers include design improvements or modifications which have enhanced the quality of radar imagery, such as those derived from interferometry, stereoscopy, polarization, and multi-spectrum transmissions. Therefore, this chapter shows how the early military SAR systems have moved to earth resource applications and begun to be used to map the lunar surface and its subsurface. Presently, SAR is contemplated to map the Venusian surface (see *Aviation Week and Space Technology*, 5 May 1972, and Kovaly (1969)).

In the first paper, L.C. Graham describes a synthetic interferometer radar which was used to obtain topographic maps. The conventional SAR with resolution in two dimensions is combined with an interferometer radar to yield the third orthogonal dimension. Also, L.C. Graham, in Graham (1975), describes a stereoscopic synthetic array which is applied to earth resource monitoring. This application has been used by earth scientists to indicate potential areas for mineralization.

R.O. Pilon and C.G. Purves discuss how a SAR was used to detect and monitor oil slicks. Early detection of oil slicks can help to reduce the potential deleterious nature of oil pollution and its effects on the ecology and beaches.

In the next three papers, a simultaneous X- and L-Band dual polarization radar is described as to its design, construction, and how it was tested. The radar imagery gathered by this SAR system has been used to classify fresh water ice, to detect and identify contour strip mines, and to study agricultural test sites. Additional experiments in agriculture assessment are described in Shuchman, Rawson, and Drake (1975).

An airborne hologram radar which combines the SAR for along-track mapping and a phased-array cross-track for receiving data is discussed by R.W. Larson, J.S. Zelenka, and E.L. Johansen in the sixth paper. It was developed as a two-dimensional analog to optical holography. Other unique SAR aspects and designs can be found in MacPhie (1973), Hansen (1974), Eichel (1975), and Harger (1976).

The last paper by L.J. Porcello et al. transcends the use of SAR's for earth mapping. This design team describes the Apollo lunar sounder radar system as it was applied to detect sub-surface geologic structures and to determine both a continuous lunar profile and an image of the moon. Other papers which discuss planetary mapping by radar are Rogers and Ingalls (1970), Shapiro (1968), and Tyler (1966).

On the design boards is a SAR system which is contemplated to map Venus' permanently hidden surface in the mid-to-late 1980's. With an anticipated resolution of 1,000 feet, this Venus orbiter is expected to map the entire surface in 243 Earth days (see pages 11 and and 61 of *Aviation Week and Space Technology*, 15 May 1972).

Synthetic Interferometer Radar For Topographic Mapping

L.C. Graham
(Goodyear Aerospace Corporation, Arizona Division, Litchfield, Park)

Proceedings of the IEEE, Vol. 62, No. 6 (June 1974)
Pages 763-768

Abstract—The production of topographic maps requires two kinds of information. First, the detail to be placed on the map sheet must be identified. Second, the positions of the various objects and features must be measured in three dimensions. Current airborne radar technology provides the means to satisfy both of these requirements in adverse weather and at any time, day or night. Radar used specifically for this purpose employs synthetic-aperture techniques to obtain fine resolution measurement in two dimensions and interferometry to obtain the third measurement.

I. Introduction

AERIAL PHOTOGRAPHY has been used for many years to collect data for the production of topographic maps. The efficiency of data collection has, however, been hampered by clouds and poor sunlight conditions. The problem is of particularly serious proportions in the cloud belt region within 15° of the equator and near the poles. The advent of fine resolution radar systems has removed these restrictions, permitting efficient scheduling of flight operations, regardless of weather and sunlight.

The radar is used in connection with other all-weather electronic systems which perform the function of navigating the aircraft and measuring its position in a geodetic coordinate system. The radar system must perform two functions. First, the various objects and features to be mapped such as roads, railroads, streams, lakes, shorelines, hills, valleys, and cities must be presented in an image with sufficient resolution to permit their ready identification. Second, a three-dimensional measurement of position, with respect to the platform, of a sufficient number of points to define the terrain surface must be made.

The fine resolution image of the terrain is provided by synthetic-aperture radar technology. The three-dimensional measurement is made by radar interferometry. The combination of these techniques is described in more detail in the following sections.

II. Radar Imaging and Geometry

The nature of the radar image, which must be understood to appreciate the radar mapping techniques, is first described with reference to a conventional- or "real"-aperture imaging radar as illustrated in Fig. 1. As shown there, a short burst, or pulse, of microwave energy is directed from a high-powered transmitter along a vertical fan-shaped beam by a directional antenna. As the resulting wave train impinges on the terrain, it becomes scattered, and a portion of it is sent back to the vehicle, where it is intercepted by the same antenna and sent to a sensitive radio receiver. It is amplified by the receiver and detected. That is, a signal is created whose amplitude depends on the magnitude of microwave energy received at any instant. This detected, or video, signal is then caused to control the brightness of a linearly moving spot of light. The moving spot is imaged on photographic film, thus forming a latent image which, when chemically processed, exhibits at any point a photographic transmission corresponding to the microwave reflectivity of terrain in the antenna beam.

Manuscript received October 3, 1973; revised January 8, 1974. The techniques described in this paper have been developed under the joint sponsorship of the U. S. Air Force Wright Air Development Center and the U. S. Army Engineer Topographic Laboratories. The resulting systems are currently deployed by the Air Force Charting and Geodetic Squadron of the Air Weather Service as tasked by the Defense Mapping Agency.

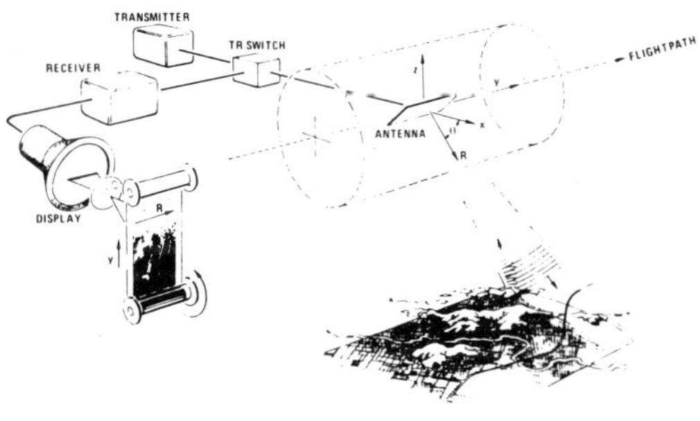

Fig. 1. Side-looking radar geometry.

As the motion of the aircraft carries the antenna along, the vertical beam is moved to new positions so that succeeding pulses sense adjacent strips of terrain. The film is moved in synchronism with the aircraft motion to record these adjacent terrain strips on adjacent film strips. The resulting map of terrain reflectivity is like a microwave photograph of the terrain, except that the measurable distance to features in the image corresponds to distances along the flight track and from the flight track, unlike the angular dependence of a photograph.

Note that two orthogonal coordinates may be measured on the radar image. The R coordinate of any object in the image is proportional to distance from the flight path to the object. The Y coordinate is proportional to distance along the flight path from some reference point to the object. These coordinates can be seen by inspection of Fig. 1 to correspond to the two linear coordinates of a cylindrical coordinate system whose axis is the flight path. A measurement of the angle θ is required to define the positions of the imaged objects in three dimensions.

III. Resolving Power

As in an optical image, the resolving power or resolution of the radar system producing the image may be taken as the 3-dB extent of the impulse response, i.e., the size (appropriately scaled) of the image of a single small scatterer. This response size is different in the two orthogonal coordinates of the image. In the range coordinate it is determined by the radar system bandwidth. If the receiver bandwidth is matched to the duration of the transmitted pulse, then range resolution is given by

$$W_R \approx \frac{c\tau}{2} \approx \frac{c}{2B} \tag{1}$$

where

W_R range resolution
τ transmitted pulse duration
B receiver bandwidth or transmitted pulse spectral width.

In the orthogonal along-track direction, the resolution is usually taken to be the 3-dB beamwidth of the antenna at the range of interest, or

$$W_A = R\beta \tag{2}$$

where

R distance from the aircraft to the scatterer
β $=\lambda/L_p$ by the well-known Rayleigh criterion

and

L_p antenna length.

IV. Radar Interferometers

The measurement of the depression angle θ may be accomplished by instrumenting the radar system with an additional aperture approximately normal to the radar-terrain line of sight and to the horizontal aperture, i.e., approximately vertical. The use of conventional angle of arrival technology, however, would result in an angular accuracy as defined by the Rayleigh criterion of:

$$\delta\theta = \frac{\lambda}{h} \tag{3}$$

where λ is the wavelength and h is the aperture size.

This uncertainty is not small enough to achieve useful accuracies with wavelengths long enough for good weather penetration and apertures small enough for airborne use. Therefore, some type of beamsplitting must be used. When beamsplitting is used, the angular uncertainty which may be achieved is given approximately by:

$$\delta\theta \approx \frac{\lambda}{2h(S/N)^{1/2}} \tag{4}$$

where S is the received signal power and N is the receiver noise power [1].

Thus the depression angle measurement uncertainty may be improved to the degree required by providing adequate transmitter power, under certain restrictions discussed later.

A convenient method of instrumenting a vertical interferometer is shown in Fig. 2 [2]. In this instrumentation, the signals backscattered from the terrain are received by two vertically separated antennas and simply added vectorially in a hybrid summing device. If the signal at the terminals of

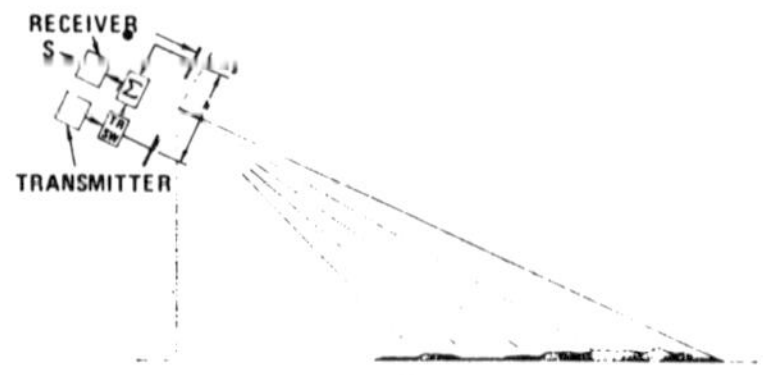

Fig. 2. Interferometer nulls.

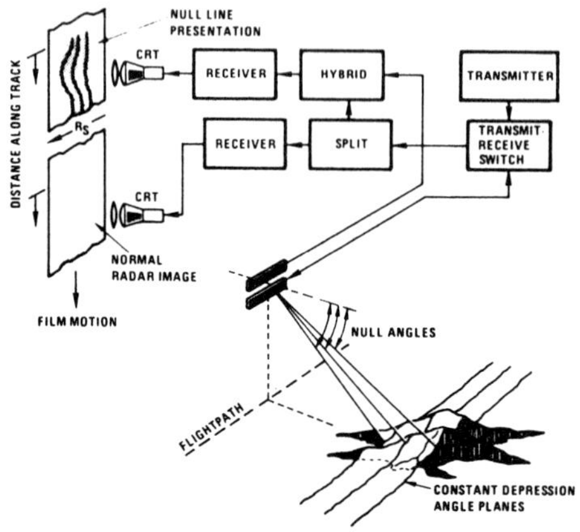

Fig. 3. Interferometer radar data acquisition.

the lower antenna is

$$e_a = A \sin \omega t$$

then the output from the upper antenna is

$$e_b = A \sin (\omega t + \phi)$$

and the sum is of the form

$$e_a + e_b = A[\sin \omega t + \sin (\omega t + \phi)] = 2A \cos \frac{\phi}{2} \sin \left(\omega t + \frac{\phi}{2}\right)$$

so that the envelope has nulls where $\phi = \pi(2N+1)$. Since the electrical angle $\phi = 2\pi a/\lambda$ nulls occur when

$$\frac{2\pi a}{\lambda} = (2n+1)\pi$$

$$a_{\text{null}} = \frac{(2n+1)\lambda}{2}.$$

But since $a = h \sin \theta$

$$\theta_{\text{null}} = \sin^{-1} \frac{(2n+1)\lambda}{2h}. \tag{5}$$

Thus nulls, or voids, will appear in the final image at certain predetermined depression angles as defined by (5). The ranges to the nulls may be measured by scaling the image, and the resulting depression angle accuracy at the range scaled will

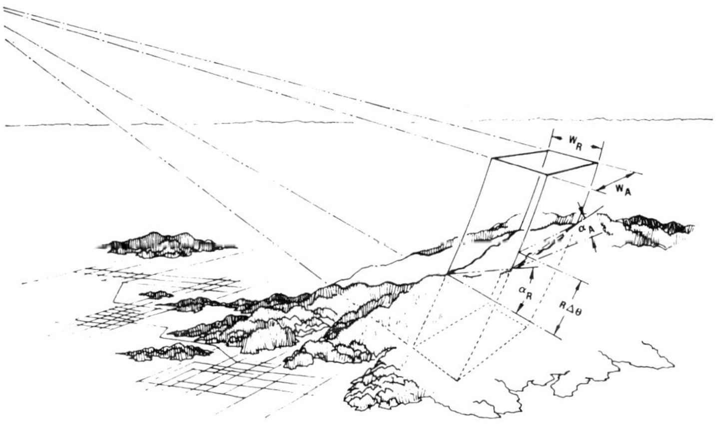

Fig. 4. Elevation ambiguity.

be defined approximately by (4). The nulls are, of course, ambiguous and the position of at least one point within the image must be determined by some other means to resolve this ambiguity.

This process is summarized in Fig. 3. In the figure, an additional recording channel is shown applied only to the lower antenna, resulting in an image as described earlier in connection with Fig. 1, in which the features to be mapped are recognized. Their locations are obtained by interpolating between the null lines at the corresponding points in the null line image. Coordinates are obtained from the transformation from cylindrical radar coordinates Y, R, and θ to Cartesian coordinates x, y, and z, viz:

$$y = Y$$
$$x = R \cos \theta$$
$$z = R \sin \theta \tag{6}$$

where x is the cross-track distance, y is the along-track distance, z is the vertical distance below the flight line, and R is the radar slant range; with appropriate image scales taken into account.

An alternative to the null line formation and measurement described here would be to measure the phase angle ϕ directly and calculate θ from

$$\theta = \sin^{-1} \frac{\lambda}{2\pi h} (\phi_m + n2\pi)$$

where ϕ_m is the measured (principal) value of the phase-angle difference. Again, the angular ambiguity would have to be resolved by knowing θ at some R to determine n. The same accuracies could be achieved, but the number of sample points could be increased beyond the number of naturally occurring null position.

V. Interferometer Restrictions

The angular measurement accuracy of (4) applies only to the case where signals arrive at the antenna aperture from one direction at a time. In the radar interferometer case, this implies that the angular uncertainty described by (4) is increased by the angular extent of signals arriving during a range resolution interval, or the angular extremes of a resolvable ground patch.

As illustrated in Fig. 4, if terrain within a resolvable ground patch were represented by a plane lying at angles α_R and α_A with respect to the range and along-track measurement directions, then the angular uncertainty ($\Delta\theta$) due to this restriction is given by

$$\Delta\theta = \frac{1}{R} (W_A \tan \alpha_A + W_R \tan \alpha_R). \tag{7}$$

Thus range and along-track resolution affect not only the range and along-track measurement accuracies but the orthogonal angular measurement accuracy as well.

VI. Mapping Accuracy

The prime performance parameter of a mapping system is the accuracy with which the various objects are located. Although there are many sources of error in a total mapping system, including such sources as the location and orientation of the airborne platform and data reduction errors, the radar sensor errors are of prime importance, and the limiting errors may be determined from (6) and (7). If ranging error due to range resolution, measurement, and timing is taken as dR, along-track error due to equivalent sources as dY, and angular accuracy due to the contributions of (3) and (7) as $d\theta$, the corresponding contributions to system measurement accuracy may be combined by differentiating (6). Thus

$$dy = dY$$
$$dx = \frac{\partial x}{\partial R} dR + \frac{\partial x}{\partial \theta} d\theta = \cos \theta \, dR + R \sin \theta \, d\theta$$
$$dz = \frac{\partial z}{\partial R} dR + \frac{\partial z}{\partial \theta} d\theta = \sin \theta \, dR + R \cos \theta \, d\theta.$$

If the terms are assumed independent error contributors and the differentials to represent small errors, then the variances may be written:

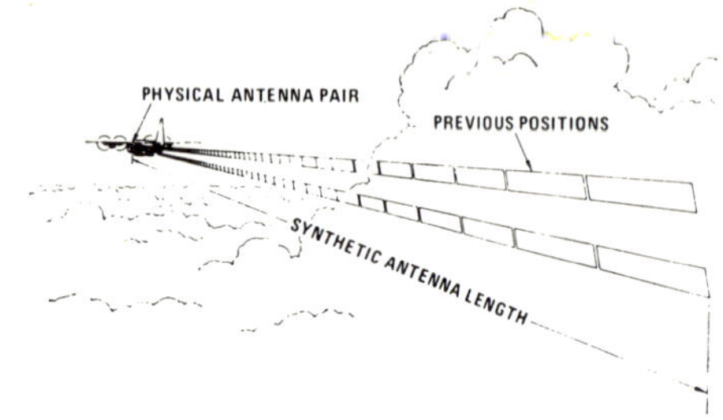

Fig. 5. Synthetic interferometer.

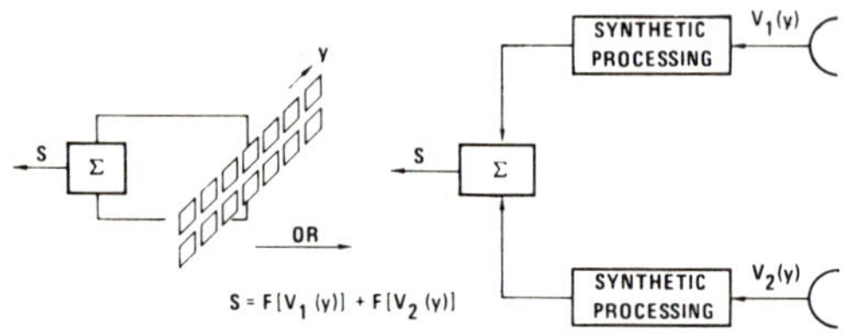

Fig. 6. Pre-summation processing.

$$\sigma_y^2 = \sigma_Y^2$$

$$\sigma_x^2 = (\sigma_R \cos\theta)^2 + (R\sigma_\theta \sin\theta)^2$$

$$\sigma_z^2 = (\sigma_R \sin\theta)^2 + (R\sigma_\theta \,\mathrm{co}\ \theta)^2. \qquad (8)$$

VII. The Synthetic Interferometer

Because of the limitation described in the previous paragraph and quantified by (7) and (8), accuracy is limited by signal to-noise ratio, range resolution (and thus bandwidth), and along-track resolution (and thus antenna length). Therefore, both accuracy and ability to recognize detail require good range and azimuth resolution. Range resolution may be improved by wide-band technology and, if necessary, pulse compression. As the radar in question is assumed to be carried on a moving aircraft, azimuth (along-track) resolution may be improved to the degree required by using synthetic apertures along the track. As the aircraft transports the physical antennas along the track, the returns from various positions are recorded and processed (vector summed) to synthesize antennas of the length desired (see Fig. 5).

Thus if two vertically spaced physical antennas are carried along parallel paths, the outputs of the two synthetic antenna systems can be combined to form a synthetic interferometer (see Fig. 6).

Functionally, this can be described as

$$S = F[V_1(y)] + F[V_2(y)]$$

where S is the system response, F is the Fresnel integral transform of the focused synthetic aperture processing, and $V_1(y)$ and $V_2(y)$ are the voltages recorded by the airborne system along the flight track.

Now, both the synthetic antenna process and the interferometer formation are linear vector summations of signals and may be interchanged to yield a more workable configuration without changing the basic character of the system with regard to resolution and accuracy (see Fig. 7).

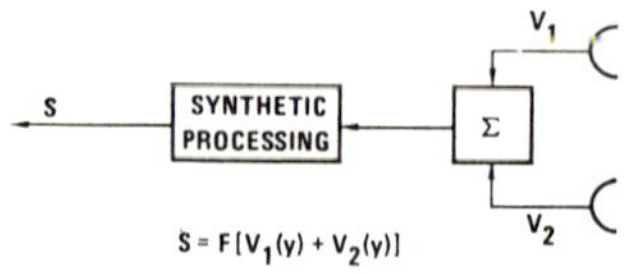

Fig. 7. Post-summation processing.

In the same functional notation, the previous relation is transformed to

$$S = F[V_1(y) + V_2(y)].$$

The second configuration is often preferred because the interferometer process is a simple summation of two signals occurring simultaneously in time, whereas the synthetic antenna processing is a summation over a large number of signals occurring sequentially in time and which must, therefore, be stored first. Thus the interferometer process is often more conveniently made linear over a wide dynamic range of input levels.

The detailed configuration of synthetic-aperture systems will not be discussed here. The interested reader may consult the available literature on the subject [3]–[7]. A simplified diagram of a synthetic interferometer using the null line approach is shown in Fig. 8. As shown there, signals from a single antenna are recorded on photographic film in a line-scan recorder similar to those described earlier. These are chemically processed and illuminated by coherent light in an analog optical processor. The synthetic aperture is formed by the action of the coherent light waves which are spatially modulated by the recorded signals, and the resulting image is recorded on a second photographic film. At the same time, signals resulting from combining the outputs of both antennas are recorded and processed in the same manner. The two resulting images are identical except that one contains the null lines whose cross-track (slant-range) positions are measured at various along-track positions. Once the depression angle for each null line is determined, the distance from the aircraft to each measured point in elevation and cross-track position is calculated from (6).

At each along-track position, the position of the aircraft in a geodetic coordinate system is simultaneously measured by a radio navigation system. These data are combined with the radar measurements to provide the positions of the imaged terrain points in three dimensions. The surface defined by these samples is used to calculate lines of equal elevation which are plotted as elevation contour lines on a final map product.

VIII. Representative System Performance

A system of this type has been built under joint Army–Air Force sponsorship and is currently operational with the Air Force Charting and Geodetic Squadron. The radar operates at a 3-cm wavelength and employs two 4 by 6 ft antennas separated about 4 ft vertically, or about 40 wavelengths. The antenna cavity is shown in Fig. 9. The antennas

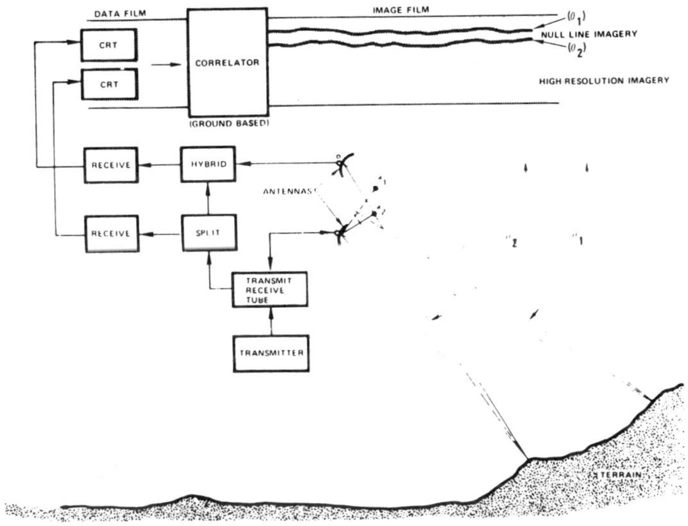

Fig. 8. Radar interferometry concept.

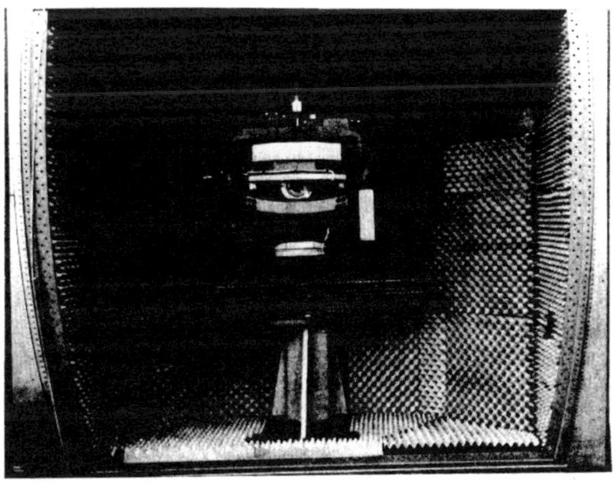

Fig. 9. Antenna and gimbal installation.

are mounted in an anechoic cavity to preclude loss of null depth through reflections into back lobes. They are mounted on a 3-axis stabilized platform to provide a vertical reference for depression angle measurement. The 40-wavelength aperture would provide a vertical beamwidth and accuracy of about 25 mrad if beamsplitting were not employed. In tests of the airborne system using synthetic interferometer beamsplitting, accuracies on individual points of 0.02, 0.07, 0.2, and 2 mrad were measured. The larger values resulted from interpolations between adjacent null lines in rough terrain. The accuracy of the entire airborne system was measured by repeated flights past a test area containing 36 identifiable points in both rough (slopes exceeding 10 percent) and smooth (slopes less than 10 percent) terrain. The results, including vertically (stability of the antenna platform), null measurement accuracy, and interpolation errors showed a 1 σ measurement accuracy of 0.2 mrad. This accuracy is more than adequate to permit compilation of 1:250 000 scale class B maps from collected data, the system requirement.

Examples of imagery from this system for both detail image or high resolution and interferometer channels are shown in Fig. 10. These images were obtained in operational flights over Puerto Rico. Note the liberal provision of range and along-track marks at 300-m intervals across both images

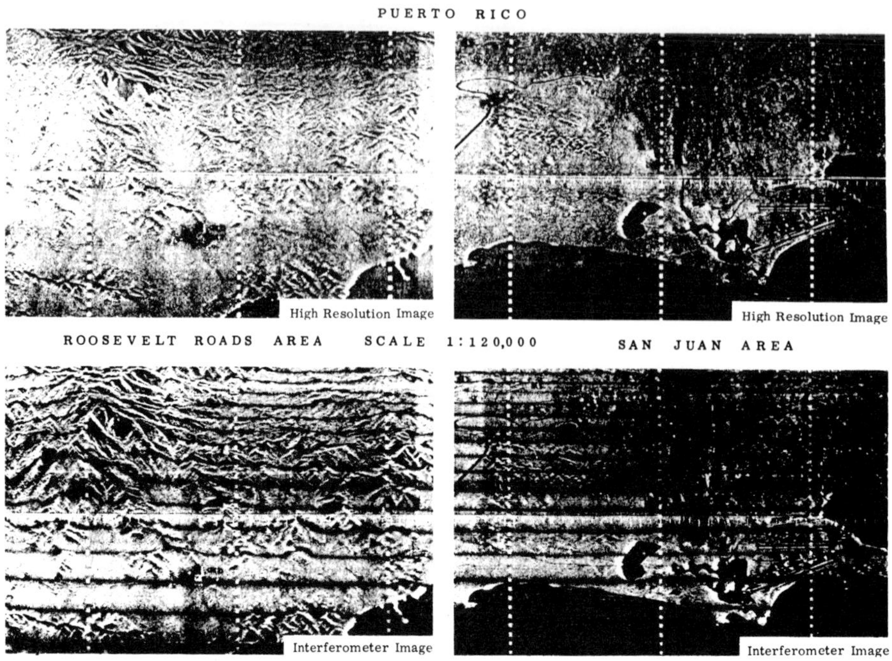

Fig. 10. Radar imagery.

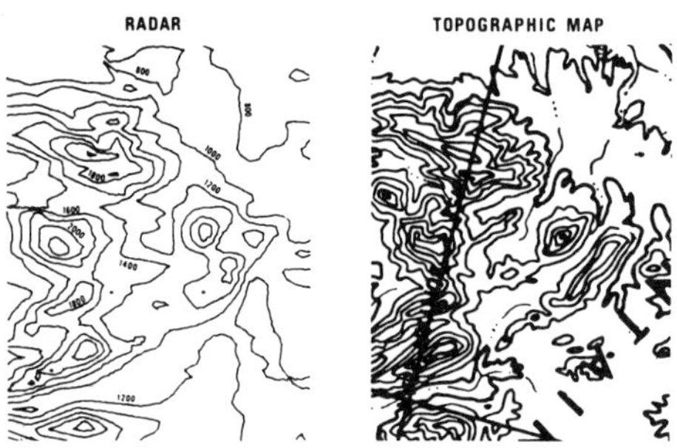

Fig. 11. Contour comparison.

to provide ready measurement and correlation with navigation system data. The images cover approximately 5 nmi in slant range and depression angles vary from about 25° to 45°. The along-track marks are *BCD* coded in length for correlation with navigation system fixes.

Data extracted from the interferometer channel during early system test are shown reduced to contour map form in Fig. 11. Although contour line shapes are not identical, the differences in elevation portrayed are within map accuracy standards.

IX. Conclusion

Systems using synthetic-aperture radar and interferometer techniques have been developed and reduced to operational status to provide all-weather collection of data from which topographic maps may be prepared. It is anticipated that future systems of improved capability will be widely used to supplement other methods of acquiring such data.

References

[1] M. L. Skolnik, *Introduction to Radar Systems*. New York: McGraw-Hill, 1962, pp. 189 and 477.
[2] Thompson *et al*, *Manual of Photogrammetry*, 3rd ed. Falls Church, Va.: American Society of Photogrammetry, 1966, p. 1030.
[3] R. O. Harger, *Synthetic Aperture Radar Systems Theory and Design*. New York: Academic Press, 1970.
[4] L. J. Cutrona, E. N. Leith, L. J. Porcello, and W. E. Vivian, "On the applications of coherent optical processing techniques to synthetic-aperture radar," *Proc. IEEE*, vol. 54, pp. 1026–1032, Aug. 1966.
[5] L. J. Cutrona, W. E. Vivian, E. N. Leith, and G. O. Hall, "A high resolution radar combat-surveillance system," *IRE Trans. Mil. Electron.*, vol. MIL-5, pp. 127–131, Apr. 1961.
[6] J. Holohan, "Synthetic aperture radar," *Space/Aeronautics*, Nov. 1963.
[7] C. W. Sherwin, J. P. Ruina, and R. D. Rawcliffe, "Some early developments in synthetic aperture radar systems," *IRE Trans. Mil. Electron.*, vol. 6, pp. 111–115, Apr. 1962.

Radar Imagery Of Oil Slicks

R.O. Pilon and C.G. Purves
(Naval Research Laboratory, Washington, DC)

IEEE Transactions on Aerospace and Electronic Systems, Vol. AES-9, No. 5 (September 1973)
Pages 630-636

Abstract

A joint agency controlled oil slick experiment, sponsored by the United States Coast Guard, was conducted in the Pacific Ocean in the Fall of 1970. The Naval Research Laboratory's synthetic aperature radar was used to detect and monitor the slicks at frequencies of 428, 1228, 4455, and 8910 MHz during the low sea state conditions encountered. At frequencies of 1228 MHz and higher, the slicks were depicted with sharp boundaries. At 428 MHz, the boundaries were indistinct. Approximately 400 liters of oil was detected as it was being discharged from the generating ship and larger spills were mapped from the initial thickness to equilibrium thicknesses of 1 micron or less. Thin streamers of oil and wind blown films were also imaged. Area growth rates were obtained for 2500 liter spills of API 26.1 crude oil and API 9.7 fuel oil on a calm sea. The respective rates, from approximately 1 to 4 hours after the spills, were 134 m^2/s and 16 m^2/s.

Introduction

The deleterious nature of oil pollution in its effects on the ecology and beaches has spurred research into various remote sensing techniques to detect, monitor, and identify oil slicks. The main problem is that of early detection. As this is the major factor in effectively dealing with the consequences of an oil spill, it is necessary to be able to search large areas of the ocean surface swiftly, with high levels of confidence in the detection of any pollution. To date, the main efforts have been centered in the ultraviolet, visible, infrared, and microwave regions of the electromagnetic spectrum. The results of investigations into the radar imagery of oil slicks are the subject of this paper.

The first radar imagery of an oil slick was obtained by the Naval Research Laboratory (NRL) in the Fall of 1969. This imagery of oil on the Chesapeake Bay is shown in Fig. 1. The large dark area to the right of center represents the main body of oil, with thin streamers visible in the right-hand quarter of the figure. The physical phenomenon that makes oil visible in radar imagery is the differing wave structure on the surface within and outside the slick. This is due to the damping effect that an oil film has on water waves. The dominant mechanism governing the backscatter of electromagnetic energy from an agitated water surface has been shown to be Bragg scattering [1], [2]. That is, water waves of length L, illuminated by radar energy of wavelength λ propagating in the same horizontal direction as that of the water wave system and at a depression angle θ from the horizontal, act as the dominant scattering centers when

$$L = \frac{\lambda}{2} \sec\theta, \quad 0 \leqslant \theta < \frac{\pi}{2}.$$

For the NRL four-frequency radar [3] (4FR), which operates in the frequency range from 428 to 8910 MHz, the resonant water wavelengths for near grazing incidence lie in the range from 1.5 to 35 cm. It is fortuitous that waves in this range are strongly damped by the presence of an oil film, and the absence of these waves markedly reduces the energy backscattered from such an area. In the imagery presented here the brightness of a particular area, within any given figure, is proportional (except in the immediate vicinity of a strongly scattering object, such as a ship) to the backscattered energy received from that area. Consequently, the oil covered areas appear as dark, nonscattering areas surrounded by the grainy gray-white image indicative of a more strongly scattering rough sea surface.

Imagery of an area is obtained by using the radar in its synthetic aperture mode [4]-[6]. This process is shown schematically in Fig. 2. In the 4FR system the antenna illuminates areas 0.19 to 3.7 km wide and 9.3 to 19 km long. During the period that each individual target area is illuminated by the radar beam, the phase of the backscattered energy from that area is recorded on film. The resulting phase histories for all the target areas viewed by the radar, when optically processed, produce the final imagery. When the system is optimized to yield the best contrast between open sea and oil, strongly scattering

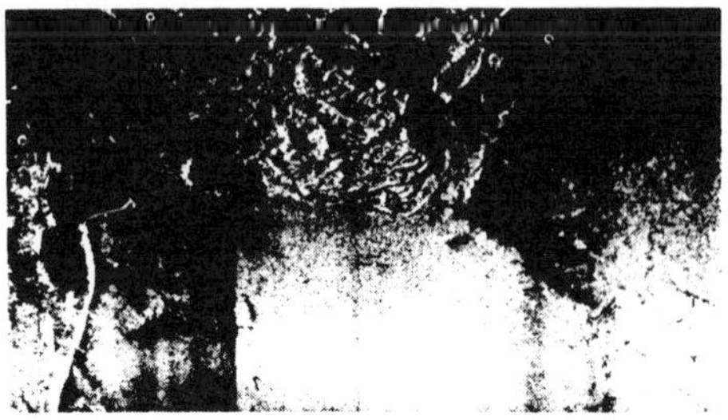

Fig. 1. Radar imagery of the Norfolk-Newport News area of Chesapeake Bay showing an oil slick in the right-hand portion of the figure. The imagery was obtained on October 16, 1969 using a synthetic aperature radar at 1228 MHz, vertical polarization. The area shown is 18 X 30 km.

Fig. 2. Schematic representation of a synthetic aperature radar system.

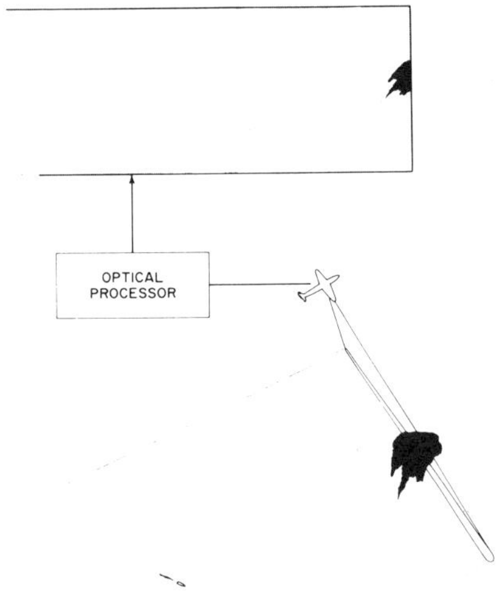

Fig. 3. Aerial photograph of a portion of the oil spill in Chadabucto Bay, Nova Scotia, February 1970. The superstructure of the sunken tanker *Arrow* appears in the lower right-hand corner.

Fig. 4. Radar imagery (4FR *L* band, 1228 MHz) of the oil slick in Chadabucto Bay and the Atlantic Ocean. Processed from the horizontal (L_{HH}) and vertical (L_{VV}) polarization channels.

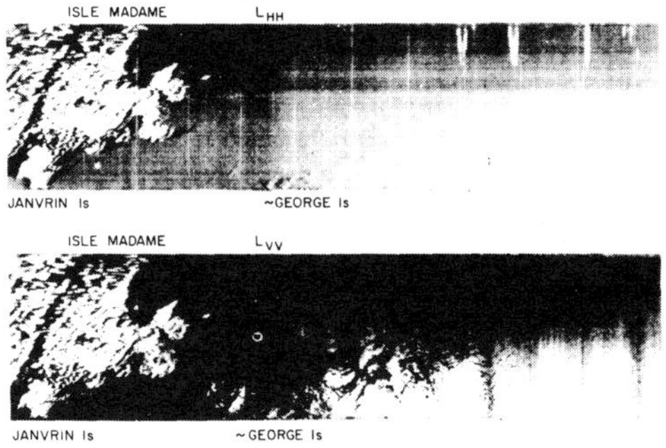

targets, such as ships, are greatly "overexposed" and appear as white areas.

The 4FR system is basically four different coherent, pulsed radars operating at UHF/*P* band (428 MHz), *L* band (1228 MHz), *C* band (4455 MHz) and *X* band (8910 MHz) at approximately 25-kW peak power. Each of the four systems is designed to operate on either horizontal or vertical polarization, or in combination, and the four frequencies may be intermixed on a pulse-to pulse basis, The resolution cell on the ocean surface measures approximately 17 meters in the along-track direction by 50 meters in range.

On February 17, 1970, at the request of the United States Coast Guard, this system was used to image the oil slick caused by the wreck of the tanker *Arrow* in the Chedabucto Bay area of Nova Scotia [7]. Approximately 5 $\times 10^6$ liters of Bunker C crude oil had spilled into the bay and surrounding waters. Fig. 3 is an aerial photograph showing a portion of the *Arrow's* superstructure and the large quantities of oil seeping from the sunken vessel.

Radar imagery was obtained at all four frequencies. An example of the *L*-band imagery is shown in Fig. 4. The bottom edge of each strip map is nearer the flight path of the aircraft. Each of the two strip maps show an area of approximately 18 X 70 km and show significant differences in the imagery obtained from the horizontally polarized channel (L_{HH}) and the vertically polarized channel (L_{VV}). The terrain of Isle Madame returns sufficient energy to yield strong imagery to the maximum range shown and beyond. The open areas of the sea, however, show the characteristic sea return pattern to ranges of only 4 km on horizontal and 11 km on vertical polarization. Beyond these ranges, the open sea appears dark due to insufficient energy return. These areas should not be confused with the dark regions due to oil at nearer ranges. As was the case in the observation of the oil slick in the Norfolk-Newport News area (Fig. 1), detection of the slick is evident when the transmitted and received signals are vertically polarized, whereas the simultaneous transmission of horizontally polarized signals shows appreciably little detection of the oil film. A ship traveling through an oil slick churns up the oil-water mixture and leaves relatively "clean" surface water in its wake. At oil-water temperatures near 0°C, as encountered, the cleared wake remains clear for some time. Such a wake would appear as a light streak, as seen in the lower left-hand corner of the L_{VV} imagery. The wake starts above "Is" of "Janvrin Is" and extends for a short distance to the right.

The limited ranges to which the open sea is imaged, and, consequently, to which a slick may be detected, is a result of the low power transmitted by the 4FR system. However, the disparity between the two channels points out the distinct advantage of using vertically polarized energy for the detection of oil slicks at sea. The typical backscatter radar cross section of the sea, as given by Guinard and Daley [8], is shown in Fig. 5. At the shallow depression angles generally used in this type of radar imagery, on the order of ten degrees, the backscattered vertically polarized signal is approximately 25 dB stronger than the horizontally polarized signal. Thus, much less transmitted power is necessary to produce the required imagery if vertically polarized signals are employed.

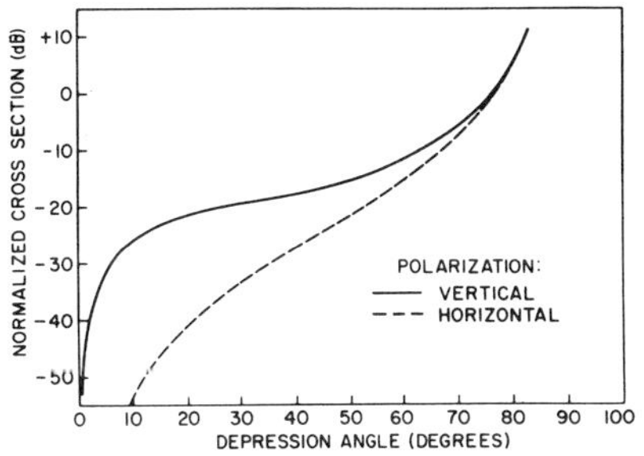

Fig. 5. Generalized variation of the radar cross section of a "clean" ocean surface with depression angle (after Guinard and Daley [8]).

Experimental Program

In late October and early November of 1970 the Naval Research Laboratory participated in a controlled oil slick experiment, also sponsored by the United States Coast Guard [9]. The purpose of the experiment was to determine the feasibility of various remote sensing techniques for the detection and identification of oil slicks. Those taking part in the experiment were the Aerojet-General Corporation, which also supplied the ship that generated the oil spills and collected ground truth data, the University of Michigan, NASA, and NRL. The sensors investigated ranged from optical sensors in the ultraviolet to infrared regions to radiometers in the sub-centimeter wavelength region and imaging radar from 3 to 70 cm wavelengths. The test was conducted in the open ocean off the coast of southern California, as shown in Fig. 6.

Fig. 6. Location of the controlled oil slick experiment off the coast of southern California.

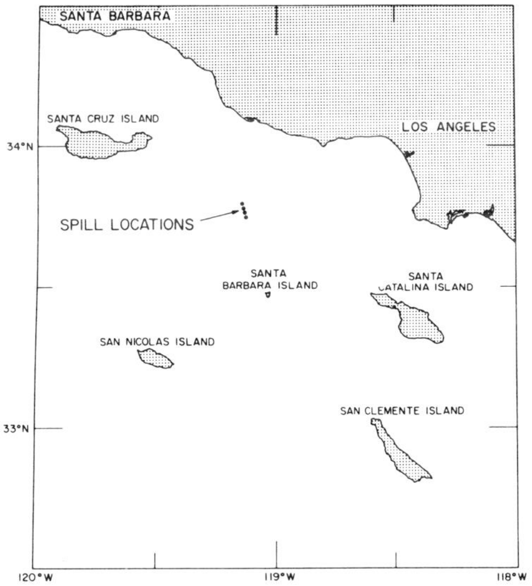

Four different weights of oil were studied: API 34.1 diesel oil, API 26.1 crude oil, API 21.6 crude oil, and API 9.7 fuel oil. The API designation for an oil is related to its specific gravity by

$$\text{API} = \frac{141.5}{\text{sp.gr.}} - 131.5$$

where sp.gr. is the oil's specific gravity at 21°C. Therefore, the specific gravities of the above oils are 0.854, 0.898, 0.924, and 1.00, respectively. Each spill was generated while the ship was lying still in the water. Upon completion of a spill, the ship would proceed on a generally northerly course for approximately 1 km, and then generate the next slick. The dates and times of each spill and pertinent sea state and synoptic weather data are given in Table I.

Fig. 7 through 10 and 12 are representative examples of the imagery obtained during this experiment. The images were processed from the vertical polarization channel and are oriented such that the bottom edge is nearer and parallel to the flight path of the aircraft. The range, from aircraft to slick, varied between 2.5 and 5 km. The rather well defined dark areas, excluding vertical banding and the "grain" of the image, represent the oil slicks generated for this experiment, while those of less clearly defined shape and extent are thought to be natural slicks.

Identification of the oil slicks and their areal extent, as measured from the radar imagery, appear in Table II. The

TABLE I
Ground Truth*

Date (1970)	Local Time	Remarks	Sea State†	Sea† (m)	Swell† (m)	Wind Dir.	Wind (m/s)
28 OCT	0800	34.1 API Spill Initiated	1	<.15	~.6-1	255°	3
	0827	34.1 Spill Completed (1250 ℓ)	1				
	0840	26.1 API Spill Initiated	1				
	0902	26.1 Spill Completed (1250 ℓ)	1				
	0908	21.6 API Spill Initiated	1				
	0935	21.6 Spill Completed (1250 ℓ)	1	<.15	~.6-1	255°	3
	0949	9.7 API Spill Initiated	1				<1
	1020	9.7 Spill Completed (1250 ℓ)	1	<.15	~.6-1	–	0
	1220		0				
	1448		0				
	1611		-	<.15	~1	310°	1-2
29 OCT	0740	The four slicks from 10/28	0				
	0840	have merged forming one	0				
	1130	large slick of irregular	0	0	–	–	0
	1457	shape ≈ 5 km x 6 km	0				
	1609		0				
6 NOV	1145	26.1 API Spill Initiated	0				
	1230	26.1 Spill Completed (2500ℓ)	0				
	1235	9.7 API Spill Initiated	0	<.15	~.6-1	225°	1-2
	1315	9.7 Spill Completed (2500ℓ)	0				
7 NOV	1230	One day aged 9.7 & 26.1 API	1-2	.3-.6	~2-3	250°	4-5
	1600	gravity oils in two discrete	1-2				
		slicks - 26.1 slick is highly					
		streaked, 9.7 slick is well					
		coalesced					

*Supplied by Aerojet-General Corporation.
†Visual Approximations.
Water temp.: 16.3±0.1 °C. Air Temp.: 17.1±0.4 °C. Relative humidity: 68±16%.
Pressure: 772±5 mm of Hg (1029±7 mb). Sky conditions: clear (low, thick, broken clouds on 6 Nov.).

TABLE II
Summary of Radar Imagery

Date (1970)	Run No.	Freq. (MHz)	Mean Local Time	Large Oil Slick: Quantity of Oil (liters)	Large Oil Slick: Area ($10^5 m^2$)	Large Oil Slick: Average Thickness (microns)	Small Oil Slick: Quantity of Oil (liters)	Small Oil Slick: Area ($10^5 m^2$)	Small Oil Slick: Average Thickness (microns)
				Diesel oil, API 34.1 Released 28 OCT, 0800-0827			Crude oil, API 26.1 Released 28 OCT, 0840-0902		
28 OCT	7	1228	0918	1250	~0.35	~35	1250	~0.19	~67
29 OCT	29	1228	0822	5000	45	1	The four spills of 28 OCT-diesel (34.1), crude (26.1), crude (21.6), #6134 fuel (9.7) - have merged into one slick		
	31	1228	0835		53	1			
	35	428	0904		40	1			
				Crude oil, API 26.1 Released 6 NOV, 1145-1230			#6175 fuel oil, API 9.7 Released 6 NOV, 1235-1315		
6 NOV	78	428	1352	2500	0.93	26	2500	~0.2	~130
	80	1228	1404		1.8	14		0.80	31
	82	1228	1415		2.7	9		0.76	33
	84	1228	1427		3.0	8		0.80	31
	87	1228	1444		3.7	7		1.1	22
	88	1228	1451		6.8	4		1.3	19
	91	8910	1517		7.8	3		1.6	16
	93	8910	1530		8.6	3		1.4	18
	95	8910	1542		10.2	2		1.6	16
	96	4455	1548		9.3	3		1.7	15
	98	4455	1600		9.3	3		1.9	13
				Unidentified Oil			Unidentified Oil		
7 NOV	108	8910	1249	2500	20	-	2500	-	-
	112	4455	1315		17	-		3.4	-
	114	4455	1328		21	-		5	-

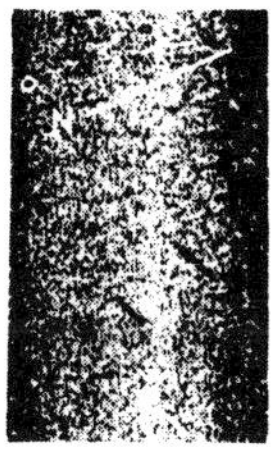

Fig. 7. Radar imagery of run 7 (1228 MHz). Two slicks are complete and a third is being generated. The area shown is 3 X 5 km.

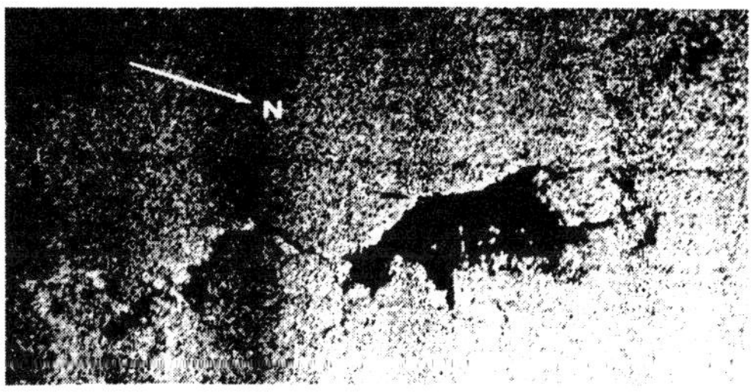

Fig. 8. Radar imagery of run 31 (1228 MHz). The four types of oil have merged to form one slick with streamers containing 5000 liters of oil. The area shown is 6 X 12 km.

Fig. 9. Radar imagery of run 35 (428 MHz). Aside from the systems induced factor of two compression in the along-track direction, the indistinct rendition of the slick boundaries compared to those in Fig. 8 is to be noted. The area shown is 5.5 X 11 km.

estimated error in the area measurements is less than 10 percent for runs 29, 31, 80, 82, 84, 91, 93, 95, and 96. The order of error increases for runs 87, 88, 98, 108, 112, and 114 due to decreased image contrast, and for runs 7 and 78 due to the small size of the image. The low value of area given for run 35 will be explained below. The average thicknesses were computed directly from the known quantity of oil in each slick and its area.

Fig. 7 (run 7) shows two completed spills, 1250 liters each of API 34.1 diesel and API 26.1 crude oil, oriented in a general west to east direction, and the beginning of a third spill of API 21.6 crude oil (approximately one third complete) to the north of the first two. The white spot at the west end of this third spill is the ship *Pacific Clipper*, in the process of generating the spill. This demonstrates that a small quantity of oil (400 liters or less) may be imaged as it is being discharged from the polluting vessel. Thus, radar is shown to have utility in policing operations. In synthetic aperature radar imagery, ships, being strongly scattering small targets, appear as white spots with dark "wings" oriented parallel to the flight path. This phenomenon, known as target capture, is quite pronounced in Figs. 7 and 8.

The imagery shown in Figs. 8 (run 31) and 9 (run 35) was obtained approximately 24 hours after that in Fig. 7. During the intervening period, the four slicks generated on October 28 had merged into one main body with streamers emanating from both ends. The values for areal extent given in Table II for runs 29 and 31 appear to indicate that the area of the slick was increasing fairly rapidly. These values, however, represent measurement error about an average. In addition to showing a systems-induced factor of two compression in the along-track direction, the 428-MHz imagery (Fig. 9) shows the expected indistinct rendition of the slick boundaries. It is believed that the "fuzziness" is due to the differing rates at which water wavelengths are attenuated upon entering an oil slick. At 1228 MHz and 428 MHz and the depression angles encountered in runs 31 and 35, the water wavelengths that act as primary radar scatters are approximately 12 cm and 35 cm, respectively. The 12-cm wavelengths damp out rapidly upon entering the oil slick, and, thus, the transition from the relatively strong scattering areas of open water to the weakly scattering oil covered areas is sharp. The 35-cm wavelength system, however, decays at a slower rate upon entering the oil and provides significant radar backscatter to a greater distance within the slick. This leads to a low measure of area. Quantitative measurements indicate that the 12-cm wavelengths decayed within 17 meters, while the 35-cm wavelengths maintained their structure to roughly 150 meters within the slick boundary.

Fig. 10 is a composite of the imagery obtained at all four frequencies over a two hour period on November 6. The area measurements from these four images and seven others are given in Table II and shown graphically in Fig. 11. This was a very calm period, zero sea state and 1 to 3 knot winds. The data, therefore, are indicative of the natural

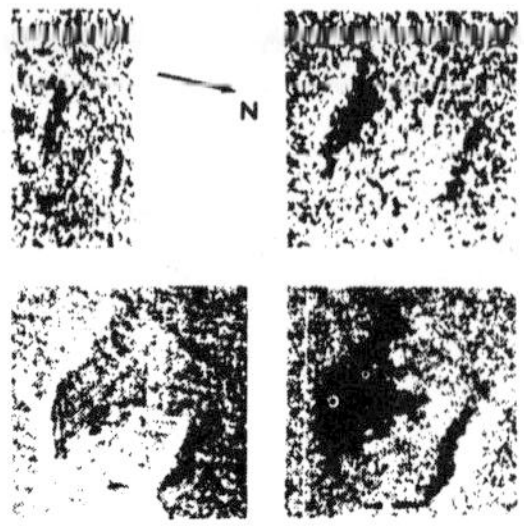

Fig. 10. Radar imagery of November 6, showing the increase in slick area with time. Reading from left to right – Top row: run 78, 428 MHz, 1352 mean local time (factor of two compression in the along-track direction); run 84, 1228 MHz, 1427 mean local time. Bottom row: run 93, 8910 MHz, 1530 mean local time; run 98, 4455 MHz, 1600 mean local time. Each image shows an area 2.5 × 2.5 km.

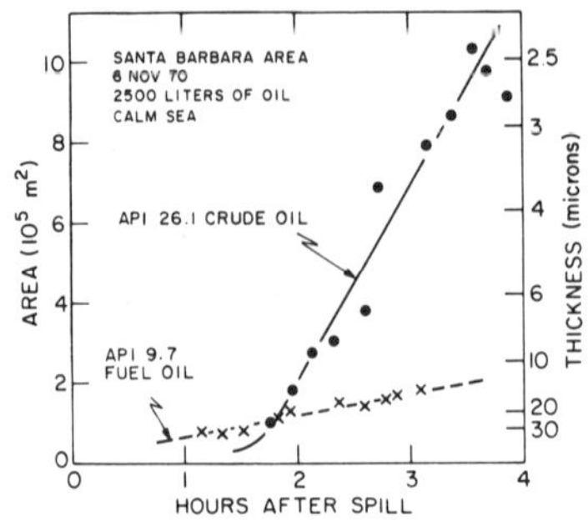

Fig. 11. The measured area and calculated average thickness of two oil slicks as a function of time.

Fig. 12. Radar imagery of run 108 (8910 MHz). Wind and wave action have reduced the radar contrast between oil and water. The area shown is 5 × 10 km.

spreading rates of API 26.1 crude oil and API 9.7 fuel oil, unaided by significant wind and wave action. The plot shows that for the two hours during which the two oil slicks were imaged the area of each slick increased approximately linearly with time. The area of the API 26.1 slick increased at the rate of 134 m^2/s, and that of the API 9.7 slick increased at the rate of 16 m^2/s. Thus, the light API 26.1 oil was spreading at a rate 8.4 times faster than the more viscous API 9.7 oil.

The imagery shown in Fig. 12 was obtained 24 hours later. During this period the sea had roughened considerably; sea states were approximated at 1 to 2 with winds of 7 to 10 knots. The slicks had started to break up and become highly streaked, and thin films, such as that extending to the left of the upper slick, were being blown off the main bodies of oil. This patchiness and wind action on the oil covered water and the interspersed regions of "clean" water increased the radar backscatter from the areas of the slicks relative to that from the open water. This reduced contrast in the radar return accounts directly for the lack of contrast between oil and open water in Fig. 12. The original imagery, which always reveals more than can be captured in a reproduction, allowed reasonable estimates of the slick areas. Sufficient data, however, were not available to reliably identify the slicks.

Conclusions

The maps obtained by the NRL 4FR system demonstrate radar imagery to be an effective tool in the detection and monitoring of oil slicks. Positive imagery of oil slicks have been obtained on all four operating frequencies, i.e., 428 MHz (*P* band), 1228 MHz (*L* band), 4455 MHz (*C* band), and 8910 MHz (*X* band). The results and conclusions arrived at are applicable to relatively low sea state and wind conditions.

Radar imagery, on vertical polarization, is capable of detecting and mapping oil from the initial thickness of the polluting spill to equilibrium thicknesses of 1 micron or less. Streamers from the main bodies of oil and wind blown films can also be imaged. At frequencies above approximately 1000 MHz, the slicks are depicted with sharp boundaries. At lower frequencies the boundaries are indistinct, possibly due to the intrusion of radar backscattering water wavelengths to a significant distance within the slick.

The detection of the third oil spill on October 28, while the oil was still being discharged from the *Pacific Clipper*, demonstrates that it is possible to image less than 400 liters of oil during the period of spillage. Thus, radar is shown to have potential in policing operations.

Area growth rates were determined for two of the oil slicks. These rates were approximately 134 and 16 m^2/s, showing that the lighter oil was spreading at a rate 8.4 times faster than the more viscous oil.

The confinement of the oil slicks generated in this experiment to rather small areas, in terms of square kilometers, necessitated only short mapping runs at a relatively low altitude. It should be pointed out that an airborne system, operating a high altitudes, could continously map a strip 200-km wide, covering thousands of square kilometers of area in a matter of a few hours. Such flights, unhampered by adverse weather, cloud cover, or darkness, could successfully monitor a coastline for man-made pollution, the appearance of natural slicks that may present a danger to beaches and the ecology, and the movements of known oil slicks, and could aid in the dispatching of men and equipment to effectively deal with such occurrences.

Acknowledgment

This study was carried out under the direction of N.W. Guinard, NRL Branch Head. The flight crew consisted of N.W. Guinard, J.T. Ransone, J.A. Burkett, W.T. Davis, W.P. Eng, N.R. Mills, and J.A. Weston. The signal film was optically processed by J.B. Thomas. Enlargements of the imagery were printed by J.O. Curtis.

References

[1] J.W. Wright, "A new model for sea clutter," *IEEE Trans. Antennas and Propagation*, vol. AP-16, pp. 217-223, March 1968.

[2] G.R. Valenzuela, M.B. Laing, and J.C. Daley, "Ocean spectra for the high-frequency waves as determined from airborne radar measurements," *J. Marine Research*, vol. 29, pp. 69-84, May 1971.

[3] N.W. Guinard, J.T. Ransone, Jr., M.B. Laing, and L.E. Hearton, "NRL terrain clutter study, phase I," Naval Research Lab., Washington, D.C., Rept. 6487, May 10, 1967.

[4] W.M. Brown, "Synthetic aperture radar," *IEEE Trans. Aerospace and Electronic Systems*, vol. AES-3, pp. 217-229, March 1967.

[5] W.M. Brown and L.J. Porcello, "An introduction to synthetic aperture radar," *IEEE Spectrum*, vol. 6, pp. 52-62, September 1969.

[6] R.O. Harger, *Synthetic Aperture Radar Systems: Theory and Design*. New York: Academic Press, 1970.

[7] N.W. Guinard and C.G. Purves, "The remote sensing of oil slicks by radar," Naval Research Lab., Rept. 5270-19; published as U.S. Coast Guard Rept., Project 714104/A/004, June 1, 1970.

[8] N.W. Guinard and J.C. Daley, "An experimental study of a sea clutter model," *Proc. IEEE*, vol. 58, April 1970.

[9] R.O. Pilon and C.G. Purves, "Remote sensing of southern California oil pollution experiment," in "Radar detection and monitoring of oil slicks," U.S. Coast Guard Rept., Project 714104, pp. 63-90, July 1971.

Roger O. Pilon was born in Cleveland, Ohio, on February 16, 1941. He received the B.A. degree in Physics from the University of Minnesota, Duluth, in 1963, and the M.S. degree in physics from the University of Minnesota, Minneapolis, in 1966.

Since 1967 he has been with the Naval Research Laboratory, Washington, D.C., where he has worked on superparamagnetism, synthetic aperture radar, and Sea Photo Analysis.

Charles G. Purves was born in Chicago, Ill., on May 27, 1919. He received the B.S. degree in mathematics from Alma College, Alma, Mich., in 1944. During World War II he completed two years of post-graduate work in meteorology at the University of Chicago, as an Aviation Cadet, and graduated as a Weather Officer in September 1943.

Since 1947 he has been a Department of Defense Research Meteorologist specializing mainly in electromagnetic wave propagation. From 1947 to 1953 he worked as a Technical Advisor to the Air Force on airborne electronic survey projects, and was a key member of the scientific staff responsible for the over-water survey of the North Atlantic guided missile range between Florida and Puerto Rico. From 1954 to 1958 he worked for the Navy Hydrographic Office as a Climatologist specializing in the military application of weather. Since 1958 he has worked at the Naval Research Laboratory, Washington, D. C., as a Meteorological Consultant doing research work in the field of electromagnetic wave propagation. He has considerable airborne experience in tropospheric atmospheric refraction work, especially in investigating elevated ducts and related transhorizon attenuation experiments in oceanic trade wind areas throughout the world. In the past few years he has participated in airborne multifrequency and polarization radar experimental research programs involving sea clutter, atmospheric back scatter, and radar imagery target evaluation studies.

The ERIM Simultaneous X- and L-Band Dual Polarization Radar

R. Rawson, F. Smith, and R. Larson
(Radar and Optics Division, Environmental Research Institute of Michigan, Ann Arbor)

The Record of the IEEE 1975 International Radar Conference (21-23 April 1975)
Pages 505-510

ABSTRACT

Design considerations, construction, and testing of a four-channel microwave radar imaging system are described. Various components of the system are described in some detail including the receiver, transmitter, recorders. Results from flight tests of the system are described. The system is unique in that images from four microwave channels are provided and the data is obtained simultaneously. Energy is radiated at both 3 cm and 25 cm wavelengths; the four receiving channels are 3 cm, like and crossed polarization and 25 cm, like and crossed polarization. The inherent large amplitude dynamic range of the radar image is too great to be linearly recorded on photographic film. The data requires special holographic display techniques to preserve this large dynamic range.

INTRODUCTION

During the late 1960's and 1970 period, several radar imaging missions were flown to provide imagery to various users under NASA sponsorship. Imagery obtained during this period was single polarization, X-band. Also, during this period, it was recognized that a multispectral, multi-polarization approach should provide imagery from which considerably more dinformation could be obtained. During the period 1970-71 the basic L-band radar was developed and operated on a non-simultaneous basis. Both X-band and L-band imagery were obtained, on separate flights. Also, antennas capable of dual polarization operation at X- and L-bands were developed and used in the data flights. This effort was conducted on Contract NAS9-11036. Users of this data were quick to point out the differences in the X- and L-band imagery. However, the differences in radar cross section could come from slight differences in the flight path as well as the difference due to wavelength or polarization. In mid-1972, an effort was initiated on Contract NAS9-1296, to instrument the X- and L-band radar to permit simultaneous operation. The first flight with the X-L band radar operating in a simultaneous mode was conducted in April 1973. This flight produced four simultaneous images of the area mapped; two at X-band, and two at L-band, both with like and cross polarization. Twenty data flights have produced imagery of many different parts of the U.S. However, most of the imagery is of the southeastern Michigan area.

Some of the imagery gathered includes,

1. Huntington agricultural test site, Huntington, Ind.
2. Melbourne, Florida coast & St. Johns River, Melbourne, Florida
3. Lake Erie Shoreline (Monroe, Michigan)
4. Lake Ice, Straights of Mackinac Michigan and Whitefish Bay Lake Superior
5. Georgia Agricultural Test Site, Athens Georgia
6. Strip mines - Kentucky

The simultaneous X-L dual polarization system consists of four major hardware components. the airborne radar system, the optical processor, the image dissector, and the digital computer. The system flow diagram is shown in Figure 1. This paper will describe the radar system, and the holographic display.

The X-L radar system is installed in a C-46 aircraft. The system weight is about 2000 lbs.

DESCRIPTION OF THE X-L RADAR SYSTEM

The X-L radar is a dual-frequency dual-polarization side-looking airborne radar designed to image a terrain swath parallel to the aircraft flight path.

The along track resolution is obtained from the synthetic aperture technique [1]. The cross-track resolution is obtained from FM pulse compression. The radar signals are recorded, during flight, in a frequency dispersed form and are optically compressed in the ground based processor.

There are, in effect, two radars in the X-L system; X-band 3 cm and L-band 25 cm. Each radar (X and L) has two channels, a like polarized channel and a cross polarization channel. In the present configuration horizontal polarization is transmitted.

The output of the airborne system is 70 mm signal film. One film with X-band like and cross polarized signals recorded on it. The other has L-band like cross polarized recorded on it.

The radar signals are recorded as astigmatic fresnel zone plates. The signal films are processed in an optical processor to produce strip maps.

The dual wavelength, dual polarization radar consists of six primary components:

1. Receiver-transmitter
2. Navigation system
3. Antenna
4. Motion compensation system
5. Recorders
6. System timing.

A block diagram of the system components is shown in Figure 2.

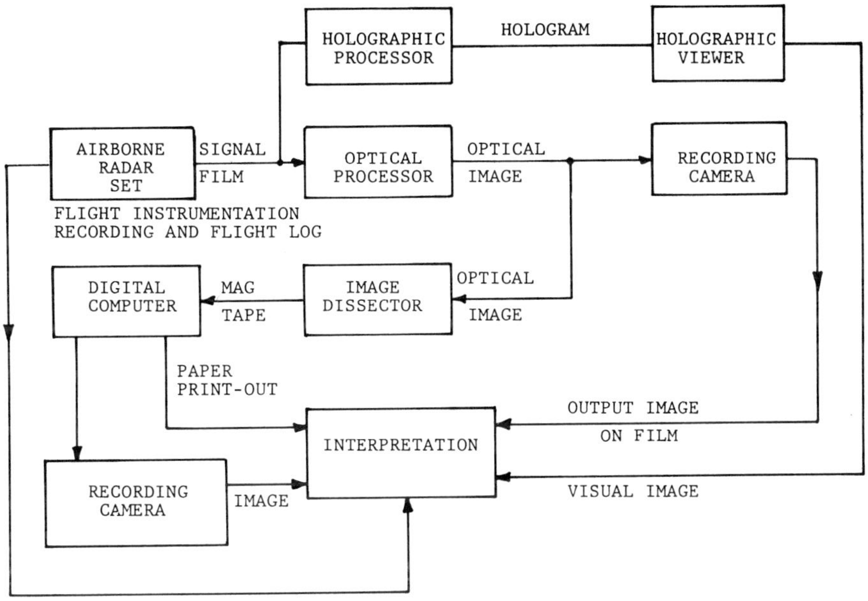

FIGURE 1. System Flow Diagram

The four channel receiver transmitter is shown in a simplified block diagram in Figure 3. Oscillator #1 is an X-band cavity stabilized reflex klystron. This is the phase reference for the receiver and transmitter. The FM generator is a backward wave oscillator that is phase locked to oscillator #1. During the transmitting time the BWO frequency is swept away from the reference oscillator. At the end of the transmitted pulse the BWO is relocked to oscillator #1. This chirp signal is divided with part of the signal going directly to the X-band transmitter TWT. The remainder of the signal is heterodyned to L-band by mixing with reference oscillator #2. It is then amplified and transmitted by the L-band TWT. The two transmitters are pulsed on alternate PRF cycles. The two transmitter tubes are connected to the proper port (horizontal or vertical) of their respective antennas. The antenna intercepts a part of the energy scattered by the scene to be imaged. Each of the four antenna channels is equipped with a low noise amplifier. In the X-band receiver, like and crossed polarization are selected on alternate X-band PRF cycles by a switch. They are amplified and phase detected with oscillator #1. The video output from the phase detector mixer is recorded on the X-band recorder.

The L-band receiver is similar to that band except after switching, the signals are heterodyned to X-band for detection and then recorded on the L-band recorder.

The two recorders are physically identical and differ only in film speed. This velocity difference is related to the wavelength difference of the radar channels.

Recording of radar data is accomplished by imaging an intensity-modulated single-line CRT onto photographic film which is moved continuously by means of the film transport system. The CRT line scan is divided into two protions each of which records a 37 μsec interval that corresponds to about a three nautical mine swath of imaged terrain. The recorded signal width of each channel on the film is about 27 mm with a 0.5 mm separation of channels at the center of the film. The portion of the film exposed for each channel is determined by the control module which supplies triggers to start the sweep and to unblank the proper portions of the sweep.

The control module determines the sequence of events. In normal operation, after four pulses all four channels have been sampled and the recording sequence starts over.

The two X-band channels are processed by a commonly-used processing technique which employs a tilted-plane decoupled mode. This type of processor has been extensively described in the optical processing literature [2].

The azimuth-to-range dimension aspect ratio is about 13.7:1 as recorded on the X-band data-storage film. The aximuth telescope is therefore set for a demagnification of 13.7 in order to produce equal scale factors in the output image. The limiting apertures at the frequency plane are adjusted to limit the signal passband in both dimen-

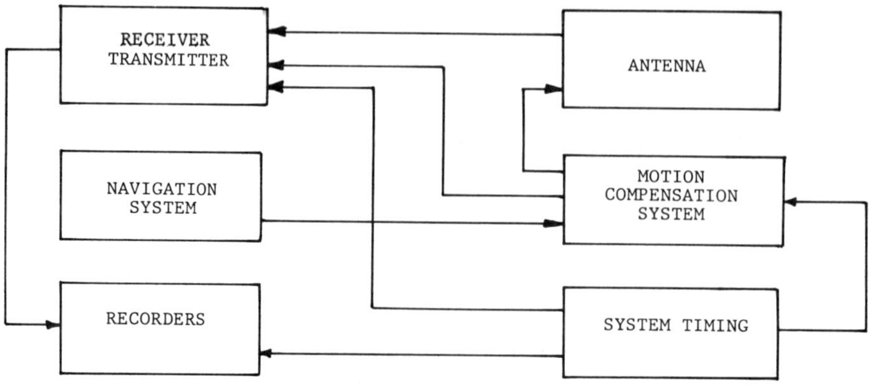

FIGURE 2. X-L Radar System Components

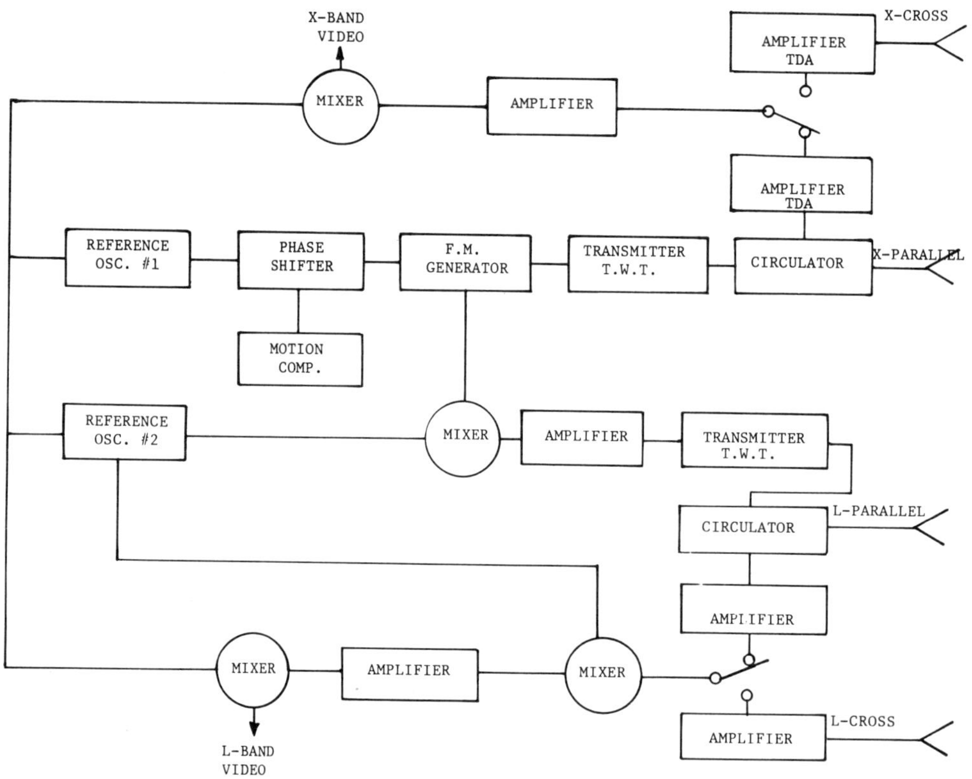

FIGURE 3. Receiver-Transmitter

sions to values commensurate with the image resolution. This is equivalent to bandpass filtering and serves to reduce extraneous noise while maintaining image resolution. The energy at the output plane of the data processor is a two-dimensional array of optical image components with a large intensity range.

In processing to obtain the L-band imagery, a tilted cylinder optical-processor configuration is used [3]. The azimuth to slant range aspect ratio on the signal storage film is 5.2. This is reduced to unity in the processor to make the output image scales equal in slant-range and cross-range.

The L-band radar imagery, like the X-band has a large dynamic range, and a choice of films must be made to produce an output image film with visible "ground painting" and yet without excessive blooming of bright-target images.

HOLOGRAPHIC DISPLAY SYSTEM

If the data user is restricted to examining film recordings of the optical processor output much of the amplitude information of the image is lost.

Radar images have been produced with as much as 70 dB variation in radar cross-section. Films have dynamic ranges of 15 to 20 dB. If the processor image plane is recorded directly, 50-55 dB of linear dynamic range is lost. To prevent the loss of amplitude integrity the holographic recording display system was developed [5].

This holographic viewer provides the observer with a simple optical system for viewing SAR imagery. The hologram is made near the processor image plane where the range to azimuth scale factor has been reduced to unity and the range dependency of azimuth focal length has been removed. In short, the new signal film (hologram) has had the astigmatism removed and the range to azimuth scale factor adjusted to unity in the image plane.

The hologram is produced by combining the undiffracted light (reference beam) with the defocused image and recording interference pattern on film. The film is bleached [6] to produce a high diffraction efficiency phase grating.

The image is formed by illuminating the hologram with collimated light, spatial filtering, and imaging the hologram with a spherical lens. The holographic projector is small, lightweight, and inexpensive compared to the anamorphic processor required to form an image from the original signal film.

INERTIAL NAVIGATION & MOTION COMPENSATION REQUIREMENTS

The airborne imaging radar uses the synthetic aperture technique to provide fine resolution in the cross-range (azimuth) direction. A synthetic aperture radar requires auxiliary systems which use inertial references to accomplish four different functions [4].

First, a desired track must be established and the aircraft must be guided along this path so that the radar beam scans over a particular strip of terrain. This is done by establishing a straight line at the altitude of the aircraft based on the desired terrain swath and the operating range of the radar. An inertial navigation system in the aircraft is used to provide guidance signals for the autopilot in order to navigate the aircraft along this line.

Second, the radar antenna must be stabilized to hold the radar beam perpendicular to the desired track and at the appropriate depression angle. The aircraft forward velocity then scans the beam along the desired terrain swath. Angular orientation signals are taken from the inertial platform to control the antenna servo so that the antenna maintains its proper orientation as the aircraft maneuvers in following the desired track.

A third function requiring inertial sensing is that of accurately correcting the radar echo signal phase for residual range-direction displacement of the radar antenna from the desired flight path. This is done by taking acceleration signals from the inertial platform, computing the component of motion along the line of sight to the target area, and shifting the phase of the radar signals (prior to recording) by an amount sufficient to cancel the motion phase shifts.

A fourth function is synchronizing the radar offset frequency, the pulse repetition frequency, and film speed of the radar recorders to the aircraft along-track velocity. Along-track velocity signals are taken from the inertial platform and applied to the offset frequency circuitry, the local time base, and the film drive servos.

A modern all-inertial navigation system (INS) was believed capable of providing all of the required stabilization and navigation signals for the operation of the radar system. An all-inertial system is typically smaller, lighter, and lower-powered than theolder Doppler-inertial system. While the INS had not been experimentally proven in this application, analysis of one such system, the LTN-51, indicated that it probably would perform adequately without extensive programming modifications. One type of navigation of which the LTN-51 is capable makes use of indicated displacement from a desired great circle track in order to guide the aircraft along the actual desired path rather than merely parallel to it. This appeared to be an attractive feature that could improve radar performance. On the basis of this analysis, then, the LTN-51 was selected for use in the aircraft with the dual-band imaging radar.

MOTION COMPENSATION

The synthetic aperture technique depends upon the predictable echo-signal phase histories which are produced as the radar antenna is transported along a straight-line path. As described above, a navigation system is provided to guide the aircraft along this straight line; but, despite this guidance, some displacements of the antenna from the desired track occur. Even small lateral displacements cause distortion of the quadratic phase rate that is utilized in the data processor to obtain fine-resolution imagery. Therefore, a motion-compensation subsystem has been designed to apply "fine-grain" phase correction to the radar. This subsystem measures the motion inertially and shifts the phase of the radar signals just enough to cancel the motion-induced phase shifts; this phase correction is accomplished before the signals are recorded on the data storage films.

A block diagram of the beam velocity computer is shown in Figure 4. The INS provides the three orthogonal acceleration components A_x, A_y, and A_V. The A_x and A_y components are resolved into the cross-track component A_{CT} by means of sin ϕ and cos ϕ, where ϕ is the angle between the desired beam orientation and the actual platform orientation. Then, the components of A_{CT} and A_V in the along-beam direction are computed by multiplication of sin θ and cos θ, where θ is the depression angle. These two components are added together to form the total along-beam acceleration A_B. Finally, A_B is integrated in order to produce the along-beam velocity V_B. Signal V_B is fed to a phase shifter located in the radar to shift the phase of the STALO frequency.

As can be seen in Figure 4, the along-beam velocity computer is lightly damped by means of a loosely-coupled clutterlock feedback loop which senses the average Doppler frequency of the radar echoes. The clutterlock output signal is zero when the average echo-signal offset is correct.

While the beam-velocity computer has a loose frequency lock because of the clutterlock feedback loop, there is no phase-lock feedback loop. Phase compensation thus depends entirely upon computing instantaneous along-beam velocity and shifting the signal phase at a rate that matches the echo phase-rate from lateral motion.

The motion compensation system for the dual-band radar uses two scale factors alternately as appropriate for the two radar wavelengths.

CONCLUSIONS

The airborne radar system produces two rolls of signal film. One X-band, two polarizations and one L-band, two polarizations. After the aircraft returns to the ground the signal film is processed in the optical processor one channel at a time. At this point an output image film may be produced or the image dissector may be used to digitize the radar image.

This radar system has proven to be a realiable instrument capable of generating four continuous maps($X_{||}$, X_X, $L_{||}$, L_X) from simultaneously collected radar data. The LTN-51 inertial navigation system has increased our number of possible flight days by navigating the aircraft to a desired area in times of total cloud cover (VFR on top) or in total darkness.

Some of the radar parameters are listed in Table I.

TABLE I

PARAMETER	X-BAND	L-BAND
Center Frequency	9.450 GHz	1.315 GHz
Resolution	Variable from 30 ft x 30	to classified
Transmitter (Peak) @ 2% duty cycle	1.2kw	6 kw
Antenna Gain	28 dB	16.5 dB
Ant. Beamwidth	1.1°	7°
Polarization Isolation	23 dB	19 dB
Depression Angle	0° to 90°	0° to 90°
Maximum Range	13 n miles	13 n miles
CRT	WX30267P11	WX30267P11
No. of Spots/Scan	8,000	4,000
Film Capacity	100 ft	100 ft
Film (Kodak)	Microfile 5460	Microfile 5460

REFERENCES

1. W. M. Brown and L. J. Porcello, "An Introduction to Synthetic Aperture Radar," IEEE Spectrum, Vol. 6, pp. 57-62, Sept. 1969.
2. A. Kozma, E. N. Leith, and N. G. Massey, Tilted Plane Optical Processor, Applied Optics, Vol. 11, No. 8, August 1972, p. 1766.

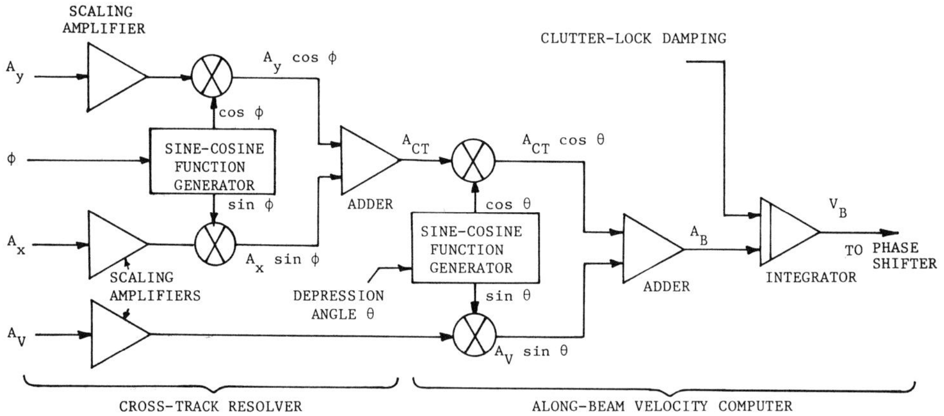

(ϕ = Angle between desired beam orientation and actual inertial platform orientation)

FIGURE 4. Cross-Track Resolver and Along-Beam Velocity Computer

3. R. A. Rendleman, et al., Multifrequency Fine-Resolution Imaging Radar Instrumentation and Data Acquisition, Report No. 198200-1-F, (NASA-CR), Environmental Research Institute of Michigan, Ann Arbor, Michigan, February 1974.
4. Liskow, C. L., Simultaneous Dual-Band Radar Development, Septemper 1974, ERIM Report No. 195100-1-F.
5. A New Technique for Processing and Viewing of Synthetic Aperture Radar Data, U of M IST, Radar and Optics Division, 1 February 1971.
6. Diffraction Efficiency of Bleached, Photographically Recorded Interference Patterns, Upatnieks and Leonard, Applied Optics Vol. 8, No. 1, January 1969.

Classification Of Freshwater Ice Using Multispectral Radar Images

M.L. Bryan and R. Larson
(Radar and Optics Division, Environmental Research Institute of Michigan, Ann Arbor)

The Record of the IEEE 1975 International Radar Conference (21-23 April 1975)
Pages 511-515

ABSTRACT

A four-channel multispectral radar was used to obtain data on freshwater ice. The study reported in this paper consists of radar data collection, (visual) analysis of these data and the continuing development of the proper ground truth techniques to be used in conjunction with such remotely sensed data. The *in situ* measurement of the electrical properties together with the determination of the physical properties of the ice (thickness, bottom and top roughness, stratification) appear to warrant further consideration. The documentation of radar energy penetration through the lake ice and its subsequent reflection from the bottom (ice-water) interface is considered to be one of the major contributions of this paper. The importance of using multichannel radar imagery for such studies is also noted.

EXPERIMENTAL APPROACH

The Environmental Research Institute of Michigan (ERIM) together with the National Aeronautics and Space Administration (NASA) Lewis Center, Cleveland conducted a field study during the winter 1973-1974. This experiment was oriented toward determining the information which is available in the data obtained from the ERIM four-channel airborne radar system. The study areas for this experiment were Whitefish Bay, Lake Superior, and the Straits of Mackinac, Michigan (Figure 1).

The approach taken in this study is outlined by BRYAN and LARSON (1973). Radar data were obtained and images of specific areas were provided to the field team as soon after the data flight was possible. In this experiment, data were provided by ERIM less than 10 hours after the completion of data flight, and using NASA aircraft delivered to the field team 12 hours after the imaging flight. Ice coring equipment, in addition to two portable instruments designed and fabricated specifically for this purpose by ERIM laboratory personnel, were used. These latter instruments measured a) the complex dielectric constant of ice (or snow) at 100 MHz, and b) the electrical properties of the ice at the two operating wavelengths of the radar system (3 cm and 25 cm).

DATA ANALYSIS

During this study, radar imagery was obtained for areas in Lake Superior and Lake Michigan. Examples of the four-channel radar imagery from one imaging pass are given in Figure 2. This covers a three-mile strip from Ile Parisienne, Ontario, Canada south to and including Nodoway Point, Michigan, USA. The wavelength and polarization of each image are as indicated.

Several straightforward interpretations are easily made:

1) Area "A", land mass of the upper peninsula of Michigan and shore line of Whitefish Bay

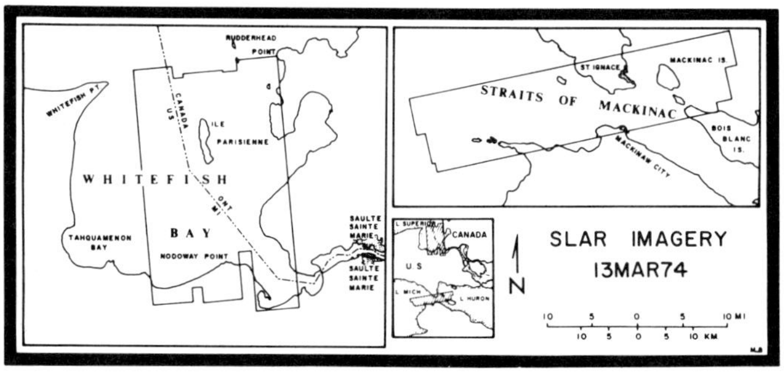

FIGURE 1. Areas of Radar Data Collection for Freshwater Ice Study

2) Area "B", island of Ile Parisienne
3) Various ice types in Whitefish Bay, smooth ice being shown by low reflective (dark tone on the image)
4) Area "C", ships' tracks, where ships have gone through ice
5) Area "D", pressure ridge.

Clearly, the response or the distribution of microwave energy reflected from the surface, as measured by the radar system, is a function of operating wavelength and polarization. This effect is the concept on which the multispectral microwave imaging system is based. Generally surface roughness (at the ice/air interface)*, as depicted in Figure 3, determines the magnitude of the backscattered energy. Thus, from Figure 3 a surface may be classified as rough at a particular wavelength and as a smooth surface at a longer wavelength.

*The term "ice/water interface" refers to the horizontal boundary between the lower surface of the lake ice and the upper surface of the water mass of the lake. Similarly, the "ice/air interface" refers to the boundary between the upper ice surface and the lower surface of the atmosphere.

ELECTRICAL MEASUREMENTS

A number of questions, based on the information obtained from the limited analysis performed, arise. For example, it has been implied that observable backscatter is received from the ice/water interface. Results from dielectric constant measurements indicate that penetration of the ice layer is possible at both X-band and L-band wavelengths. Also, the two-way energy loss in the ice is not too excessive at L-band so that energy backscattered from the ice/water interface should be received by the system antenna. However, adequate empirical data to substantiate these conclusions and to determine the reflection coefficients from the ice/water interface are not presently available.

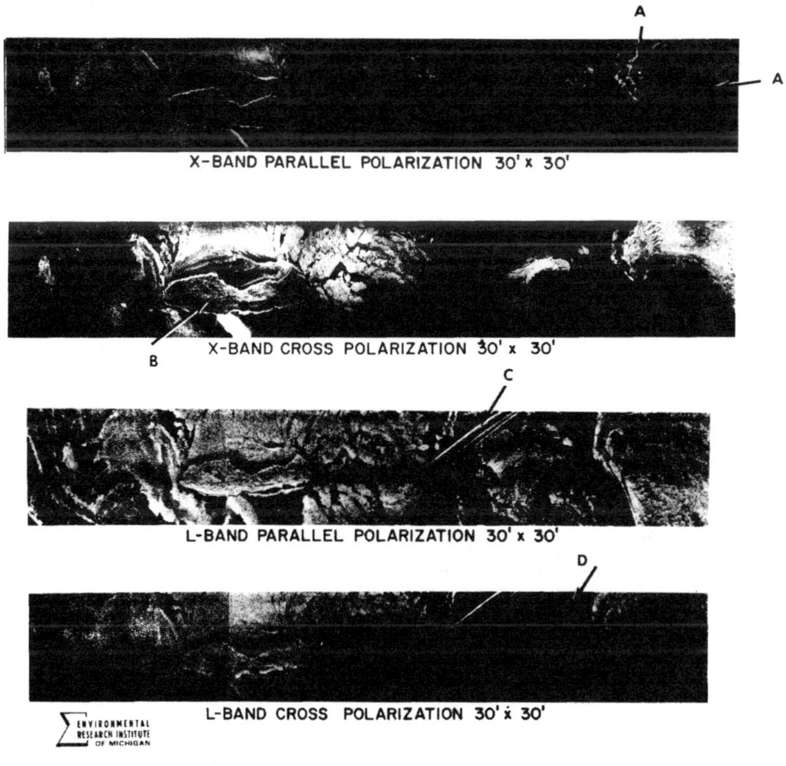

FIGURE 2. Simultaneously Obtained, Dual Frequency Radar Imagery, 13 APR 74, Whitefish Bay, Michigan (See Text for Discussion of Symbols).

Estimates of the ratio of the backscattered energy from the air/ice interface (P_i) and that from the ice/water interface (P_w) can be calculated. First, the reflection coefficients are approximated by assuming equal roughness factors for the two boundaries. Then, normalized by the surface roughness factor, the air/ice reflection coefficient is approximately -10 db. Using published values (CUMMINGS, 1952) for the dielectric constant of water, the ice/water reflection coefficient is approximately -3 to -5 db.

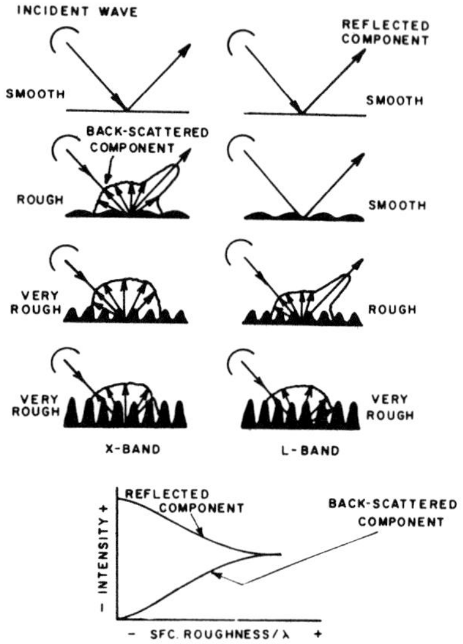

FIGURE 3. Effect of Surface Roughness on Radar Backscatter.

Reflection coefficients at the ice/air and ice/water interfaces can be either measured directly or calculated using measured values of dielectric constants. (Obviously, the direct measurements are preferred due to the difficulty in calculating the interface roughness.) However, lacking these empirical data, and using a) the calculated reflection coefficients for these interfaces, b) the loss tangent (tanδ) for the ice, c) a radar depression angle of 20° and d) X-band and L-band wavelength values, the values of P_w and P_i as a function of ice thickness may be calculated. Such calculations show that $P_w \approx P_i$ for 40-50 cm of ice for λ = 3.0 cm (X-band) and 3-4 m of ice for λ = 25.0 cm (L-band).

For ice thicknesses less than these values, it can be assumed that the backscattered energy is coming primarily from the ice/water interface - if, of course, the ice/air interface is considered smooth at the operating wavelength. Similarly, for ice thicknesses greater than the calculated values, the scattering of the ice/air interface would be the predominate component of the backscattered energy.

It must be emphasized that, in the example, it has been assumed that both scattering surfaces have an equal roughness factor. Clearly, in many situations this is not true, and corrections to account for differences in the boundary roughness would be required. This obviously points out the need for additional and meticulous ground truth efforts to obtain a more complete understanding of the data available in the radar imagery.

The data concerning dielectric constant (ε_i'), ice thickness and loss tangent (tan δ) for five locations in the study area are given in Figure 4. These data were collected using a 7.62 cm (3") diameter SIPRE corer. Dielectric measurements were taken at a frequency of 100 MHz (BRYAN and LARSON, 1973). For the wavelengths used in the radar system, the dielectric constant was calculated to be ε_i' = 3.0 or slightly higher than that measured at 100 MHz.

These data confirm the comments previously found in the literature. Specifically, there is every possibility that there was considerable penetration into the lake ice by the microwave energy from the imaging radar and therefore the effects observed in the imagery, that is, the brightnes of the image tone, are not necessarily a direct function of the roughness of the air/ice interface.

CLASSIFICATION OF ICE

A small southern portion of the imaged area included in Figures 1 & 2 was selected for intense study (Figure 5). In this area the ice was classed into nine separate units as illustrated in the small location map. Area 1 is composed of black (i.e. new) ice with small pieces of brash ice (less than 1 m). In this area, the ice/air interface was rough at both wavelengths. It is also assumed that the ice/water interface was equally rough. In area 2, composed of brash ice, the upper surface was also rough. This area is extremely bright in the radar imagery due to the multiple scattering of the microwave energy. Smooth brash ice, illustrated by Area 3 (Figure 5), is similar to Area 1 noted above. The somewhat brighter returns for this area (when compared to Area 1) are thought to be due to the specular reflectors at the ice/water interface.

Ice foots (BRYAN and MARCUS, 1972) which are 3-4 meters in height, are easily identified in Area 4, while pressure ridges, another type of linear feature, are seen as item 5 on Figure 4. Area 6 is one of the areas in which the bottom roughness was determined, and in this area, the ice/air surface is smooth at both wavelengths; for 6a, in which the ice is 25-35 cm thick, the ice/water interface is smooth at both wavelengths, but, for Area 6b the ice thickness varies between 25-35 cm and 65-70 cm thick. The thicker ice resulted from the imbedding of pancake ice into a matrix of the new and thinner black ice and formed the rough bottom interface. It is in these areas where scattering from the ice/water interface is most pronounced.

Area 7 is very similar to that noted in Area 1, except that the ice/water interface is somewhat rougher. Area 8, for which both interfaces are smooth at both wavelengths, should have presented very low returns to the radar. However, the brighter returns are from the snowline and the drifted snow along the shore. Finally, Area 9 is

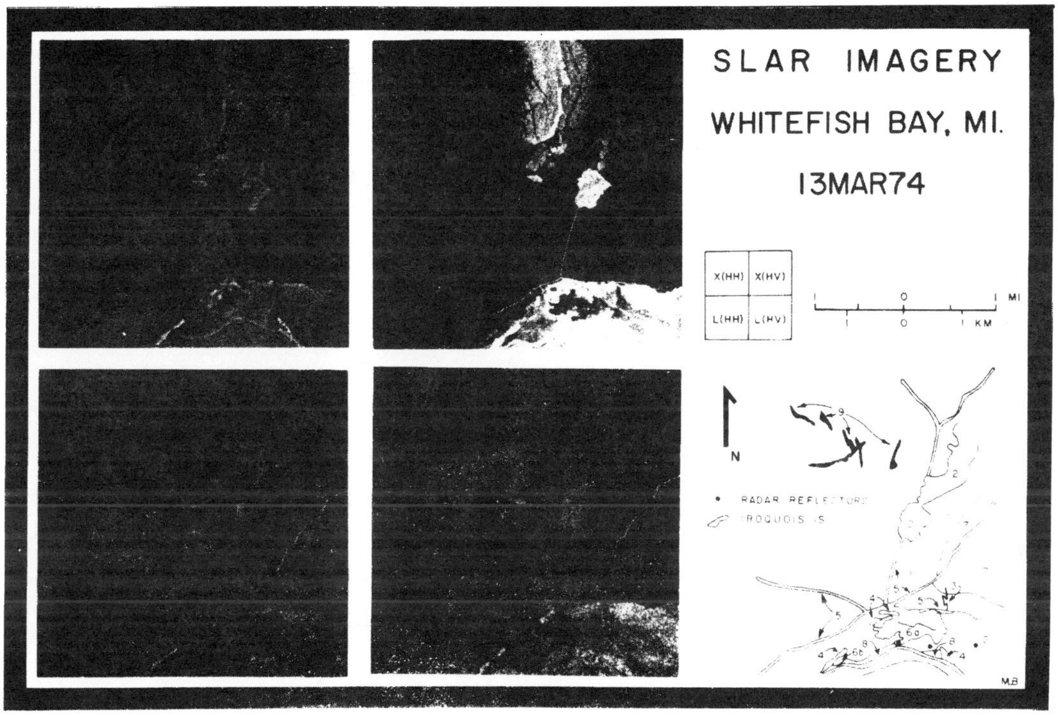

FIGURE 5. Detail of Multichannel Imagery of Fresh Water Ice Presented in Figure 2. Shoreline of the Upper Peninsula of Michigan is Seen at the Lower (Southern) Edge of the Figure.

an example of what open water (leads or polynyas) would look like, should there have been any open water in the image. This emphasizes the difficulty in identifying the areas of open water from those having moderately thin ice (which allow penetration through the entire thickness and reflection from the ice/water interface) in which both interfaces are smooth at the operating wavelengths of the radar.

CONCLUSION

From the results obtained during this study, it has been verified that surface roughness of the freshwater ice on the scales of 3 cm and 25 cm can be delineated. Also, based on on-site measurements of the electrical and physical properties of the cie, it is determined that radiation at 25 cm penetrates up to 1 meter of ice and a reflection from the ice/water boundary can be seen.

Clearly, there is a need for much additional work before ice classification can be accomplished on an operational basis. In addition, carefully controlled work is required in which both ground measurements and radar imagery are simultaneously obtained. In addition, such studies can be used to guide both visual and machine assisted analysis and interpretation. At the present time, however, analysis is conducted primarily through visual image interpretation.

REFERENCES

Bryan, M.L. and M.G. Marcus, 1972, "Physical Characteristics of Near-Shore Ice Ridges," ARCTIC 25:3 pp. 182-192.

Bryan, M.L. and R. W. Larson, 1973, "Applications of Dielectric Constant Measurements to Radar Imagery Interpretation," pp. 529-547 in: F. SHAHROKHI, (ed) REMOTE SENSING OF EARTH RESOURCES, Vol. II, TULLAHOMA, TN, MARCH (Also available: NTIS #E73-10429)

Cummings, W.A., 1952. "The Dielectric Properties of Ice and Snow at 3.2 Centimeters," J. APPLIED PHYSICS, 23:7, pp. 768-773.

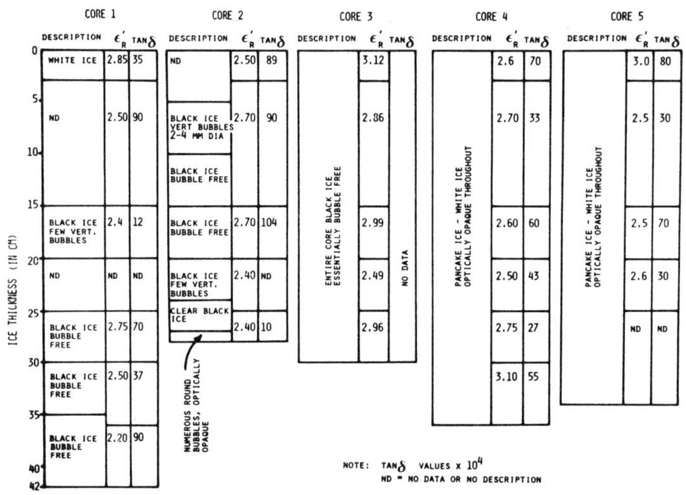

FIGURE 4. Physical and Electrical Properties for Five Cores of Freshwater Ice, Whitefish Bay, Michigan, March 1974.

Contour Strip Mine Detection And Identification With Imaging Radar

R.A. Shuchman, C.F. Davis, and P.L. Jackson
(Radar and Optics Division, Environmental Research Institute of Michigan, Ann Arbor)

The Record of the IEEE 1975 International Radar Conference (21-23 April 1975)
Pages 516-521

ABSTRACT

Sixty strip mine areas out of sixty-three mine areas present within a Kentucky test site were correctly identified using dual wavelength (3.2 and 25 cm), dual polarization Synthetic Aperture Radar (SAR) imagery and existing 1:24,000 USGS topographic maps. Three basic mining categories were identified on the multi-channel radar: 1) Active mine areas; 2) Reclaimed strip mines; and 3) Unreclaimed strip mines ("orphan" areas). More detailed information was inferred from the imagery after subsequent field observation. Radar is especially sensitive to surface roughnesses, subtle height variations, and orientation of surfaces. In addition, its all weather capability and its resolution (essentially independent of range) makes radar capable of large area enforcement and monitoring of strip mine activities.

INTRODUCTION

Initial field observations in southeastern Kentucky indicate that three basic mining categories can be identified on multi-channel radar imagery:

1. Active strip mines,
2. Reclaimed strip mines,
3. Unreclaimed strip mines ("orphan" areas)

On April 18, 1974, fine resolution radar imagery was collected over a region containing 63 strip mining areas. The radar used was the Environmental Research Institute of Michigan (ERIM) multichannel synthetic aperture radar (SAR) system [1,2, 3]. This system simultaneously gathers four separate images which are produced at X-band (3.2 cm) and L-band (25 cm) wavelengths while using parallel (HH) and cross (HV) polarizations.

Using only this imagery and existing 1:24,000 USGS topographic maps (to yield elevation contour information), we identified 60 of these areas correctly. Subsequent field observations in southeastern Kentucky confirmed the identifications and indicated that detailed information about the mining areas could be determined from this radar imagery.

It was possible in some cases to determine the density of vegetation, on the benches, roughness of unvegetated benches, presence of highwalls, slumping of outslopes, and water impoundment.

To make some of the above discriminations it was necessary to study all four radar channels. As each image is obtained with different sensing parameters, the differences among the four separate images aid in interpretation.

Approximately 9.3 meter resolution radar imagery was interpreted in this study. This resolution is adequate in imagery used for the identification of relatively small Kentucky strip mines which cannot be resolved with ERTS-type satellite imagery. Radar imagery obtainable through clouds and rain, conditions which prevent monitoring by other airborne or satellite imaging systems makes radar capable of large area enforcement and monitoring of strip mine activities.

BACKGROUND

The ERIM radar is a two wavelength SAR system that simultaneously images the terrain with X-band (3.2 cm) and L-band (25 cm) radar wavelengths. Alternate X- and L-band horiztally polarized radar pulses are transmitted. Both like- and cross-polarized energy is reflected back to the transmitting antenna for each wavelength; thus four simultaneous channels of radar imagery are obtainable [1]. The radar characteristics of the flight are as follows:

1. Airplane altitude above mean sea level: 2,590 meters;
2. The depression angles: near range 43°, far range 14°;
3. Slant range swath width was 6,096 meters;
4. Radar look direction true azimuth: 81°;
5. Slant range distance to mean edge of imagery: 3,353 meters.

The test area, approximately 58 kilometer (36 miles) long and 5.6 km (3.5 miles) wide, is shown in Figure 1. The area runs NNW through Whitley and Knox Counties, beginning near the Kentucky-Tennessee borner 24 km (15 miles) SSW of Middlesboro, Kentucky ending 3.2 km (2 miles) east of Corbin, Kentucky.

The imagery was interpreted primarily with the unaided eye; however, a densitometer was used for detailed signature comparison of the benches of reclaimed areas. Identification parameters used in the interpretation were tone and texture return, size, shape and context. Field observations consisted of intensive inspection of 33 mining areas, 8 active sites, 18 reclaimed areas, and 7 orphan areas. Field observations could not be conducted until 3 months after the flight; thus the exact progress of active mines in respect to location on the imagery and of some reclaimed benches is indefinite in some cases.

TYPES OF MINING

Two basic strip mining techniques are employed within the test area; contour and contour area or mountain top removal. However, the dominant strip mining operation within the test area is contour mining, employed where coal deposits follow a topographic elevation in rolling or mountainous terrain. After removing the overburden (material above the coal seam), coal is mined until the ratio of the thickness of overburden to coal becomes too great for profitable operation.

Contour mining creates a shelf of "bench" which is bordered on the inner edge by a highwall. The highwall in eastern Kentucky generally ranges from 10-38 m in height, is usually almost vertical, and consists of sandstone, shales, and unconsolidated material covering the coal seam. The coal under the highwall is not removed.

Below the outer rim of the bench lies the outslope, consisting of spoil material (the overburden rock and soil) that has been cast down the slope. The angle of repose of the outslope is generally 37° from horizontal, the original slopes in the area prior to mining averaged 24°. Contour mining is permitted only on slopes 27° or less from the horizontal (Kentucky Division of Reclamation).

Mountain top removal, a specialized form of contour mining is employed where coal seams lie near the tops of mountains. Sometimes referred to as contour area mining, mountain top removal is contour mining with a very wide bench. Many cuts are made into the mountain side until the entire top is removed.

RESULTS

With no previous knowledge of the number of distribution of mining sites within the test area, 60 strip mines were identified by the authors using the radar imagery and 1:24,000 USGS topographic maps, the USGS maps being used to obtain contour information.

Prior to field observation it was not known whether the radar could discriminate between the three basic mining types:

1. Active strip mines,
2. Reclaimed strip mines,
3. Unreclaimed strip mines ("orphan" areas).

Upon completion of the field inspection of 33 mining sites within the test area, it was found that the radar imagery could be used not only to differentiate the above mining types, but also to determine whether reclaimed benches were vegetated or non-vegetated. Non-vegetated benches are defined by the Kentucky Division of Reclamation as benches with less than 70% vegetation. The field inspection also revealed that one could use radar imagery to identify individual features of mines, including the highwall, bench, outslope, spoils, sediment ponds, and mining machinery.

Figure 1 and 2 illustrate the appearance of small strip mines on the radar imagery. The sixty strip mines in the larger test area were initially identified on this type of imagery. Figures 1 and 2 represent a section of the test area that includes all three basic mining types. Figure 1, an X-band parallel polarization radar image, illustrates the identification of strip mines on radar imagery, and acts as a location and identification key for Figure 2. Figure 2 consists of all four channels of SAR imagery that were generated. The Rayleigh resolution of the four images (X-HH, X-HV, L-HH, and L-HV) is approximately 9.3 m in both range and azimuth.

Table 1 summarizes the radar returns exhibited on the radar imagery of the various types of mines and mining features. Examining mining types and features found within the test area, an area longer than that shown in Figs. 1 and 2, we made the following basic observations and conclusions.

A. Active Areas

Both types of active mine areas can be recognized and mapped on the imagery. At one active mine area (See A, Figure 1) the spoil piles are clearly seen. The mining areas have a distinct highwall return when oriented near perpendicular to the radar look direction and a characteristic rough return from the spoils area.

Contour area mines or mountain top removal mines are generally differentiated from contour mines due to elevation, contour location, size, and shape. Unfortunately, an example of a mountain top removal mine is not present on Figure 1 or 2.

B. Reclaimed Areas

As mentioned previously reclaimed areas are formerly mined areas where the bench has been leveled. Reclaimed benches within the test area vary in width from 45-244 m (150-800 ft) wide and in length from less than a mile to many miles.

The bench and outslope are both planted with vegetation. Rye, barley, and feskew are the most common plant species on vegetated benches and outslopes. They range at maturity from .5-1.2 m, in height. Occasionally locust trees are planted along with the grains which are usually planted by a hydroseeder and hence are not organized in rows. Current reclamation laws allow the retention of the highwall (left from active mining). Reclaimed and partially reclaimed areas are clearly discerned on the imagery when oriented on the proper side of the mountain with respect to the radar look direction. The acreage of these reclaimed areas can be estimated down to the resolution limit of the radar imagery.

It appears from a close examination of the imagery that relative bench vegetation densities (vegetated vs. non-vegetated) of some reclaimed areas can be estimated by the close examination of the radar return from the benches. The differences in return between the four channels of the radar system are used for this interpretation.

Reclaimed benches generally appear dark on L-band in both the like and cross polarizations. It is presumed that the L-band is penetrating the vegetation, if present, on the bench and reflecting away from the soil which appears smooth to L-band, (25 cm) hence the no return (dark).

The return of the X-band is generally moderate to bright from vegetated benches, while a darker return results from the reclaimed benches lacking in vegetation. The moderate to bright return indicates a roughness range to which the X-band (2.3 cm) energy responds. The vegetated reclaimed mine site labeled B at the lower right corner of Figure

1 is imaged as dark on the L-band, bright on X-band, demonstrating how the differential return between the wavelengths is diagnostic.

The total effect of polarization as a diagnostic in differentiating mining features has not been fully explored. Some problems arise in respect to the study of subtle polarization changes on this imagery.

1. The radar system is not calibrated, thus an apparent polarization difference could be an artifact in the system.
2. Lack of timely ground truth of the mine sites. (i.e. moisture content and exact vegetation status unknown) The area of polarization differences caused by relative moisture differences of objects looks particularly proimsing. Polarization returns are influenced by roughness, dielectric constant, and EM attenuation of the surface.

The highwall present at each reclaimed site is clearly visible on all four radar channels, but exhibits the strongest return on X-band. As mentioned previously, the strength of radar return from highwalls is orientation to the radar look direction dependent. In general, all mining highwalls act as very strong radar reflectors, the highwall and bench of each mine site forming approximately a dihedral corner reflector to the radar. Breaks in the highwall are visible when it is oriented nearly perpendicular to the radar look direction. The debris of highwall strata often found at the base of the highwall apparently does not affect the radar return. In some cases the imagery clearly displayed standing bodies of water at the base of the highwalls and drainage ditches cut into the reclaimed bench.

Settling ponds used to collect water run off from the reclaimed bench, usually sediment ladened, are also visible on the X-band imagery at some mine sites. These settling ponds are located either on the bench or in the hollows between strip areas.

Outslope slump or bench subsidence occurs near the outer rim, due to mass wasting of the outslope. Two slumped areas appeared as dark returns on both X- and L-band. This dark return was a result of the orientation of the slumped areas with respect to the radar look direction causing a shadowing effect. One reclaimed area briefly examined at the southern end (Kentucky-Tennessee border) of the test site has severe sliding and slippage of the outslope. The sliding was so great that tree areas below the slide were affected. The imagery shows a triangular area where these slides occurred. The tree line on top of the highwall (as well as the tree line at the base of the outslope) is clearly defined.

C. Orphan Lands

The term "orphan land" refers to areas mined before enactment of reclamation laws, and consequently not reclaimed. Thus only natural revegetation occurred. The highwall, bench and spoil pile of orphan areas were usually left ungraded. The bench of reclaimed areas is on the order of 15 to 30 ft wide while the spoil area is another 25-50 ft. Often the bench accummulates standing water due to the depression of the bench in relation to the spoil pile and highwall. Orphan lands are discerned on all four radar images. However, the individual mining features constituting the orphan areas are not generally recognizable on the L-band imagery. The radar on X-band picks up the highwall as a bright return while the bench appears dark. The spoils generally have vegetation growing on them and thus blend into the surrounding vegetation return. Orphan areas have narrower benches that current reclaimed land. Because these benches are wider than road systems and are located adjacent to the bright highwall return on X-band imagery, no problem is presented in identification. Often, orphan areas are not visible from tree-top level and cannot be mapped by conventional photographic means. Radar imagery has been shown to reveal ground contours from the study of returns from the tops of brush and trees [4]. In one case, an orphan mine, unknown to the authorities, was discovered with radar imagery.

DISCUSSION

The preceding section has demonstrated the value of SAR in locating and identifying various strip mining activities. Imaging radar can contribute to strip mine monitoring and enforcement in a unique way.

Radar images are obtainable through cloud cover and rainfall up to a rate of approximately 1 in/hr. It is the only imaging device whose resolution is essentially independent of distance to the object being imaged. Although it is essentially independent of distance, the resolution is approximately 25% coarser at the near edge of the swath than at the far edge. This difference in resolution is caused by the difference in depression angles at which the swath edges are viewed ($\cos 20^\circ/\cos 40^\circ = 1:1.23$). Thus the radar resolutions considered here with respect to low flying aircraft should also be obtainable from satellites. Radar is especially sensitive to surface roughnesses, subtle height variations, and orientation of surfaces; these characteristics are particularly useful for strip mine detection and identification. Thus radar imagery has the distinct advantages of all-weather capability, adequate uniform resolution from aircraft or satellite, and availability to be machine processed. Other applicable remote sensors are not only restricted to essentially cloud-free weather (in SE Kentucky cloud cover is common) but, from satellites, restricted to resolution cells longer than many contour strip mines.

Various strip mine problems are especially suited to solution using radar imagery:

1) Mapping of Orphan Lands

Because of vegetation cover on orphan lands radar is most suited to delineate the underlying surface contours. The surface contours being inferred by the study of radar returns from the tops of vegetation. Future large strip-mining activities could be undertaken at these orphan sites. It should again be noted that only a 15 to 30 ft wide strip of coal seam was removed in these old orphan areas. Thus using modern techniques these orphan areas could be remined to obtain more coal and proper reclamation could then be implemented.

2) Monitoring wildcat operations

There appears to be a great potential for the use of all weather imaging radar to monitor regulated (permitted) mining vs. wildcat operation. Wildcat operations are mines illegally operated where no reclamation bond or permit is filed. Due to the relatively high price for coal, increasingly stringent strip mining regulations, and cost of filing a permit, wildcat operations are on the increase.

3) Highwall Detection

Highwalls, very prominent on radar imagery, will play a more significant role in the future if the federal government enacts the anticipated law requiring mined land returned to original elevation contours, thus prohibiting highwalls on reclaimed land. The radar imagery could act as a monitor to ensure highwall removal.

4) Reclamation Enforcement

Current Kentucky reclamation laws require mine operators to file permits and post bonds to engage in any kind of surface mining. After mining, the site is reclaimed by leveling the mined bench and spoil with either a downward or upward slope of the bench with respect to the highwall and "manicuring" of the outslope. In addition, the law requires vegetation, usually grasses but sometimes trees, to be planted on the bench and outslope to prevent erosion. At present the highwall can remain, but the bench elevation must be at least 2m (5 ft) above the coal seam (Kentucky Division of Reclamation). The final texture and elevation of the benches are left up to the discretion of the reclamation inspector, and thus differ between reclaimed benches.

In the enforcement of reclaimed areas various problems necessitate the use of remote sensing:

A. On site examination of individual mines, and particularly older mines, is hindered by (1) a lack of adequate mine map coverage; (2) deeply eroded, non existent, or blocked access roads; (3) lack of accurate or adequate records; (4) the great total area of the strip mining; (5) strip mine reclamation planting along roads that obscures adjacent barren land; and (6) dated or non-existent aerial photographic coverage [5].

B. Overburdened reclamation inspectors cannot inspect mines frequently enough for effective enforcement. A 500% increase in mining has occurred since 1966 [6]. Remote sensing techniques could determine problem mines, and only these mines would have to be individually inspected.

C. Inconsistencies exist between individual inspectors interpretation of reclamation regulations[6]. Remote sensing can be used for reference in controversial interpretations.

Certain inherent limitations of SAR presented themselves in this study. Namely the rolling hill topography of southeastern Kentucky (test area) with elevations differences ranging between 60-152 meters, the look direction of the radar becomes quite important. One radar look direction provided a significant amount of information; however, some mined areas were excessively dark on the imagery, particularly at the far range of each image. This darkness was due to radar shadowing and fading of the radar return in the far range. For future strip mining flights, two perpendicular looks of the same area are recommended. The depression angle for future flights should be as large as possible to avoid excessive shadowing.

SUMMARY

Using four channel imaging radar the recognition of three basic types of contour strip mining has been shown. Active, reclaimed and unreclaimed "orphan" mines were detected and identified. Highwalls, benches, relative bench vegetation and texture, outslope, spoils, sediment ponds, and mining machinery were recognized on the imagery. In some cases all four radar channels are needed for recognition of these features.

Radar is especially sensitive to surface roughnesses, subtle height variations, and orientations of surfaces. In addition, its all-weather capability and its resolution (essentially independent of distance), both of which are unique in imaging devices, make imaging radar particularly useful for strip mine monitoring.

Small contour mines and unreclaimed "orphan" areas were resolved by radar under cloudy conditions. Other remote sensors are hampered by often persistent cloud cover, and current ERTS type imaging systems are not able to resolve these small mines from space.

Imaging radar is most useful for the following:

1) Mapping of unreclaimed "orphan" areas,
2) Monitoring of wildcat (illegal) mining operations,
3) Highwall detection,
4) Reclamation evaluation for enforcement purposes.

ACKNOWLEDGEMENTS

We wish to thank Frank Ruskey of the U.S. Bureau of Mines for helpful suggestions and encouragement. We also wish to thank William Boyd, Liaison Officer for the States of Kentucky and W. Virginia, U.S. Bureau of Mines as well as Herchel Moyers, Bill Harris, Ron McCarty, and Joe Hensley of the Middlesboro Regional Office of the Kentucky Division of Reclamation for their great help in the ground observation effort. This research was supported in part by the U.S. Bureau of Mines under Contract No. S0241056.

REFERENCES

1. Smith, F., and R. Rawson, 1974, Four Channel Simultaneous X-L Imaging SLAR Radar, Proceedings of the Ninth International Symposium on Remote Sensing of the Environment, Ann Arbor, Mi.
2. Brown, W.M., and L.J. Porcello, 1969, An Introduction to Synthetic Aperture Radar, IEEE Spectrum, Vol. 6, No. 9, pp. 52-62.
3. Harger, R.O., 1960, Synthetic Aperture Radar System, Theory and Design, Academic Press, NY.
4. Westinghouse, 1967, Side Look Radar, Brochure produced by Westinghouse Elec. Corp., pp. 1-46.
5. Rogers, H., L.E. Reed, and W.A. Pettijohn, 1973, Automated Strip Mine and Reclamation Mapping from ERTS. Paper presented at Third ERTS Symposium, pp. 1.
6. ARC, 1974. Design of Surfacing Mining Systems in Eastern Kentucky, Volume I, Summary, Appalachian Regional Commission, Report ARC-71-66-T1, p. 82.

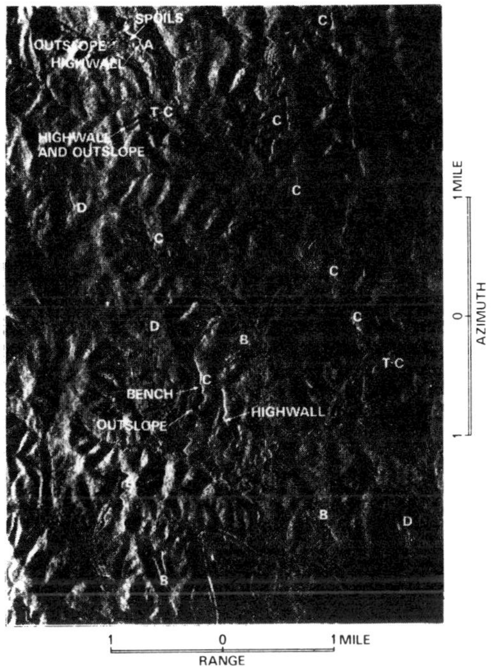

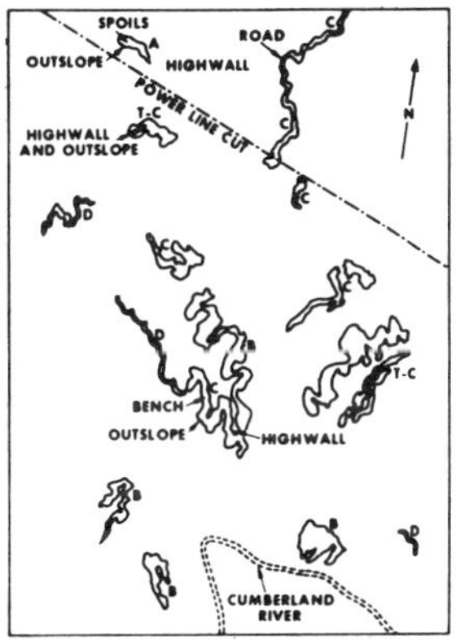

FIGURE 1. Identification Key for Strip Mines Types and Individual Mining Features Found Within Test Site

A - Active Mine Site

B - Reclaimed (Vegetated)

C - Reclaimed (Non-vegetated)

D - Unreclaimed ("Orphan Area")

T - Double Tiered (Bileveled Areas)

Table 1. Radar Tonal and Textural Characteristics of Mining Features

MINING TYPE	FEATURE WITHIN MINING AREA	X-BAND (HH)[1]		X-BAND (HV)[1]		L-BAND (HH)		L-BAND (HV)	
		TONE[2]	TEXTURE[3]	TONE	TEXTURE	TONE	TEXTURE	TONE	TEXTURE
ACTIVE	A. HIGHWALL	B	R	B	R	M	R	M	R
	B. BENCH	D-M	M-S	D-M	M-S	ND	ND	ND	ND
	C. SPOILS	M	R	M	R	M-B	R	B	R
	D. OUTSLOPE	D-M	R	D-M	R	ND	ND	ND	ND
	E. MACHINERY	B	PR	B	PR	B	PR	B	PR
	F. GENERAL (NA)								
RECLAIMED	G. HIGHWALL	B	R	B	R	B	R	B	R
	H. BENCH-NON. VEG.	D-M	M	M-D	M	M-D	M-R	D	M-R
	I. BENCH-VEG.	M	M-R	M	M-R	D	M-R	D	M-R
	J. OUTSLOPE	D-B	M-R	D-B	M-R	M	R	M	R
	K. SETTLING PONDS	D	S	D	S	ND	ND	ND	ND
	L. GENERAL (NA)								
UNRECLAIMED "ORPHAN" AREAS	M. HIGHWALL	B-M	R	M	R	ND	ND	ND	ND
	N. BENCH-VEG.	M-D	M	M-D	M	D	S-M	D	S-M
	O. SPOILS	M-D	M	M-D	M	D	S-M	D	S-M
	P. OUTSLOPE	ND	ND	ND	ND	ND	ND	ND	ND
	Q. GENERAL (NA)								

1. HH REFERS TO NORMAL-POLARIZATION, TRANSMITTED AND RECEIVED HORIZONTALLY
   HV REFERS TO CROSS-POLARIZATION, TRANSMITTED HORIZONTALLY AND RECEIVED VERTICALLY

2. B-BRIGHT
   M-MODERATE
   D-DARK

3. R-ROUGH
   M-MODERATE
   S-SMOOTH
   PR-POINT REFLECTOR
   ND-NOT DISCERNIBLE
   NA-NOT APPLICABLE

TONE IS THE GRADIATION (BLACK TO WHITE) OF GRAY WITHIN AN AREA REPRESENTING THE INTENSITY OF BACKSCATTERED ENERGY (GREATER RADAR REFLECTION = BRIGHTER TONE).

TEXTURE IS THE FREQUENCY OF TONAL CHANGE WITHIN AN AREA.

Near Range Far Range

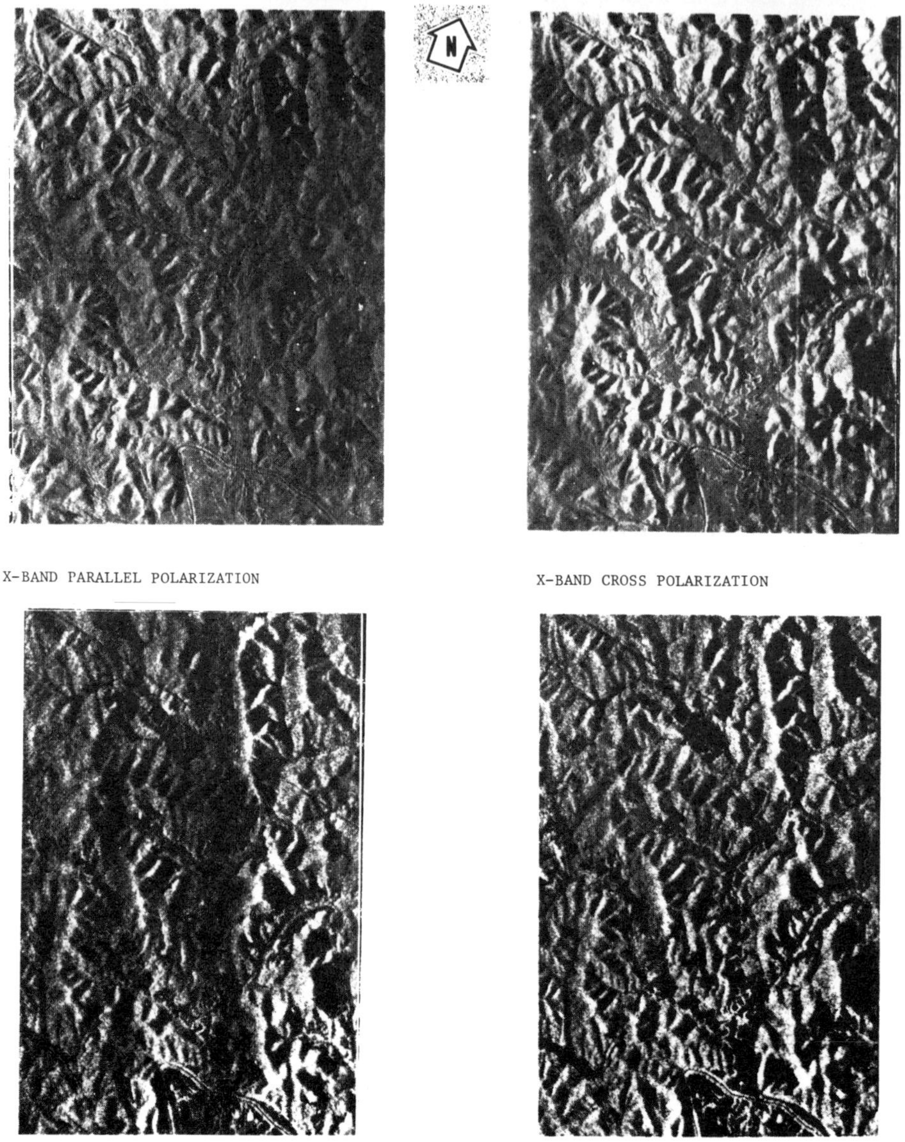

X-BAND PARALLEL POLARIZATION

X-BAND CROSS POLARIZATION

L-BAND PARALLEL POLARIZATION

L-BAND CROSS POLARIZATION

FIGURE 2. Simultaneously Obtained Dual Wavelength and Polarization Radar Imagery of a Strip Mining Region in Southeastern Kentucky. 9.3 x 9.3 Resolution, All Four Channels.

A Microwave Hologram Radar System

R.W. Larson, J.S. Zelenka, and E.L. Johansen
(Institute of Science and Technology, University of Michigan, Ann Arbor)

IEEE Transactions on Aerospace and Electronic Systems, Vol. AES-8, No. 2 (March 1972)
Pages 208-217

Abstract

An airborne microwave hologram radar system has been developed which is a two-dimensional analog to optical holography. The field of view of the radar is directly below and to either side of the aircraft. Resolution is realized in the along-track direction by utilizing the synthetic aperture technique, and in the cross-track direction by means of a phased receiving array. The theory of operation is summarized, the demonstration system is described, and results for both the normal and contouring modes of operation are presented.

I. Introduction

During the past twenty years considerable advances have been made in radar imaging systems. This work has included the application of optical processing techniques to synthetic aperture side-looking radar to obtain fine resolution maps of the ground [1]-[3] and moving targets [4]. It has been shown that the azimuth channel of synthetic aperture radar systems is an analog to optical holography [2], [3], [5]-[7]. With synthetic aperture radars fine range resolution is obtained by the use of short pulsewidths or pulse compression techniques.

Recently an airborne microwave hologram radar system has been developed which is an analog to optical holography in two dimensions. The field of view of the radar is directly below and to either side of the aircraft (Fig. 1); this field of view is similar to that of airborne infrared and photographic systems. Resolution is realized in the along-track direction by utilizing the synthetic aperture technique, and in the cross-track direction by means of a phased array. The hologram radar system operates in a CW mode and radiation from the transmitting antenna in the aircraft illuminates the ground area. The backscattered energy is coherently detected and recorded on film. Signal processing is then accomplished using two-dimensional optical techniques similar to those developed for side-looking synthetic aperture systems. Synthetic aperture compression is performed in the along-track direction and antenna beam forming is accomplished simultaneously in the across-track direction. Photographic film serves to record the image at the output plane of the optical processor.

A unique feature of the microwave hologram radar is the range contouring capability. When the frequency of the transmitter is switched rapidly between two values f_1 and f_2, the received signals will interfere either constructively or destructively at slant range increments separated by the velocity of light divided by twice the frequency separation of f_1 and f_2.

This paper summarizes the theoretical and experimental hologram radar program conducted at The University of Michigan, which started with theoretical studies in 1964. An experimental hologram radar system was built, and ground tests were performed in 1967 and 1968 [4]. Fourteen imaging flights in a C-131 aircraft were made in 1968 and 1969.

II. Basic Theory

Fig. 2 displays the geometry of the hologram radar system for a point target. The point target is located at x, y in the ground plane (x-y plane). Above this plane at a height h there is a receiving array which is parallel to the y axis. This array is of length $2a$ and it moves in the x direction with velocity V. At the center of the array, $(x_0, 0, h)$, there is a source of coherent radiation illuminating the ground plane. To simplify this initial discussion we shall imagine that there are an infinite number of receivers

Manuscript received July 26, 1971; revised September 30, 1971.

This work was supported by the Air Force Avionics Laboratory, Wright-Patterson AFB, Ohio, under Contract F33615-67-C-1814.

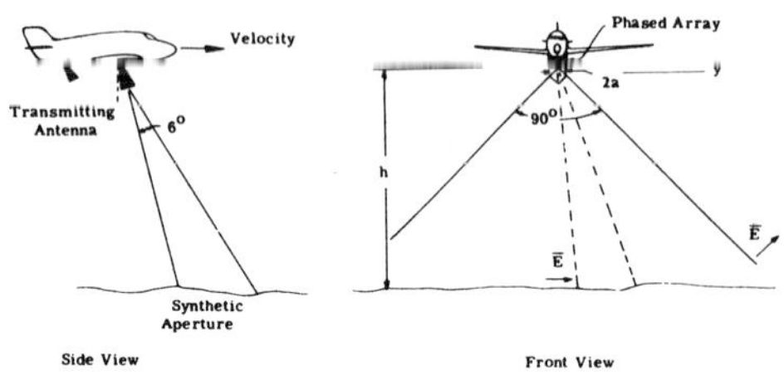

Fig. 1. **Hologram radar geometry.**

Fig. 2. Hologram radar geometry for point target.

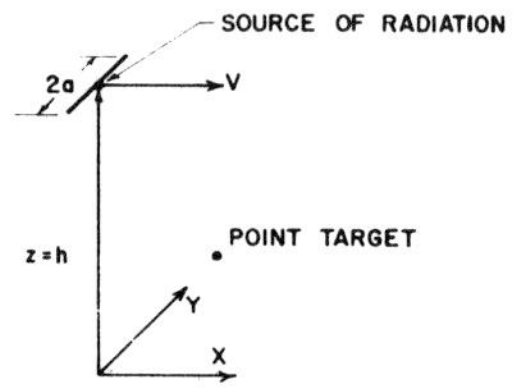

closely spaced along the array so that the return from the point target can be continuously recorded along the length of the array.

Suppose the transmitted signal S_t is in a sine wave

$$S_t = \exp j\omega t; \tag{1}$$

then the received signal S_R at a point on the array (x_0, y_0, h) has the form

$$S_R = K \exp j\omega(t - \tau) \tag{2}$$

at arbitrary time t where t is equal to x/V. Here K is a constant which accounts for transmitter power, antenna gain, range to the target, etc., and τ is the time required for the transmitted signal to reach the target and return to a point on the array. In our analysis we shall assume that the array velocity V is much less than the velocity of light so that the array moves only a small fraction of a wavelength during τ:

$$\tau = \frac{R_t + R_r}{c} \tag{3}$$

with

$$R_t = [(x_0 - x)^2 + y^2 + h^2]^{1/2} \tag{4}$$

$$R_r = [(x_0 - x)^2 + (y_0 - y)^2 + h^2]^{1/2}. \tag{5}$$

After synchronous detection, the phase of the return becomes

$$\phi_1 = -\frac{2\pi}{\lambda_r}(R_t + R_r). \tag{6}$$

Here λ_r is the radar wavelength. It may be shown that ϕ_1 can be written as the sum of two terms. The first term ϕ_2 represents the phase history recorded by a perfect hologram, and the second $\Delta\phi$ can be considered as an error term that causes aberrations in the reconstructed image; this term arises from the relative motion between the object and the illuminating source. Let

$$R_0 = \left[(x_0 - x)^2 + \left(\frac{y_0}{2} - y\right)^2 + h^2\right]^{1/2}; \tag{7}$$

then

$$\phi_1 = -\frac{4\pi}{\lambda_r} R_0 + \Delta\phi. \tag{8}$$

To record the signal history on film we must introduce a bias term S so that signal values can be recorded. It is also necessary to scale down the dimensions of the diffraction pattern in space so that the pattern will fit on the film. In the x dimension the scaling factor P_x is the ratio of film speed v to array speed V:

$$P_x = v/V. \tag{9}$$

In the y direction the scaling factor P_y is the ratio of film width w to array width $2a$:

$$P_y = \frac{w}{2a}. \tag{10}$$

By substituting x/P_x for x and y/P_y for y in R_0, the phase term ϕ_2 recorded on film becomes

$$\phi_2 = -\frac{4\pi}{\lambda_r P_x}\left[(x - P_x x_0)^2 + \left(\frac{P_x y}{2P_y} - P_x y_0\right)^2 + P_x{}^2 h^2\right]^{1/2} \tag{11}$$

It has been shown that if a hologram is constructed at a longer wavelength than the reconstruction wavelength, the hologram should be scaled down approximately by the wavelength ratio before the image is viewed [8]-[10]. Without such scaling the lateral and longitudinal magnifications of the image are unequal and aberrations are introduced. Reference [11] discusses these problems in greater detail. The motion of the hologram radar array introduces an additional factor of 2 which complicates the reconstruction. A more detailed analysis shows that $\Delta\phi$ contains a quadratic phase error in the y dimension. This phase error is a function of target location; however, it can be eliminated by setting

$$\frac{P_x}{\sqrt{2}} = P_y = \frac{\lambda_l}{\lambda_r} \tag{12}$$

where λ_l is the wavelength of light used in the reconstruction. If this relationship is satisfied, the along-track or x

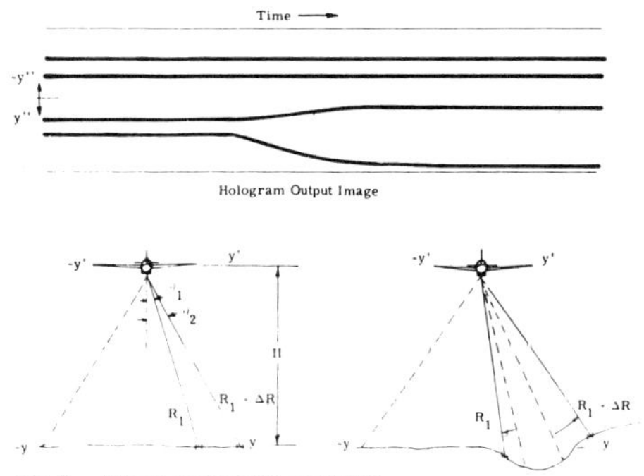

Fig. 3. Range-contouring geometry.

dimension of the image will be larger than the cross-track or y dimension by the factor $\sqrt{2}$. If we choose

$$P_x = 2P_y = \frac{2\lambda_l}{\lambda_r} \tag{13}$$

the output image will have unity aspect ratio; however, the length of the array ($2a$) cannot be too large or the quadratic phase error in $\Delta\phi$ will be excessive [11]. The quadratic phase error can be compensated whenever

$$h > \frac{4a^2}{\lambda_r}(1 - \cos^3 \beta_0/2) \tag{14}$$

where β_0 denotes the cross-track beamwidth.

When (13) and the distance criteria are satisfied, ϕ_2 is a good approximation to the signal available for recording on film. A linear phase term is introduced before recording so that the desired output image can be separated from the conjugate image and the undiffracted light in the processor. When collimated, monochromatic light is incident on the hologram, a demagnified image similar to the scene illuminated by the hologram radar will appear.

With the hologram radar it is impractical to utilize the ideal demagnification ratio from radar to reconstruction space ($2\lambda_l/\lambda_r$) because this ratio is too small, ranging in practice from 10^{-5} to 10^{-3}. Current recorders cannot pack information on film densely enough to permit the ideal scaling, and even if the ideal scaling could be realized, practical limitations would restrict the length of the array so that the size of the scaled hologram would be too small for convenient viewing. A lens system is therefore needed to reconstruct the image and magnify it to a reasonable size.

In the along-track direction standard synthetic aperture theory applies, and for nearly vertical viewing the theoretically achievable resolution ρ_x is

$$\rho_x = \frac{\lambda_r}{4 \sin(\phi_x/2)} \tag{15}$$

where ϕ_x is the antenna beamwidth in this direction. In the cross-track direction the array length determines the resolution ρ_y,

$$\rho_y = \frac{A\lambda_r}{2a} h \sec^3 \theta_y . \tag{16}$$

Here A is a constant near unity, which depends on the aperture weighting, θ_y is the angle measured from the vertical, and h is the height.

III. Range Contouring

In optical holography, two-frequency light can be used to generate holograms which contain contour lines [12]. The same principle applies to the hologram radar. When the frequency of the transmitted radiation is switched rapidly between two values f_1 and f_2, it is clear that the received signals will interfere either constructively or destructively when received from range increments given by

$$\Delta R = \frac{c}{2(f_2 - f_1)} \tag{17}$$

where c is the velocity of light. Fig. 3 shows the contouring geometry. Assume that signals received along the path R_1 at the frequencies f_1 and f_2 interfere; similarly, signals received along path $R_1 + \Delta R$, at transverse angle θ_2, also will interfere. Lines of minimum intensity will appear in the output image corresponding to the transverse angle associated with range R_1 and $R_1 + \Delta R$.

It is clear that the indications obtained showing terrain relief are not "map contour lines," but slant range contour lines referenced to the aircraft. Since the aircraft geometry is known, knowledge of the aircraft altitude and ground elevation references will permit the surface contour to be obtained from the range contour lines on the microwave hologram imagery.

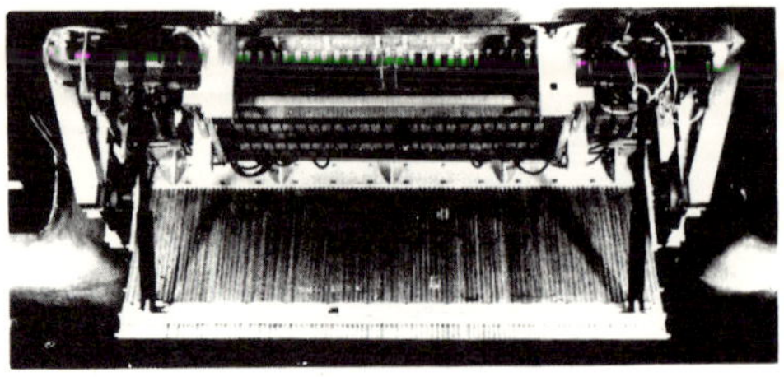

Fig. 4. View looking aft toward the forward radome showing the 108-element antenna and one row of receivers.

IV. A Description of the Demonstration System

The basic requirements for the construction of the microwave hologram are that 1) the terrain be illuminated by a microwave signal, 2) the backscattered wavefront must be adequately sampled in two dimensions, 3) the phase information of the return must be recovered by coherent detection, and 4) the phase information of the return must be recorded on a film or some other recording medium. The University of Michigan has constructed a demonstration system meeting these requirements. This radar operates at low altitudes with a relatively short receiving array.

In the theoretical discussion, the transmitting antenna was located at the center of the receiving array; in the demonstration system, however, the transmitting horn was moved 10 meters behind the receiving array to isolate the transmitter from the receiving elements. A receiving array of 100 elements served to sample the received field across the aircraft. The elements were equally spaced; the spacing of slightly over half a wavelength was sufficient to prevent grating lobes from appearing on the output image. Fig. 4 shows the receiving array; it consists of 108 *H*-plane horns mounted side by side. The outer four elements on each end were not utilized because their patterns differed from those of the other elements.

Fig. 1 shows the antenna beam coverage. In the along-track direction the beam is narrow, but across track a broad 90° beam is used for wide angular coverage. The transmitting antenna was also an *H*-plane horn with the same beamwidth.

An operating frequency of 16.8 GHz was selected because suitable components could not be obtained at a reasonable cost at a higher frequency. A high operating frequency is, of course, desirable for good cross-track resolution. The space available in the radome of the C-131 test aircraft determined the length of the array.

Fig. 5 is a simplified block diagram of the complete system. Table I lists the operating parameters. A cavity stabilized klystron generates the basic operating frequency of 16.8 GHz. Power output from the klystron is divided in two with each output line driving the input of a single-sideband generator. Offset frequencies used to drive the single-sideband (SSB) generators are obtained by multiplying the output of a 5-MHz crystal oscillator by factors of 12 and 13. The difference between the output frequencies of the two SSB generators is equal to the intermediate frequency of the receiver amplifier. Power output from one SSB generator is amplified in a TWT and radiated as the transmitted signal; power output from the other SSB generator is also amplified in a TWT which drives the local oscillator distribution system.

The radar return is received by 100 individual receiving elements simultaneously. Each return is converted down to an IF of 15 kHz, amplified, demodulated, and then sent to a multiplexer. The required synchronous demodulation frequency is obtained from the basic 5-MHz crystal oscillator through a countdown circuit dividing by 300. The multiplexer output is recorded on film. The 100 channels are recorded across the film as the multiplexer switches from one channel to the next and the Doppler signal histories are recorded along the film as it passes through the recorder. Fig. 6 is an enlarged section of hologram signal film.

Alignment of the system consisted of adjusting all receivers to indicate signals of equal phase when the equivalent of a plane wave was incident on the array. Amplitude weighting of the receiving array was accomplished as a part of the same operation, by adjusting the gains of the receivers. Practical difficulties were encountered when using a far-field source to provide an incident plane wave with long-term stability across the array. To avoid this difficulty, a small source that could be moved across the face of the array was used instead of a plane wave. Results of a number of phase measurements indicate that the phases can be aligned within 10 degrees peak-to-peak error for the 100 receivers. Once aligned, phase tolerances of 20 degrees peak-to-peak over a period of several months were realized. Considerable design effort and temperature cycling testing were necessary in order to arrive at the final phase-stable circuitry.

When the radar was being designed, provision was made for pulsed operation in the event of inadequate isolation

TABLE I

Hologram Radar Parameters

Transmitter	
Radiated frequency	16.8 GHz (λ = 1.78 cm)
Polarization	horizontal
Transmitter power (maximum)	400 mW
Frequency difference in contouring mode	5 MHz
Receiving Array	
Total array elements	108
Active array elements	100
Across-track spacing of array elements	0.56 λ
Across-track array length	1.01 meters
Along-track width (both array and elements)	0.15 meters
Receivers	
Intermediate frequency	15 kHz
Bandwidth (receivers)	1800 Hz
Receiver system noise figure (average)	17 dB
Receiver sampling rate (multiplexer)	500 kHz
Average array sampling rate	3800 Hz
Peak-to-peak phase variation (maximum)	± 20°
Amplitude taper across array	Taylor, for 20-dB sidelobes
Recording Film	
Width of recorded strip	5 mm
Average film speed (at 150 knots)	40 mm/s
Scaling factors:	
P_y(across-track)	6.8 X 10^{-3}
P_x(along-track)	5.2 X 10^{-4}
Image	
Ground swath width (across-track)	Twice aircraft altitude above terrain
Across-track resolution	0.018 radians
Along-track resolution	0.076 meters
Range-contour intervals	30 meters (100 feet)

Fig. 5. Simplified block diagram of the experimental system.

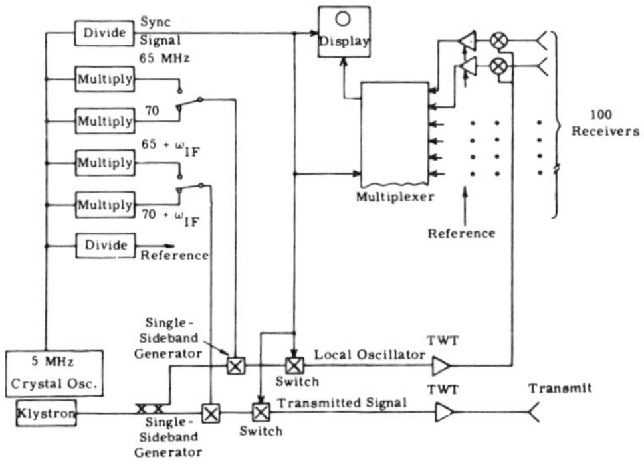

Fig. 6. Section of microwave hologram signal film. Receiver channel numbers are in the vertical direction, and time is in the horizontal direction. Actual film width is 5 millimeters.

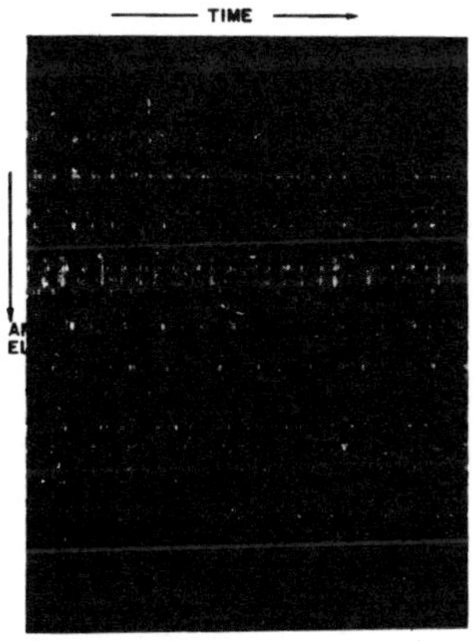

between the transmitter and receiver. Fortunately, the 10-meter separation of the transmitter horn gave sufficient isolation and many of the maps were made in the CW mode.

So that range contours can be generated, provisions have been made to permit operation in the two-frequency mode. This is accomplished by switching the SSB generators, as shown in Fig. 5. Alternate pulses consist of two transmitted frequencies differing by 5 MHz. The contours that result denote slant range intervals of 30 meters.

V. Processing

As mentioned in Section II, recorder limitations do not permit the microwave wavefront to be scaled down by the ideal wavelength ratio, resulting in phase errors across the aperture. An additional complication is usually introduced by choosing along-track and cross-track scale factors which differ greatly in order to record data on film efficiently. As a result the holographic image does not come to a common focus in the along-track and cross-track directions. Relatively simple optical systems will serve to compensate for practical scaling factors and to adjust the reconstructed image to a suitable size. Standard spatial filtering techniques in the processor permit the removal of both the dc bias and the unwanted sideband. There are several optical configurations capable of generating an image from a microwave hologram, but we shall only discuss the processor in Fig. 7, which was the processor used most frequently in the program.

In Fig. 7, the microwave hologram is placed in plane H where it is illuminated with a collimated monochromatic source. Following the hologram plane are two telescopes in tandem. The first is a spherical telescope composed of lenses L_1 and L_2, which have respective focal lengths f_1 and f_2. The second is a cylindrical telescope with lenses L_3 and L_4.

The spherical telescope transports desired wavefronts emanating from the hologram to the plane H'. The unwanted wavefronts (i.e., the undiffracted wave resulting from the dc bias together with one of the first-order waves resulting from having the signal on a spatial carrier) are easily removed by placing the appropriate slit in the frequency plane F. Of course, any spacial filtering of the signal can be readily accomplished by placing the proper optical element, usually a transparency or lens, in plane F.

The cylindrical telescope is used to compensate for the usual mismatch in scaling factors. This telescope does not affect any wavefront appreciably in a direction parallel to the axes of its lenses, say the along-track dimension. In the other dimension, that is, cross-track, it operates on the virtual wavefront formed by the spherical telescope at plane H' to form a scaled-down wavefront at the same location. For the cylindrical telescope to form a rescaled wavefront at H', it is required that

$$x_2' = f_4 \frac{3 + f_4/f_3}{1 - (f_4/f_3)} \,. \tag{18}$$

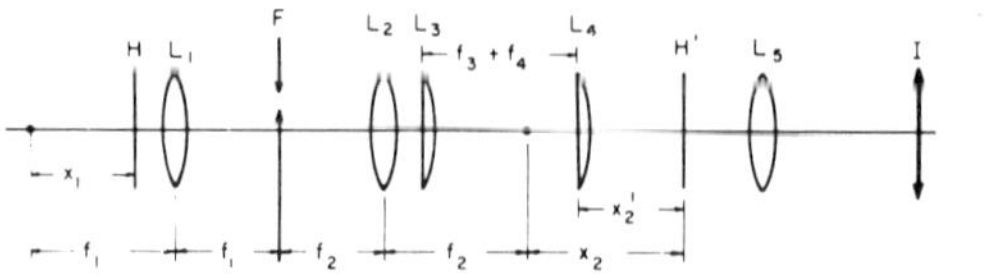

Fig. 7. Decoupled telescopic processor.

The complex wavefront at H' is identical to one of the first-order waves emanating from H, except that it has been rescaled or demagnified by an amount

$$M_a = f_2/f_1 \tag{19}$$

in the along-track direction and

$$M_c = (f_2/f_1)(f_4/f_3) \tag{20}$$

in the cross-track direction.

We can view the reconstruction of the target field directly from the H' plane. As described below, this image can be three-dimensional in nature providing two conditions are met. First, the aperture associated with the demagnified wavefront at H' must be large enough to accept both the viewer's eyes. Second, the reconstruction has to form a reasonable distance behind the hologram, say one-half to ten feet. When these processors are operated in the mapping mode, it is usually convenient to have an additional spherical lens L_5 in the vicinity of H'. This lens can be easily used for adjusting the final image to some specified size.

The processors are tracking processors when operating in the mapping mode. That is, all targets formed in the output plane move in synchronism with the input data. Tracking processors have two distinct advantages over other types, both advantages arising from the fact that the output film can be exposed via an adjacent slit several resolution elements wide. First, such a wide slit reduces the exposure time required to make the map, since for nontracking processors, this slit can be no wider than a resolution cell. Second, the peak output noise resulting from dust particles, lens imperfections, etc., is greatly reduced by averaging.

The processor depicted in Fig. 7 is suitable for modifying scaling factors from 4 to 25 times. Another processor must be used for scale factors higher than 25 [11]. Compensating for scaling factors less than 4 can be accomplished with the processor simply by placing a cylindrical shift lens in the frequency plane F. When the axis of this lens is at right angles to those of the cylindrical telescope, the along-track focus position H' is shifted further down the system axis. Thus, more space becomes available for the use of a cylindrical lens L_4 with a sufficiently long focal length.

With the experimental hologram radar the intensity of the imagery varied considerably with slant range and antenna gain, and at the edges of the swath the attenuation was 10 dB higher than at the center. Since this attenuation is a function of the cross-track position of the targets, we can easily compensate for it in the frequency plane of the

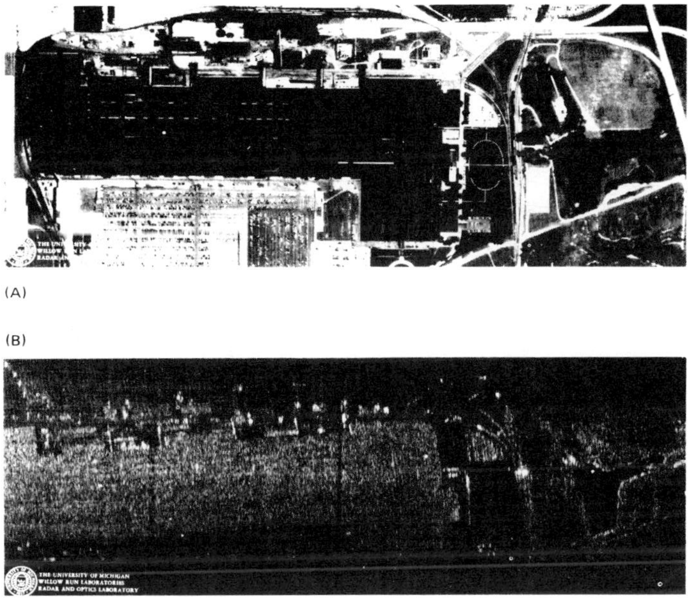

(A)

(B)

Fig. 8. (A) Photograph of Willow Run Airport, Ypsilanti, Mich., and factory complex (300-meter altitude). (B) Microwave hologram image of factory complex (CW operation; 4-mW average power; 300-meter altitude).

optical processor. The reflectivity of the terrain is dependent upon the depression angle from the radar, and variations in image intensity may also result from this factor.

Although hologram radars have the capability of generating three-dimensional imagery, the experimental system was not well suited for this purpose. Since it had only 100 cross-track receiving elements and an along-track beam of 6°, an ideal microwave hologram generated from an altitude of 300 meters would have dimensions of only 0.036 by 2.2 mm. Even if such a hologram were enlarged ten times, it presents a very small aperture for direct viewing with both eyes. Some experiments on the generation of three-dimensional imagery were seen with enlarged holograms; however, only those people having considerable experience with stereo viewers were able to see the three-dimensional image associated with tall objects. Although the generation of three-dimensional images has been experimentally verified, the resulting images are not of any practical use. Hologram apertures an order of magnitude larger in both dimensions would give more satisfactory results.

VI. Imagery

The flight test program consisted of fourteen mapping flights, each one hour or less in duration, and imagery was obtained on all flights. All flights were made at a low altitude, 300 meters above ground level or less, in order to obtain the best possible resolution. No motion compensation was used during any flight; consequently, the generation of a good synthetic array was dependent upon a stable flight path.

Fig. 8(A) is an aerial photograph of a factory complex near the Willow Run Airport, and 8(B) is a hologram radar image of the same area when the flight path was directly over the factory. A large number of features in the image can be identified in the photograph. For instance, there is a lake on the right side next to a group of railroad tracks.

Fig. 9(B) is a hologram radar image of a ramp area at the Willow Run Airport taken from a 150-meter altitude, and 9(A) is a photograph of the boxed area in the image. A large hanger facing the ramp is at the bottom of the figure. Five C-46 aircraft can be seen in the center of the ramp area, and a twin-engine beechcraft is at the left end of the line of parked aircraft. A single aircraft, a four-engine Electra turbo prop, is at the top of the ramp. There are a large number of stored parts, boxes, and crates at the right of the ramp.

The radar system was also operated in the two-frequency mode to produce range contours in the output image. With a 5-MHz frequency difference, the slant range difference between contours is 30 meters. Fig. 10(A) is a photograph of an industrial plant and gravel pit on opposite sides of a river. The constant range lines in the image [Fig. 10(B)] show the steep slopes in the gravel pit (*A*), the downward slope from the building toward the river on its right side

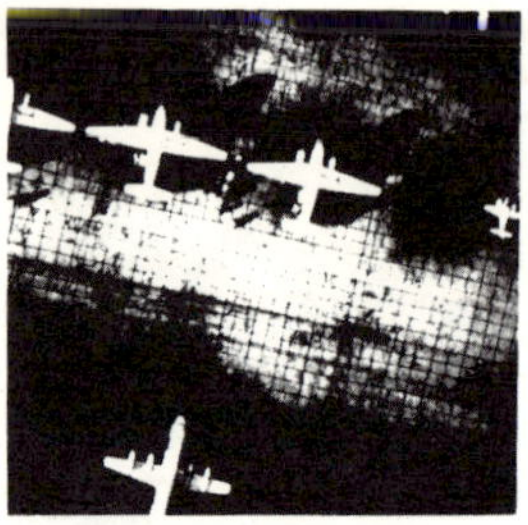

(A)

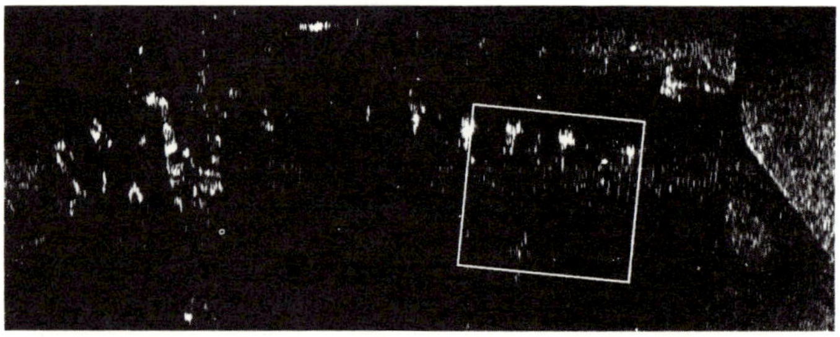

(B)

Fig. 9. (A) Photograph of air-freight parking ramp, Willow Run Airport. (B) Microwave hologram of air-freight parking ramp and hanger, Willow Run Airport, at 150-meter altitude.

Fig. 10. (A) Photograph of an industrial plant and a gravel pit northwest of Ann Arbor. (B) Microwave hologram image of industrial plant and gravel pit at 300-meter altitude, range contour mode.

(A)

(B)

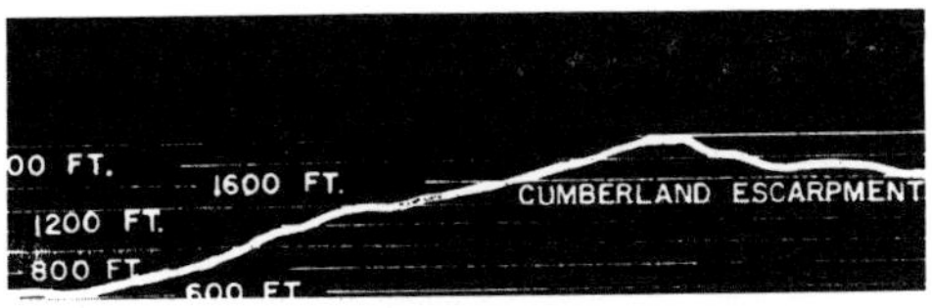

Fig. 11. Microwave hologram of hill west of Chattanooga, Tenn. Minimum altitude is 300 meters; maximum is 510 meters.

(*B*), and the sudden shift in range contour due to the height of the building (*C*). Within the gravel pit, depressions (*D*) and mounds (*E*) can be seen though the two cannot be differentiated as such without additional information, and signal returns from many objects in the pit are also evident. It can reasonably be deduced from the image that the slope from the building to the river is downhill. It may be observed that the lines converge toward the center of the images, becoming closer to the aircraft's ground track as it is moved from the left side of the image toward the river. Since the slant range is constant for a given line, the terrain should slope downward in that direction. The closed line in the center of the image is ambiguous and could be caused by either a hill or a depression beneath the aircraft. The shift in the range contour line which goes over the building can be measured, and using the measured value, the vertical height of the building was found by simple geometry to be 5 meters. The structure is, in fact, a one-story factory building about 5 to 6 meters high.

Fig. 11 is an interesting example of range contour imagery obtained from a flight over a wooded section of mountains in Tennessee along a westward line from Chattanooga to Monteagle. An elevation change of 390 meters in the vicinity of the Sequatchie Valley is shown on the profile. The line intervals in the image represent 30-meter range increments, and 14 of these lines can be counted.

VII. Conclusions

The objective of the hologram radar program was to study the applicability of holographic techniques to airborne terrain-imaging systems operating in the microwave region and to demonstrate feasibility. The theory underlying such a hologram system was examined, and an experimental system designed, constructed, and operated successfully. Visible hologram images of good quality have been generated from airborne microwave radar signals. Synthetic-aperture beam compression and phased-array beam forming were accomplished simultaneously in an optical processor. Measured system resolutions were in good agreement with theoretically predicted values. The capability of a hologram radar to generate range contours on terrain images by using a two-frequency interference technique has also been successfully demonstrated.

These hologram images show the terrain beneath the aircraft and out to 45° on either side of the aircraft's ground track, and they reveal the microwave appearance of terrain and targets from a perspective unlike that of traditional pulse-ranging radars. Because this type of radar system produces microwave images with a perspective which nearly matches that of infrared and photographic sensors, it may be particularly well suited to applications in the new field of multispectral sensing.

The present experimental hologram radar system is the first to be constructed and it was intended only to demonstrate the principles of microwave holography, not to compete with state-of-the-art side-looking radars. Cross-track resolution could be improved in future hologram radar systems by using longer arrays or transmitter switching techniques in various combinations. For arrays with a greater number of elements, thinning techniques (either the random removal of elements or incorporation of point transmitters along the array to remove sections of redundant elements) should be utilized. Improvement in across-track resolution for a given array aperture size can be realized by increasing the operating frequency. With the present aperture size of 1 meter, a system designed to operate at a wavelength of 3 mm would have an array beamwidth of 0.003 radian.

There should be considerable interest in the range-contouring capability of such radar systems. The ground swath can be increased by flying at higher altitudes, and the range-contour intervals can be varied by changing the difference between the two frequencies used. Designers of future systems should also consider the possible employment of fringe sharpening to reduce the amount of image blanking or loss resulting from destructive interference. This can be accomplished, for example, by transmitting more than two equally spaced frequencies. Simultaneous imagery with and without range contouring, on two output films, could be obtained with minor modifications to The University of Michigan system.

Acknowledgment

The authors wish to acknowledge the assistance in this program of many other members of the Radar and Optics Division of the Institute of Science and Technology. The meticulous efforts of F. Smith in constructing and operating the equipment merit special attention. Dr. K. A. Haines was responsible for the first theoretical analysis of the radar and R. A. Rendleman directed the flight test operations.

References

[1] L.J. Cutrona *et al.*, "A high resolution radar combat-surveillance system," *IRE Trans. Military Electronics*, vol. MIL-5, pp. 127-131, April 1961.

[2] E.N. Leith, "Quasi-holographic techniques in the microwave region," *Proc. IEEE*, vol. 59, pp. 1305-1318, September 1971.

[3] W.M. Brown and L.J. Porcello, "An introduction to synthetic aperture radar," *IEEE Spectrum*, vol. 6, pp. 52-62, September 1969.

[4] R.W. Larson *et al.*, "Microwave holography," *Proc. IEEE*, vol. 57, pp. 2162-2164, December 1969.

[5] E. Leith, "Synthetic aperture radar viewed as a holographic process," Radar Lab., The University of Michigan, Ann Arbor, Unpublished Internal Memo., 1956.

[6] ——, "Pulse compression in optics," in *Introduction to Optical Data Processing,* University of Michigan Summer Course Notes, Ann Arbor, sec. 2, pt. 2, July 25-August 5, 1966.
[7] D.E. Duffy, "Optical reconstruction from microwave holograms," *J. Opt. Soc. Am.*, vol. 56, p. 832, 1966.
[8] E.B. Champagne, "Nonparaxial imaging, magnification, and aberration properties in holography," *J. Opt. Soc. Am.*, vol. 57, 1968.
[9] E.N. Leith, J. Upatnieks, and K.A. Haines, "Microscopy by wavefront reconstruction," *J. Opt. Soc. Am.*, vol. 55, pp. 981-986, 1965.
[10] R.W. Meier, "Magnification and third-order aberrations in holography," *J. Opt. Soc. Am.*, vol. 55, pp. 987-992, 1965.
[11] J.S. Zelenka, "Microwave holography," in *Principles of Imaging Radars,* The University of Michigan, Ann Arbor, Class Notes from Intensive Short Course, July 20-31, 1970.
[12] B.P. Hildebrand and K.A. Haines, "Multiple-wavelength and multiple-source holography applied to contour generation," *J. Opt. Soc. Am.*, vol. 57, p. 155, 1967.
[13] R.W. Larson *et al.*, "Results obtained from The University of Michigan hologram radar," presented at the Seventh Internatl. Symp. on Remote Sensing of the Environment, Ann Arbor, Mich., May 17-21, 1971.

Richard W. Larson (S'50–A'51–M'58) was born in Chicago, Ill. He received the B.S. degree in electrical engineering from The University of Illinois, in 1952, and the M.S. degree in electrical engineering from The University of Michigan, Ann Arbor, in 1958.

He has been active in projects relating to microwave radar systems or related components for 19 years. After three years in the U.S. Army Signal Corps, he was with the Electron Physics Laboratory of The University of Michigan from 1955 to 1958. He then worked for a year in the Microwave Branch of Raytheon, Santa Barbara, Calif. For the past ten years he has been a Research Engineer with the Radar and Optics Laboratory, Willow Run Laboratories, Institute of Science and Technology, at The University of Michigan. Work during this period has included various aspects of fine resolution radar systems design such as propagation characteristics, oscillation stability, and microwave hologram techniques. He has also been a Lecturer in the Electrical Engineering Department since 1965.

Jerry S. Zelenka (S'57–M'60) was born in Cleveland, Ohio, on January 27, 1936. He received the B.S., M.S., and Ph.D. degrees in electrical engineering from The University of Michigan, Ann Arbor, in 1958, 1959, and 1966, respectively.

He joined the Bendix Research Division, Southfield, Mich., in 1959, and worked for two years in the Radar Group as a systems engineer. Since 1961 he has been with the Radar and Optics Laboratory, Willow Run Laboratories, Institute of Science and Technology, of The University of Michigan, where he has worked on systems analysis of side-looking radar, coherent optical processing techniques, and applications of holography. He is currently Head of the Electro-Optics Section.

Dr. Zelenka is a member of the Optical Society of America, Sigma Xi, Tau Beta Pi, and Eta Kappa Nu.

Elmer L. Johansen (S'54–A'55–M'60) was born in Lake Forest, Ill., on June 28, 1930. He received the B.A. degree from Harvard University, Cambridge, Mass., in 1952, and the M.S.E.E. and Ph.D. degrees from The University of Michigan, Ann Arbor, in 1954 and 1964, respectively.

For two years following discharge from the U.S. Army in 1956 he was engaged in radar countermeasures for the Cook Research Laboratories, Morton Grove, Ill. Since 1958 he has been at The University of Michigan, where most of his time has been spent on the staff of the Radar and Optics Laboratory, Willow Run Laboratories, Institute of Science and Technology, working on synthetic aperture radar and target reflectivity.

Dr. Johansen is a member of Sigma Xi and Phi Kappa Phi.

The Apollo Lunar Sounder Radar System

L.J. Porcello, R.L. Jordan, J.S. Zelenka, G.F. Adams, R.J. Phillips, W.E. Brown, Jr., S.H. Ward, and P.L. Jackson
(Porcello, Zelenka, Adams, and Jackson — Environmental Research Institute of Michigan, Ann Arbor; Jordan, Phillips, and Brown — Jet Propulsion Laboratory, Pasadena, California; Ward — The University of Utah, Salt Lake City)

Proceedings of the IEEE, Vol. 62, No. 6 (June 1974)
Pages 769-783

***Abstract*—The objectives of the Apollo 17 Lunar Sounder Experiment (ALSE) were to detect subsurface geologic structures, to generate a continuous lunar profile, and to image the moon a radar wavelengths. The first objective is generally impossible on Earth, but is possible on the moon because of the very low EM attenuation found in lunar rocks. A three-wavelength synthetic-aperture radar (SAR) operating at 60 m, 20 m, and 2 m (5, 15, and 150 MHz) was designed to attain these objectives. The design choices reflected a balance of scientific requirements versus Apollo mission and hardware constraints. The most difficult choice was that of wavelength, since longer wavelengths have the potential for deeper subsurface penetration at the expense of depth resolution. Shorter wavelengths enhance the ability to generate surface images and accurate profiles. The sounding requirement led to a system requirement for large dynamic range and precise sidelobe control.**

The radar data from the lunar mission were recorded on photographic film in a conventional SAR format, and were returned to Earth for processing. A combination of optical and digital processing and exploitation techniques was applied to the scientific interpretation of the data. Some preliminary results from the lunar mission have been obtained. The system design and data exploitation techniques developed in support of the ALSE experiment show considerable promise for eventual application to earth-resource survey radar systems.

I. Introduction and Objectives

SIDELOOKING airborne radar (SLAR) has been used over the past decade to generate radar images of the Earth's surface. A radar image is simply a pictorial representation of the reflection properties of the terrain at the radar wavelength. The image is determined by both the roughness and electrical properties of the surface. The relatively high moisture content and consequent high conductivity of vegetation, soil, and rock precludes (except for a few special cases) substantial surface penetration, although it is possible to infer subsurface structures from observed surface features. Glacial ice, however, can be penetrated with radar waves, which have been used to probe glaciers successfully for a number of years. Glacial ice attenuates radar waves much less than soil or rocks. Ice with no free water has low conductivity and low attenuation.

Lunar regolith (soil) and exposed lunar rock exhibit very low electrical loss for wavelengths near 1 m. The absence of free water in the regolith is probably the principal reason for much smaller conductivities than those of typical terrestrial soils. If lunar subsurface structures have large electrical discontinuities, then an imaging radar has the potential to detect the structures. Electrical discontinuities can arise from geologic contacts expected from such features as the boundaries of lava flows, regolith layers, and crater ejecta blankets. The late stages of near-surface igneous activity (which apparently ceased on the moon some three billion years ago) and the details of ejecta deposition would be better understood if a subsurface electrical map were added to the available geophysical, geological, and geochemical data on the lunar surface.

Manuscript received March 11, 1974.

With this broad scientific objective in mind, a three-wavelength orbital imaging radar system was flown on the Apollo-17 lunar mission. The Apollo Lunar Sounder Experiment (hereafter ALSE, assigned by NASA the scientific experiment designation S-209) had the following objectives.

Primary: The detection and location of subsurface discontinuities ("sounding").

Secondary: Generation of a lunar surface profile; generation of surface images; and measurement of galactic noise in the lunar environment.

Surface profiling is related to the global shape of the moon as well as to local topographic forms. The global shape, in particular its disequilibrium figure, is related to the present stress distribution in the interior as well as to lunar thermal evolution. On the last three Apollo missions, profiling was performed using a laser altimeter with a 30- to 43-km sampling interval [1]. The ALSE system, however, could generate a continuous lunar profile yielding detailed shapes of lunar land forms to aid in geologic interpretation. Examples of useful information to be gained from the profile include dips of fault scarps, disposition of lava flow units, and cross sections of rilles. The accompanying image will place the profile in context with more familiar "map" views of lunar features. In addition, the images yield important information in their own right, such as three-dimensional crater shape, slope information, and delineation of areas of differing radar brightness. The galactic noise measurement is an attempt to use the ALSE receiver in the listen-only mode. Relatively uncontaminated galactic-noise flux could be measured in the HF band.

The emphasis in this paper is on the system aspects of subsurface detection, profiling, and surface imaging. The basic ALSE system is described, and some examples of preliminary results are provided. Since much remains to be done on the scientific analysis of the data, scientific results *per se* are deferred for future publication. The measurement of galactic noise is not treated in this paper.

II. Experiment Design

The design of the ALSE was dominated by the primary objective of subsurface-feature detection. A major impediment to the detection of subsurface features is the strong return from the lunar surface against which the weak subsur-

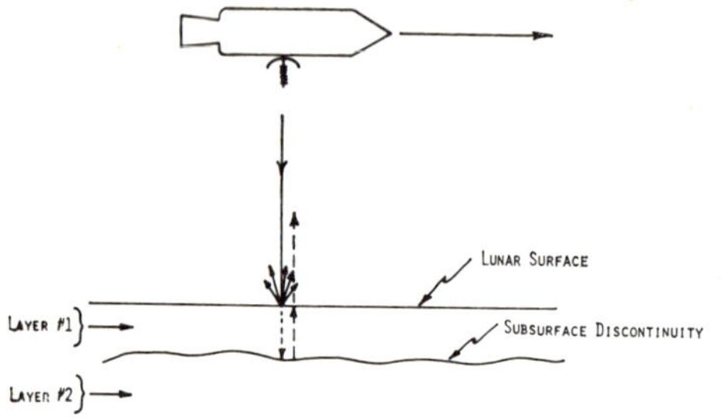

Fig. 1. Electromagnetic sounder.

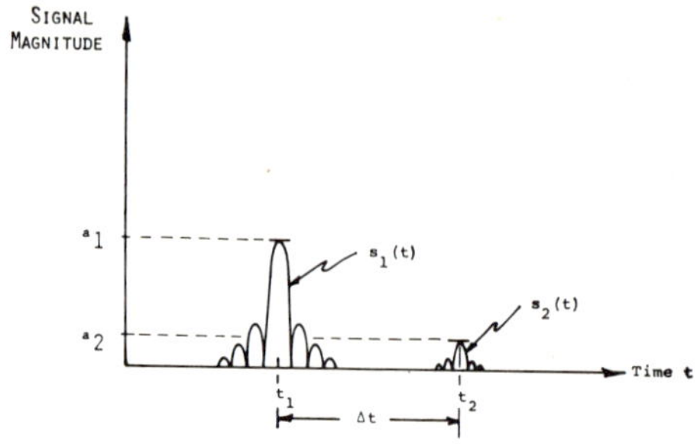

Fig. 2. Resolution of signals with widely different amplitudes.

face returns must be observed. The surface return, undesirable for subsurface sounding, is precisely the desired signal required to meet the secondary objectives of profiling and surface imaging. Data for sounding, profiling, and surface imaging must be gathered simultaneously. Data interpretation, however, is different for each of the three objectives. The surface return is composed of two components, a specular or smooth surface component and a diffuse or rough surface component. The specular component is caused by normal reflection from the smooth portion of the surface and the diffuse component is caused by surface irregularities. A simplified electromagnetic sounder specular model is illustrated in Fig. 1. For the expected range of lunar relative dielectric constants, approximately a quarter of the energy of the incident wave will be reflected from the surface. A small portion of the wave energy which penetrates the surface layer will be reflected back toward the orbiting spacecraft from the first subsurface discontinuity. The reflected energy from this interface is reduced with respect to the transmitted energy by the transmission loss through the surface layer, and the three reflections which the wave must undergo. In order to detect the subsurface discontinuity, the sounder must be able to detect the presence of the weak second reflection in the presence of the strong first reflection. The two reflected signals are separated in time by an amount

$$\Delta t = \frac{2\Delta_1 n_1}{c}$$

where

- Δ_1 the thickness of the surface layer,
- n_1 the refractive index of the surface layer,
- c the free-space velocity of light.

As for a conventional radar system, the sounder system transmits a modulated waveform, and the reflected signal is processed to yield fine time resolution. The problem of resolving two closely spaced signals with widely different amplitudes is illustrated in Fig. 2.[1] Let the signal reflected from the first surface induce at the output of the signal processor the waveform

$$s_1(t) = a_1 f(t - t_1).$$

Then the signal reflected from the subsurface boundary induces the waveform

$$s_2(t) = a_2 f(t - t_1 - \Delta t)$$

where $a_2 \ll a_1$ and noise is neglected. As Δ_t becomes small, $s_2(t)$ becomes buried in the sidelobes of $s_1(t)$. The lower the sidelobe level of $s_1(t)$, the smaller we can make Δt for fixed (a_2/a_1) and still detect $s_2(t)$. Clearly, subsurface sounding requires both large system dynamic range and a system impulse response with low sidelobe levels.

The ALSE system provides for depth resolution by pulse ranging. Some surface scatterers will have radar ranges identical to ranges of the subsurface layers. Subsurface returns cannot be directly distinguished from such surface reflections when they are at identical along-track position. Their separation relies upon methods of inference. Along-track scatter resolution is provided by processing the radar signal Doppler spectrum.

The objective of the profiling mode is to accurately measure the nadir altitude. The process is simple as long as the nadir contributes the first return to the radar receiver. This case, which is illustrated in Fig. 3(a), is a reasonable representation when the line to the nadir is normal to the surface and the surface is smooth. However, two situations which cause the first return to come from a point other than the nadir complicate the profiling process. Fig. 3(b) shows the problem associated with profiling over an inclined smooth surface. Fig. 3(c) shows the problem associated with profiling a surface which has large changes of elevation close to the nadir. In practice, the ellipticity of the orbits and the slopes of the mare surfaces add to the difficulty of unambiguous profiling. Auxiliary photographic and cartographic data are useful in resolving these ambiguities. We should add that dynamic range is not a major consideration for profiling.

The principal surface imaging requirements are those of adequately resolving along-track and cross-track structural components of the surface itself. Dynamic range for imaging is more demanding than for profiling, but it is far less demanding than for sounding. Shadow regions become more distinct from low-reflecting surface areas as the system dynamic range increases. A desirable feature of the imaging mode is sufficient overlap on successive orbits so that stereo techniques may be employed in the interpretive procedures.

A. General Technical Approach

The general technical approach taken in the design of the ALSE experiment was to employ a multiwavelength synthetic-aperture radar (SAR) having large dynamic range and

[1] For a detailed discussion of this problem, see [2].

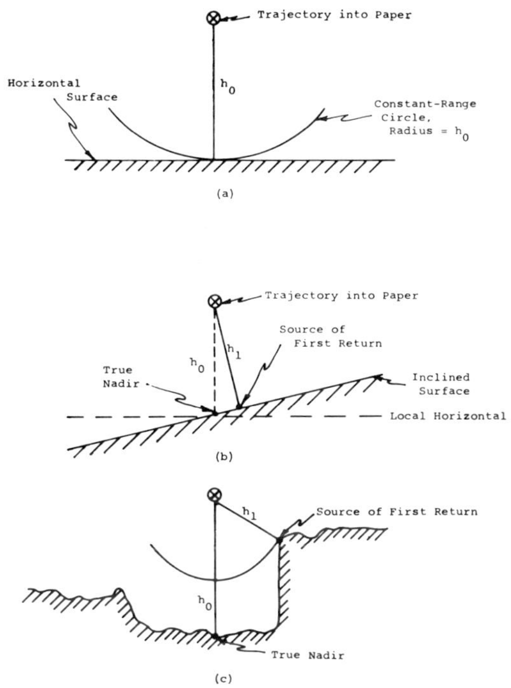

Fig. 3. Profiling. (a) Over a smooth horizontal surface. (b) Over a smooth inclined surface. (c) Over a highly irregular surface.

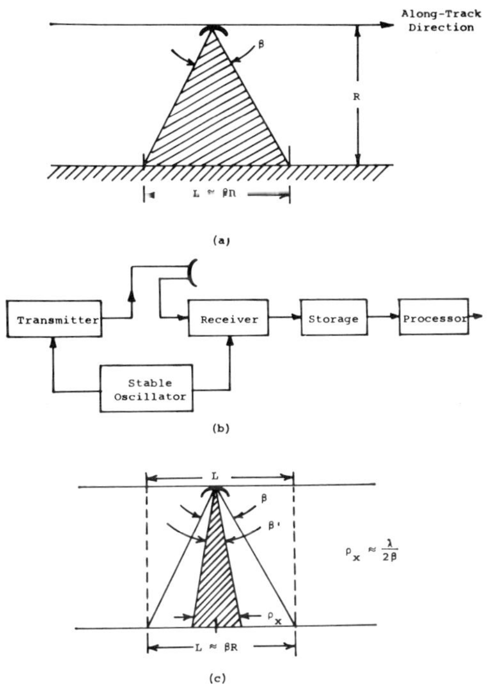

Fig. 4. Synthetic-aperture radar. (a) Physical beamwidth. (b) Simplified block diagram. (c) Along-track resolution ρ_x.

low sidelobes. The SAR technique, described in the literature,[2] was chosen to provide fine along-track resolution. The multiwavelength feature was chosen to provide varying degrees of surface penetration. Hardware and mission constraints limited the system to three wavelengths, with the side condition that one could operate the two longer wavelengths simultaneously (the "HF" mode) and the shortest wavelength only by itself (the "VHF" mode). The multiwavelength approach designed into the experiment reflected the following considerations:

1) The uppermost lunar layers were believed to have a small loss tangent.

2) Lunar subsurface dielectric contrasts were expected to be essentially independent of wavelength.

3) The loss tangent was expected to decrease with increasing wavelength. Therefore, longer wavelengths should allow sounding to greater depths.

4) At longer wavelengths, the off-track diffuse surface return was expected to be smaller than at shorter wavelengths, thereby introducing less surface return to interfere with anticipated subsurface returns.

5) At longer wavelengths, the relatively fine rubble of the regolith layer was expected to be in the Rayleigh scattering region, in which case it would contribute relatively small backscattered energy to occlude the returns from deeper dielectric discontinuities.

Estimates of the likely range of values for the loss tangents of lunar materials were obtained from the literature [5], as were estimates of the likely values of lunar surface reflectivity for the wavelengths under consideration [6]. On the basis of these estimates of loss and reflectivity, scattering models were constructed to yield estimates of the signal-to-clutter ratios (as functions of depth, dielectric contrast, and wavelength) associated with postulated subsurface discontinuities.

To realize along-track resolution ρ_x much finer than that dictated by the angular beamwidth of the radar antenna, a typical SAR system employs a coherent radar (i.e., one in which both transmitter and receiver are phase-referenced to a stable master oscillator) in conjunction with data storage and processing. The geometry and configuration of such a system are illustrated in Fig. 4. A given terrain element at range R is observed by a radar while the latter moves along a path subtended by an angle β. Stored data collected during the observation time are processed to synthesize an antenna aperture of length L, yielding improved along-track resolution ρ_x:

$$\rho_x \approx \frac{\lambda}{2\beta}.$$

In theory, a SAR system can achieve very fine along-track resolution independent of range and operating wavelength.

[2] An extensive literature on SAR is available. See, for example, [3] for a detailed treatment of SAR theory, or [4] for a tutorial introduction to the SAR concept. SAR theory is not treated in this paper, except as required for specific discussion of the ALSE system design.

In practice, however, system errors and design restrictions limit the maximum realizable synthetic-aperture length L.

Fine temporal resolution is obtained via pulse-compression techniques. This provides us with fine resolution of depth for sounding, of height for profiling, and of cross-track location for surface imaging. Since the theory and practice of pulse-compression radars are also well known, they are not repeated here.[3] For our purposes, it is adequate to recall the central features of this class of radar, in which a transmitted waveform of duration T, bandwidth W, and time-bandwidth product $TW \gg 1$ is radiated. Upon reception, the reflected signal is processed (for example, with a matched filter) to yield a narrow pulse from each scatterer, with effective pulsewidth $T' \approx 1/W \ll T$. As mentioned previously, the specular surface return and its far-range sidelobe structure combine with the diffuse return to form the surface clutter against which subsurface returns must be observed. This combined return problem is asymmetric; high sidelobes on the near-range side of the main lobe are of little consequence as they precede the surface return, while far-range sidelobes obscure potential subsurface returns. Therefore, heavy emphasis on sidelobe control was required. Sidelobe control was accomplished by combining the shaping of the radiated waveform in the transmitter with spectral weighting in the signal processor. Furthermore, since subsurface returns were expected to be much weaker than specular returns from the surface, large dynamic range was also required.

Finally, some serious design constraints were placed on the ALSE system, especially on the flight hardware. These included limitations placed on the following: weight, volume, and available power; operating time (caused by constraints on the Apollo mission itself and by crew duties); data storage capacity; thermal output of the hardware (which, together with recorder constraints, eventually precluded having all three ALSE wavelengths operate simultaneously); data link (a data link for real-time wide-band data relay was not available); and antenna structure (which precluded the use of polarization-diversity techniques). The ALSE system design discussed in the following sections represents a balance of mission and engineering constraints against scientific requirements; this balance is a ubiquitous feature of spacecraft hardware.

B. Parameter Selection

Since it is difficult to provide large bandwidth at a long wavelength, time resolution deteriorates as wavelength increases. Therefore, it becomes increasingly difficult to resolve obscure subsurface returns from the specular surface return. The most critical experiment parameter is wavelength, and the wavelength selection process is a delicate balance of penetration versus resolution considerations.

1) Wavelength Selection: To illustrate the effects of lunar material on the detection of the return from a dielectric discontinuity at a depth of X meters, we refer to Fig. 5. Consider a linearly swept (chirped) signal specularly reflected from a smooth surface of dielectric constant ϵ. If this waveform is passed through an ideal pulse-compression filter, the system response (after suitable weighting) is expected to have the form shown in Fig. 5. A dielectric discontinuity at a depth of X meters below the surface will give rise to a second weaker

[3] See, for example, [7].

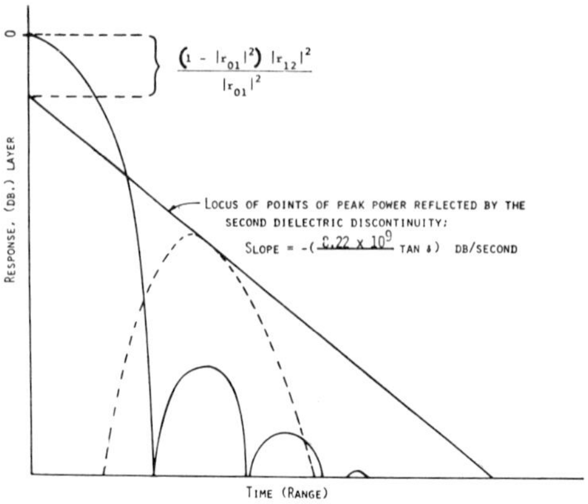

Fig. 5. Amplitude responses of surface and subsurface features.

return, with a shape similar to the first but delayed in time. This second return is depicted as the dashed line in Fig. 5. The intensity of multiple reflections is so low that they are neglected. The subsurface return will add vectorially to the surface return producing constructive or destructive interference. Before the second return is clearly visible, its peak magnitude must be greater than the competing structure of the first return by an adequate margin, perhaps 5 dB.

The peak power P_{r_1} of the specular return is given by

$$P_{r_1} = P_t \left(\frac{G\lambda}{8\pi H} \right)^2 | r_{01} |^2$$

where

- P_t the peak effective transmitter power,
- G the antenna gain,
- λ the radar wavelength,
- H the altitude,
- r_{01} the reflection coefficient at the lunar surface/free-space boundary.

The power P_{r_2} reflected from a subsurface specular dielectric interface is given by

$$P_{r_2} = P_t \left(\frac{G\lambda}{8\pi(H+X)} \right)^2 (1 - | r_{01} |^2)^2 | r_{12} |^2 \cdot 10^{-2.74\tau f \tan \delta}$$

where

- r_{12} the subsurface interface reflection coefficient,
- X the depth of the subsurface interface,
- τ the time delay between the arrival of both returns,
- $\tan \delta$ the lunar material loss tangent,
- f the radiated frequency.

For fixed reflection coefficients, the power return from the subsurface interface decreases rapidly with the thickness of the intervening material. The derivative of the power return with respect to the delay time due to thickness is

$$10 \frac{d}{d\tau} (\log_{10} P_{r2}) = - \left(\frac{8.22 \times 10^9}{\lambda} \tan \delta \right) \quad \text{dB/s}$$

for $X \ll H$. This derivative, also shown in Fig. 5, establishes the rate at which the far-range sidelobes of the system impulse response must decay in order to permit the detection of subsurface features in a context characterized by r_{01}, r_{12}, X, and tan δ.

Measurements on lunar material brought to earth during previous Apollo missions indicated loss tangents on the order of 0.025 at a frequency of 5 MHz, 0.01 at 15 MHz, and 0.003 at 150 MHz. For these values of loss tangents, the energy transmission through the lunar material is expected to decay at a rate of 3.4 dB/μs at 5 MHz, 4.1 dB/μs at 15 MHz, and 12.3 dB/μs at 150 MHz.

The minimum depth of exploration is determined by the smallest delay at which a second return is distinguishable from the first. The range resolution in the lunar material is given by

$$X_{\min} = \frac{c}{2W} \frac{K}{\sqrt{\epsilon}}$$

where

- W the bandwidth of the transmitted signal,
- ϵ the dielectric constant,
- K a factor dependent on the weighting function utilized during data processing to reduce range sidelobes,
- c the velocity of light.

(For the case of sinc-squared weighting between the first nulls of the spectrum, K is approximately 2.3.) It is apparent from this equation that to minimize the detectable depth of an interface, it is desirable to extend the bandwidth of the system as much as possible to improve resolution. Decreasing the weighting function constant K increases the specular surface return sidelobes which, in turn, reduces the maximum depth of detection.

The maximum depth of exploration is determined by the weakest detectable return as limited by the loss tangent, system noise, sidelobe level, or surface clutter. For a given lunar model, this depth is determined by either the system dynamic range or the system sidelobe response.

On the basis of models investigated, such as those previously discussed, the ALSE system was provided with wavelengths of 60, 20, and 2 m (5, 15, and 150 MHz).

2) Range/Depth Resolution: At each frequency, a system bandwidth of one-tenth the center frequency was selected, as was a time–bandwidth product of 128 for the transmitted waveforms. As a result of this choice, depth resolution was proportional to wavelength. The lowest frequency (hereafter HF1) had the greatest potential for penetration but the poorest depth resolution.[4] The next higher frequency (HF2) had smaller penetration but three times finer depth resolution. Finally, the highest frequency (VHF) had still smaller penetration, but offered the potential to resolve structures ten times finer than at HF2.

3) Along-Track Resolution: The ALSE system was designed to be operated from the Apollo 17 Command Module while in a near-circular orbit at an altitude of 110 km. Synthetic-aperture techniques were employed with each of the three

[4] In free space, the range resolution ρ_r for HF1 is on the order of 300 m without weighting and 700 m with weighting; the depth resolution within the lunar structure is finer and is given by ρ_r/n, where n is the refractive index at HF1.

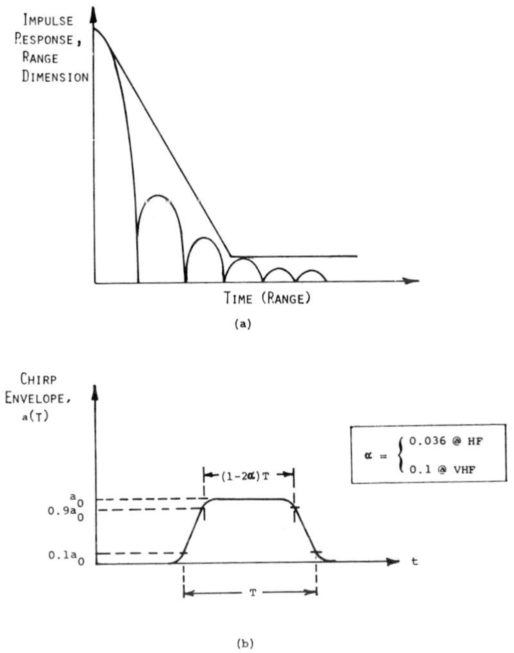

Fig. 6. (a) Sidelobe structure of impulse response. (b) Form of chirp envelope.

TABLE I

SYSTEM REQUIREMENTS FOR ACCEPTABLE SIDELOBE LEVELS

Amplitude linearity	0.1%
Phase linearity	0.001 rad
Sweep linearity	1.0%
2nd harmonic	−20 dB
3rd harmonic	−30 dB

wavelengths. The ALSE system was designed to permit integration over effective beamwidths of 0.1 rad at each wavelength. Therefore, the free-space along-track resolution was approximately 5λ for all three channels.

4) Sidelobe Control: As discussed in Section II, extreme control of the far-range sidelobes of the system's impulse response is required. This sidelobe structure was specified such that the peak values of the sidelobes would descend rapidly, as shown in Fig. 6(a), and would remain at least 45 dB below the mainlobe peak for all sidelobes beyond the third. To provide this performance, spectral weighting with very good linearity was specified. Sinc-squared amplitude weighting was to be applied in the signal processor. In order to control sidelobe effects of ripples in the spectrum, which result from sharp cutoff of the transmitted chirp pulse, amplitude weighting as shown in Fig. 6(b) was specified.

The linearity requirements listed in Table I were specified for all three wavelengths.

5) Coverage: In the along-track dimension, coverage at each wavelength was determined by both the length of mission time which could be allocated to the ALSE and the availability of storage for the raw data. In the range dimension, there are two distinct considerations:

1) In order to use recorder capacity efficiently, we would like to turn on the data recorder shortly before the return from the first surface reaches the radar. This need is common to the sounding, profiling, and imaging modes.

2) In order to meet specific sounding, profiling, and imaging requirements, we need sufficient recorder capacity to store returns from all significant scatterers at each wavelength.

With respect to 1), a delay variation of ± 100 μs results from a spacecraft altitude variation of ± 10 km and a surface elevation variation of ± 5 km. The recording system must provide either enough storage capacity to cope with the 200-μs uncertainty, or an adaptive sensor which can decide when to turn on the recorder after each pulse is transmitted.

Range-dimension storage requirements over and above the 200-μs uncertainty differ from one mode to the next:

1) For sounding, we need to store the echoes from a delay interval which is long enough to include returns both from the surface and from the deepest detectable discontinuities. The delays required for this purpose were not expected to exceed 300 μs at HF1, 100 μs at HF2, and 20 μs at VHF, including the lengths of the three transmitted waveforms.

2) For profiling, we need to store data adequate to monitor, on a continuous basis, the elevation variations of the nadir whenever the spacecraft is operating within its nominal altitude range.

3) For surface imaging, the cross-track width of the imaged swath is dependent upon flight geometry, antenna elevation, beamwidth, and data storage capacity. Imaging was a design consideration at VHF but not at HF1 and HF2. For the VHF mode, a swath width of 40 km was desired. With the spacecraft altitude of 110 km, this corresponds to a delay interval of 70 μs.

6) Gain and Calibration: To preserve large dynamic range in the receiving and signal-processing system, it is necessary to match the received signal level to signal-input levels. However, the return signal level is strongly influenced by the degree of flatness of the lunar surface. To cope with this problem, an automatic gain control (AGC) was specified. In addition, the nadir return was expected to be much stronger than that from off nadir. For VHF, therefore, where surface imaging was of high interest, a requirement was established to automatically increase the system gain by 12 dB, 13 μs after detection of the initial return for each pulse.

Amplitude calibration was specified for the experiment in order to permit quantitative estimates of surface reflectivity. The calibration is discussed in Section III-D.

C. Data-Processing Considerations

SAR systems store data during flight and subsequently process the data to yield fine along-track resolution. SAR systems traditionally have employed coherent optical systems to perform the necessary along-track processing operations. Three basic reasons motivated the early choice of coherent optical processing for SAR:

1) The signal emerges from the radar receiver with relatively large bandwidth, and the quantity of data is very large. Recording the data on film with an intensity-modulated CRT was compatible with radar bandwidth and data-storage requirements [8].

2) Coherent optical systems can be configured to perform one-dimensional and/or two-dimensional linear integral transform operations [9]. In particular, one- and two-dimensional Fourier transforms are easily displayed. The along-track processing in a SAR system may be implemented using coherent optics [8], [10].

3) It is possible to configure the optics for range-pulse compression simultaneous with the along-track processing. When the range pulse is a linear-FM waveform, the pulse compression is particularly straightforward [11].

Coherent optical processing was adopted in the ALSE system as the fundamental method for performing along-track processing. Variations on the processing exist for the various interpretive modes.

1) For sounding purposes, it is of interest to perform the synthetic-aperture processing to achieve fine along-track resolution. However, flexibility is also desired to carry out Fourier-transform analysis in place of the synthetic-aperture processing—which, in itself, is a modified Fourier transform. In some instances, the dispersed range pulse is to be compressed; in others, it is desirable to leave the pulse uncompressed in range while performing the azimuth processing, and then perform the range-pulse compression in a digital computer where phase errors could be estimated and compensated digitally. To provide this flexibility, a coherent optical processor was equipped with several alternate readouts including a photograph, hologram, and a digital tape. Since the optical system has sufficient flexibility, it may be programmed to perform along-track and/or range processing and to select the output best suited for subsequent analysis. The hologram is a photographic format which is observed in a special viewer [12]. This format has the advantages that it preserves very large dynamic range (40 to 50 dB) in the visual presentation and the user can adjust the Doppler filtering in the "frequency plane" of the viewer.

2) For profiling purposes, synthetic-aperture processing to realize fine along-track resolution as well as the best available range resolution is desired. The possibility exists of using the phase structure of the dispersed and/or compressed pulse to refine the elevation estimate in view of the high signal-to-noise ratio of the specular return from the surface. For this case, a desirable optical processing mode is to compress along-track and then digitize for subsequent computer processing to improve the accuracy of the elevation estimate.

3) For imaging, two-dimensional compression (i.e., synthetic-aperture processing with range-pulse compression) is desired. Photographic hard copy and holograms are the primary output forms for this purpose.

Section IV of this paper is devoted to a treatment of the various data processing and display techniques employed in the ALSE experiment.

III. Flight Hardware

The S-209 flight hardware consisted of two distinct coherent radar subsystems, one operating at two frequencies in the HF band and the second at a single VHF frequency, which shared a common data-storage device. The HF subsystem radiated the two HF signals, HF1 and HF2, on alternate pulses via a center-fed dipole antenna. The VHF subsystem radiated via a Yagi antenna. The system could operate at either HF or VHF, simultaneously. The on-board data storage device was an "optical recorder" which employed an intensity-modulated CRT with photographic film to store the received radar data.

The parameter requirements discussed in Section II-B, together with constraints which arose during the hardware

TABLE II

LUNAR SOUNDER SYSTEM CHARACTERISTICS

	HF Mode		VHF Mode
	HF1	HF2	
Wavelength (m)	60	20	2
Estimated depth of penetration (m)[a]	1300	800	160
Center frequency (MHz)	5.266	15.8	158
RF bandwidth (MHz)	0.5333	1.6	16.0
Pulsewidth (μs)	240	80	8.0
Time-bandwidth product	128	128	128
Range resolution, free space (m)	300	100	10
Transmitter peak power (W)	130	118	95
Transmitter average power (W)	12.4	3.7	1.5
Effective antenna gain (dB one-way, including efficiency)	−0.8	−0.7	+7.3
Nominal specular power (dBm)	−53	−57	−73
Noise figure (dB)	11.4	11.4	10.0
Pulse repetition rate (s^{-1})	397[b]	397[b]	1984
Recording duration (μs)	600	600	70
AGC gain range (dB)	12.1	12.1	13.9
Echo tracker	no	no	yes

[a] Based on model analysis.
[b] Interlaced on HF1 and HF2.

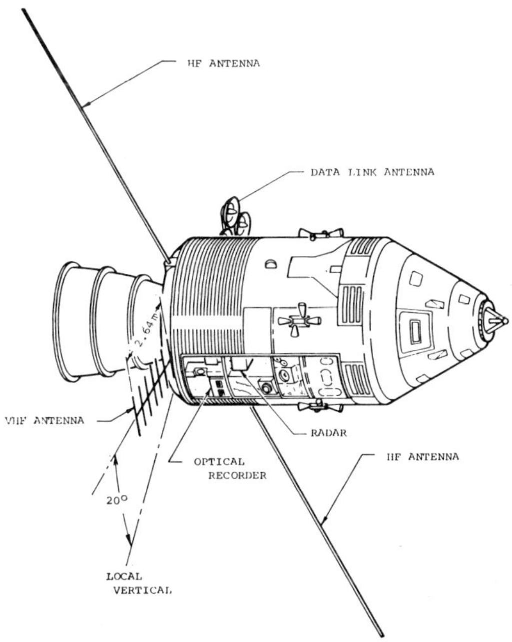

Fig. 7. Lunar sounder configuration.

design and fabrication process, led to an ALSE hardware package with the characteristics listed in Table II.

The approximate physical configuration of the hardware in the Apollo spacecraft is shown in Fig. 7. The radar was mounted in the Scientific Instrument Module (SIM) Bay of the spacecraft, along with the recorder. The VHF Yagi antenna and the two HF dipole antenna elements were mounted at the aft end of the Service Module, forward of the SPS engine. Radar operation, as well as HF antenna deployment and retraction, were controlled by the astronaut in the Command Module. Since no internal access from the Command Module to the optical recorder was available, the photographic film on which the recorder stored the radar data was retrieved by an astronaut during an EVA; this was carried out toward the end of the Apollo 17 mission, during the trans-Earth phase.

Sections III-A, III-B, and III-C describe, respectively, the radar, antenna system, and recorder. The amplitude-calibration techniques used in the system are described in Section III-D. The prime contract to fabricate the S-209 hardware and to integrate it into the Apollo spacecraft was held by Rockwell International. The radar, antenna system, and recorder were fabricated by RCA, Spar Aerospace Corporation, and Goodyear Aerospace Corporation, respectively, each under subcontract from North American.

A. Radar

The basic parameters of the ALSE radar were given in Table II. Each of the three frequency bands utilizes a linear FM transmitter pulse; pulse compression is performed optically in the data processor. An FM bandwidth of 10 percent of the carrier frequency is employed at each frequency band, as limited by the antenna subsystems. A greater relative bandwidth would result in poor transmitter efficiency and difficulty in meeting the stringent antenna-phase-error requirements. The linear FM pulse is generated by a voltage-controlled oscillator which, at the start of each pulse, is phase-locked to a stable local oscillator. The duration of the FM sweep in each system is set to maintain a time–bandwidth product of 128. Upon reception of the echo, the signals are offset in frequency while maintaining phase coherence. The transmitted waveform envelope had a linear rise and fall for all three channels, as shown in Fig. 6.

AGC was incorporated at both HF and VHF for the reasons cited in Section II-B6). To avoid undue complexity in the data processing, the gain setting was allowed to change no more often than once every 30 s and then in discrete steps. The gain could be varied by up to 12 dB at each frequency. During the flight of Apollo 17, the power return did follow predicted specular values on the average; however, significant variations in return power were observed over some regions of the lunar surface.

As stated in Section II-B6), the VHF subsystem incorporated a 12-dB increase in gain 13 μs after arrival of the specular pulse. The initial gain of the radar system was set so that the specular pulse would have a received amplitude close to the saturation point of the optical recorder in order to achieve a wide dynamic range. However, shortly after arrival of the specular return, the level of the received signal was expected to drop significantly as a result of weak diffuse off-nadir scattering. Consequently, additional gain was required in order to obtain a radar image of the lunar surface. The gain change was delayed for 13 μs after the arrival of the leading edge of the specular return to avoid compromising the sounding aspects of the experiment. By this time, no portion of any sounding data was expected to be present. The additional gain was limited to 12 dB because of the system noise.

The combined HF/VHF radar weighed approximately 49 kg and required a prime input power of 103 W.

B. Antenna System

The Apollo Lunar Sounder antenna system consisted of two separate antennas, one for both HF1 and HF2 and the other for VHF. The HF antenna was composed of two ele-

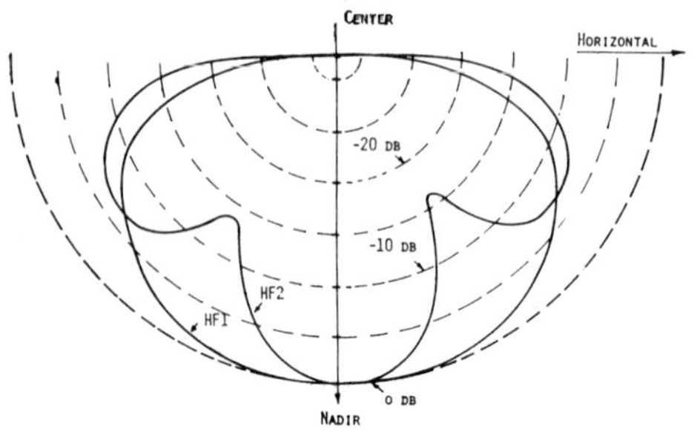

Fig. 8. HF radiation pattern from scale model.

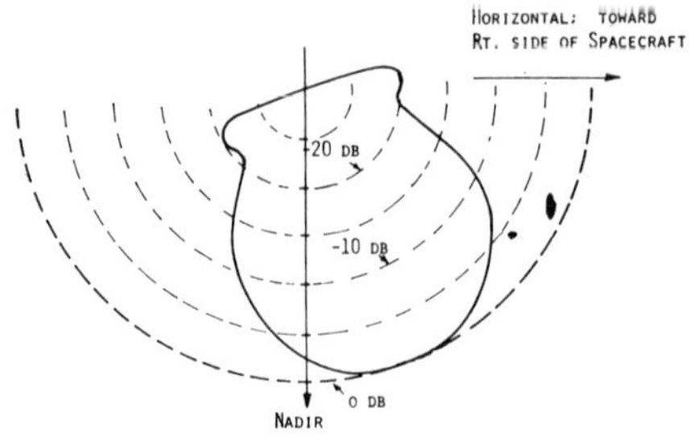

Fig. 10. *E*-plane pattern of VHF antenna.

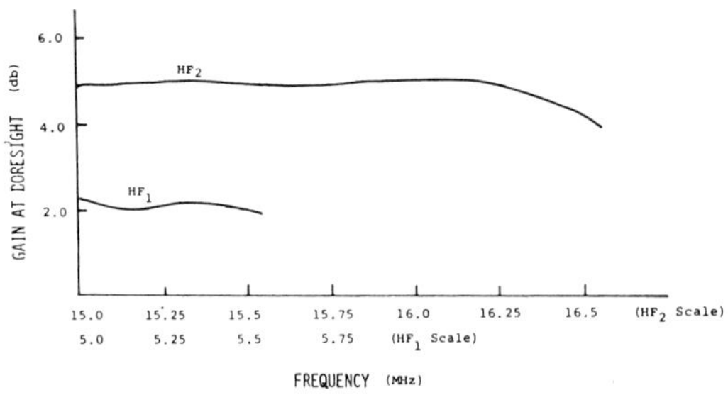

Fig. 9. HF antenna gain versus frequency at boresight.

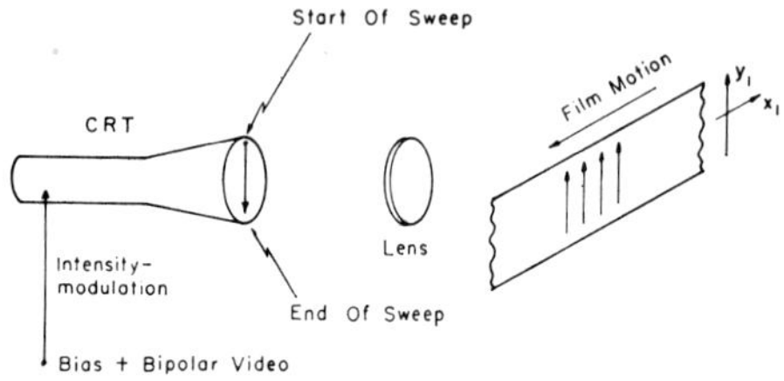

Fig. 11. Optical recorder, simplified.

ments 10.4 m in length driven by a matching network. The VHF antenna was a seven-element Yagi. The HF dipole was oriented parallel to the lunar surface, and perpendicular to the vehicle track. The VHF Yagi was pointed toward the surface, 20° off nadir toward the north, in a plane perpendicular to the vehicle track.

1) HF Antenna Subsystem: The HF antenna system was designed to have high efficiency and maximum gain in the spacecraft nadir direction for both HF1 and HF2. The gain was to be preserved over a 10-percent bandwidth across both HF bands, with amplitude and phase error small enough to keep the system impulse response sidelobes below the system specification. A further requirement was that the antenna was to be driven by a single source whose frequency band was to be switched on successive pulses.

The desired maximum radiation gain was achieved with two retractable, center-fed, 10.4-m elements driven by a suitable matching network. The matching network was required since the antenna structure had a resonant frequency of approximately 6 MHz. Selection of the antenna length was based on 1) pattern studies obtained from a computer simulation of a wire grid model of the spacecraft and antenna elements and 2) impedance properties of the various antenna element lengths simulated. The simulated radiation system models were verified by pattern and impedance measurements on a $\frac{1}{10}$ scale model of the spacecraft and antenna elements. The (scale model) radiation patterns at the center of the 5-MHz band and at the center of the 15-MHz band are shown in Fig. 8. The variation of *E*-plane radiation gain with frequency at boresight (spacecraft nadir) is shown in Fig. 9. Overall antenna radiation efficiencies of 54 and 32 percent were obtained in the 5- and 15-MHz bands, respectively.

2) VHF Antenna Subsystem: The VHF antenna was designed to have maximum gain on the right side of the spacecraft, with the one-way gain in the nadir direction 3 dB below the peak gain. As with HF, the VHF subsystem had a 10-percent bandwidth. Furthermore, adequate phase and amplitude control to assure meeting the sidelobe specification on the system impulse response was again of prime importance.

The VHF antenna was a seven-element Yagi operating in the 150- to 166-MHz band. The antenna was mounted on a hinged structure so that, during ascent, it could be stored next to the CSM main propulsion engine and deployed. The antenna consisted of one driven element, a reflector, and five director elements.

The elements were driven by a modified Roberts balun, with an overall efficiency greater than 90 percent with a center-band, *E*-plane, 3-dB beamwidth of 50° and *H*-plane, 3-dB beamwidth of 60°. The *E*-plane pattern is shown in Fig. 10.

C. Recorder

The optical recorder was required to accept input data over the video bandwidths of the radar and to have adequate storage capacity to permit the radar to operate for nearly six orbits of the moon. The video bandwidth is largest, 16 MHz, during VHF operation. The decision to use a CRT/film-type optical recorder was based both upon the ease with which such a recorder would meet the input and storage requirements and upon the compatibility of the film-storage medium with coherent optical processing [6]–[8]. The recorder consisted of a CRT, relay lens, photographic film, and associated transport, as shown in Fig. 11. The video output from the radar intensity-modulates the CRT about a constant bias level while an image of the CRT spot is swept across a photographic film at a constant velocity. Hence there is a linear correspondence between time delays of individual target echoes and the spatial position across the film. Returns from

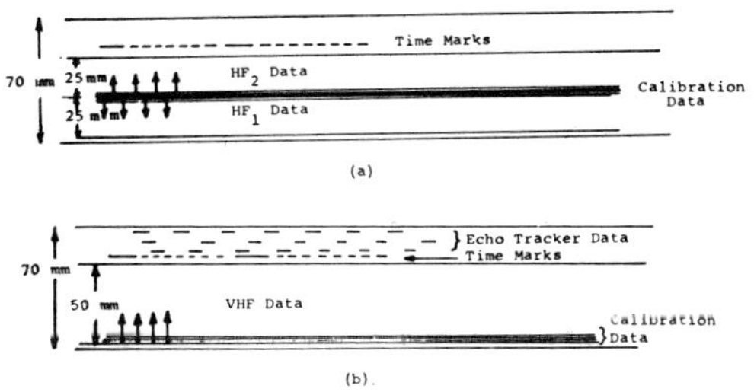

Fig. 12. (a) Optical recorder film format, HF mode. (b) Optical recorder film format, VHF mode.

successive pulses are recorded side by side along the film as it is transported at a constant speed past the CRT image. Thus the location along the film is linearly related to the orbital position of the spacecraft.

HF1 and HF2 radar data were recorded on two separate 25-mm-wide strips along the 70-mm-wide film. VHF data were recorded across a single 50-mm-wide strip.

At VHF, the system bandwidth of 16 MHz coupled with the available recorder time–bandwidth product resulted in a serious constraint on data storage capability. Specifically, the delay interval over which returns could be stored was limited to 70 μs. Since the delay uncertainty could be on the order of $\pm 100\,\mu$s, the recorder was equipped with an "echo tracker," which monitored the receiver output to estimate the round-trip propagation time for the first return, and then turned on the recorder just prior to the arrival of the first return. The receiver itself operated for an interval of 275 μs, and the echo tracker operated within this gate.

During both modes of operation, time marks denoting the number of seconds, minutes, hours, and days from liftoff were recorded every 9 s along the edge of the signal film with a binary light source. To calibrate the optical recorder, six cycles of a square wave also were continuously recorded across one edge of each data channel. The spatial frequency of this calibration signal was about four line-pairs per millimeter. In addition, during VHF operation, the echo-tracker range-bin status was recorded every 2 s along the edge of the film. The recording film formats are shown in Fig. 12.

The recorder's physical parameters are given in Table III. The 10 h of recording time was adequate to record two orbits of VHF data, two orbits of HF data, and nearly two orbits of individual sounding targets in either sounder mode. The recorder's signal-storage parameters are listed in Table IV.

A film speed of 5 mm/s was used for both modes of operation. Operating this recorder in the VHF mode only, it was possible to record over 10^{10} pixels (resolution cells) of information.

TABLE III

PHYSICAL PARAMETERS OF RECORDER

Primary power	100 W dc 17 VA 400 Hz 3 ϕ ac
Weight	55 kg
Film	
Length	195 m
Width	70 mm
Type	Kodak 3400
Film capacity (operating time)	10 h

TABLE IV

ALSE RECORDER-SIGNAL-STORAGE PARAMETERS

Parameter (Units)	Channel HF$_1$	HF$_2$	VHF
Film speed (mm/s)	5	5	5
CRT sweep speed (mm/s)	4.03×10^4	4.03×10^4	7.14×10^5
Sweep duration (μs)	620	620	70
Calibration frequency (cycles/mm)	3.75	3.75	3.6
Dynamic range (dB)	30	25	20
Temporal range bandwidth (MHz)	0.533	1.6	16
Range spatial frequencies (cycles/mm)	5.62 to 18.55	21.8 to 60.6	4.32 to 27.4
Number of 5λ range resolution cells	331	992	1120
Temporal along-track (Doppler) bandwidth (Hz)	± 43	± 100	± 100
Along-track spatial frequencies recorded (cycles/mm)	± 8.6	± 20	± 20

D. System Calibration

In order to obtain quantitative estimates of the various return levels from the lunar structures, an effort was made to incorporate certain amplitude-calibration features into the system. For calibration purposes, it was assumed that the output of the optical processor would be digitized for further analysis. Fig. 13 shows a simplified signal-flow diagram extending from the transmitter to the ultimate user. A dispersed pulse generated by the radar transmitter is radiated by the system antenna. The return echo is amplified and coherently down-converted by the receiver, and the resulting

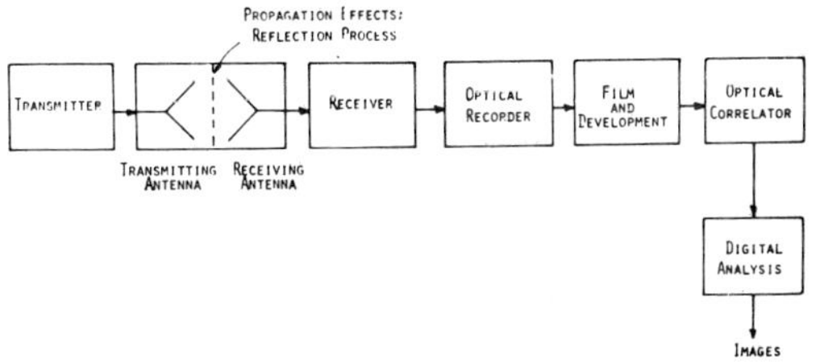

Fig. 13. Signal flow for calibration.

signal is then relayed to the optical recorder, which exposes a data film. This film is developed after the mission and then photocopied. The photocopy serves as the input to an optical processor whose output is in turn digitized for further analysis with a digital computer. Nonlinearities and uncertainties throughout the signal flow cause difficulties in maintaining calibration. To estimate overall calibration, the approach taken was to incorporate subsystem calibration to the maximum extent feasible as follows:

1) Transmitter Power: Transmitted power was monitored on the spacecraft through use of a circuit which had been calibrated on the ground prior to the mission. The output of this circuit was telemetered to Earth. As a cross check, transmitter performance was calibrated as a function of voltage and temperature for normal operation, and voltage and temperature observed during radar operation were relayed to Earth.

2) Antenna: The antenna posed a difficult calibration problem for two reasons: pointing angles were not known precisely

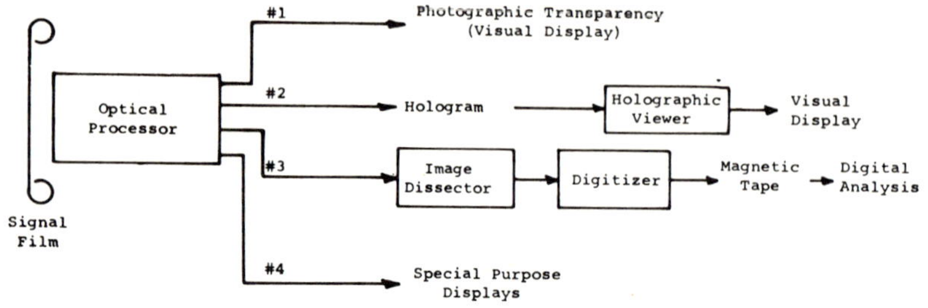

Fig. 14. Basic processor/display system.

and no antenna range existed which could measure the gain patterns of the actual flight hardware. Consequently, the overall antenna gain was estimated by these techniques:

a) Measurements were obtained of the driving source impedances of the transmitter, cables, matching network, and the antenna-terminal characteristics.

b) Patterns on a $\frac{1}{10}$-scale model of the CSM and antennas were measured and compared to theoretical patterns obtained with a digital simulation of the CSM and antenna structures (wire-grid model).

c) Measurements were made of the equivalent network representing the antenna, matching device, and receiver.

d) It was assumed that the HF and VHF beams had main lobes much wider than the pointing inaccuracies which then could be ignored.

When the pattern of a full-scale monopole over a ground plane was measured, the HF gains were found to be within 1 dB of the values estimated via the above procedures.

3) Receiver: Receiver-amplitude transfer functions were obtained for each stage over the temperature range at different frequencies in their passband. The receiver gain step and temperature were monitored on the spacecraft and telemetered to Earth.

4) Optical Recorder, Film Development, Optical Processor and Digitizer: To calibrate the signal chain, recorder, and processor, a small square wave was injected at the receiver gate and recorded on a strip along the signal film. Thus the small-signal gain of the electronics, the electrooptical interface, the optical processor, and the photographic processes could be measured.

IV. Signal Processing and Display

The signal processing and display system was designed with several criteria in mind, including the following:

a) flexibility to optimize sounding, profiling, or surface imaging per the objectives of the user geoscientist;
b) preservation of large dynamic range, primarily for sounding purposes;
c) availability of optical or digital outputs to satisfy needs for both visual study and computer analysis;
d) flexibility for digital compression of reflected range pulses to reduce effects of system phase errors across the signal spectrum;
e) ability to apply amplitude weighting for sidelobe control;
f) flexibility in the optical processor to provide either compressed or uncompressed data for digitizing, per the requirements of the user.

Since the input data to the processing/display system consisted of photographic film, an optical processor was the natural first element in the signal processing chain. The overall processing system was configured as shown in Fig. 14.

Four basic types of output exist:

1) photographs, for direct visual inspection;
2) holograms, for visual inspection through a holographic viewer;
3) digital data records on magnetic tape, for subsequent digital computer analysis;
4) special-purpose displays, such as records of time-varying Doppler spectrum.

For each type of output, the processor may be adjusted for different processing modes. Section IV-A of this paper describes the processing operations and hardware depicted in Fig. 14. Section IV-B provides a brief discussion of the digital processing techniques employed to analyze the ALSE data.

A. Optical Processing

Photo duplicates of the raw-data signal film were processed using coherent optical-processing techniques. The optical-processing techniques used in this experiment have been developed over the past two decades for use with synthetic-aperture side-looking radar data and have been reported in the literature [8], [9]. The coherent optical processor is an analog computer which carries out the linear integral transform operations required in the processing of SAR data. The sounder, as for any other SAR, requires that two independent operations be carried out with the data, one to compress the coded range pulses and the other to compress the along-track data records. For the data storage format described in Section III-C, these two independent operations reduce to a single two-dimensional operation in an orthogonal coordinate system.

1) Photographic Imagery: A simplified diagram of an optical processor [8] is presented in Fig. 15(a). A photograph of the ERIM Precision Optical Processor, the specific instrument used for this purpose, is shown in Fig. 15(b). The input data are positioned in plane P_1 and illuminated by a coherent light beam derived from a laser source. The data film is immersed in a "liquid gate" to minimize the effects of random film-thickness variations on the coherent light beam. The data, recorded on film in variable density format, modulate the light-beam intensity so that an astigmatic radar image of the lunar surface is formed; azimuth focus occurs at some plane P_a, while range focus occurs at some plane P_r. The pair of spherical lenses (S_1, S_2) operate as a unity-magnification tele-

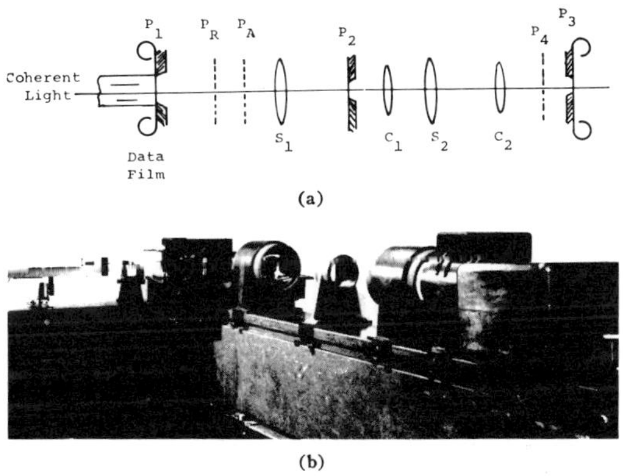

Fig. 15. (a) Optical processor configuration. (b) Photograph of precision optical processor.

scope in the range dimension and transfer the range image plane to the output plane P_3. The spherical, cylindrical lens pairs (S_1, C_1; S_2, C_2) operate as separate demagnifying telescopes in the azimuth dimension and both demagnify the azimuth focus image and transfer it from P_a to P_3. The output film then records the image at plane P_3.

The spherical lens (S_1) produces a display of the two-dimensional Fourier transform of the input data in its rear focal plane P_2. Various frequency-filtering functions can be carried out in this spatial plane. Simple limiting apertures placed here act as sharp cutoff bandpass filters. Such apertures were used to eliminate both noise outside the data band and the conjugate image which forms an out-of-focus background in the image plane, as well as to restrict the processed azimuth or Doppler bandwidth. Weighting filters were placed in this frequency plane to reduce the range sidelobe levels. These filters were simple intensity-modulation masks. On occasion a phase filter, composed of a set of lenses,[5] was placed at plane P_2 to correct for the range curvature problem [8] encountered in the VHF data. This filter allows the processor to compress large azimuth bandwidth data without appreciably degrading the image quality.

Photographic images were made of the data for scientific analysis and interpretation. The entire range bandwidth was used for all systems, resulting in range resolution better than 5λ without weighting and 10λ with weighting. The HF channels were processed with full Doppler bandwidth so that theoretically an along-track resolution better than 5λ could be achieved. The VHF data were processed, with curvature correction, to yield 5λ resolution.

2) *Holographic Imagery:* Two severe limitations of photographic imagery restrict its scientific usefulness. First, the very large compression ratios of synthetic-aperture coded-pulse systems result in a very large dynamic range in the processed data, as much as 50–60 dB. Since photographic film typically stores only 20–25 dB, much of the dynamic range of the image is thereby lost. Second, when an image is stored on film, its phase information is destroyed, restricting subsequent data manipulation. By way of example: the Doppler bandwidth which contributes to the image is fixed and the Doppler

[5] The design used was originated by Goodyear Aerospace Corporation and Perkin-Elmer Corporation, and is proprietary.

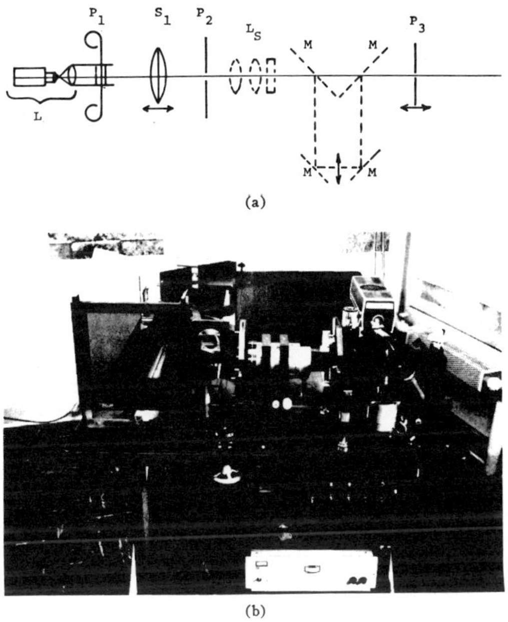

Fig. 16. (a) Holographic viewer configuration. (b) Photograph of holographic viewer.

signature lost when the image is recorded; the user cannot view or filter the spectrum as he views the image; and he cannot adjust the image focus.

These limitations are removed if one employs holographic techniques to record and later view the output of the optical processor [12]. Therefore, holographic viewing systems were developed for use with the ALSE data to provide scientific users with this desirable flexibility. Holograms were produced by using a coherent optical processor with a configuration similar to that shown in Fig. 16(a); the major difference was that the output hologram was recorded in a plane P_4, in which the signal film was imaged. The primary function of the processor in producing holograms was to demagnify the data film in azimuth and to introduce desired range-pulse spectral weighting.

Hologram viewers were designed and constructed. The viewers, configured essentially as shown in Fig. 16(a), are quite flexible and allow an operator wide latitude in data manipulation. Each viewer consists of a small argon laser and a beam-expanding lens system. A continuous film drive P_1 is used to transport the hologram through a liquid gate. A spherical lens S_1 acts as both a projection lens and a Fourier-transforming lens. Bandpass filtering is achieved in the frequency plane P_2 by use of simple rectangular apertures. The viewer system includes a variety of spherical and cylindrical lenses (L_5), a beam-path folding mirror system M, and a liquid-crystal rear-projection viewing screen. A photograph of an ALSE holographic viewer is shown in Fig. 16(b). While viewing the processed image on the screen, an operator can adjust focus and Doppler-filter bandwidth. He also has access to a zoom capability. A simple lens manipulation is provided to project the two-dimensional Fourier transform onto the viewing screen so that the data spectrum and Doppler-filter

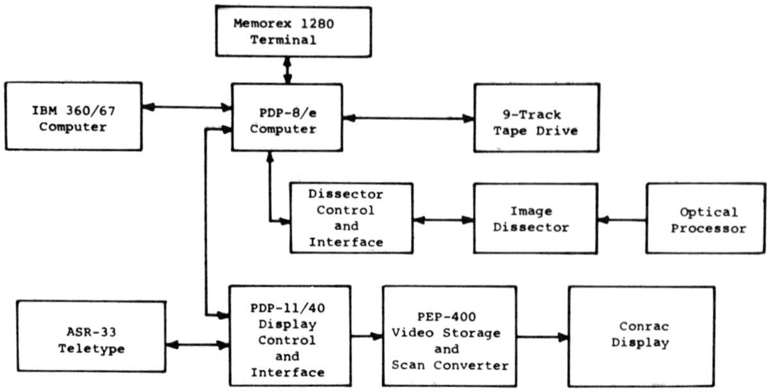

Fig. 17. Functional diagram of dissector and digitizer.

TABLE V
DISSECTOR CHARACTERISTICS

Resolution	24 line pairs/mm at 50% 50 line pairs/mm at 0%
Output	8-bit words
Dynamic range	24-dB linear mode 40-dB log mode
Sensitivity	55 mA/W at 4200 Å
Integration time	10 to 990 μs
Photocathode	S-20 phosphor
Line spots	3000
Cathode area	30 mm long 0.5 mm wide
Scanning aperture	10 μ long, up to 500 μ wide

placement can be observed. By reconfiguring the lens manipulation, the operator can display an image composed of compressed range pulses in one dimension and Doppler-frequency signatures in the other.

3) Digital Tape Records: Digital tape records of the data are required for subsequent processing in a digital computer. A line-scan image dissector was developed for placement at the output of an optical processor to produce digital tape records of output of the optical processor [12]. The dissector system developed for use with the Apollo Lunar Sounder data is shown in the functional block diagram of Fig. 17.

The image dissector is basically a photomultiplier tube using magnetic deflection to scan the photocathode area. The dissector output current is directly proportional to image intensity. Dissector characteristics are summarized in Table V. The dissector is operated in successive line-scan mode for optimum stability and linearity, with the data film being transported in the processor.

A special-purpose electronic system was built to digitize the analog data and to interface the output to a PDP-8/e digital computer which in turn feeds the data to a nine-track magnetic-tape recorder. A display monitor system was also implemented so that the digitized data can be observed on a TV monitor. Either digital tapes or real-time inputs can be fed into the display.

The signal film records contain approximately 5 billion data points, enough to fill over 1000 nine-track digital tapes. Therefore, the photographic outputs and the holographic displays are used to prescreen the data, and high-interest areas are digitized for further study. Data may be digitized with the range pulse either compressed or dispersed, with the azimuth data either compressed or dispersed, and with the signal spectrum either weighted or unweighted.

4) Special-Purpose Imagery: Special-purpose optical displays of the sounder data are useful for analyzing the data; two of these will be described briefly. The first is a display of range-compressed and azimuth-Fourier-transformed data. Data from lunar specular layers having very small inclinations will result in energy peaks in such a display. Such energy peaks will be relatively stationary while a region of data is scanned. These data essentially perform the function of a scatterometer.

A second type of useful special data display is a record of the time-varying Doppler spectrum. Such a display is obtained by recording, on a continuous strip of film, a Doppler-frequency display across the film with data time along the film. The characteristic signature of each point target on such a film will be an "S" shape pattern along the film, while a reflecting plane yields a constant Doppler line.

5) Processor Response with Weighting: At several points in the preceding discussion, reference was made to the weighting necessary for sidelobe control. The weighting was accomplished through generation of a filter, having a sinc-squared amplitude transfer characteristic, which was then applied to weight the range-pulse spectrum. The upper curve in Fig. 18 shows the effect of introducing this weighting function on the optical system impulse response. Note that the sidelobe structure drops rapidly to approximately 40 dB below the mainlobe peak. The lower curve shows the ideal response of a sinc-squared weighting in the absence of phase and amplitude errors.

B. Digital Signal Processing

Two types of digital analysis were planned for the ALSE data:

1) Digital processing of radar data with uncompressed range pulses. This mode is used to correct for systematic phase errors across the range spectrum (as required for precise side-

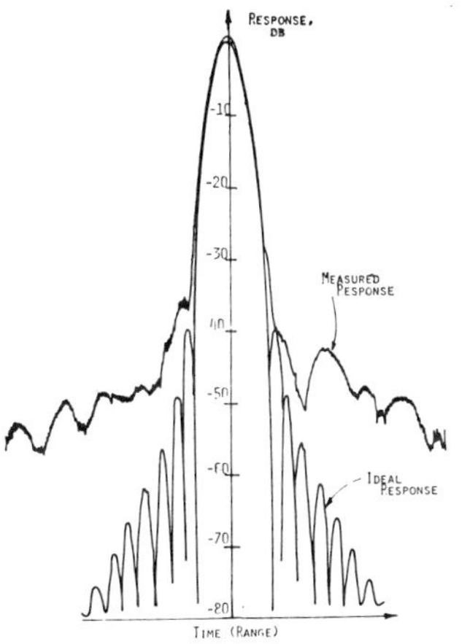

Fig. 18. Sidelobe control with weighting.

lobe control), while preserving dynamic range. It is also used for amplitude calibration.

2) Digital analysis of compressed data as part of the geophysical data interpretation.

In this section we restrict our attention to 1) above, for which the following procedure is followed:

a) The signal film is first placed in the optical processor where it is compressed in the azimuth dimension only. The signal remains dispersed in range, about a small spatial carrier frequency. The resulting "image" is projected onto an image-dissector tube where it is digitized and recorded on digital magnetic tape.

b) The data are then examined in a digital computer to estimate the system transfer function in range.

c) Via a digital deconvolution process, systematic phase errors in the data are compensated and range-pulse compression is carried out.

d) The compressed signal is then displayed using photographs, printouts, and computer plots. The final optical recording process is computer-controlled to achieve a logarithmic amplitude image dynamic range compression.

V. Preliminary Results

Two sets of flight data exist, one from an aircraft test program in which a modified version of the ALSE prototype was flown primarily to confirm the validity of the hardware, and the other from the Apollo 17 mission.

A. *Prototype Flight Data*

The modified ALSE prototype was flown during August and September 1972 in a KC-135 aircraft with the objective of validating the hardware and exercising the data analysis system. Flight paths over terrain areas which might offer some penetration capability were included in the plan. Data runs were made over the deserts of the Southwestern United States and culminated in a flight over the Greenland ice fields. No HF1 data could be collected from the KC-135. The ALSE HF1 transmitter was designed to generate a relatively long pulse of 240 μs consistent with Apollo altitudes (110 km)

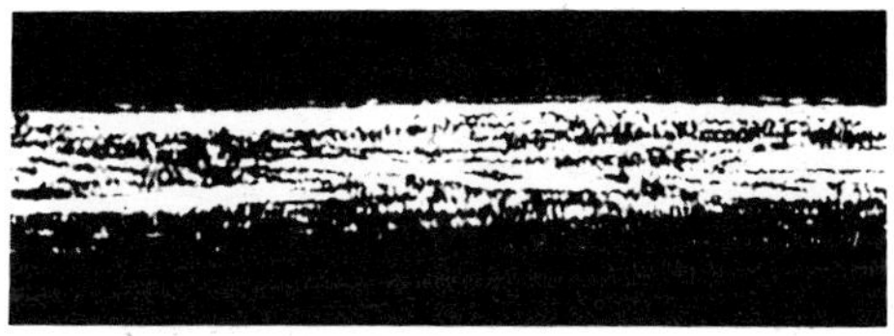

Fig. 19. VHF image of subsurface layers in Greenland ice cap. 7× vertical exaggeration. Surface represented by upper bright horizontal line. Lowest subsurface reflection approximately 120 m deep.

where the average round-trip propagation delay of 730 μs was much longer than the corresponding delay of 70–80 μs at KC-135 altitudes. The HF2 pulse length of 80 μs requires a terrain clearance of 12 km to avoid saturation of the receiver by the transmitter. Minor modifications were carried out to permit HF2 operation. VHF operation was best matched to the test-flight geometry.

Fig. 19 shows a VHF image with 7× vertical exaggeration of an area in the Greenland icecap. The profile generated along the nadir of the aircraft is evident in the image, as are subsurface reflections. If we assume that the dielectric constant of the ice is on the order of 3.1, then the maximum depths of the features shown in Fig. 19 are approximately 120 m. The Greenland data gave confidence that the ALSE system would perform to specification during the lunar mission.

B. *Lunar Mission Data*

The Apollo 17 mission was carried out in December 1972, and ALSE data began to flow to the experimenter team in January 1973. Several weeks were required to generate "survey" processings of the data at the three wavelengths, and to provide these preliminary images to the team to aid in selection of candidate sites for further processing. At this time (March, 1974), visual interpretation of the data has progressed for all three wavelengths, through use of the holographic displays and the other special display techniques. One major observation is that the lunar surface reflectivity at HF is higher than was predicted during the early model analysis; the high surface return leads to HF images which are more photogenic than had been expected, but it also produces a higher background level against which the faint subsurface returns must be detected. The digital analysis of the data, which holds the greatest hope for controlling range sidelobe structure, is in an early stage and has not yet yielded definitive results.

Interpretation to date has concentrated on the VHF imagery, which is contributing to the understanding of the deformational structures in the circular maria. Most certainly, many of the features we see are the result of partial isostatic adjustment of the excess mass (mascons) of surface volcanic flows. Fig. 20 shows a VHF image of the crater Condorcet on the edge of Mare Crisium. This image dramatically illustrates both the profiling and imaging capabilities of the ALSE system at 2 m. Fig. 21 shows a VHF image of the eastern part of Crisium basin. The mascon boundary is associated with the 300-m surface displacement near the outer margin of the basin. The abrupt change of slope to the right of the displacement suggests elastic, but unrecovered, bending before structure failure. The same geometry appears at the

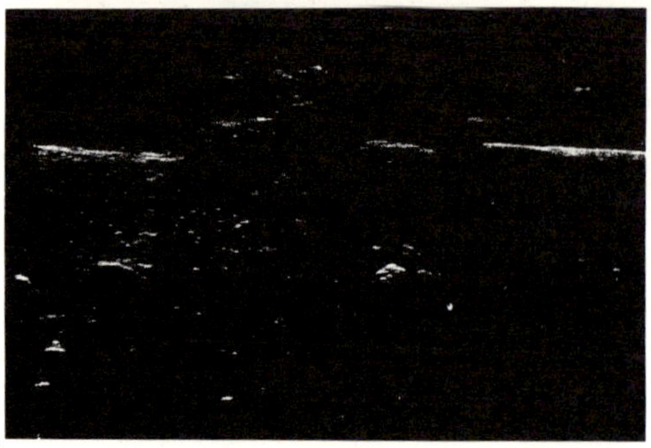

Fig. 20. VHF image of floor and rim of crater Condorcet on eastern edge of Mare Crisium. The strong lineal returns in the upper right are typical of mare surfaces and smooth crater floors; seeming to hover above the floor is the rugged rim which is closer in range but off to the side of the groundtrack, as diagrammed in Fig. 3. The small crater Condorcet Y in the upper center lies on the rim of the larger crater. The mare floor on the upper left is rougher than the crater floor, both of which are pockmarked by small craters. Figure is 75 km by 50 km.

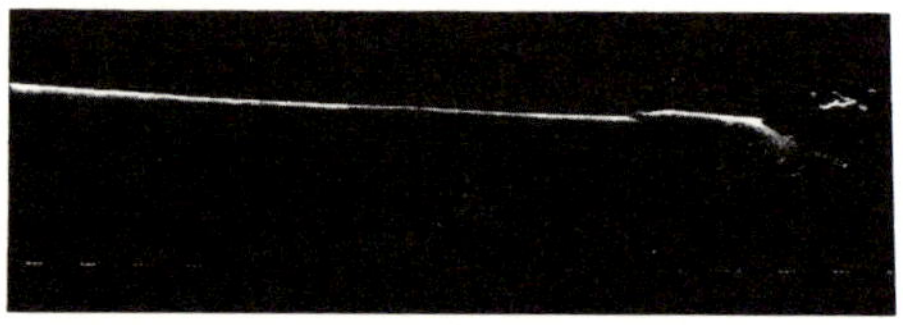

Fig. 21. Lunar sounder image of Eastern Mare Crisium. In this image, east is to the right and the scene is imaged to the north. Time mark spacing shown at the bottom of lunar sounder images represents a ground distance of approximately 15 km. The general linear trend in surface elevation over each image is the uncorrected effect of the elliptical orbit of the spacecraft.

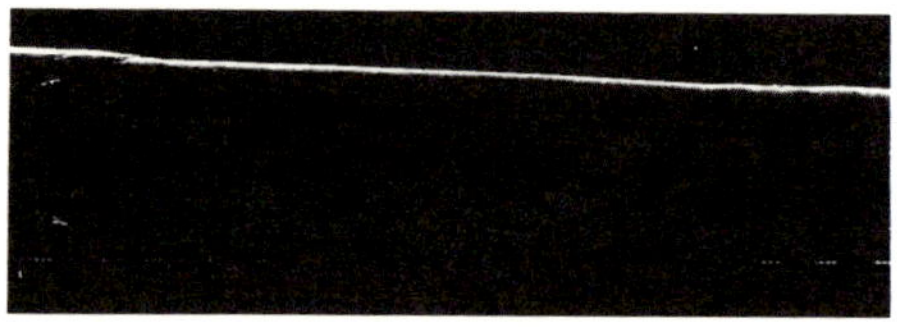

Fig. 22. Image of Western Mare Serenitatis. Here east is to the left and the scene is imaged to the north.

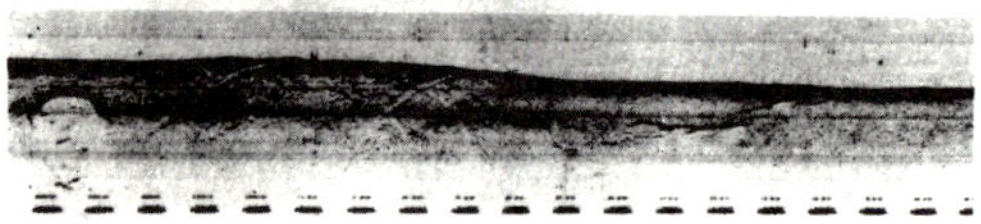

Fig. 23. VHF image of the Tranquillitatis–Serenitatis boundary. Here east is to the left and the scene is imaged to the north. The crosscutting rille is actually an image superposition from the south.

mascon boundary in Western Mare Serenitatis (Fig. 22). Here a wrinkle ridge is clearly associated with a 200-m displacement. A VHF image of the boundary between Mare Serenitatis and Mare Tranquillitatis is shown in Fig. 23. The crater Dawes is on the extreme left. (Note lack of rectification.) The Plinius rille system (left-center) is a possible

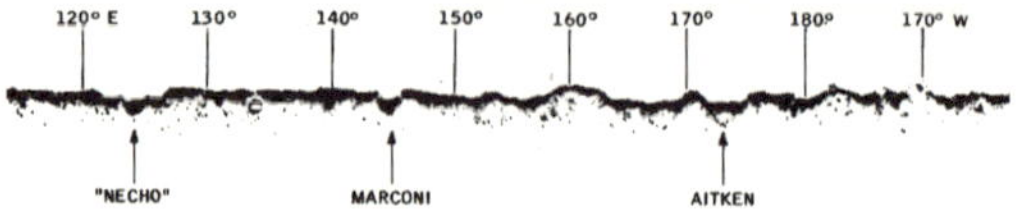

Fig. 24. HF1 image of region of the back side of the moon.

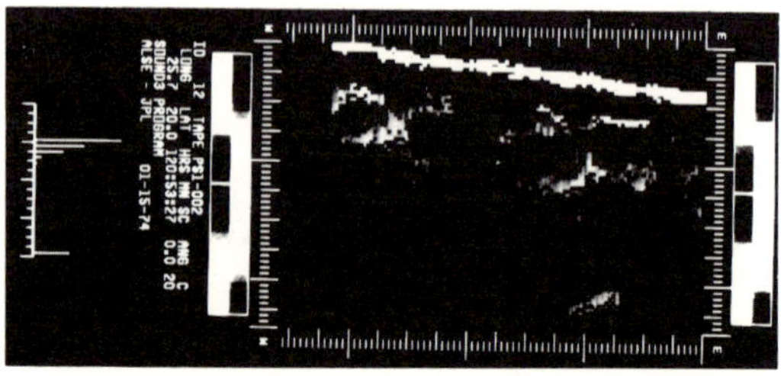

Fig. 25. Digital image display for subsurfacing processing of data. Horizontal scale (E–W along ground track) about 1 km/div. Vertical scale is 120 m/div (free space). There is about 5 dB per tonal shade with the faintest features about 35–40 dB weaker than the bright surface return. The blockiness of the data represents a decimation by 10 sample averaging. The data have been deconvolved with the Lunar Sounder system transfer function and convolved with a simple smoothing operator, following decimation. The total smoothing length is about 10 km, along a direction parallel to the local horizontal.

tension feature associated with stress imparted by the Serenitatis mascon. The floor of Tranquillitatis stands 900 m above the outer bench of Serenitatis. This is interpreted as reflecting either the greater original depth of the Serenitatis basin or a down-faulting of the basin caused by the mascon load. The 200-m displacement and associated wrinkle ridge at the mascon boundary may also be seen.

A typical example of HF imagery is shown in Fig. 24. This figure shows an HF1 image profile on the lunar farside, unadjusted for the ellipticity of the orbit. As a scale of reference, the crater Aitken is about 80 km wide under the orbital track and 5–6 km from rim to floor. We observe that the profile from 114° to 147°E is relatively flat compared to the remainder of the profile farther east. This flat region is deeply notched by two large depressions. One is the 75-km diameter crater Marconi which was profiled roughly half way from the center toward the southern rim. Part of the floor shows in the profile and the southern rim appears in the imagery. The other depression has been recognized by El-Baz (1972) [13] in Apollo 16 orbital photography; he recommended the name Necho for it. This depression is approximately 180 km in diameter and the profile in Fig. 24 is a nearly central section of it. The crater is very old; roughly 50 percent of its interior is overlapped by the interiors of superposed craters larger than 20 km in diameter. The moderately rough plane between Marconi and "Necho" appears to be very primitive in the sense of being a little-dissected remnant of an old plane situated between many 25- to 100-km craters. High points along the profile in Fig. 24 near 160°E, 177.5°W, and 170°W may represent the high parts of very ancient multiringed basins tentatively recognized during the preparation of geologic maps for planning the Lunar Sounder observations. Fig. 25 is a typical output frame from the subsurface (digital) processing program for a region in Eastern Mare Serenitatis. The bright line is the strong specular return from the lunar surface. Fainter lines are reflections from the lunar subsurface or from linear surface features away from and approximately parallel to the spacecraft ground track. The key issue is to distinguish real subsurface reflections from these linear surface features, which arise principally from mare scarps and ridges. The data may now be viewed in a regional context (i.e., multiple complete passes across a mare). In particular, the availability of multiple passes aids in a geometrical separation of pervasive subsurface features from linear surface features. An intensive study to correlate results of the type in Fig. 25 with high-resolution photography has begun. The aim is to place the subsurface data in a geologic framework.

V. Summary Remarks

In summary, the ALSE system represents a unique combination of active microwave sensing and data-processing techniques which were applied to the problem of lunar exploration. The three-wavelength, Apollo-qualified, synthetic-aperture radar system was required to have certain features not normally available in conventional aircraft-borne SAR systems. In retrospect, the specifications which were directed toward control of the sidelobe structure of the overall system impulse response and toward maximization of the available dynamic range presented the greatest challenge, as did the attempt to incorporate amplitude calibration at all three wavelengths. A novel feature was the introduction of hybrid optical/digital processing techniques into the data-analysis system. Coherent optical processors have long been used in SAR systems because of their natural propensity for evaluating Fourier transforms and other related linear-integral transforms, as well as their ability to handle huge amounts of data stored at video rates. Two-dimensional transforms, expensive to evaluate digitally, are readily available in these systems. Digital computers, on the other hand, have well-deserved reputations for precision, accuracy, and flexibility of programs which are difficult to realize optically. The hybrid configuration which utilizes the best features of each method appears to offer much promise. Much was learned in system design, data analysis, and interpretation methods which will be valuable for planning and using future radar remote sensing systems.

Acknowledgment

The authors of this paper are but a small subset of the many individuals whose contributions were essential not only to the design, fabrication, and flight of the hardware itself, but also to the configuration of the processing system and the analysis of data. These include NASA personnel; the engineering staffs of North-American Rockwell, RCA, Spar Aerospace of Canada, and Goodyear Aerospace Corporation; other contributing staff members at the authors' institutions (ERIM, JPL, and the University of Utah) and at the U. S. Geological Survey; and finally, the crew of astronauts, especially Ron Evans for both his role in operating the S-209 hardware and his EVA to retrieve the data film from the SIM Bay of the spacecraft.

References

[1] W. R. Wollenhaupt and W. L. Sjogren, "Apollo 16 laser altimeter," *Apollo 16 Preliminary Science Rep.*, NASA Doc. SP-315, 1972.

[2] B. D. Steinberg, "The effects of relative source strength and signal-to-noise ratio on angular resolution of antennas," this issue, pp. 758–762.

[3] R. O. Harger, *Synthetic Aperture Radar Systems: Theory and Design.* New York: Academic Press, 1970.

[4] W. M. Brown and L. J. Porcello, "An introduction to synthetic-aperture radar," *IEEE Spectrum*, vol. 6, pp. 52–62, Sept. 1969.

[5] T. J. Katsube, and L. S. Collett, "Electrical properties of Apollo 11 and Apollo 12 lunar samples," in *Proc. 2nd Lunar Science Conf.*, *Geochim. et Cosmochim.*, *Acta Suppl. 2*, vol. 3. Cambridge, Mass.: M.I.T. Press, 1971.

[6] G. L. Tyler, "Oblique-scattering radar reflectivity of the lunar surface: Preliminary results from Explorer 35," *J. Geophys. Res.*, vol. 73, Dec. 15, 1968.

[7] C. E. Cook and M. Bernfeld, *Radar Signals: An Introduction to Theory and Applications.* New York: Academic Press, 1967.

[8] L. J. Cutrona, E. N. Leith, L. J. Porcello, and W. E. Vivian, "On the application of coherent optical processing techniques to synthetic-aperture radar," *Proc. IEEE*, vol. 54, pp. 1026–1032, Aug. 1966.

[9] L. J. Cutrona, E. N. Leith, C. J. Palermo, and L. J. Porcello, "Optical data processing and filtering systems," *IRE Trans. Inform. Theory*, vol. IT-6, pp. 386–400, June 1960.

[10] A. Kozma, E. N. Leith, and N. G. Massey, "Tilted-plane optical processor," *Appl. Opt.*, vol. 11, Aug. 1972.

[11] E. N. Leith, "Optical processing techniques for simultaneous pulse compression and beam sharpening," *IEEE Trans. Aerosp. Elec. Syst.*, vol. AES-4, pp. 879–885, Nov. 1968.

[12] J. S. Zelenka, W. D. Hall, and N. G. Massey, "Optical processors and displays for synthetic aperture radar systems," in *Proc. Electro-Optical Systems Design Conf.* (New York, Sept. 1972).

[13] F. El-Baz, "Discovery of two lunar features," *Apollo 16 Preliminary Science Rep.*, NASA Doc. SP-315, 1972.

REFERENCES

Ausherman, D.A.; et al.: "Radar Data Processing and Exploitation Facility," *The Record of the IEEE 1975 International Radar Conference,* pp. 493-498, 21-23 April 1975.

Bickmore, R.W.: "Time Versus Space in Antenna Theory," Chapter 4 in Vol. III of *Microwave Scanning Antennas* (R.C. Hansen, ed.), Academic Press, New York and London, 1966.

Brown, W.M.; and Palermo, C.J.: "Theory of Coherent Systems," *IRE Transactions on Military Electronics, Vol. MIL-6, No. 2,* pp. 187-196, April 1962.

Brown, W.M.; and Palermo, C.J.: "Effects of Phase Errors on Resolution," *IEEE Transactions on Military Electronics, Vol. MIL-9, pp. 4-9,* January 1965.

Brown, W.M.; and Fredricks, R.J.: "Range-Doppler Imaging with Motion through Resolution Cells," *IEEE Transactions on Aerospace and Electronic Systems, Vol. AES-5, No. 1,* pp. 98-102, January 1969.

Cary, R.H.J.; and Thraves, J.: "Antennas for Sideways Looking Airborne Radar," *IEE Conference Publication 77,* pp. 37-42, 1971.

Cheng, D.K.; "Degradation Effects of Arbitrary Phase Errors on High-Resolution Radar Performance," *Proceedings of the IEE, Vol. III, No. 8,* pp. 1375-1378, August 1964.

Clarke, J.; "Reduction of Storage Requirement in Synthetic Aperture Radar," *IEEE Transactions on Aerospace and Electronic Systems, Vol. AES-7, No. 6,* pp. 1213-1215, November 1971.

Collin, R.E.; and Zucker, F.J.: *Antenna Theory, Part II,* Chapter 27, pp. 626-643, McGraw-Hill, New York, 1969.

Cutrona, L.J.; Leith, E.N.; Palermo, C.J.; and Porcello, L.J.: "Optical Data Processing and Filtering Systems," *IRE Transactions on Information Theory, Vol. IT-6, No. 3,* pp. 386-400, June 1960.

Cutrona, L.J.: "Synthetic Aperture Radar," Chapter 23 in *Radar Handbook* (M.I. Skolnik, ed.), McGraw-Hill, New York, 1970.

Develet, J.A.: "The Influence of Random Phase Errors on the Edge Response of Synthetic Aperture Mapping Radar Systems," *MIL-E-CON,* pp. 255-258, 14-16 September 1964.

Dulberger, L.: "Targeting for Air Attack," *Space/Aeronautics,* pp. 84-95, November 1965.

Eichel, L.A.: "An Inexpensive Side-Looking Radar with a Novel Display," *The Record of the IEEE 1975 International Radar Conference,* pp. 522-526, 21-23 April 1975.

Ellis, A.B.E.: "Parallel Processing for a Synthetic Aperture Radar," *IEE Conference Publication 105,* pp. 311-317, 23-25 October 1973.

Graham, L.C.: "Stereoscopic Synthetic Array Application in Earth Resource Monitoring," *Proceedings of the IEEE 1975 NAECON,* pp. 125-132.

Hansen, R.C.: "The Segmented Aperture Synthetic Aperture Radar (SASAR)," *IEEE Transactions on Aerospace and Electronic Systems, Vol. AES-10, No. 6,* pp. 800-804, November 1974.

Harger, R.O.; and Crimmins, T.R.: "The Solution of a Nonlinear Design Problem for Synthetic-Aperture Radars," *IEEE Transactions on Aerospace and Electronic Systems, Vol. AES-4, No. 2,* pp. 316-317, March 1968.

Harger, R.O.: "Harmonic Radar Systems for Near-Ground In-Foliage Nonlinear Scatterers," *IEEE Transactions in Aerospace and Electronic Systems, Vol. AES-12, No. 2,* pp. 230-245, March 1976.

Hermann, G.; Kerr, D.; and Tammy, M.: "Interactive SAR Interpretation Facility," *The Record of the IEEE 1975 International Radar Conference,* pp. 488-492, 21-23 April 1975.

Holahan, J.: "Synthetic Aperture Radar," *Space/Aeronautics,* pp. 88-93, November 1963.

Johansen, E.L.: "The Synthetic-Array Radar Image of a Flat Plate," *IEEE Transactions on Aerospace and Electronic Systems, Vol. AES-6, No. 3,* pp. 395-398, May 1970.

Kelly, P.M.; et al.: "Data Processing for Synthetic Aperture Radar," *Proceedings of the IEEE 1964 NAECON,* pp. 194-200.

Kirk, John C. Jr.: "Digital Synthetic Aperture Radar Technology," *The Record of the IEEE 1975 International Radar Conference,* pp. 482-487, 21-23 April 1975.

Kock, W.E.: "Holography Can Help Radar Find New Performance Horizons," *Electronics,* pp. 80-88, 12 October 1970.

Kock, W.E.: "A Holographic (Synthetic Aperture) Method for Increasing the Gain of Ground-to-Air Radars," *Proceedings of the IEEE, Vol. 59, No. 3,* pp. 426-427, March 1971.

Kovaly, J.J.: "Satellite High Resolution Radar Mapping Techniques," *Annals of the New York Academy of Sciences, Vol. 163, Art. 1,* pp. 154-170, 4 September 1969.

Leith, E.N.; Friesem, A.A.; and Funkhouser, A.T.: "Optical Simulation of Radar Ambiguities," *IEEE Transactions on Aerospace and Electronic Systems, Vol. AES-6, No. 6,* pp. 832-840, November 1970.

MacPhie, R.H.: "The Circular Synthetic Radar," *IEEE Transactions on Aerospace and Electronic Systems, Vol. AES-9, No. 4,* pp. 608-611, July 1973.

Mitchell, R.L.: "Models of Extended Targets and their Coherent Radar Images," *Proceedings of the IEEE, Vol. 62, No. 6,* pp. 754-758, June 1974.

Moore, R.K.; and Rouse, J.W. Jr.: "Electronic Processing for Synthetic Aperture Array Radar," *Proceedings of the IEEE (Letters), Vol. 55, No. 2,* pp. 233-234, February 1967.

Nuttall, J.A.: "Dynamic Range Study of Correlators used in Conjunction with Side-Looking Synthetic Radar Systems," *Supplement to IEEE Transactions on Aerospace and Electronic Systems, Vol. AES-2, No. 6,* pp. 385-394, November 1966.

Revillon, G.: "Synthetic Aperture Antennas and their Applications to Side-Looking Radars," *L'Onde Electrique 45, No. 458,* pp. 561-567, May 1965. (In French)

Rihaczek, A.W.: "An Example: Synthetic Aperture Radar," *Chapter 13 in Principles of High-Resolution Radar,* McGraw-Hill, New York, pp. 441-483, 1969.

Rogers, E.E.; and Ingalls, R.P.: "Radar Mapping of Venus with Interferometric Resolution of the Range-Doppler Ambiguity," *Radio Science, Vol. 5, No. 2,* pp. 425-433, February 1970.

Ruina, J.P.; and Angulo, C.M.: "Antenna Resolution as Limited by Atmospheric Turbulence," *IEEE Transactions on Antennas and Propagation, Vol. AP-11, No. 2,* pp. 153-161, March 1963.

Shapiro, I.I.: "Planetary Radar Astronomy," *IEEE Spectrum,* pp. 70-79, March 1968.

Shuchman, R.A.; Rawson, R.F.; and Drake, Ben: "A Dual Frequency and Dual Polarization Synthetic Aperture Radar System and Experiments in Agriculture Assessment," *Proceedings of the IEEE 1975 NAECON,* pp. 133-140.

Steinberg, B.D.: "The Effects of Relative Source Strength and Signal-to-Noise Ratio on Angular Resolution of Antennas," *Proceedings of the IEEE, Vol. 62, No. 6,* pp. 758-762, June 1974.

Tyler, G.L.: "The Bistatic, Continuous-Wave Radar Method for the Study of Planet Surfaces," *Journal of Geophysical Research, Vol. 71, No. 6,* pp. 1559-1567, 15 March 1966.

Books on Synthetic Aperture Radar

Harger, R.O.: *Synthetic Aperture Radar Systems: Theory and Design,* Academic Press, New York, 1970.

Reutov, A.P.; and Mikhaylov, B.A.: *Side-Looking Radar,* Radiolokatsionnyye Stantsii Bokvogo Obzora, Moscow, Soviet Radio, 1970. (Available in translation AD787070 from US Department of Commerce Clearinghouse for Federal Scientific and Technical Information)

Books on Related Subjects

Levine, D.: *Radargrammetry,* McGraw-Hill, Inc., New York, 1960.

Mitchell, R.L.: *Radar Signal Simulation,* Artech House, Dedham, Massachusetts, 1976.